SECOND EDITION

SAFETY
PROFESSIONAL'S
REFERENCE & STUDY GUIDE

SECOND EDITION

SAFETY PROFESSIONAL'S
REFERENCE & STUDY GUIDE

W. DAVID YATES

CRC Press
Taylor & Francis Group
Boca Raton   London   New York

CRC Press is an imprint of the
Taylor & Francis Group, an **informa** business

CRC Press
Taylor & Francis Group
6000 Broken Sound Parkway NW, Suite 300
Boca Raton, FL 33487-2742

First issued in paperback 2017

© 2015 by Taylor & Francis Group, LLC
CRC Press is an imprint of Taylor & Francis Group, an Informa business

No claim to original U.S. Government works

ISBN-13: 978-1-4822-5665-9 (hbk)
ISBN-13: 978-1-138-89297-2 (pbk)

Visit the Taylor & Francis Web site at
http://www.taylorandfrancis.com

and the CRC Press Web site at
http://www.crcpress.com

Printed and bound by CPI Group (UK) Ltd, Croydon, CR0 4YY

With enormous gratitude to my wonderful and supportive wife, Sharon, who has been there during the late nights, and to our children, Joseph, Jason, Katie, Cody, and Jesse; our grandchildren, Jonathon Elisha, Jordan Nicole, Jacob Michael, Bailey Addison, Mason Tanner, and Madison Olivia, I dedicate this book.

Contents

Preface

During the course of my professional career, I have had the opportunity to read and utilize numerous technical resources. As a practicing safety professional in comprehensive practice, I have yet to find a single-source reference that includes the majority of information that I encounter on a daily basis. That is not to say that there are no useful references available. However, it has been my experience that maintaining a library suitable for comprehensive practice becomes very costly and requires a wide variety of topics to obtain the information that is needed. The majority of useful references available focus primarily on the practicing industrial hygienist.

It is for this reason that I set out to publish the first edition of the book. It was intended to serve several purposes as outlined below:

- To function as a "quick desk reference" for the experienced, practicing safety professional in comprehensive or specialized practice
- To be utilized by university students at all levels as a useful reference tool to supplement more in-depth textbooks
- To serve as a primary study resource for those individuals preparing to take the Associate Safety Professional (ASP), Certified Safety Professional (CSP), Occupational Health and Safety Technologist (OHST), and the Construction Health and Safety Technologist (CHST) examinations

The first edition had a very positive response from my fellow safety professionals, as well as students and prospective ASP/CSP Candidates. In this edition, I have added new chapters on behavior-based safety programs, safety auditing procedures and techniques, environmental management, measuring health and safety performance, OSHA's laboratory safety standard, process safety management standard, and BCSP Code of Ethics. In addition to the new chapters, this book will serve as a primary study guide for the examinations listed above. It includes such topics as History of the Safety Profession, Regulations, OSHA Record Keeping, Particulates and Gases, Toxicology, Industrial Hygiene, Ventilation, Noise, Illumination, Biological Hazards, Thermal Stressors, Personal Protective Equipment, Math Review, Statistics for the Safety Professional, Fire Protection and Prevention, Mechanics, Hydrostatics and Hydraulics, Electrical Safety, Engineering Economy, Training, and Worker's Compensation Issues. It is my belief that the reader of this book will determine it to be an "invaluable" resource at any level of their professional safety career.

Acknowledgments

In preparing a work of this nature, there are undoubtedly people whose contributions must be acknowledged. These people are professional colleagues, friends, and family members. I would like to personally thank Dr. K. Bhamidipaty, PE, CSP, Fellow, ASME International, for his review of the first edition and for preparing an errata. His review was obviously time-consuming and provided great assistance in improving the second edition.

I would also like to once again thank Tim Hallmark, CIH, CSP, who provided significant contributions to various aspects of the first edition, which are also included within this book. I would like to thank the various authors of cited materials for their work on the safety professional and further for their copyright permission responses in a timely manner.

Last, but most definitely not the least, I would like to thank my wife, Sharon, for her continued support and dedication to this work.

Author

W. David Yates was born in Morton, Mississippi, and lived most of his childhood in Crystal Springs, Mississippi. He has earned a BS in Health Care Services from Southern Illinois University, Carbondale, Illinois; an MS in Hazardous Materials Management from Jackson State University, Jackson, Mississippi; and a PhD in Environmental Science from Jackson State University, Jackson, Mississippi. Dr. Yates served 10 years in the United States Navy as a Preventive Medicine Technician. Dr. Yates currently serves in the Army National Guard as a safety and occupational health officer. In his civilian career, Dr. Yates has operated his own professional consulting firm; served as the safety and mission assurance manager for Stennis Space Center, Mississippi; served as the corporate safety director for Bodine Services of the Midwest, Decatur, Illinois; and held lead safety positions in mining and manufacturing. Dr. Yates currently serves as the safety manager for FMC Corporation, Green River, Wyoming, which has both Mine Safety and Health Administration (MSHA) and Occupational Safety and Health Administration (OSHA) facilities. He is a certified safety professional (CSP) with the Board of Certified Safety Professionals (BCSP).

Dr. Yates has extensive knowledge and experience in hazardous materials management, safety programs management, indoor air quality, ventilation, noise and sound, and industrial hygiene sampling and analysis. Dr. Yates' email address is wdyates@gmail.com.

1

The Safety Profession and Preparing for the ASP/CSP Exam

The safety profession has a long and distinguished history tracing back to Hammurabi (ca. 1728–1686 BC), who was the sixth king of Babylon. Hammurabi is best known for his codification of laws, which included some, if not the first, set of worker's compensation laws known. The safety profession has greatly changed since the days of Hammurabi. On December 29, 1970, Public Law 91-596 (The Williams–Steiger Occupational Safety and Health Act of 1970) was signed into law. This legislation focused on controlling workplace hazards and ensuring safe and healthful working conditions for working men and women; by authorizing enforcement of the standards developed under the Act; by assisting and encouraging the States in their efforts to assure safe and healthful working conditions; by providing for research, information, education, and training in the field of occupational safety and health; and for other purposes. Under the Act, the Occupational Safety and Health Administration was created within the Department of Labor. The passage of this legislation highlighted the need for educated and knowledgeable professionals in the area of safety and health. Over the past 39 years, the safety profession has matured, as evidenced by universities offering undergraduate and advanced degrees in safety and health, placement of safety professionals at the highest levels of management, and certification of safety professionals. Today's safety profession requires a broad range of knowledge, including biology, chemistry, mathematics, business, and management. Your decision to become a candidate for the Certified Safety Professional (CSP) designation is an important step in your professional career. This book is written to assist you in achieving that ultimate designation as a safety professional.

Board of Certified Safety Professionals

NOTE: Information provided in this section is derived from the "Safety Fundamentals Examination Guide, Fifth Edition, April 2008." This information is derived from copyrighted materials that are owned by the Board of Certified Safety Professionals.

The Board of Certified Safety Professionals (BCSP) was organized in 1969 as a peer certification board. Its purpose is to certify practitioners in the safety profession. The specific functions of the Board are to

- Evaluate the academic and professional experience qualifications of safety professionals
- Administer examinations
- Issue certifications to those professionals who meet the Board's criteria and successfully pass required examinations

In 1968, the American Society of Safety Engineers studied the issue of certification for safety professionals and recommended the formation of a professional certification program. This recommendation led to establishing BCSP in July 1969. The BCSP governing Board consists of 13 directors who represent the breadth and depth of safety, health, and environmental practice, as well as the public. Six of the directors are nominated to a pool by professional membership organizations affiliated with BCSP. The professional membership organizations currently affiliated with BCSP are the following:

- American Industrial Hygiene Association
- American Society of Safety Engineers
- Institute of Industrial Engineers
- National Fire Protection Association
- National Safety Council
- Society of Fire Protection Engineers
- System Safety Society

BCSP has issued the CSP credential to more than 20,000 people and more than 11,000 currently maintain their certification.

The CSP credential meets or exceeds the highest national and international accreditation and personnel certification standards for certification bodies. International Accreditation is ISO/IEC 17024 and National Accreditation is the National Commission for Certifying Agencies.

Definitions

NOTE: The definitions in this section are derived from the "Safety Fundamentals Examination Guide, Fifth Edition, April 2008." These definitions are derived from copyrighted materials that are owned by the Board of Certified Safety Professionals.

A *safety professional* is one who applies the expertise gained from a study of safety science, principles, practices, and other subjects from professional safety experience to create or develop procedures, processes, standards, specifications, and systems to achieve optimal control or reduction of the hazards and exposures that may harm people, property, or the environment.

Professional safety experience, as interpreted by BCSP, must be the primary function of a position and account for at least 50% of the position's responsibility. Professional safety experience involves analysis, synthesis, investigation, evaluation, research, planning, design, administration, and consultation to the satisfaction of peers, employers, and clients in the prevention of harm to people, property, and the environment. Professional safety experience differs from nonprofessional experience in the degree of responsible charge and the ability to defend analytical approaches and recommendations for engineering or administrative controls.

A *Certified Safety Professional or CSP* is a safety professional who has met and continues to meet all requirements established by BCSP and is authorized by BCSP to use the Certified Safety Professional title and the CSP credential.

An *Associate Safety Professional or ASP* is a temporary designation awarded by BCSP. This designation describes an individual who has met the academic requirements for the CSP credential and has passed the Safety Fundamentals Examination, the first of two examinations leading to the CSP credential.

A *Graduate Safety Practitioner or GSP* is a temporary designation awarded by BCSP. This designation describes an individual who has graduated from an independently accredited academic program meeting BCSP's standards.

Minimum Qualifications to Sit for the CSP Examination

The minimum qualifications to sit for the CSP examination include academic requirements, professional safety experience requirements, successfully passing the ASP (or obtain a waiver), and CSP examinations. BCSP uses a point system for determining eligibility to sit for the examinations. To qualify for the ASP examination, you must have a total of 48 points. The CSP examination qualification requires a total of 98 points. The following details are only a summary of the requirements. Should you have specific questions on requirements, you can visit http://www.bcsp.org.

Academic Requirements

Academic requirements to obtain the CSP credential includes a bachelor's degree or higher in any field or an associate degree in safety and health. The degree(s) must have been awarded from an accredited university or institution. Unaccredited degrees will not be accepted to satisfy the academic requirements.

Professional Safety Experience Requirements

Professional safety experience is required to qualify for the ASP/CSP credential. The exact number of years depends on the level of education. All professional safety experience claimed on the application must meet all five of the following criteria to be acceptable by the BCSP:

- Professional safety must be the primary function of the position.
- Primary responsibility of the position must be the prevention of harm to people, property, or the environment, rather than responsibility for responding to harmful events.
- Professional safety functions must be at least 50% of the position duties.
- The position must be at the professional level, which is determined by the degree of responsibility.
- The position must have breadth of professional safety duties.

ASP Process

The process of earning the interim ASP designation is summarized as follows:

- Complete and submit application materials (requires application fee).
- Receive evaluation and letter of authorization from BCSP to take the Safety Fundamentals Examination.
- Register to take the Safety Fundamentals Examination (examination fee required).
- Make an appointment to take your examination at a test center near you, and sit for the examination at the scheduled time.
- Complete all requirements for the ASP credential. Once you pass the examination, BCSP will award your ASP credential.
- Pay an annual renewal fee to BCSP (fee varies).

CSP Process

The process for earning the CSP credential is summarized as follows:

- Complete and submit application materials (requires application fee).
- Receive evaluation and letter of authorization from BCSP to take the Comprehensive Practice Examination.
- Register to take the Comprehensive Practice Examination (examination fee required).
- Make an appointment to take your examination at a test center near you, and sit for the examination at the scheduled time.
- Complete all requirements for the CSP credential. Once you pass the examination, BCSP will award your CSP credential.
- Pay an annual renewal fee to BCSP (fee varies).

Safety Fundamentals Examination Blueprint

There are four domains on the Safety Fundamentals Examination, which represent the major areas of practice for safety professionals at the ASP level. Within each domain are knowledge topics. These domains and topics are listed below.

Domain 1: Recognizing Safety, Health, and Environmental Hazards (35.4% of the exam)
- Topic 1: Biological Hazards
- Topic 2: Chemical Hazards
- Topic 3: Electrical Hazards
- Topic 4: Natural Hazards
- Topic 5: Radiation Hazards
  - Ionizing radiation
  - Nonionizing radiation
- Topic 6: Structural and Mechanical Hazards
- Topic 7: Hazards Related to Fires and Explosions
- Topic 8: Hazards Related to Human Factors and Ergonomics
  - Fitness for duty
  - Manual materials handling
  - Organizational, behavioral, and psychological influences

- Physical and mental stressors
- Repetitive activities
- Workplace violence

Domain 2: Measuring, Evaluating, and Controlling Safety, Health, and Environmental Hazards (30.9% of the exam)

- Topic 1: Measurement and Monitoring
  - Methods and techniques for measurement, sampling, and analysis
  - Uses and limitations of monitoring equipment
- Topic 2: Engineering Controls
  - Dust control
  - Equipment and material handling
  - Excavation shoring
  - Facility physical security
  - Fall protection
  - Fire prevention, protection, and suppression
  - Hazardous energy control
  - Human factors and ergonomic design
  - Mechanical and machine guarding
  - Segregation and separation
  - Substitution and selection of alternative design strategies
  - Ventilation
- Topic 3: Administrative Controls
  - Accountability
  - Behavior modification
  - Decontamination process
  - Exposure limitation
  - Fitness for duty
  - Housekeeping
  - Labels
  - Material safety data sheets
  - Safe work permits
  - Training and education
  - Warnings and signs
  - Work zone establishment
  - Written plans, procedures, and work practices

- Topic 4: Personal Protective Equipment
  - Assessment of need for personal protective equipment
  - Selection and testing of personal protective equipment
  - Usage of personal protective equipment
  - Maintenance of personal protective equipment

Domain 3: Safety, Health, and Environmental Training and Management (20.6% of the exam)

- Topic 1: Training and Communication Methods
  - Adult learning techniques
  - ANSI/ASSE Z490
  - Behavior modification
  - Methods of training delivery
  - Methods of training evaluation
  - Presentation tools
- Topic 2: Management Processes
  - Emergency/crisis/disaster planning and response
  - Identification of expert resources
  - Incident data collection and analysis
  - Techniques for performing incident investigation and root cause analysis
- Topic 3: Inspections and Auditing
  - Elements of an inspection and auditing program
  - Reasons to perform inspections and audits
  - Purpose and objective of ISO 19011
- Topic 4: Group Dynamics
  - Conflict resolution
  - Methods of facilitating teams
  - Multidisciplinary teamwork
- Topic 5: Project Management
  - Evaluation of cost, schedule, performance, and risk
  - Project management terminology
  - Review of specifications and designs against requirements
- Topic 6: Risk Management
  - The risk management process
  - Risk analysis methods

- Topic 7: Safety, Health, and Environmental Management Systems
  - Purpose and objective of ANSI/AIHA Z10
  - Purpose and objective of ISO 14000 series of environmental management system standards
  - Purpose and objective of OHSAS 18000 series of occupational health and safety management system standards
  - Purpose and objective of the OSHA VPP

Domain 4: Business Principles, Practices, and Metrics in Safety, Health, and Environmental Practice (13.1% of the exam)

- Topic 1: Basic Financial Principles
  - Cost–benefit analysis
  - Definition and use of the life cycle cost
  - Definition and use of net present value
  - Definition and use of return on investment
- Topic 2: Probability and Statistics
  - Concepts of probability
  - Normal distribution, description, calculations, and interpretations
  - Poisson distribution
  - Descriptive statistics: description, calculations, and interpretations (mean, mode, median, standard deviation, standard error of measurement, variance)
  - Inferential statistics: description, calculations, and interpretations (t test, z test, chi-square test, Pearson product-moment correlation, Spearman's rank correlation, linear regression techniques, confidence intervals, control limits)
- Topic 3: Performance Metrics and Indicators
  - Lagging indicators (incidence rates, lost time, direct costs)
  - Leading indicators (inspection frequency, number of safety interventions, training frequency)
  - Economic effects of losses
  - Relationship between cost of losses and the effect on profitability

Comprehensive Practice Examination Blueprint

The Comprehensive Practice Examination is divided into three major categories (domains), which also represent the major functions performed by safety professionals at the CSP level. Each domain has a list of knowledge and skills areas that can be viewed at http://www.bcsp.org.

Domain 1: Collecting Safety, Health, Environmental, and Security Risk Information (28.6% of the exam)

- Task 1: Identify and characterize hazards, threats, and vulnerabilities using equipment and field observation methods in order to evaluate safety, health, environmental, and security risk.

- Task 2: Design and use data management systems for collecting and validating risk information in order to evaluate safety, health, environmental, and security risk.

- Task 3: Collect and validate information on organizational risk factors by studying culture, management style, business climate, financial conditions, and the availability of internal and external resources in order to evaluate safety, health, environmental, and security risk.

- Task 4: Research applicable laws, regulations, consensus standards, best practices, and published literature using internal and external resources to develop benchmarks for assessing an organization's safety, health, environmental, and security performance and to support the evaluation of safety, health, environmental, and security risk.

Domain 2: Assessing Safety, Health, Environmental, and Security Risk (36.6% of the exam)

- Task 1: Evaluate the risk of injury, illness, environmental harm, and property damage to which the public or an organization is exposed associated with the organization's facilities, products, systems, processes, equipment, and employees by applying quantitative and qualitative threat, vulnerability, and risk assessment techniques.

- Task 2: Audit safety, health, environmental, and security management systems using appropriate auditing techniques to compare an organization's management systems against established standards for identifying the organization's strengths and weaknesses.

- Task 3: Analyze trends in leading and lagging performance indicators related to safety, health, environmental, and security management systems using historical information and statistical methods to identify an organization's strengths and weaknesses.

Domain 3: Managing Safety, Health, Environmental, and Security Risk (34.8% of the exam)

- Task 1: Design effective risk management methods using the results of risk assessments to eliminate or reduce safety, health, environmental, and security risks.

- Task 2: Educate and influence decision makers to adopt effective risk management methods by illustrating the business-related benefits associated with implementing them to eliminate or reduce safety, health, environmental, and security risks.
- Task 3: Lead projects to implement the risk management methods adopted by decision makers using internal and external resources to eliminate or reduce safety, health, environmental, and security risks.
- Task 4: Promote a positive organizational culture that is conscious of its safety, health, environmental, and security responsibilities by communicating these responsibilities to all stakeholders and by training all stakeholders as part of the organization's overall risk management program.

Preparing for the ASP/CSP Examinations

Now that you understand the process of qualifying for the examination and the information that you will be tested on, you will need to develop a plan of action to prepare to take the examinations. The methods and techniques used for preparing for the examination(s) is an individual decision. The methods listed in this chapter have been developed over the years and have been determined to be highly successful. There are no shortcuts to preparing for these examinations. Preparation for this examination started in elementary and high school, by taking the required courses in math, science, economics, and business. With this being said, this chapter will guide you through some time-proven methods and techniques that will assist you in successfully passing the examinations and obtaining the professional designations.

Knowing Your Strengths and Weaknesses

One of your first steps in preparing for the ASP/CSP examinations is to determine where your strengths and weaknesses are. This can be achieved in a number of fashions. One way is to do a self-evaluation, by rating yourself on how well you know each subject area listed in the domains. Simply look at the domains and individual topics or tasks and rate yourself on how well you are familiar with the topic or task. A second method of determining how well you know the topic is to take a self-assessment examination. These self-assessment examinations can be purchased from BSCP for approximately $95.00 each. They are useful in determining your initial status and for studying for the examination. A rating form to assist you in rating yourself is available on the BCSP website by downloading the examination guides. There is no cost for this document.

Developing an Examination Preparation Plan

Based on the results of your self-evaluation, the next step is to develop a plan to prepare for the examination. There is no set period to prepare for the examination. The time required to prepare is strictly a personal decision. Personal experience has shown that the average person with a solid background in mathematics (trigonometry and algebra), basic physics, chemistry, and business will require an average of 6 months to prepare. As stated in Knowing Your Strengths and Weaknesses, the first step is to determine your strengths and weaknesses. One of the easiest things to do, and probably one of the worst mistakes made when preparing for an examination, is to spend a lot of time studying those topics that you are strongest in. Let's face it. It's the easy thing to do. However, it will not serve you well to spend most of your time doing this. Preparing for the examination is hard work. Focus on those areas that you are weakest in.

Use the self-assessment and evaluation to develop your study plan. Determine how much time you will need to study each topic and develop a schedule to help you meet these goals. Stick to the schedule, as best as you can. Cramming a lot of information at the last minute may work for a few individuals, but for the majority of people, this method does not work. When studying a particular topic, don't just rely on remembering details of the subject. Try to gain a solid understanding of each topic. The examination questions will not be questions directly from the recommended resources. They are designed to test your overall knowledge and understanding.

References and Resources

This book is designed to cover the majority of topics listed on the examination, focusing primarily on the examination reference sheet and the equations. A list of other references is available on the BCSP website, should you need more in-depth knowledge of a particular subject. A number of professional organizations and private companies offer ASP/CSP review courses and materials. These courses can be helpful but should not be considered to be all-inclusive or to provide you with a complete set of knowledge of the requirements of the ASP/CSP skill set in a week's time. If you are going to spend the time and money to attend one of these courses, it is my recommendation that you do it early on in your preparation. This allows you to review and modify your preparation plans. There are study materials offered by the American Society of Safety Engineers, SPAN International, and DataChem Software, Inc., in the form of study books and compact disks with example tests. These materials average approximately $400 to $650 each. As you can see, a person preparing for the examinations has a lot of options. All of these are viable options. You must determine what works best for you, as an individual, before proceeding.

Test-Taking Strategy

Both the Safety Fundamental Examination and the Comprehensive Practice Examination contain 200 questions each. Each question is multiple-choice, with four possible answers. Only one answer is correct. Each item is independent and does not rely on the correct answer to any other item. Data necessary to answer items are included in the item or in a scenario shared by several items. Your score is based on the number of scored items you correctly answer. There are no penalties assigned for wrong answers. Therefore, it is to your advantage to answer all of the questions, even if you are guessing on some. However, only correct answers count toward reaching the passing score.

In order for you to improve your chances of passing the examination, you develop a strategy for actually taking the examination. Therefore, it is helpful to understand an item (question) construction. A four-choice, objectively scored examination item contains an item stem and four possible answers. The premise, or lead-in statement or question, is called the stem. One of the choices is correct and three are not. As mentioned previously, there are no penalties for incorrect answers, so for some items you may have to guess. If you do not know the answer to an item or are not sure about it, you should guess intelligently. Look for choices that you know are incorrect or do not appear as plausible as others. Choose your answer from among the remaining choices. This increases your chance of selecting a correct answer. Above all else, read the items carefully. Consider the item from the viewpoint of an examination item writer. Look for the item focus. Each item evaluates some subject or area of knowledge. Try to identify what knowledge the item is trying to test. Avoid reading things into an item. The item can only test on the information actually included. Recognize that the stems for some items may include information that is not needed for correctly answering them. Next, consider the context of the item. Often an item is framed around a particular industry or situation. Even if you do not work in that industry or have not experienced a particular situation, the item may be testing knowledge that you have. Avoid dismissing an item because of the context or the industry in which it is framed.

Use your examination time wisely. When taking your examination, complete those items first that you know or can answer quickly. Then go back to items that were difficult for you or required considerable time to read, analyze, or compute. This approach allows you to build your score as quickly as possible. You may want to go back over skipped or marked items several times. Complete the skipped items. After you have gone through the examination once, or if you are running out of time, look for items that you have not answered. Select an answer for any skipped or incomplete item. By chance alone, you can get one of every four correct. Many times, a later item may contain the answer or at least a trigger to allow you to answer some previous question.

Go back to troublesome items. It is a good idea to mark items that you are not sure about or items that are difficult for you. After you have worked through the entire examination, go back to marked items. Reread the items and study the choices again. You may recall some knowledge or information that you had not considered earlier and be able to answer the item correctly. You may also be able to eliminate a choice that is not correct and increase your chance of guessing the answer.

One of the single most important items that you are allowed to bring with you when taking the exam is your calculator. It is recommended that you bring two in case the battery in one is used up. In the latest version of the BCSP guidelines, you are authorized to bring any of the following types of calculators:

- Casio models (FX-115, FX-250, FX-260, FX-300)
- Hewlett-Packard models (HP 9, HP10, HP12, HP30)
- Texas Instruments models (TI30, TI-34, TI-35, TI36)

My personal preference is the TI30, since I am very familiar with the model. It is recommended that regardless of the calculator you use, you should spend a good deal of time practicing with it before taking the exam. This will help in saving valuable test-taking time and will also allow you to know the specific functions and capabilities of the calculator.

2

Regulations

In the course of this chapter, we will discuss the basic requirements of the more common regulations that the safety professional encounters on a routine basis. This chapter is not intended as a substitute for reading the complete regulation for specific details. It is intended only as a summary of the standards. However, the information provided will provide a good basic understanding and familiarize the examination candidate sufficiently enough to answer questions on the Associate Safety Professional/Certified Safety Professional examinations. This chapter will cover the following topics: General Duty Clause, Occupational Safety and Health Administration (OSHA) Record Keeping, Hazardous Communication Standard, Bloodborne Pathogens, Lockout/Tagout, Confined Space Entry, Respiratory Protection, Powered Industrial Trucks and Aerial Lifts, Personal Protective Equipment (PPE), and Fall Protection.

Occupational Safety and Health Act

The Occupational Safety and Health Act (Public Law 91-596) was passed into law on December 29, 1970. It may also be referred to as the Williams–Steiger Occupational Safety and Health Act of 1970. The purpose of the law is "to assure safe and healthful working conditions for working men and women; by authorizing enforcement of the standards developed under the Act; by assisting and encouraging the States in their efforts to assure safe and healthful working condition; by providing for research, information, education, and training in the field of occupational safety and health; and for other purposes." The OSHA was created within the Department of Labor. The primary responsibilities assigned to OSHA under the Act are as follows:

- Encourage employers and employees to reduce workplace hazards and to implement new or improve existing safety and health standards.
- Provide for research in occupational safety and health and develop innovative ways of dealing with occupational safety and health problems.
- Establish "separate but dependent responsibilities and rights" for employers and employees for the achievement of better safety and health conditions.

- Maintain a reporting and record keeping system to monitor job-related injuries and illnesses; establish training programs to increase the number of competent occupational safety and health personnel.
- Develop mandatory job safety and health standards and enforce them effectively.

Who Is Covered under the Occupational Safety and Health Act?

Basically, all private sector employers with one or more workers in all 50 states and US territories are governed under the OSH Act. OSHA regulations do not apply to all employers in the public sector (municipal, county, state, or federal government agencies), self-employed individuals, family members operating a farm, or domestic household workers.

Horizontal and Vertical Standards

Standards are referred to as either horizontal or vertical. *Horizontal standards* are those standards that apply to all industries and employers. For example, fire prevention and protection standards are horizontal standards. *Vertical standards* are those standards that apply only to particular industries and employers. Standards that apply only to the construction industry are an example of vertical standards.

General Duty Clause

Each standard promulgated by OSHA cannot cover every specific detail. Therefore, OSHA has implemented a "general duty clause" into the regulations. The general duty clause states that an employer shall furnish "a place of employment which is free from recognized hazards that are causing or are likely to cause death or serious physical harm to its employees." Where there is no specific standard, OSHA will use the general duty clause for the issuance of citations and fines.

The general duty clause can be found in Section 5 (a)(1) of the Occupational Safety and Health Act of 1970.

Employer Rights and Responsibilities

Besides meeting the intent of the general duty clause, the employer must

- Examine workplace conditions to make sure they comply with applicable standards.
- Minimize or reduce hazards.
- Use color codes, poster, labels, or signs when needed to warn employees of potential hazards.

- Provide training required by applicable OSHA standards.
- Keep OSHA-required records.
- Provide access to employee medical records and exposure records to employees or their authorized representatives.

Employee Rights and Responsibilities

OSHA requires workers to comply with all safety and health standards that apply to their actions on the job. Employees should

- Read the OSHA poster.
- Follow the employer's safety and health rules and wear or use all required gear and equipment.
- Follow safe work practices for your job, as directed by your employer.
- Report hazardous conditions to a supervisor or safety committee.
- Report hazardous conditions to OSHA, if employers do not fix them.
- Expect safety and health on the job without fear of reprisal.

Communications and Correspondence with OSHA

There are two trains of thought among safety professionals when communicating and dealing with OSHA. The first train of thought is the belief that OSHA is the enemy and any communication or correspondence with them will result in a negative way for the company. The second thought is that OSHA has a purpose to protect the health and safety of employees and also serve as a valuable resource and partner for American businesses. Throughout my professional career, I follow the second train of thought. I have always found it beneficial to develop a working relationship with regulatory agencies. The large majority of OSHA personnel are experienced and knowledgeable professionals and can serve as a useful resource in making decisions and often have ideas that you may not have thought of when trying to solve a particular problem within your facility or operation. With this being said, I would urge caution in providing information on an official basis. The safety professional must remember that a large portion of OSHA's responsibility lies in enforcing the regulations. Therefore, information provided to them can be used to levy penalties against your organization. When confronted on an official basis with a potential citation or penalty, it is best to provide only those specific documents that are requested. This, of course, is only a personal opinion, based on my experiences.

OSHA Inspections and Process

Whenever an OSHA inspection occurs, the employer must

- Be advised by the compliance officer of the reason for the inspection
- Require identification of the OSHA compliance officer
- Accompany the compliance officer on the inspection
- Be assured of the confidentiality of any trade secrets observed by an OSHA compliance officer during an inspection

Under the OSH Act, OSHA is authorized to conduct workplace inspections during normal operating hours. Inspections are based on the following priorities:

- Imminent danger situations
- Catastrophes and fatal accidents
- Employee complaints
- Programmed high-hazard inspections
- Follow-up inspections

The inspection process starts even before the compliance officer visits your site. The compliance officer will prepare himself or herself by becoming familiar with your particular industry and business through research. He or she will be familiar with the potential hazards and processes involved with your particular business. Once on site, the compliance officer begins by presenting his or her credentials. *Note: The compliance officer has the right, under the law, for timely admission to the facility. Any unnecessary delay or refusal of admittance may prompt the compliance officer to obtain a warrant for inspection purposes.* Once the authorized credentials have been presented, the compliance officer will hold an opening conference. In the opening conference, the compliance officer will explain the purpose of the visit, how your particular facility was selected for inspection, the scope of the inspection, and the standards that will apply. The compliance officer will ask the employer to select an employer representative to accompany him or her on the inspection. The compliance officer also gives an authorized employee representative the opportunity to attend the opening conference and accompany the compliance officer during the inspection. The Act does not require an employee representative to accompany the compliance officer. After the opening conference, the compliance officer conducts a walkthrough inspection of the facility. During the walkthrough inspection, the compliance officer determines which route to take and to which employees he or she will talk with. The compliance officer may review records, collect air samples, measure noise readings, or photograph and videotape certain areas. Once the compliance officer has completed the walkthrough inspection, he or she will hold a closing conference. During the closing conference, the compliance officer gives the employer and all other interested parties a copy of the Employer Rights and Responsibilities Following an OSHA Inspection (OSHA 3000) for their review and discussion. The compliance officer discusses with the employer all unsafe or unhealthful conditions observed during the inspection and indicates the violations for which he or she may recommend a citation and a proposed penalty. At this time, the compliance officer will also inform the employer of the appeal process.

OSHA Citations

After the compliance officer files his or her report with the Area Director, it is the Area Director who determines whether he or she will issue a citation or propose penalties. The Area Director will send all citations via certified mail. Once the employer has received the citation, they must post the citation for *3 days or until the violation has been abated*, whichever is longer.

OSHA Citation Penalties

The categories of violations that are cited and the penalties that may be proposed are as follows:

Other-than-Serious Violation: A violation that has a direct relationship to job safety and health but probably would not cause death or serious physical harm. OSHA may assess a penalty from $0 to $1000 for each violation. The agency may adjust a penalty for an other-than-serious violation downward by as much as 95%, depending on the employer's good faith, history of previous violations, and size of business.

Serious Violation: A violation where there is a substantial probability that death or serious physical harm could result. OSHA assesses the penalty for a serious violation from $1500 to $7000 depending on the gravity of the violation. OSHA may adjust a penalty for a serious violation downward on the basis of the employer's good faith, history of previous violations, and size of business.

Willful Violation: A violation that the employer intentionally and knowingly commits. The employer is aware that a hazardous condition exists, knows that the condition violates a standard or other obligation of the Act, and makes no reasonable effort to eliminate it. OSHA may propose penalties of up to $70,000 for each willful violation. The minimum willful penalty is $5000. An employer who is convicted in a criminal proceeding of a willful violation of a standard that has resulted in the death of an employee may be fined up to $250,000 (or $500,000 if the employer is a corporation) or imprisoned up to 6 months, or both. A second conviction doubles the possible term of imprisonment.

Repeated Violation: A violation of any standard, regulation, rule, or order where, upon reinspection, a substantially similar violation is found and the original citation has become a final order. Violations can bring a fine of up to $70,000 for each such violation with the previous 3 years. To calculate repeated violations, OSHA adjusts the initial penalty for the size and then multiplies by a factor of 2, 5, or 10, depending on the size of the business.

Failure to Abate Violation: Failure to correct a prior violation may bring a civil penalty of up to $7000 for each day that the violation continues beyond the prescribed abatement date.

Potential Other Penalties: Additional violations for which OSHA may issue citations and proposed penalties are as follows:

- Falsifying records, reports, or applications can, upon conviction, bring a criminal fine of $10,000 or up to 6 months in jail, or both.

- Violating posting requirements may bring a civil penalty of $7000.

- Assaulting a compliance officer or otherwise resisting, opposing, intimidating, or interfering with a compliance officer in the performance of his or her duties is a criminal offense and is subject to a fine of not more than $5000 and imprisonment for not more than 3 years.

NOTE: Citations and penalty procedures may differ slightly in states with their own occupational safety and health programs.

Appeals

Once a citation or non-citation is issued, an employee or an employer may appeal the decision by the Area Director. The appeals process is different for the employer than it is for the employee.

Employee Appeals

If an employee complaint initiates an inspection, the employee or authorized employee representative may request an informal review of any decision not to issue a citation. Employees may not contest citations, amendments to citations, proposed penalties, or lack of penalties. They may, however, contest the time allowed for abatement of a hazardous condition. They also may request an employer's "Petition for Modification of Abatement," which requests an extension of the proposed abatement period. Employees must contest the petition within 10 working days of its posting or within 10 working days after an authorized employee receives a copy. Employees may request an informal conference with OSHA to discuss any issues raised by an inspection, citation, notice of proposed penalty, or employer's notice of intention to contest.

Employer Appeals

Within 15 working days of receiving a citation, an employer who wishes to contest must submit a written objection to OSHA. The OSHA Area Director forwards the objection to the Occupational Safety and Health Review Commission (OSHRC), which operates independently of OSHA. The OSHRC is a commission of three member administrative law judges appointed by the President of the United States, with the consent of congress. Each judge appointed serves a term of 6 years. Initial appointments were 2 years for the first judge, 4 years for the second judge, and 6 years for the third judge. Each succeeding judge is appointed for a term of 6 years. This provides some administrative consistency within the commission.

When issued a citation and notice of proposed penalty, an employer may request an informal meeting with OSHA's Area Director to discuss the case. OSHA encourages employers to have informal conferences with the Area Director if the employer has issues arising from the inspection that he or she wishes to discuss or provide additional information. The Area Director is authorized to enter into settlement agreements that revise citations and penalties to avoid prolonged legal disputes and result in speedier hazard abatement. (Alleged violation contested before OSHRC do not need to be corrected until the contest is ruled upon by OSHRC.)

Petition for Modification of Abatement

After receiving a citation, the employer must correct the cited hazard by the abatement date unless he or she contests the citation or abatement date. Factors beyond the employer's control, however, may prevent the completion of corrections by that date. In such a situation, the employer who has made a good-faith effort to comply may file a petition to modify the abatement date.

The written petition must specify the steps taken to achieve compliance, the additional time needed to comply, the reasons additional time is needed, and interim steps taken to safeguard employees against the cited hazard during the intervening period. The employer must certify that he or she posted a copy of the petition in a conspicuous place at or near each place where a violation occurred and that the employee representative received a copy of the petition.

Notice of Contest

If the employer decides to contest either the citation, the abatement period, or the proposed penalty, he or she has *15 working days* from the time the citation and proposed penalty are received to notify the OSHA Area Director in writing. Failure to do so results in the citation and proposed penalty becoming a final order of the OSHRC without further appeal. An orally expressed disagreement will not suffice. This written notification is called a "Notice of Contest."

Although there is no specific format for the Notice of Contest, it must clearly identify the employer's basis for filing—the citation, notice of proposed penalty, abatement period, or notification of failure to correct violations.

The employer must give a copy of the Notice of Contest to the employees' authorized representative. If any affected employees are not represented by a recognized bargaining agent, the employer must post a copy of the notice in a prominent location in the workplace or give it personally to each unrepresented employee.

Review Procedure

If the employer files a written Notice of Contest within the required 15 working days, the OSHA Area Director forwards the case to OSHRC. The commission is an independent agency not associated with OSHA or the Department of Labor. The commission assigns the case to an administrative law judge.

OSHRC may schedule a hearing at a public place near the employer's workplace. The employer and the employee have the right to participate in the hearing; the OSHRC does not require them to be represented by attorneys. Once the administrative law judge has ruled, any party to the case may request a further review by OSHRC. Any of the three OSHRC commissioners may also, at his or her own motion, bring a case before the commission for review. Employers and other parties may appeal commission ruling to the appropriate US Court of Appeals.

NOTE: The sections included above from OSHA Inspection Process through Review Procedure were taken from an OSHA web pamphlet (OSHA 2098 Rev. 2002) that requires no copyright permissions because it is in the public domain. However, it is the author's preference to give credit for work that is not his own.

States with their own occupational safety and health programs have a state system for review and appeal of citations, penalties, and abatement periods. The procedures are generally similar to Federal OSHAs, but a state review board or equivalent authority hears cases.

Hazard Communication Standard (29 CFR 1910.1200)

NOTE: This standard has undergone tremendous changes since the initial publication of this book. Therefore, at the end of this chapter, I have elected to republish OSHA's Comparison of Hazard Communication Requirements between OSHA Hazard Communication Standard 29 CFR 1910.1200 (HCS) and Globally Harmonized System (GHS).

Purpose

This standard is still known as the Hazard Communication Standard but is unofficially known as Globally Harmonized System (GHS). However, it has undergone significant changes in the last few years. The purpose of this section is to ensure that the hazards of all chemicals produced or imported are classified and that information concerning the classified hazards is transmitted to employers and employees. The requirements of this section are intended to be consistent with the provisions of the *United Nations Globally Harmonized System of Classification and Labeling of Chemicals (GHS), Revision 3.* The transmittal of information is to be accomplished by means of comprehensive hazard communication programs, which are to include container labeling and other forms of warning, safety data sheets (SDSs), and employee training.

Scope and Application

This standard requires chemical manufacturers or importers to classify the hazards of chemicals that they produce or import and all employers to provide information to their employees about the hazardous chemicals to which they are exposed, by means of a hazard communication program, labels and other forms of warning, SDSs, and information and training. In addition, this standard requires distributors to transmit the required information to employers. (Employers who do not produce or import chemicals need only focus on those parts of the rule that deal with establishing a workplace program and communicating information to their workers.)

This standard applies to any chemical that is known to be present in the workplace in such a manner that employees may be exposed under normal conditions of use or in a foreseeable emergency.

Written Hazard Communication Standard

Employers are required to develop, implement, and maintain at each workplace a written hazard communication program that at least describes how the program will be managed and operated, including the requirements for labeling and other forms of warning, SDSs, and employee information and training, and how these requirements will be met, which also includes the following:

- A list of the hazardous chemicals known to be present using a product identifier that is referenced on the appropriate SDS (the list may be compiled for the workplace as a whole or for individual work areas)
- The methods the employer will use to inform employees of the hazards of nonroutine tasks (e.g., the cleaning of reactor vessels) and the hazards associated with chemicals contained in unlabeled pipes in their work areas
- *Multi-employer workplaces.* Employers who produce, use, or store hazardous chemicals at a workplace in such a way that the employees of other employers may be exposed shall additionally ensure that the hazard communication programs developed and implemented are transmitted to those employees and employers.

Label and Other Forms of Warning

The chemical manufacturer, importer, or distributor shall ensure that each container of hazardous chemicals leaving the workplace is labeled, tagged, or marked. Hazards not otherwise classified do not have to be addressed on the container. Where the chemical manufacturer or importer is required to label, tag, or mark, the following shall be provided:

- Product identifier
- Signal word
- Hazard statement(s)
- Pictogram(s)
- Precautionary statement(s)
- Name, address, and telephone number of the chemical manufacturer or other responsible party

Safety Data Sheets

Chemical manufacturers and importers shall obtain or develop an SDS for each hazardous chemical they produce or import. Employers shall have an SDS in the workplace for each hazardous chemical that they use. The chemical manufacturer or importer preparing the SDS shall ensure that it is in English (although the employer may maintain copies in other languages as well) and includes at least the following section numbers and headings, and associated information under each heading, in the order listed below: (See Appendix D to 29 CFR 1910.1200—Safety Data Sheets, for the specific content of each section of the SDS.)

- Section 1: Identification
- Section 2: Hazard identification
- Section 3: Composition/information on ingredients
- Section 4: First-aid measures
- Section 5: Fire-fighting measures
- Section 6: Accidental release measures
- Section 7: Handling and storage
- Section 8: Exposure controls/personal protection
- Section 9: Physical and chemical properties
- Section 10: Stability and reactivity information
- Section 11: Toxicological information

NOTE 1: To be consistent with the GHS, an SDS must also include the following headings in this order:

- Section 12: Ecological information
- Section 13: Disposal considerations
- Section 14: Transport information
- Section 15: Regulatory information

Employee Information and Training

Employee training and information is at the core of this standard. Employers shall provide employees with effective information and training on hazardous chemicals in their work area at the time of their initial assignment and whenever a new physical or health hazard the employees have not previously been trained about is introduced into their work area. Information and training may be designed to cover categories of hazards or specific chemicals. Chemical-specific information must always be available through labels and material SDSs.

Training

Employee training shall include at least the following information:

- Methods and observations that may be used to detect the presence or release of hazardous chemicals in the work area
- The physical and health hazards of the chemicals in the work area
- The measure employees can take to protect themselves from these hazards, including specific procedures the employer has implemented to protect employees from exposure to hazardous chemicals
- The details of the hazard communication program developed by the employer, including the explanation of the labeling system and the material SDS, and how employees can obtain and use the appropriate hazard information

Bloodborne Pathogen Standard (29 CFR 1910.1030)

Scope, Application, and Definitions

The information provided in this standard applies to all occupation exposure to blood or other potentially infectious materials present in the workplace. Definitions within this standard are as follows:

Blood means human blood, human blood components, and products made from human blood.

Bloodborne pathogens means pathogenic microorganisms that are present in human blood and cause disease in humans. These pathogens include, but are not limited to, hepatitis B virus (HBV) and human immunodeficiency virus (HIV).

Other potentially infectious materials means the following human body fluids: semen, vaginal secretions, cerebrospinal fluid, synovial fluid, pleural fluid, pericardial fluid, peritoneal fluid, amniotic fluid, saliva in dental procedures, any body fluid that is visibly contaminated with blood, and all body fluids in situations where it is difficult or impossible to differentiate between body fluids or any unfixed tissue or organ from a human.

Exposure Control Plan

Each employer having an employee (or employees) with occupational exposure or potential exposure to bloodborne pathogens shall establish a written Exposure Control Plan designed to eliminate or minimize employee exposure. The Exposure Control Plan shall contain at least the following information:

- The exposure determination
- The schedule and method of implementation for Methods of Compliance, HIV and HBV Research Laboratories and Production Facilities, Hepatitis B Vaccination and Postexposure Evaluation and Follow-up, Communication of Hazards to Employees, and Record Keeping

The Exposure Control Plan shall be reviewed and updated at least annually and whenever necessary to reflect new or modified tasks and procedures that affect occupational exposure and to reflect new or revised employee positions with occupational exposure.

Hepatitis B Vaccination and Postexposure Follow-Up

The employer shall make available the hepatitis B vaccine and vaccination series to all employees who have occupational exposure and postexposure evaluation and follow-up to all employees who have had an exposure incident. Should an employee refuse to take the hepatitis B vaccine, the employer is required to obtain a written statement of his or her refusal.

Communication of Hazards

Labels and signs shall be affixed to containers of regulated waste, refrigerators and freezers containing blood or other potentially infectious materials, and other containers used to store, transport, or ship blood or other potentially infectious materials.

Record Keeping

Medical Records

The employer shall establish and maintain an accurate record for each employee with occupational exposure. This record shall include the following:

- The name and social security number of the employee
- A copy of the employee's hepatitis B vaccination status including the dates of all the hepatitis B vaccinations and any medical records relative to the employee's ability to receive vaccination
- A copy of all results of examinations, medical testing, and follow-up procedures
- The employer's copy of the health care professional's written opinion
- A copy of the information provided to the health care professional

The employer shall ensure that employee medical records and information are maintained in the strictest of confidence. The information contained in the medical records may not be disclosed or reported without the employee's express written consent, except as required by this standard. The employer shall maintain the records required under this standard for at least the duration of employment plus 30 years.

Training Records

Training records shall include the following information:

- The dates of the training sessions
- The contents or a summary of the training sessions
- The names and qualification of persons conducting the training
- The names and job titles of all persons attending the training sessions

Training records shall be maintained for 3 years from the date on which the training occurred.

Control of Hazardous Energy Standard (29 CFR 1910.147)

Scope, Application and Purpose

This standard covers the servicing and maintenance of machines and equipment in which the unexpected energization or start-up of the machines or equipment or the release of stored energy could cause injury to employees.

This standard establishes minimum performance requirements for the control of such hazardous energy. This standard applies to the control of energy during servicing or maintenance of machines and equipment.

Definitions

Affected employee. An employee whose job requires him or her to operate or use a machine or equipment on which servicing or maintenance is being performed under lockout or tagout, or whose job requires him or her to work in an area in which such servicing or maintenance is being performed.

Authorized employee. A person who locks or tags out machines or equipment in order to perform servicing or maintenance on a machine or equipment. An affected employee becomes an authorized employee when that employee's duties include performing servicing or maintenance covered by this standard.

Capable of being locked out. An energy-isolating device capable of being locked out if it has a hasp or other means of attachment to which a lock can be affixed or it has a locking mechanism built into it. Other energy-isolating devices are capable of being locked out, if lockout can be achieved without the need to dismantle, rebuild, or replace the energy-isolating device or permanently alter its energy control capability.

Energized. Connected to an energy source or containing a residual or stored energy.

Energy-isolating device. A mechanical device that physically prevents the transmission or release of energy, including but not limited to the following: a manually operated electrical circuit breaker; a disconnect switch; a manually operated switch by which the conductors of a circuit can be disconnected for all ungrounded supply conductors, and, in addition, no pole can be operated independently; a line valve; a block; and any similar device used to block or isolate energy. Push buttons, selector switches, and other control circuit-type devices are not energy-isolating devices.

Energy source. Any source of electrical, mechanical, hydraulic, pneumatic, chemical, thermal, or other energy.

Hot tap. A procedure used in the repair, maintenance, and services activities that involves welding on a piece of equipment under pressure, in order to install connections or appurtenances.

Lockout. The placement of a lockout device on an energy-isolating device, in accordance with established procedure, ensuring that the energy-isolating device and the equipment being controlled cannot be operated until the lockout device is removed.

Lockout device. A device that utilizes a positive means, such as a lock, either key or combination type, to hold an energy-isolating device in a safe position and prevent the energizing of a machine or equipment.

Normal production operations. The utilization of a machine or equipment to perform its intended production function.

Servicing or maintenance. Workplace activities such as constructing, installing, setting up, adjusting, inspecting, modifying, and maintaining or servicing machines or equipment.

Setting up. Any work performed to prepare a machine or equipment to perform its normal production operation.

Tagout. The placement of a tagout device on an energy-isolating device, in accordance with established procedure, to indicate that the energy-isolating device and the equipment being controlled may not be operated until the tagout device is removed.

Tagout device. A prominent warning device, such as a tag and a means of attachment, which can be securely fastened to an energy-isolating device in accordance with established procedure, to indicate that the energy-isolating device and the equipment being controlled may not be operated until the tagout device is removed.

Energy Control Program

The employer shall establish a program consisting of energy control procedures, employee training, and periodic inspections to ensure that before any employee performs any servicing or maintenance on a machine or equipment where the unexpected energizing, start-up, or release of stored energy could occur and cause injury, the machine or equipment shall be isolated from the energy source and rendered inoperative.

Periodic Inspection

The employer shall conduct a periodic inspection of the energy control procedure at least annually to ensure that the procedure and the requirements of this standard are being followed. The periodic inspection shall be performed by an authorized employee other than the one(s) utilizing the energy control procedure being inspected. The periodic inspection shall be conducted to correct any deviations or inadequacies identified. The employer shall certify that the periodic inspections have been performed. The certification shall identify the machine or equipment on which the energy control procedure was being utilized, the date of the inspection, the employees included in the inspection, and the person performing the inspection.

Training and Communication

The employer shall provide training to ensure that the purpose and function of the energy control program are understood by employees and that the knowledge and skills required for the safe application, usage, and removal of the energy controls are acquired by employees. The training shall include the following:

- Recognition of applicable hazardous energy sources, the type and magnitude of the energy available in the workplace, and the methods and means necessary for energy isolation and control
- Purpose and use of the energy control procedure
- Prohibition relating to attempts to restart or reenergize machines or equipment that are locked or tagged out

Retraining shall be provided for all authorized and affected employees whenever there is a change in their job assignments, a change in machines, equipment, or processes that present a new hazard, or when there is a change in the energy control procedures. Additional retraining shall also be conducted whenever a periodic inspection reveals inadequacies in the program or procedures. The employer shall certify that employee training has been accomplished and is being kept up to date. The certification shall contain each employee's name and dates of training.

Confined Space Entry Standard (29 CFR 1910.146)

Scope and Application

This standard contains requirements for practices and procedures to protect employees in general industry from the hazards of entry into permit-required confined spaces. This standard does not apply to agriculture, construction, or shipyard employment.

Definitions

Acceptable entry conditions means the conditions that must exist in a permit space to allow entry and to ensure that employees involved with a permit-required confined space entry can safely enter into and work within the space.

Attendant means an individual stationed outside one or more permit spaces who monitors the authorized entrants and who performs all attendant's duties assigned in the employer's permit space program.

Authorized entrant means an employee who is authorized by the employer to enter a permit space.

Confined space means a space that (1) is large enough and so configured that an employee can bodily enter and perform assigned work, (2) has limited or restricted means for entry or exit, and (3) is not designed for continuous employee occupancy.

Engulfment means the surrounding and effective capture of a person by a liquid or finely divided (flowable) solid substance that can be aspirated to cause death by filling or plugging the respiratory system or than can exert enough force on the body to cause death by strangulation, constriction, or crushing.

Entry means the action by which a person passes through an opening into a permit-required confined space.

Entry permit means the written or printed document that is provided by the employer to allow and control entry into a permit space and that contains the information required in the standard.

Entry supervisor means the person responsible for determining if acceptable entry conditions are present at a permit space where entry is planned, for authorizing entry and overseeing entry operations, and for terminating entry as required in the standard.

Hazardous atmosphere means an atmosphere that may expose employees to the risk of death, incapacitation, impairment of ability to self-rescue, injury, or acute illness.

Immediately dangerous to life or health means any condition that poses an immediate or delayed threat to life or that would cause irreversible adverse health effects or that would interfere with an individual's ability to escape unaided from a permit space.

Isolation means the process by which a permit space is removed from service and completely protected against the release of energy and material into the space.

Oxygen-deficient atmosphere means an atmosphere containing less than 19.5% oxygen by volume.

Oxygen-enriched atmosphere means an atmosphere containing more than 23.5% oxygen by volume.

Permit-required confined space means a confined space that has one or more of the following characteristics: (1) contains or has a potential to contain a hazardous atmosphere, (2) contains a material that has the potential for engulfing an entrant, (3) has an internal configuration such that an entrant could be trapped or asphyxiated by inwardly converging walls or by a floor that slopes downward and tapers to a smaller cross section, or (4) contains any other recognized serious safety or health hazard.

Rescue service means the personnel designated to rescue employees from permit spaces.

General Requirements

The employer shall evaluate the workplace to determine if any spaces are permit-required confined spaces. If the workplace contains permit spaces, the employer shall inform exposed employees by posting danger signs, or by any other equally effective means, of the existence and location of and the danger posed by the permit spaces.

Confined Space Entry Program

If the employer decides that its employees will enter permit spaces, the employer shall develop and implement a written permit space program that complies with this standard. The written program shall be available for inspection by employees and their authorized representatives.

Under the confined space entry program, the employer shall

- Implement the measures necessary to prevent unauthorized entry
- Identify and evaluate the hazards of permit spaces before the employee enters them
- Develop and implement the means, procedures, and practices necessary for safe permit space entry operations

Entry Permits

The entry permit that documents compliance with this standard and authorizes entry to a permit space shall include the following:

- The permit space to be entered
- Purpose of the entry
- Date and the authorized duration of the entry permit
- Authorized entrants by name
- Attendant name
- Entry supervisor (by name)
- Measures to isolate the permit space and to eliminate or control permit space hazards before entry

Employees must have the opportunity to observe the monitoring under this standard.

Training

The employer shall provide training so that all employees whose work is regulated by this standard acquire the understanding, knowledge, and skills necessary for the safe performance of their duties. Training shall be provided to each affected employee before the employee is first assigned duties under this standard. Whenever there is a change in assigned duties, the employer determines that there is a discrepancy in the program, or there is a change in permit space operations that present a hazard about which an employee has not previously been trained, and thus additional training is required. The employer shall certify that the training required has been accomplished. The certification shall contain each employee's name, the signatures or initial of the trainers, and the dates of training. The certification shall be available for inspection by employees and their authorized representatives.

Personal Protective Equipment (29 CFR 1910.132)

Application

Protective equipment covered in this standard includes PPE for eyes, face, head, and extremities; protective clothing; respiratory devices; and protective shield and barriers, which shall be provided, used, and maintained in a sanitary and reliable condition wherever it is necessary by reason of hazards of processes or environment, chemical hazards, radiological hazards, or mechanical irritants encountered in a manner capable of causing injury or impairment in the function of any part of the body through absorption, inhalation, or physical contact.

NOTE: OSHA's Final PPE Rule, which was effective February 13, 2008, requires employers to provide PPE, at no cost to the employees. The Final Rule does not require an employer to provide normal safety boots or shoes, but does require the employer to provide specialty boots.

Employee-Owned Equipment

Where employees provide their own protective equipment (of their own choice), the employer shall be responsible to assure its adequacy, including proper maintenance and sanitation of such equipment.

Hazard Assessment and Equipment Selection

The employer shall assess the workplace to determine if hazards are present, or are likely to be present, which necessitate the use of PPE. If such hazards are present, or likely to be present, the employer shall (a) select and have each affected employee use the types of PPE that will protect the affected employee from the hazards identified in the hazard assessment, (b) communicate selection decisions to each affected employee, and (c) select PPE that properly fits each affected employee. The employer shall verify that the required workplace hazard assessment has been performed through a written certification that identifies the workplace evaluated; the person certifying that the evaluation has been performed; the date(s) of the hazard assessment; and, which determines the document as a certification of the hazard assessment.

Training

The employer shall provide training to each employee who is required under this standard to use PPE. Each employee shall be trained to know at least the following:

- When PPE is necessary
- What PPE is necessary
- How to properly don, doff, adjust, and wear PPE
- Limitations of the PPE
- Proper care, maintenance and useful life, and disposal of the PPE

Each affected employee shall demonstrate an understanding of the training and the ability to use PPE properly, before being allowed to perform work requiring the use of PPE. When the employer has reason to believe that any affected employee who has already been trained does not have the understanding and skill required, the employer shall retrain each such employee.

Respiratory Protection Standard (29 CFR 1910.134)

Purpose

The purpose of the Respiratory Protection Standard is to control those occupational diseases caused by breathing air contaminated with harmful dusts, fogs, fumes, mists, gases, smokes, sprays, or vapors. This shall be accomplished as far as feasible by accepted engineering control measures. When effective engineering control measures are not feasible, or while they are being instituted, appropriate respirators shall be used.

Respirators shall be provided by the employer when such equipment is necessary to protect the health of the employee. The employer shall provide the respirators that are applicable and suitable for the purpose intended. The employer shall be responsible for the establishment and maintenance of a respiratory protection program, which shall include the requirements of this standard.

Definitions

Air-purifying respirator means a respirator with an air-purifying filter, cartridge, or canister that removes specific air contaminants by passing ambient air through the air-purifying element.

Assigned protection factor means the protection factor assigned to the respirator type.

Atmosphere-supplying respirator means a respirator that supplies the respirator user with breathing air from a source independent of the ambient atmosphere and includes supplied-air respirators and self-contained breathing apparatus units.

Fit test means the use of a protocol to qualitatively or quantitatively evaluate the fit of a respirator on an individual.

Powered air-purifying respirator means an air-purifying respirator that uses a blower to force the ambient air through air-purifying elements to the inlet covering.

Qualitative fit test means a pass/fail test to assess the adequacy of respirator fit that relies on the individual's response to the test agent.

Quantitative fit test means an assessment of the adequacy of respirator fit by numerically measuring the amount of leakage into the respirator.

Self-contained breathing apparatus means an atmosphere-supplying respirator for which the breathing air source is designed to be carried by the user.

Supplied-air respirator or airline respirator means an atmosphere-supplying respirator for which the source of breathing air is not designed to be carried by the user.

Respiratory Protection Program

This standard requires the employer to develop and implement a written respiratory protection program with required worksite-specific procedures and elements for required respirator use. The program must be administered by a suitably trained program administrator. In addition, certain program elements may be required for voluntary use to prevent potential hazards associated with the use of the respirator. The employer shall include in the written program the following information:

- Procedures for selecting respirators for use in the workplace
- Medical evaluations of employees required to use respirators

- Fit testing procedures
- Procedures for proper use of respirators
- Procedures and schedules for cleaning, disinfecting, storing, inspecting, repairing, discarding, and otherwise maintaining respirators
- Procedures to ensure adequate air quality, quantity, and flow of breathing air for atmosphere-supplying respirators
- Training required for respirator usage
- Procedures for evaluating the effectiveness of the program

Training and Information

This standard requires the employer to provide effective training to employees who are required to use respirators. The training must be comprehensive, must be understandable, and must recur annually, and more often, if necessary. Training must ensure that each employee can demonstrate knowledge and understanding of the topic, and include the following:

- Why respirator protection is necessary and how improper wearing or use can compromise the protection received
- Limitations and capabilities of the respirator and cartridge (filter)
- Inspection and maintenance procedures
- Cleaning, disinfecting, and storage procedures
- Proper wear of the respirator

Retraining shall be administered annually, or when a new process or procedure is implemented that the employee has not been previously trained.

Fall Protection Standard (29 CFR 1926.500–503) (Subpart M)

Scope and Application

This standard sets forth the requirements and criteria for fall protection in construction work areas covered by this standard. Exception: The provisions of this standard do not apply when employees are making inspection, investigation, or assessment of workplace conditions prior to the actual start of construction work or after all construction work has been completed.

Definitions

Anchorage means a secure point of attachment for lifelines, lanyards, or deceleration devices. Anchorage points must be rated to 5000 lb per person attached.

Body harness means straps that may be secured about the employee in a manner that will distribute the fall arrest forces over at least the thighs, pelvis, waist, chest, and shoulders with means for attaching it to other components of a personal fall arrest system.

Connector means a device that is used to couple parts of the personal fall arrest system and positioning device systems together.

Controlled access zone means an area in which certain work may take place without the use of guardrail systems, personal fall arrest systems, or safety net systems and access to the zone is controlled.

Deceleration device means any mechanism, such as a rope grab, rip-stitch lanyard, specially woven lanyard, tearing or deforming lanyards, automatic self-retracting lifelines/lanyards, and so on, that serves to dissipate a substantial amount of energy during a fall arrest, or otherwise limit the energy imposed on an employee during a fall arrest.

Deceleration distance means the additional vertical distance a falling employee travels, excluding lifeline elongation and free fall distance, before stopping, from the point at which the deceleration device begins to operate.

Free fall means the act of falling before a personal fall arrest system begins to apply force to arrest the fall.

Guardrail system means a barrier erected to prevent employees from falling to lower levels.

Lanyard means a flexible line of rope, wire rope, or strap that generally has a connector at each end for connecting the body harness to a deceleration device, lifeline, or anchorage.

Leading edge means the edge of a floor, roof, or formwork for a floor or other walking/working surface that changes location as additional floor, roof, decking, or formwork sections are placed, formed, or constructed. A leading edge is considered to be an "unprotected side and edge" during periods when it is not actively and continuously under construction.

Lifeline means a component consisting of a flexible line for connection to an anchorage at one end to hang vertically (vertical lifeline), or for connection to anchorages at both ends to stretch horizontally (horizontal lifeline), and which serves as a means for connecting other components of a personal fall arrest system to the anchorage.

Personal fall arrest system means a system used to arrest an employee in a fall from a working level. It consists of an anchorage, connectors, and body harness and may include a lanyard, deceleration device, lifeline, or suitable combinations of these.

Walking/working surface means any surface, whether horizontal or vertical, on which an employee walks or works, including, but not limited to, floors, roofs, ramps, bridges, runways, formwork, and concrete reinforcing steel, but not including ladders, vehicles, or trailers, on which employees must be located in order to perform their duties.

Duty to Have Fall Protection

The employer shall determine if the walking/working surfaces on which its employees are to work have the strength and structural integrity to support employees safely. Employees shall be allowed to works on those surfaces only when the surfaces have the requisite strength and structural integrity.

Each employee on a walking/working surface (horizontal and vertical surface) with an unprotected side or edge which is 6 ft or more above a lower level shall be protected from falling by the use of guardrail systems, safety net systems, or personal fall arrest system.

Training

The employer shall provide a training program for each employee who might be exposed to fall hazards. The program shall enable each employee to recognize the hazards of falling and shall train each employee in the procedures to be followed in order to minimize these hazards. The employer shall assure that each employee has been trained, as necessary, by a competent person qualified to teach the following information:

- Nature of fall hazards in the work area
- Correct procedures for erecting, maintaining, disassembling, and inspecting fall protection systems to be used
- Use and operation of guardrail systems, personal fall arrest systems, safety net systems, warning line systems, safety monitoring systems, controlled access zones, and other protection to be used
- The role of each employee in the safety monitoring system in which this system is used
- The limitations on the use of mechanical equipment during the performance of roofing work on low-sloped roofs

- Correct procedures for the handling and storage of equipment and materials and the erection of overhead protection
- The role of employers in fall protection plans
- The specific requirements of the standard

The employer shall verify compliance with the standard by preparing a written certification record. The written certification record shall contain the name or other identity of the employee trained, the date(s) of the training, and the signature of the person who conducted the training or the signature of the employer. The latest training certificate shall be maintained. When the employer has reason to believe that any affected employee who has already been trained does not have the understanding and skill required to work safely, the employer shall retrain each such employee or whenever new equipment or systems are installed.

Record Keeping

OSHA 300, 300-A, and 301 Forms (29 CFR 1910.29)

Employers must use OSHA 300, 300-A, and 301 forms, or equivalent forms, for recordable injuries and illnesses. The OSHA 300 form is called the Log of Work-Related Injuries and Illnesses, the 300-A form is the Summary of Work-Related Injuries and Illnesses, and the OSHA 301 form is called the Injury and Illness Incident Report.

How quickly must each injury or illness be recorded?

An employer must enter each recordable injury or illness on the 300 and 301 incident report with *seven calendar days* of receiving information that a recordable injury or illness has occurred.

How are "privacy cases" listed on the forms?

The following cases are considered to be privacy cases and therefore the employee's name is not entered on the OSHA 300 form. Instead, enter "PRIVACY CASE" in place of the employee's name.

- An injury or illness to an intimate body part or the reproductive system
- An injury or illness resulting from a sexual assault
- Mental illnesses
- HIV infection, hepatitis, or tuberculosis

- Needlestick injuries and cuts from sharp objects that are contaminated with another person's blood or other potentially infectious materials
- Other illnesses if the employee voluntarily requests that his or her name not be entered on the log

Annual Summary

An annual summary is created on the OSHA 300 log and is certified by a company owner or designated representative. The OSHA 300-A form is signed by a company owner or designated representative and posted in a conspicuous location no later than February 1 of the year following the year covered by the records. The OSHA 300-A form shall remain posted until April 30th of the year following the year covered by the records. OSHA 300 logs (and separate privacy case files, if required) shall be maintained for a period of 5 years following the end of the calendar year that these records cover.

Providing Records to Government Representatives

When an authorized government representative asks for the records you keep under 1904, you must provide copies of the records within four business hours. Authorized government representatives under this standard include

- A representative of the Secretary of Labor
- A representative of the Secretary of Health and Human Services
- A representative of a State agency responsible for administering a state plan approved under the Act

Reporting Fatalities and Multiple Hospitalizations (29 CFR 1904.39)

OSHA requires an employer to notify them within 8 h after the death of any employee from a work-related incident or multiple hospitalizations of employees. This reporting must be orally by telephone or in person to the Area Office of the OSHA nearest the location of the incident. The toll free number to report this is 1-800-321-6742. One exception to this is a fatality or hospitalizations resulting from a motor vehicle accident occurring on a public road. These do not have to be reported.

Determining Recordable Injuries or Illnesses

In general, an employer must consider an injury or illness to be recordable, if it results in any of the following:

- Death
- Days away from work

- Restricted work or transfer to another job
- Medical treatment beyond first aid
- Loss of consciousness
- Injury or illness diagnosed by a physician or other licensed health care professional

NOTE: Determining whether an injury or illness is work related may appear simple on the surface. However, this is not always the case. In the case where an employer believes that the injury or illness is personal and not work related, he or she must record the injury or illness on the OSHA 300 and 301 forms, until such time as it is definitively determined to be non-work related.

Calculating Total Case Incident Rates

To calculate a company's total case incident rate (TCIR), use the following equation:

$$TCIR = \frac{\text{No. of injury or illness cases} \times 200{,}000}{\text{Total number of hours worked}}.$$

Example

A company has two recordable injury cases and one days away or restricted case for a total of three cases. The company has worked a total of 278,942 h for the year. Calculate the TCIR for this company.

$$TCIR = \frac{3 \times 200{,}000}{278{,}942 \text{ h}}$$

$$TCIR = \frac{600{,}000}{278{,}942}$$

$$TCIR = 2.15.$$

The company's TCIR for the year is 2.15. This rate can be compared to the Bureau of Labor and Statistics average rating for your particular Standard Industry Code category. The constant of 200,000 is based on 100 employees working 2000 h/year. Therefore, this rate is stating that for every 100 employees, 2.15 of them have sustained an injury or illness as a result of a work-related accident.

Calculating Days Away, Restricted, or Transfer Rates

To calculate the days away, restricted, or transfer (DART) rate, use only those injury cases (included in the TCIR) that resulted in days away, restricted, or transfer from job. The equation is as follows:

$$DART = \frac{No.\ of\ DART\ cases \times 200{,}000}{Total\ number\ of\ hours\ worked}.$$

Example

As in the previous example, use 1 DART case and 278,942 total hours work to calculate the DART rate.

$$DART = \frac{1\ case \times 200{,}000}{278{,}942}.$$

$$DART = \frac{200{,}000}{278{,}942}$$

$$DART = 0.72.$$

Calculating Severity Rates

To calculate severity rates, use the following equation:

$$Severity\ rate = \frac{No.\ of\ lost\ work\ days \times 200{,}000}{Total\ number\ of\ hours\ worked}.$$

Example

Company XYZ had two recordable injuries with one of them resulting in 52 days of lost time. The total number of hours worked were 278,942 h.

$$Severity\ rate = \frac{52 \times 200{,}000}{278{,}942}$$

$$Severity\ rate = \frac{10{,}400{,}000}{278{,}942}$$

$$Severity\ rate = 37.28.$$

Key Information to Remember on Regulations

1. The Occupational Safety and Health Act (Public Law 91-596) was passed into law on December 29, 1970.

2. OSHA regulations do not apply to all employers in the public sector (municipal, county, state, or federal), self-employed individuals, family members operating a farm, or domestic household workers.

3. Horizontal standards are those standards that apply to all industries and employers.

4. Vertical standards are those standards that apply only to particular industries and employers.

5. Section 5(a)(1) of the OSH Act of 1970 is the General Duty Clause.

6. Once an employer receives a citation, he or she must post the citation in a conspicuous location for a period of 3 days or until the violation has been abated, whichever is longer.

7. If an employer decides to contest a citation or abatement period, or the proposed penalty, he or she has 15 working days from the time the citation or proposed penalty is received to notify the OSHA Area Director in writing.

8. If an employee who has received an exposure to bloodborne pathogens refuses to take the hepatitis B vaccination, he or she must sign a refusal statement, which is maintained on file with the employer.

9. Employee medical records, under the Bloodborne Pathogen standard, must be maintained on file for the duration of employment plus 30 years.

10. A work-related recordable injury must be recorded on the OSHA 300 and 301 forms within seven working days of receiving notification of the injury or illness.

11. When an authorized government representative asks for records required in 29 CFR 1904, an employer must provide copies within 4 h.

12. A work-related fatality or multiple employee hospitalizations must be reported to OSHA within 8 h.

Comparison of Hazard Communication Requirements

OSHA HAZARD COMMUNICATION STANDARD 29 CFR 1910.1200 (HCS) AND GLOBALLY HARMONIZED SYSTEM (GHS)

(https://www.osha.gov/dsg/hazcom/ghoshacomparison.html)

INTRODUCTION

The Globally Harmonized System (GHS) is not in itself a regulation or a model regulation. It is a framework from which competent authorities may select the appropriate harmonized classification and communication elements. Competent authorities will decide how to apply the various elements of the GHS within their systems based on their needs and the target audience.

The GHS includes the following elements:

a. Harmonized criteria for classifying substances and mixtures according to their health, environmental, and physical hazards
b. Harmonized hazard communication elements, including requirements for labeling and material safety data sheets

The harmonized elements of the GHS may be seen as a collection of building blocks from which to form a regulatory approach. While the full range is available to everyone, and should be used if a country or organization chooses to cover a certain effect when it adopts the GHS, the full range does not have to be adopted. This constitutes the GHS building block approach.

Competent authorities, such as OSHA, will determine how to implement the elements of the GHS within their systems. This document compares the GHS elements to the OSHA Hazard Communication Standard (HCS) elements. The competent authority allowances/decision points and the selection of building blocks are addressed in Section VI.

This Comparison of Hazard Communication Requirements document includes the following segments:

- General provisions comparison
- Health hazard comparison
- Physical hazard comparison
- Label comparison
  - GHS and transport pictograms
  - Label examples
- MSDS comparison
- GHS competent authority allowances and building block discussion

Purpose

Comparison

The purpose of the HCS and that of the GHS are consistent. The HCS is one of the major existing systems that was to be harmonized by the GHS.

OSHA HCS 29 CFR 1910.1200	GHS
29 CFR 1910.1200 (a)(1) Purpose	**1.1.1 Purpose**
The purpose of this section is to ensure that the hazards of all chemicals produced or imported are evaluated and that information concerning their hazards is transmitted to employers and employees. This transmittal of information is to be accomplished by means of comprehensive hazard communication programs, which are to include container labeling and other forms of warning, material safety data sheets, and employee training.	1.1.1.1 The use of chemical products to enhance and improve life is a widespread practice worldwide. But alongside the benefits of these products, there is also the potential for adverse effects to people or the environment. As a result, a number of countries or organizations have developed laws or regulations over the years that require information to be prepared and transmitted to those using chemicals, through labels or Safety Data Sheets (SDSs). Given the large number of chemical products available, individual regulation of all of them is simply not possible for any entity. Provision of information gives those using chemicals the identities and hazards of these chemicals and allows the appropriate protective measures to be implemented in the local use settings.
(a)(2)	1.1.1.2 While these existing laws or regulations are similar in many respects, their differences are significant enough to result in different labels or SDSs for the same product in different countries. Through variations in definitions of hazards, a chemical may be considered flammable in one country but not in another. Or it may be considered to cause cancer in one country but not in another. Decisions on when or how to communicate hazards on a label or SDS thus vary around the world, and companies wishing to be involved in international trade must have large staffs of experts who can follow the changes in these laws and regulations and prepare different labels and SDSs. In addition, given the complexity of developing and maintaining a comprehensive system for classifying and labeling chemicals, many countries have no system at all.
This occupational safety and health standard is intended to address comprehensively the issue of evaluating the potential hazards of chemicals, and communicating information concerning hazards and appropriate protective measures to employees and to preempt any legal requirements of a state, or political subdivision of a state, pertaining to this subject. Evaluating the potential hazards of chemicals, and communicating information concerning hazards and appropriate protective measures to employees, may include, for example, but is not limited to,	1.1.1.3 Given the reality of the extensive global trade in chemicals, and the need to develop national programs to ensure their safe use, transport, and disposal, it was recognized that an internationally harmonized approach to classification and labeling would provide the foundation for such programs. Once countries have consistent and appropriate information on the chemicals they import or produce in their own countries, the infrastructure to control chemical exposures and protect people and the environment can be established in a comprehensive manner.

(Continued)

Purpose

Comparison

The purpose of the HCS and that of the GHS are consistent. The HCS is one of the major existing systems that was to be harmonized by the GHS.

OSHA HCS 29 CFR 1910.1200	GHS
provisions for developing and maintaining a written hazard communication program for the workplace, including lists of hazardous chemicals present; labeling of containers of chemicals in the workplace, as well as of containers of chemicals being shipped to other workplaces; preparation and distribution of material safety data sheets to employees and downstream employers; and development and implementation of employee training programs regarding hazards of chemicals and protective measures. Under Section 18 of the Act, no state or political subdivision of a state may adopt or enforce, through any court or agency, any requirement relating to the issue addressed by this Federal standard, except pursuant to a Federally approved state plan.	1.1.1.4 Thus, the reasons for setting the objective of harmonization were many. It is anticipated that, when implemented, the GHS will a. Enhance the protection of human health and the environment by providing an internationally comprehensible system for hazard communication b. Provide a recognized framework for those countries without an existing system c. Reduce the need for testing and evaluation of chemicals d. Facilitate international trade in chemicals whose hazards have been properly assessed and identified on an international basis 1.1.1.5 The work began with examination of existing systems and determination of the scope of the work. While many countries had some requirements, the following systems were deemed to be the "major" existing systems and were used as the primary basis for the elaboration of the GHS: a. Requirements of systems in the United States of America for the workplace, consumers, and pesticides b. Requirements of Canada for the workplace, consumers, and pesticides c. European Union directives for classification and labeling of substances and preparations d. The United Nations Recommendations on the Transport of Dangerous Goods 1.1.1.6 The requirements of other countries were also examined as the work developed, but the primary task was to find ways to adopt the best aspects of these existing systems and develop a harmonized approach. This work was done based on agreed principles of harmonization that were adopted early in the process:

a. The level of protection offered to workers, consumers, the general public, and the environment should not be reduced as a result of harmonizing the classification and labeling systems.

b. The hazard classification process refers principally to the hazards arising from the intrinsic properties of chemical elements and compounds and mixtures thereof, whether natural or synthetic.

c. Harmonization means establishing a common and coherent basis for chemical hazard classification and communication, from which the appropriate elements relevant to means of transport, consumer, worker, and environment protection can be selected.

d. The scope of harmonization includes both hazard classification criteria and hazard communication tools, for example, labeling and chemical safety data sheets, taking into account especially the four existing systems identified in the ILO report.

e. Changes in all these systems will be required to achieve a single globally harmonized system; transitional measures should be included in the process of moving to the new system.

f. The involvement of concerned international organizations of employers, workers, consumers, and other relevant organizations in the process of harmonization should be ensured.

g. The comprehension of chemical hazard information, by the target audience, for example, workers, consumers, and the general public, should be addressed.

h. Validated data already generated for the classification of chemicals under the existing systems should be accepted when reclassifying these chemicals under the harmonized system.

i. A new harmonized classification system may require adaptation of existing methods for testing of chemicals.

j. In relation to chemical hazard communication, the safety and health of workers, consumers, and the public in general, as well as the protection of the environment, should be ensured while protecting confidential business information, as prescribed by the competent authorities.

Scope

Comparison

The GHS scope clarification is consistent with the HCS exemptions and labeling exceptions. Consumer products and pharmaceuticals are specifically addressed in the GHS scope. The HCS includes the GHS scope. The HCS includes laboratories, sealed containers, and distributors, while as a framework for systems, the GHS does not include these specific issues.

The GHS addresses testing in the scope section. The HCS addresses testing under hazard determination. The GHS and HCS do not require testing for health hazards. All the physical hazards in the HCS are not linked to specific test methods (as is the case in the GHS) and testing for physical hazards is not required.

OSHA HCS 29 CFR 1910.1200	GHS
29 CFR 1910.1200 (b) Scope and Application	**1.1.2 Scope**
(b)(1)	1.1.2.1 The GHS includes the following elements:
This section requires chemical manufacturers or importers to assess the hazards of chemicals that they produce or import, and all employers to provide information to their employees about the hazardous chemicals to which they are exposed, by means of a hazard communication program, labels and other forms of warning, material safety data sheets, and information and training. In addition, this section requires distributors to transmit the required information to employers. (Employers who do not produce or import chemicals need only focus on those parts of this rule that deal with establishing a workplace program and communicating information to their workers. Appendix E of this section is a general guide for such employers to help them determine their compliance obligations under the rule.)	a. Harmonized criteria for classifying substances and mixtures according to their health, environmental, and physical hazards; and
	b. Harmonized hazard communication elements, including requirements for labeling and safety data sheets
	1.1.2.2 This document describes the classification criteria and the hazard communication elements by type of hazard (e.g., acute toxicity, flammability). In addition, decision logics for each hazard have been developed. Some examples of classification of chemicals in the text, as well as in Annex 7, illustrate how to apply the criteria. There is also some discussion about issues that were raised during the development of the system where additional guidance was thought to be necessary to implement the system.
(b)(2)	1.1.2.3 The scope of the GHS is based on the mandate from the 1992 United Nations Conference on Environment and Development (UNCED) for development of such a system as stated in paragraphs 26 and 27 of Agenda 21, Chapter 19, Program Area B, reproduced below:
This section applies to any chemical that is known to be present in the workplace in such a manner that employees may be exposed under normal conditions of use or in a foreseeable emergency.	*"26. Globally harmonized hazard classification and labeling systems are not yet available to promote the safe use of chemicals, inter alia, at the workplace or in the home. Classification of chemicals can be made for different purposes and is a particularly important tool in establishing labeling systems. There is a need to develop harmonized hazard classification and labeling systems, building on ongoing work;*
(b)(3)	
This section applies to laboratories only as follows:	
(b)(3)(i)	
Employers shall ensure that labels on incoming containers of hazardous chemicals are not removed or defaced.	

27. *A globally harmonized hazard classification and compatible labeling system, including material safety data sheets and easily understandable symbols, should be available, if feasible, by the year 2000."*

This mandate was later analyzed and refined in the harmonization process to identify the parameters of the GHS. As a result, the following clarification was adopted by the Interorganization Program for the Sound Management of Chemicals (IOMC) Coordinating Group to ensure that participants were aware of the scope of the effort:

"The work on harmonization of hazard classification and labeling focuses on a harmonized system for all chemicals, and mixtures of chemicals. The application of the components of the system may vary by type of product or stage of the life cycle. Once a chemical is classified, the likelihood of adverse effects may be considered in deciding what informational or other steps should be taken for a given product or use setting. Pharmaceuticals, food additives, cosmetics, and pesticide residues in food will not be covered by the GHS in terms of labeling at the point of intentional intake. However, these types of chemicals would be covered where workers may be exposed, and, in transport if potential exposure warrants. The Coordinating Group for the Harmonization of Chemical Classification Systems (CG/HCCS) recognizes that further discussion will be required to address specific application issues for some product use categories which may require the use of specialized expertise."

1.1.2.5 In developing this clarification, the CG/HCCS carefully considered many different issues with regard to the possible application of the GHS. There were concerns raised about whether certain sectors or products should be exempted, for example, or about whether or not the system would be applied at all stages of the life cycle of a chemical. Three parameters were agreed in this discussion and are critical to application of the system in a country or region. These are described below:

a. **Parameter 1: The GHS covers all hazardous chemicals. The mode of application of the hazard communication components of the GHS (e.g., labels, safety data sheets) may vary by product category or stage in the life cycle. Target audiences for the GHS include consumers, workers, transport workers, and emergency responders**

(Continued)

(b)(3)(ii)
Employers shall maintain any material safety data sheets that are received with incoming shipments of hazardous chemicals and ensure that they are readily accessible during each work shift to laboratory employees when they are in their work areas.

(b)(3)(iii)
Employers shall ensure that laboratory employees are provided information and training in accordance with paragraph (h) of this section, except for the location and availability of the written hazard communication program under paragraph(h)(2)(iii) of this section.

(b)(3)(iv)
Laboratory employers that ship hazardous chemicals are reconsidered to be either a chemical manufacturer or a distributor under this rule, and thus must ensure that any containers of hazardous chemicals leaving the laboratory are labeled in accordance with paragraph (f)(1) of this section and that a material safety data sheet is provided to distributors and other employers in accordance with paragraphs (g)(6) and (g)(7) of this section.

(b)(4)
In work operations where employees only handle chemicals in sealed containers that are not opened under normal conditions of use (such as are found in marine cargo handling, warehousing, or retail sales), this section applies to these operations only as follows:

(b)(4)(i)
Employers shall ensure that labels on incoming containers of hazardous chemicals are not removed or defaced.

(b)(4)(ii)
Employers shall maintain copies of any material safety data sheets that are received with incoming shipments of the sealed containers of hazardous chemicals, shall obtain a material safety data sheet as soon as possible for sealed containers of hazardous chemicals received without a material safety data sheet if an employee requests the material safety data sheet, and shall ensure that the material safety data sheets are readily accessible during each work shift to employees when they are in their work area(s).

Scope

Comparison

The GHS scope clarification is consistent with the HCS exemptions and labeling exceptions. Consumer products and pharmaceuticals are specifically addressed in the GHS scope. The HCS includes laboratories, sealed containers, and distributors, while as a framework for systems, the GHS does not include these specific issues.

The GHS addresses testing in the scope section. The HCS addresses testing under hazard determination. The GHS and HCS do not require testing for health hazards. All the physical hazards in the HCS are not linked to specific test methods (as is the case in the GHS) and testing for physical hazards is not required.

OSHA HCS 29 CFR 1910.1200	GHS
(b)(4)(iii) Employers shall ensure that employees are provided with information and training in accordance with paragraph (h) of this section (except for the location and availability of the written hazard communication program under paragraph (h)(2)(iii) of this section), to the extent necessary to protect them in the event of a spill or leak of a hazardous chemical from a sealed container. **(b)(5)** This section does not require labeling of the following chemicals: **(b)(5)(i)** Any pesticide as such term is defined in the Federal Insecticide, Fungicide, and Rodenticide Act (7 U.S.C. 136 et seq.), when subject to the labeling requirements of that Act and labeling regulations issued under that Act by the Environmental Protection Agency. **(b)(5)(ii)** Any chemical substance or mixture as such terms redefined in the Toxic Substances Control Act (15 U.S.C. 2601 et seq.), when subject to the labeling requirements of that Act and labeling regulations issued under that Act by the Environmental Protection Agency. **(b)(5)(iii)** Any food, food additive, color additive, drug, cosmetic, or medical or veterinary device or product, including materials intended for use as ingredients in such products (e.g., flavors and fragrances), as such terms are defined in the Federal Food, Drug, and Cosmetic Act (21 U.S.C. 301	i. Existing hazard classification and labeling systems address potential exposures to all potentially hazardous chemicals in all types of use situations, including production, storage, transport, workplace use, consumer use, and presence in the environment. They are intended to protect people, facilities, and the environment. The most widely applied requirements in terms of chemicals covered are generally found in the parts of existing systems that apply to the workplace or transport. It should be noted that the term *chemical* is used broadly in the UNCED agreements and subsequent documents to include substances, products, mixtures, preparations, or any other terms that may be used in existing systems to denote coverage. ii. Since all chemicals and chemical products in commerce are made in a workplace (including consumer products), handled during shipment and transport by workers, and often used by workers, there are no complete exemptions from the scope of the GHS for any particular type of chemical or product. In some countries, for example, pharmaceuticals are currently covered by workplace and transport requirements in the manufacturing, storage, and transport stages of the life cycle. Workplace requirements may also be applied to employees involved in the administration of some drugs, or cleanup of spills and other types of potential exposures in health care settings. SDSs and training must be available for these employees under some systems. It is anticipated that the GHS would be applied to pharmaceuticals in a similar fashion.

et seq.) or the Virus–Serum–Toxin Act of 1913 (21 U.S.C. 151 et seq.), and regulations issued under those Acts, when they are subject to the labeling requirements under those Acts by either the Food and Drug Administration or the Department of Agriculture.

(b)(5)(iv)
Any distilled spirits (beverage alcohols), wine, or malt beverage intended for nonindustrial use, as such terms are defined in the Federal Alcohol Administration Act (27 U.S.C. 201 et seq.) and regulations issued under that Act, when subject to the labeling requirements of that Act and labeling regulations issued under that Act by the Bureau of Alcohol, Tobacco, and Firearms.

(b)(5)(v)
Any consumer product or hazardous substance as those terms are defined in the Consumer Product Safety Act (15 U.S.C. 2051 et seq.) and Federal Hazardous Substances Act (15 U.S.C. 1261 et seq.), respectively, when subject to a consumer product safety standard or labeling requirement of those Acts, or regulations issued under those Acts by the Consumer Product Safety Commission.

(b)(5)(vi)
Agricultural or vegetable seed treated with pesticides and labeled in accordance with the Federal Seed Act (7 U.S.C. 1551 et seq.) and the labeling regulations issued under that Act by the Department of Agriculture.

(b)(6)
This section does not apply to the following:

(b)(6)(i)
Any hazardous waste as such term is defined by the Solid Waste Disposal Act, as amended by the Resource Conservation and Recovery Act of 1976, as amended (42 U.S.C. 6901 et seq.), when subject to regulations issued under that Act by the Environmental Protection Agency.

iii. At other stages of the life cycle for these same products, the GHS may not be applied at all. For example, at the point of intentional human intake or ingestion, or intentional application to animals, products such as human or veterinary pharmaceuticals are generally not subject to hazard labeling under existing systems. Such requirements would not normally be applied to these products as a result of the GHS. (It should be noted that the risks to subjects associated with the medical use of human or veterinary pharmaceuticals are generally addressed in package inserts and are not part of this harmonization process.) Similarly, products such as foods that may have trace amounts of food additives or pesticides in them are not currently labeled to indicate the presence or hazard of those materials. It is anticipated that application of the GHS would not require them to be labeled as such.

b. **Parameter 2: The mandate for development of a GHS does not include establishment of uniform test methods or promotion of further testing to address adverse health outcomes.**

i. Tests that determine hazardous properties, which are conducted according to internationally recognized scientific principles, can be used for purposes of a hazard determination for health and environmental hazards. The GHS criteria for determining health and environmental hazards are test method neutral, allowing different approaches as long as they are scientifically sound and validated according to international procedures and criteria already referred to in existing systems for the hazard class of concern and produce mutually acceptable data. While the OECD is the lead organization for development of harmonized health hazard criteria, the GHS is not tied to the OECD Test Guidelines Program. For example, drugs are tested according to agreed criteria developed under the auspices of the World Health Organization (WHO). Data generated in accordance with these tests would be acceptable under the GHS. Criteria for physical hazards under the

(Continued)

Scope

Comparison

The GHS scope clarification is consistent with the HCS exemptions and labeling exceptions. Consumer products and pharmaceuticals are specifically addressed in the GHS scope. The HCS includes laboratories, sealed containers, and distributors, while as a framework for systems, the GHS does not include these specific issues.

The GHS addresses testing in the scope section. The HCS addresses testing under hazard determination. The GHS and HCS do not require testing for health hazards. All the physical hazards in the HCS are not linked to specific test methods (as is the case in the GHS) and testing for physical hazards is not required.

OSHA HCS 29 CFR 1910.1200	GHS
	UNSCETDG are linked to specific test methods for hazard classes such as flammability and explosivity.
(b)(6)(ii)	ii. The GHS is based on currently available data. Since the
Any hazardous substance as such term is defined by the Comprehensive	harmonized classification criteria are developed on the basis
Environmental Response, Compensation and Liability Act (CERCLA)	of existing data, compliance with these criteria will not
(42 U.S.C. 9601 et seq.) when the hazardous substance is the focus of	require retesting of chemicals for which accepted test data
remedial or removal action being conducted under CERCLA in	already exist.
accordance with the Environmental Protection Agency regulations.	**c. Parameter 3: In addition to animal data and valid in vitro**
	testing, human experience, epidemiological data, and clinical
(b)(6)(iii)	**testing provide important information that should be**
Tobacco or tobacco products.	**considered in application of the GHS.**
	Most of the current systems acknowledge and make use of
(b)(6)(iv)	ethically obtained human data or available human experience.
Wood or wood products, including lumber that will not be processed,	Application of the GHS should not prevent the use of such data,
where the chemical manufacturer or importer can establish that the only	and the GHS explicitly acknowledges the existence and use of all
hazard they pose to employees is the potential for flammability or	appropriate and relevant information concerning hazards or the
combustibility (wood or wood products that have been treated with a	likelihood of harmful effects (i.e., risk).
hazardous chemical covered by this standard and wood that may be	1.1.2.6 Other scope limitations
subsequently sawed or cut, generating dust, are not exempted).	1.1.2.6.1 The GHS is not intended to harmonize risk assessment
	procedures or risk management decisions (such as establishment
(b)(6)(v)	of a permissible exposure limit for employee exposure), which
Articles (as that term is defined in paragraph (c) of this section).	generally require some risk assessment in addition to hazard
	classification. In addition, chemical inventory requirements in
(b)(6)(vi)	various countries are not related to the GHS.
Food or alcoholic beverages that are sold, used, or prepared in a retail	
establishment (such as a grocery store, restaurant, or drinking place),	
and foods intended for personal consumption by employees while in the	
workplace.	

(b)(6)(vii)
Any drug, as that term is defined in the Federal Food, Drug, and Cosmetic Act (21 U.S.C. 301 et seq.), when it is in solid, final form for direct administration to the patient (e.g., tablets or pills); drugs that are packaged by the chemical manufacturer for sale to consumers in a retail establishment (e.g., over-the-counter drugs); and drugs intended for personal consumption by employees while in the workplace (e.g., first-aid supplies).

(b)(6)(viii)
Cosmetics that are packaged for sale to consumers in a retail establishment and cosmetics intended for personal consumption by employees while in the workplace.

(b)(6)(ix)
Any consumer product or hazardous substance, as those terms are defined in the Consumer Product Safety Act (15 U.S.C. 2051 et seq.) and Federal Hazardous Substances Act (15 U.S.C. 1261 et seq.), respectively, where the employer can show that it is used in the workplace for the purpose intended by the chemical manufacturer or importer of the product, and the use results in a duration and frequency of exposure that is not greater than the range of exposures that could reasonably be experienced by consumers when used for the purpose intended.

(b)(6)(x)
Nuisance particulates where the chemical manufacturer or importer can establish that they do not pose any physical or health hazard covered under this section.

(b)(6)(xi)
Ionizing and nonionizing radiation.

(b)(6)(xii)
Biological hazards.

Application

Comparison

Implementing the GHS will require the HCS to make decisions concerning the application of the building blocks for physical and health hazard classes and hazard categories. See the individual health and physical endpoints for details. For other competent authority decisions, see GHS competent authority allowances and building block discussion section.

OSHA HCS 29 CFR 1910.1200	GHS
Comment: The HCS is a performance-oriented regulation. The GHS is specification oriented. Implementation of the GHS will require changes to the performance-oriented nature of the HCS. These changes will include required label elements and a required MSDS format, as well as criteria changes. Although the GHS has the building block approach, changes to the HCS are expected. The HCS may not implement all hazard classes, for example, hazardous for the environment. Even within some hazard classes, the HCS may not regulate all hazard categories, for example, acute toxicity. Many hazard classes will require some type of change to the HCS.	**1.1.3 Application of the GHS** 1.1.3.1 Harmonization of the application of the GHS 1.1.3.1.1 The goal of the GHS is to identify the intrinsic hazards found in chemical substances and mixtures and to convey hazard information about these hazards. The criteria for hazard classification are harmonized. Hazard statements, symbols, and signal words have been standardized and harmonized and are now part of a foreman-integrated hazard communication system. The GHS will allow the hazard communication elements of the existing systems to converge. Competent authorities will decide how to apply the various elements of the GHS on the basis of the needs of the competent authority and the target audience. (See also Hazard Communication: Labeling (Chapter 1.4, paragraph 1.4.10.5.4.2) and Consumer Product Labeling Based on the Likelihood of Injury, Annex 4.) 1.1.3.1.2 For transport, it is expected that application of the GHS will be similar to application of current transport requirements. Containers of dangerous goods will be marked with pictograms that address acute toxicity, physical hazards, and environmental hazards. As is true for workers in other sectors, workers in the transport sector will be trained. The elements of the GHS that address such elements as signal words and hazard statements are not expected to be adopted in the transport sector.

1.1.3.1.3 In the workplace, it is expected that all of the GHS elements will be adopted, including labels that have the harmonized core information under the GHS and safety data sheets. It is also anticipated that this will be supplemented by employee training to help ensure effective communication.

1.1.3.1.4 For the consumer sector, it is expected that labels will be the primary focus of GHS application. These labels will include the core elements of the GHS, subject to some sector-specific considerations in certain systems. (See also Hazard Communication: Labeling (Chapter 1.4, paragraph 1.4.10.5.4.2) and Consumer Product Labeling Based on the Likelihood of Injury, Annex 4.)

1.1.3.1.5 Building block approach.

1.1.3.1.5.1 Consistent with the building block approach, countries are free to determine which of the building blocks will be applied in different parts of their systems. However, where a system covers something that is in the GHS, and implements the GHS, that coverage should be consistent. For example, if a system covers the carcinogenicity of a chemical, it should follow the harmonized classification scheme and the harmonized label elements.

1.1.3.1.5.2 In examining the requirements of existing systems, it was noted that coverage of hazards may vary by the perceived needs of the target audience for information. In particular, the transport sector focuses on acute health effects and physical hazards but has not to date covered chronic effects owing to the types of exposures expected to be encountered in that setting. But there may be other differences as well, with countries choosing not to cover all of the effects addressed by the GHS in each use setting.

(*Continued*)

Application

Comparison

Implementing the GHS will require the HCS to make decisions concerning the application of the building blocks for physical and health hazard classes and hazard categories. See the individual health and physical endpoints for details. For other competent authority decisions, see GHS competent authority allowances and building block discussion section.

OSHA HCS 29 CFR 1910.1200	GHS
	1.1.3.1.5.3 The harmonized elements of the GHS may thus be seen as a collection of building blocks from which to form a regulatory approach. While the full range is available to everyone, and should be used if a country or organization chooses to cover a certain effect when it adopts the GHS, the full range does not have to be adopted. While physical hazards are important in the workplace and transport sectors, consumers may not need to know some of the specific physical hazards in the type of use they have for a product. As long as the hazards covered by a sector or system are covered consistently with the GHS criteria and requirements, it will be considered appropriate implementation of the GHS. Notwithstanding the fact that an exporter needs to comply with importing countries' GHS implementation, it is hoped that the application of the GHS worldwide will eventually lead to a fully harmonized situation.

Definitions

Comparison

The HCS will need to add definitions and some existing HCS definitions will need to be changed. The above definitions illustrate some of the key changes to be considered. Substance, chemical, liquid, and gas are some differences. In the HCS, physical hazards have definitions. The GHS has criteria for physical hazards.

OSHA HCS 29 CFR 1910.1200	GHS
"Chemical" means any element, chemical compound, or mixture of elements or compounds.	*Substance* means chemical elements and their compounds in the natural state or obtained by any production process, including any additive necessary to preserve the stability of the product and any impurities deriving from the process used, but excluding any solvent that may be separated without affecting the stability of the substance or changing its composition.
"Chemical name" means the scientific designation of a chemical in accordance with the nomenclature system developed by the International Union of Pure and Applied Chemistry (IUPAC) or the Chemical Abstracts Service (CAS) rules of nomenclature, or a name that will clearly identify the chemical for the purpose of conducting a hazard evaluation.	*Chemical identity* means a name that will uniquely identify a chemical. This can be a name that is in accordance with the nomenclature systems of the International Union of Pure and Applied Chemistry (IUPAC) or the Chemical Abstracts Service (CAS), or a technical name.
"Specific chemical identity" means the chemical name, Chemical Abstracts Service (CAS) Registry Number, or any other information that reveals the precise chemical designation of the substance.	
"Combustible liquid" means any liquid having a flagship at or above 100°F (37.8°C) but below 200°F (93.3°C), except any mixture having components with flash points of 200°F (93.3°C) or higher, the total volume of which make up 99% or more of the total volume of the mixture.	See flammable liquid hazard categories.
"Common name" means any designation or identification such as code name, code number, trade name, brand name, or generic name used to identify a chemical other than by its chemical name.	No GHS definition.

(Continued)

Definitions

Comparison

The HCS will need to add definitions and some existing HCS definitions will need to be changed. The above definitions illustrate some of the key changes to be considered. Substance, chemical, liquid, and gas are some differences. In the HCS, physical hazards have definitions. The GHS has criteria for physical hazards.

OSHA HCS 29 CFR 1910.1200	GHS

"Compressed gas" means
 	
   i. A gas or mixture of gases having, in a container, an absolute pressure exceeding 40 psi at 70°F (21.1°C)
 	
  ii. A gas or mixture of gases having, in a container, an absolute pressure exceeding 104 psi at 130°F (54.4°C) regardless of the pressure at 70°F (21.1°C)
 	
  iii. A liquid having a vapor pressure exceeding 40 psi at 100°F (37.8°C) as determined by ASTM D-323-72
	Compressed gas: A gas that when packaged under pressure is entirely gaseous at −50°C, including all gases with a critical temperature ≤ −50°C.

No HCS definition.	*Corrosive to metal* means a substance or a mixture that, by chemical action, will materially damage, or even destroy, metals.
"Explosive" means a chemical that causes a sudden, almost instantaneous release of pressure, gas, and heat when subjected to sudden shock, pressure, or high temperature.	*Explosive article* means an article containing one or more explosive substances. *Explosive substance* means a solid or liquid substance (or mixture of substances) that is in itself capable, by chemical reaction, of producing gas at such a temperature and pressure and at such a speed as to cause damage to the surroundings. Pyrotechnic substances are included even when they do not evolve gases.
"Flammable" means a chemical that falls into one of the following categories: i. "Aerosol, flammable" means an aerosol that, when tested by the method described in 16 CFR 1500.45, yields a flame projection exceeding 18 inches at full valve opening, or a flashback (a flame extending back to the valve) at any degree of valve opening.	*Flammable gas:* A gas having a flammable range with air at 20°C and a standard pressure of 101.3 kPa. *Flammable liquid:* A liquid having a flash point of not more than 93°C. *Flammable solid:* A solid that is readily combustible or may cause or contribute to fire through friction.

ii. "Gas, flammable" means
  a. A gas that, at ambient temperature and pressure, forms a flammable mixture with air at a concentration of 13% by volume or less
  b. A gas that, at ambient temperature and pressure, forms a range of flammable mixtures with air wider than 12% by volume, regardless of the lower limit

iii. "Liquid, flammable" means any liquid having a flash point below 100°F (37.8°C), except any mixture having components with flash points of 100°F (37.8°C) or higher, the total of which make up 99% or more of the total volume of the mixture.

iv. "Solid, flammable" means a solid, other than a blasting agent or explosive as defined in 1910.109(a), that is liable to cause fire through friction, absorption of moisture, spontaneous chemical change, or retained heat from manufacturing or processing, or which can be ignited readily and when ignited burns so vigorously and persistently as to create a serious hazard. A chemical shall be considered to be a flammable solid if, when tested by the method described in 16 CFR 1500.44, it ignites and burns with a self-sustained flame at a rate greater than one-tenth of an inch per second along its major axis.

"Flash point" means the minimum temperature at which a liquid gives off a vapor in sufficient concentration to ignite when tested as follows:
  i. Tagliabue Closed Tester (see American National Standard Method of Test for Flash Point by Tag Closed Tester, Z11.24-1979(ASTM D 56-79)) for liquids with a viscosity of less than 45 Saybolt Universal Seconds (SUS) at 100°F (37.8°C) that do not contain suspended solids and do not have a tendency to form a surface film under test

Aerosols means any nonrefillable receptacles made of metal, glass, or plastics containing a gas compressed, liquefied, or dissolved under pressure, with or without a liquid, paste or powder, and fitted with a release device allowing the contents to be ejected as solid or liquid particles in suspension in a gas, as a foam, paste, or powder or in a liquid state or in a gaseous state. Aerosol includes aerosol dispensers.

Readily combustible solid: Readily combustible solids are powdered, granular, or pasty substances that are dangerous if they can be easily ignited by brief contact with an ignition source, such as a burning match, and if the flame spreads rapidly.

Flash point means the lowest temperature (corrected to a standard pressure of 101.3 kPa) at which the application of an ignition source causes the vapors of a liquid to ignite under specified test conditions.

(Continued)

Definitions

Comparison

The HCS will need to add definitions and some existing HCS definitions will need to be changed. The above definitions illustrate some of the key changes to be considered. Substance, chemical, liquid, and gas are some differences. In the HCS, physical hazards have definitions. The GHS has criteria for physical hazards.

OSHA HCS 29 CFR 1910.1200	GHS

ii. Pensky–Martens Closed Tester (see American National Standard Method of Test for Flash Point by Pensky–Martens Closed Tester, Z11.7-1979 (ASTM D 93-79)) for liquids with a viscosity equal to or greater than 45 SUS at 100°F (37.8°C), or that contain suspended solids, or that have a tendency to form a surface film under test
iii. Setaflash Closed Tester (see American National Standard Method of Test for Flash Point by Setaflash Closed Tester (ASTM D3278-78)).
 	
No HCS definition.
	Gas means a substance that (i) at 50°C has a vapor pressure greater than 300 kPa or (ii) is completely gaseous at 20°C at a standard pressure of 101.3 kPa.
	
	Dissolved gas means a gas that, when packaged under pressure, is dissolved in a liquid phase solvent.
	
	Liquefied gas means a gas that, when packaged under pressure, is partially liquid at temperatures above −50°C. A distinction is made between the following:
	  i. High-pressure liquefied gas: a gas with a critical temperature between −50°C and +65°C
	  ii. Low-pressure liquefied gas: a gas with a critical temperature above +65°C
	
	Refrigerated liquefied gas: A gas that when packaged is made partially liquid because of its low temperature.

"Health hazard" means a chemical for which there is statistically significant evidence based on at least one study conducted in accordance with established scientific principles that acute or chronic health effects may occur in exposed employees. The term *health hazard* includes chemicals that are carcinogens, toxic or highly toxic agents, reproductive toxins, irritants, corrosives, sensitizers, hepatotoxins, nephrotoxins, neurotoxins, agents that act on the hematopoietic system, and agents that damage the lungs, skin, eyes, or mucous membranes. Appendix A provides further definitions and explanations of the scope of health hazards covered by this section, and Appendix B describes the criteria to be used to determine whether or not a chemical is to be considered hazardous for purposes of this standard. (See appendix for criteria.)

"Hazard warning" means any words, pictures, symbols, or combination thereof appearing on a label or other appropriate forms of warning that convey the specific physical and health hazard(s), including target organ effects, of the chemical(s) in the container(s). (See the definitions for "physical hazard" and "health hazard" to determine the hazards that must be covered.)

Hazard category means the division of criteria within each hazard class; for example, oral acute toxicity includes five hazard categories and flammable liquids includes four hazard categories. These categories compare hazard severity within a hazard class and should not be taken as a comparison of hazard categories more generally.

Hazard class means the nature of the physical, health, or environmental hazard, for example, flammable solid carcinogen and oral acute toxicity.

Hazard statement means a statement assigned to a hazard class and category that describes the nature of the hazards of a hazardous product, including, where appropriate, the degree of hazard.

Signal word: A signal word means a word used to indicate the relative level of severity of hazard and alert the reader to a potential hazard on the label. The GHS uses "Danger" and "Warning."

Pictogram: A pictogram means a composition that may include a symbol plus other graphic elements, such as a border, background pattern, or color, that is intended to convey specific information.

Precautionary statement: A precautionary statement means a phrase (and/or a pictogram) that describes recommended measures that should be taken to minimize or prevent adverse effects resulting from exposure to a hazardous product or improper storage or handling of a hazardous product.

(*Continued*)

Definitions

Comparison
The HCS will need to add definitions and some existing HCS definitions will need to be changed. The above definitions illustrate some of the key changes to be considered. Substance, chemical, liquid, and gas are some differences. In the HCS, physical hazards have definitions. The GHS has criteria for physical hazards.

OSHA HCS 29 CFR 1910.1200	GHS
	Supplemental label element: A supplemental label element means any additional nonharmonized type of information supplied on the container of a hazardous product that is not required or specified under the GHS. In some cases, this information may be required by other competent authorities or it may be additional information provided at the discretion of the manufacturer/distributor.
	Symbol: A symbol means a graphical element intended to succinctly convey information.
"Label" means any written, printed, or graphic material displayed on or affixed to containers of hazardous chemicals.	***Label*** means an appropriate group of written, printed, or graphic information elements concerning a hazardous product, selected as relevant to the target sector(s), that is affixed to, printed on, or attached to the immediate container of a hazardous product or to the outside packaging of a hazardous product.
	Label element means one type of information that has been harmonized for use in a label, for example, pictogram or signal word.
No HCS definition.	***Liquid*** means a substance or mixture that at 50°C has a vapor pressure of not more than 300 kPa (3 bar), that is not completely gaseous at 20°C and at a standard pressure of 101.3 kPa, and that has a melting point or initial melting point of 20°C or less at a standard pressure of 101.3 kPa. A viscous substance or mixture for which a specific melting point cannot be determined shall be subjected to the ASTM D 4359-90 test or to the test for determining fluidity (penetrometer test) prescribed in Section 2.3.4 of Annex A of the European Agreement concerning the International Carriage of Dangerous Goods by Road (ADR).

Mixture means a mixture or a solution composed of two or more substances in which they do not react.

Alloy means a metallic material, homogeneous on a macroscopic scale, consisting of two or more elements so combined that they cannot be readily separated by mechanical means. Alloys are considered to be mixtures for the purpose of classification under the GHS.

Oxidizing gas means any gas that may, generally by providing oxygen, cause or contribute to the combustion of other material more than air does.

Oxidizing liquid means a liquid that, while in itself not necessarily combustible, may, generally by yielding oxygen, cause, or contribute to, the combustion of other material.

Oxidizing solid means a solid that, while in itself not necessarily combustible, may, generally by yielding oxygen, cause, or contribute to, the combustion of other material.

Organic peroxide means a liquid or solid organic substance which contains the bivalent -O-O- structure and may be considered a derivative of hydrogen peroxide, where one or both of the hydrogen atoms have been replaced by organic radicals. The term also includes organic peroxide formulation (mixtures).

Product identifier: A product identifier means the name or number used for a hazardous product on a label or in the SDS. It provides a unique means by which the product user can identify the substance or mixture within the particular use setting, for example, transport, consumer, or workplace.

Pyrophoric liquid: A pyrophoric liquid is a liquid that, even in small quantities, is liable to ignite within 5 min after coming into contact with air.

Pyrophoric solid: A pyrophoric solid is a solid that, even in small quantities, is liable to ignite within 5 min after coming into contact with air.

Pyrotechnic substance: A substance or mixture of substances designed to produce an effect by heat, light, sound, gas, or smoke or a combination of these as a result of nondetonative self-sustaining exothermic chemical reactions.

(Continued)

"Mixture" means any combination of two or more chemicals if the combination is not, in whole or in part, the result of a chemical reaction.

"Oxidizer" means a chemical other than a blasting agent or explosive as defined in 1910.109(a) that initiates or promotes combustion in other materials, thereby causing fire either of itself or through the release of oxygen or other gases.

"Organic peroxide" means an organic compound that contains the bivalent -O-O- structure and which may be considered to be a structural derivative of hydrogen peroxide where one or both of the hydrogen atoms have been replaced by an organic radical.

"Identity" means any chemical or common name that is indicated on the material safety data sheet (MSDS) for the chemical. The identity used shall permit cross-references to be made among the required list of hazardous chemicals, the label, and the MSDS.

"Pyrophoric" means a chemical that will ignite spontaneously in air at a temperature of 130°F (54.4°C) or below.

Definitions

Comparison

The HCS will need to add definitions and some existing HCS definitions will need to be changed. The above definitions illustrate some of the key changes to be considered. Substance, chemical, liquid, and gas are some differences. In the HCS, physical hazards have definitions. The GHS has criteria for physical hazards.

OSHA HCS 29 CFR 1910.1200	GHS
No HCS definition.	*Technical name*: A name that is generally used in commerce, regulations, and codes to identify a substance or mixture, other than the IUPAC or CAS name, and that is recognized by the scientific community. Examples of technical names include those used for complex mixtures (e.g., petroleum fractions or natural products), pesticides (e.g., ISO or ANSI systems), dyestuffs (Color Index system), and minerals.
	CBI means "confidential business information."
"Trade secret" means any confidential formula, pattern, process, device, information, or compilation of information that is used in an employer's business that gives the employer an opportunity to obtain an advantage over competitors who do not know or use it. Appendix D sets out the criteria to be used in evaluating trade secrets.	
"Unstable (reactive)" means a chemical that in the pure state, or as produced or transported, will vigorously polymerize, decompose, condense, or will become self-reactive under conditions of shock, pressure, or temperature.	*Self-heating substance*: A self-heating substance is a solid or liquid substance, other than a pyrophoric substance, which, by reaction with air and without energy supply, is liable to self-heat; this substance differs from a pyrophoric substance in that it will ignite only when in large amounts (kilograms) and after long periods of time (hours or days).
"Pyrophoric" means a chemical that will ignite spontaneously in air at a temperature of 130°F (54.4°C) or below.	
"Unstable (reactive)" means a chemical that in the pure state, or as produced or transported, will vigorously polymerize, decompose, condense, or will become self-reactive under conditions of shock, pressure, or temperature.	*Self-reactive substance*: Self-reactive substances are thermally unstable liquid or solid substances liable to undergo a strongly exothermic decomposition even without participation of oxygen (air). This definition excludes substances or mixtures classified under the GHS as explosive, organic peroxides, or oxidizing.
"Water reactive" means a chemical that reacts with water to release a gas that is either flammable or presents a health hazard.	*Substances that, in contact with water, emit flammable gases* are solid or liquid substances that, by interaction with water, are liable to become spontaneously flammable or to give off flammable gases in dangerous quantities.

Hazard Determination/Classification

Comparison

A significant difference between the HCS and GHS is the evaluation of mixtures. The GHS criteria for mixtures vary by hazard class. See individual endpoints for details. The HCS allows test data on mixtures to be used for all hazard classes. The GHS allows test data on carcinogens, mutagens, and reproductive toxins on a case-by-case basis. The GHS expectation of physical test data for mixtures is another difference.

The HCS "floor" of hazardous chemicals is a difference and one that is likely helpful to small businesses. Guidance on how IARC, NTP, and OSHA carcinogens fit with the GHS cancer classification scheme could also be useful guidance.

OSHA HCS 29 CFR 1910.1200	GHS
1910.1200(d) Hazard Determination	**1.3.2 General considerations on the GHS**
(d)(1)	1.3.2.1 Scope of the system
Chemical manufacturers and importers shall evaluate chemicals produced in their workplaces or imported by them to determine if they are hazardous. Employers are not required to evaluate chemicals unless they choose not to rely on the evaluation performed by the chemical manufacturer or importer for the chemical to satisfy this requirement.	1.3.2.1.1 The GHS applies to pure chemical substances, their dilute solutions, and mixtures of chemical substances. "Articles" as defined in the Hazard Communication Standard (29 CFR 1910.1200) of the US Occupational Safety and Health Administration, or by similar definition, are outside the scope of the system.
(d)(2)	1.3.2.1.2 One objective of the GHS is for it to be simple and transparent with a clear distinction between classes and categories in order to allow for "self-classification" as far as possible. For many hazard classes, the criteria are semiquantitative or qualitative and expert judgment is required to interpret the data for classification purposes. Furthermore, for some hazard classes (e.g., eye irritation, explosives, or self-reactive substances), a decision tree approach is provided to enhance ease of use.
Chemical manufacturers, importers, or employers evaluating chemicals shall identify and consider the available scientific evidence concerning such hazards. For health hazards, evidence that is statistically significant and that is based on at least one positive study conducted in accordance with established scientific principles is considered to be sufficient to establish a hazardous effect if the results of the study meet the definitions of health hazards in this section. Appendix A shall be consulted for the scope of health hazards covered, and Appendix B shall be consulted for the criteria to be followed with respect to the completeness of the evaluation and the data to be reported.	1.3.2.2 Concept of "classification"
	1.3.2.2.1 The GHS uses the term *hazard classification* to indicate that only the intrinsic hazardous properties of substances or mixtures are considered.

(Continued)

Hazard Determination/Classification

Comparison

A significant difference between the HCS and GHS is the evaluation of mixtures. The GHS criteria for mixtures vary by hazard class. See individual endpoints for details. The HCS allows test data on mixtures to be used for all hazard classes. The GHS allows test data on carcinogens, mutagens, and reproductive toxins on a case-by-case basis. The GHS expectation of physical test data for mixtures is another difference.

The HCS "floor" of hazardous chemicals is a difference and one that is likely helpful to small businesses. Guidance on how IARC, NTP, and OSHA carcinogens fit with the GHS cancer classification scheme could also be useful guidance.

OSHA HCS 29 CFR 1910.1200	GHS
(d)(3) The chemical manufacturer, importer, or employer evaluating chemicals shall treat the following sources as establishing that the chemicals listed in them are hazardous:	1.3.2.2.2 Hazard classification incorporates only three steps, that is: a. Identification of relevant data regarding the hazards of a substance or mixture b. Subsequent review of those data to ascertain the hazards associated with the substance or mixture c. A decision on whether the substance or mixture will be classified as a hazardous substance or mixture and the degree of hazard, where appropriate, by comparison of the data with agreed hazard classification criteria
(d)(3)(i) 29 CFR part 1910, subpart Z, Toxic and Hazardous Substances, Occupational Safety and Health Administration (OSHA).	1.3.2.2.3 As noted in IOMC Description and Further Clarification of the Anticipated Application of the GHS text in the Purpose, Scope and Application (Chapter 1.1, paragraph
(d)(3)(ii) "Threshold Limit Values for Chemical Substances and Physical Agents in the Work Environment," American Conference of Governmental Industrial Hygienists (ACGIH) (latest edition). The chemical manufacturer, importer, or employer is still responsible for evaluating the hazards associated with the chemicals in these source lists in accordance with the requirements of this standard.	1.3.2.3 Classification criteria (mixtures) The classification criteria for substances and mixtures are presented in Parts 2 and 3 of this document, each of which is for a specific hazard class or a group of closely related hazard classes. The recommended process of classification of mixtures is based on the following sequence: a. Where test data are available for the complete mixture, the classification of the mixture will always be based on that data.
(d)(4) Chemical manufacturers, importers, and employers evaluating chemicals shall treat the following sources as establishing that a chemical is a carcinogen or potential carcinogen for hazard communication purposes:	
(d)(4)(i) National Toxicology Program (NTP), "Annual Report on Carcinogens" (latest edition)	

(d)(4)(ii)

International Agency for Research on Cancer (IARC) "Monographs" (latest editions)

(d)(4)(iii)

29 CFR part 1910, subpart Z, Toxic and Hazardous Substances, Occupational Safety and Health Administration

(d)(5)

The chemical manufacturer, importer, or employer shall determine the hazards of mixtures of chemicals as follows:

(d)(5)(i)

If a mixture has been tested as a whole to determine its hazards, the results of such testing shall be used to determine whether the mixture is hazardous.

(d)(5)(ii)

If a mixture has not been tested as a whole to determine whether the mixture is a health hazard, the mixture shall be assumed to present the same health hazards as do the components that comprise 1% (by weight or volume) or greater of the mixture, except that the mixture shall be assumed to present a carcinogenic hazard if it contains a component in concentrations of 0.1% or greater, which is considered to be a carcinogen under paragraph (d)(4) of this section.

(d)(5)(iii)

If a mixture has not been tested as a whole to determine whether the mixture is a physical hazard, the chemical manufacturer, importer, or employer may use whatever scientifically valid data are available to evaluate the physical hazard potential of the mixture.

b. Where test data are not available for the mixture itself, then bridging principles included and explained in each specific chapter should be considered to see whether they permit classification of the mixture.In addition, for the health and environmental classes,

c. If (i) test data are not available for the mixture itself and (ii) the available information is not sufficient to allow application of the abovementioned bridging principles, then the agreed method(s) described in each chapter for estimating the hazards based on the information known will be applied to classify the mixture.

1.3.2.4 UNCETDG/ILO Working Group on Physical Hazards

The UNCETDG/ILO Working Group for Physical Hazards used a similar process to the OECD Task Force on HCL. The work involved a comparison of the major classification systems, identification of similar or identical elements, and, for the elements that were dissimilar, development of a consensus on a compromise. For physical hazards, however, the transport definitions, test methods, and classification criteria were used as a basis for the work since they were already substantially harmonized. The work proceeded through examination of the scientific basis for the criteria, gaining consensus on the test methods, data interpretation, and the criteria. For most hazard classes, the existing schemes were already in place and being used by the transport sector. On this basis, a portion of the work focused on ensuring that workplace, environment, and consumer safety issues were adequately addressed.

See the individual health hazard classes for mixture considerations.

Labels

Comparison

The HCS label requirements will now have to be specified. GHS labels have pictograms, as well as specified signal words, and hazard statements. See pictogram table and label element comparison. The HCS provision for no component disclosure on labels is accommodated in the GHS. The use of transport pictograms in nontransport settings is an option, as is the use of a black border for GHS pictograms in domestic settings. See competent authority allowance discussions.

OSHA HCS 29 CFR 1910.1200	GHS
1910.1200(f) Labels and Other Forms of Warning	1.4.6.2 Application of standardization in the harmonized system
(f)(1)	For labels, the hazard symbols, signal words, and hazard statements have
The chemical manufacturer, importer, or distributor shall ensure that each	all been standardized and assigned to each of the hazard categories.
container of hazardous chemicals leaving the workplace is labeled,	These standardized elements should not be subject to variation and
tagged, or marked with the following information:	should appear on the GHS label as indicated in the chapters for each
	hazard class in this document. For safety data sheets, the chapter *Hazard*
(f)(1)(i)	*Communication: Safety Data Sheets* (Chapter 1.5) provides a standardized
Identity of the hazardous chemical(s)	format for the presentation of information. While precautionary
	information was considered for standardization, there was insufficient
(f)(1)(ii)	time to develop detailed proposals. However, there are examples of
Appropriate hazard warnings	precautionary statements and pictograms in Annex 3 and it remains a
	goal to develop them into fully standardized label elements.
(f)(1)(iii)	1.4.6.3 Use of nonstandardized or supplemental information
Name and address of the chemical manufacturer, importer, or other	1.4.6.3.1 There are many other label elements that may appear on a label
responsible party	that have not been standardized in the harmonized system. Some of
	these clearly need to be included on the label, for example,
(f)(4)	precautionary statements. Competent authorities may require
If the hazardous chemical is regulated by OSHA in a substance-specific	additional information, or suppliers may choose to add
health standard, the chemical manufacturer, importer, distributor, or	supplementary information on their own initiative. In order to
employer shall ensure that the labels or other forms of warning used are	ensure that the use of nonstandardized information does not lead
in accordance with the requirements of that standard.	

(f)(6)

The employer may use signs, placards, process sheets, batch tickets, operating procedures, or other such written materials in lieu of affixing labels to individual stationary process containers, as long as the alternative method identifies the containers to which it is applicable and conveys the information required by paragraph (f)(5) of this section to be on a label. The written materials shall be readily accessible to the employees in their work area throughout each work shift.

(f)(7)

The employer is not required to label portable containers into which hazardous chemicals are transferred from labeled containers, and which are intended only for the immediate use of the employee who performs the transfer. For purposes of this section, drugs that are dispensed by a pharmacy to a health care provider for direct administration to a patient are exempted from labeling.

(f)(11)

Chemical manufacturers, importers, distributors, or employers who become newly aware of any significant information regarding the hazards of a chemical shall revise the labels for the chemical within 3 months of becoming aware of the new information. Labels on containers of hazardous chemicals shipped after that time shall contain the new information. If the chemical is not currently produced or imported, the chemical manufacturer, importer, distributor, or employer shall add the information to the label before the chemical is shipped or introduced into the workplace again.

to unnecessarily wide variation in information or undermine GHS information, the use of supplementary information should be limited to the following circumstances:

    a. The supplementary information provides further detail and does not contradict or cast doubt on the validity of the standardized hazard information.

    b. The supplementary information provides information about hazards not yet incorporated into the GHS.

In either instance, the supplementary information should not lower standards of protection.

1.4.6.3.2 The labeler should have the option of providing supplementary information related to the hazard, such as physical state or route of exposure, with the hazard statement rather than in the supplementary information section on the label, see also paragraph 1.4.10.5.4.1.

1.4.10 Labeling Procedures

1.4.10.1 Scope

The following sections describe the procedures for preparing labels in the GHS, comprising the following:

    a. Allocation of label elements
    b. Reproduction of the symbol
    c. Reproduction of the hazard pictogram
    d. Signal words
    e. Hazard statements
    f. Precautionary statements and pictograms
    g. Product and supplier identification
    h. Multiple hazards and precedence of information
    i. Arrangements for presenting the GHS label elements
    j. Special labeling arrangements

(Continued)

Labels

Comparison

The HCS label requirements will now have to be specified. GHS labels have pictograms, as well as specified signal words, and hazard statements. See pictogram table and label element comparison. The HCS provision for no component disclosure on labels is accommodated in the GHS. The use of transport pictograms in nontransport settings is an option, as is the use of a black border for GHS pictograms in domestic settings. See competent authority allowance discussions.

OSHA HCS 29 CFR 1910.1200	GHS
	1.4.10.2 Label elements
	The tables in the individual chapters for each hazard class detail the label elements (symbol, signal word, hazard statement) that have been assigned to each of the hazard categories of the GHS. Hazard categories reflect the harmonized classification criteria. A summary of the allocation of label elements is provided in Annex 1. There are special arrangements that apply to the use of certain mixture concentrations in the GHS to take account of the information needs of different target audiences. These are further described in paragraph 1.4.10.5.4
	1.4.10.5.2 *Information required on a GHS label*
	a. *Signal words*
	A signal word means a word used to indicate the relative level of severity of hazard and alert the reader to a potential hazard on the label. The signal words used in the GHS are "Danger" and "Warning." "Danger" is used for the more severe hazard categories (i.e., in the main for hazard categories 1 and 2), while "Warning" is used for the less severe. The tables in the individual chapters for each hazard class detail the signal words that have been assigned to each of the hazard categories of the GHS.
	b. *Hazard statements*
	A hazard statement means a phrase assigned to a hazard class and category that describes the nature of the hazards of a hazardous product, including,

where appropriate, the degree of hazard. The tables of label elements in the individual chapters for each hazard class detail the hazard statements that have been assigned to each of the hazard categories of the GHS.

c. *Precautionary statements and pictograms*

A precautionary statement means a phrase (and/or pictogram) that describes recommended measures that should be taken to minimize or prevent adverse effects resulting from exposure to a hazardous product, or improper storage or handling of a hazardous product. The GHS label should include appropriate precautionary information, the choice of which is with the labeler or the competent authority. Annex 3 contains examples of precautionary statements, which can be used, and also examples of precautionary pictograms, which can be used where allowed by the competent authority.

d. *Product identifier*

  i. A product identifier should be used on a GHS label and it should match the product identifier used on the SDS. Where a substance or mixture is covered by the UN Model Regulations on the Transport of Dangerous Goods, the UN proper shipping name should also be used on the package.

 ii. The label for a substance should include the chemical identity of the substance. For mixtures or alloys, the label should include the chemical identities of all ingredients or alloying elements that contribute to acute toxicity, skin corrosion or serious eye damage, germ cell mutagenicity, carcinogenicity, reproductive toxicity, skin or respiratory sensitization, or Target Organ Systemic Toxicity (TOST), when these hazards appear on the label. Alternatively, the competent authority may require the inclusion of all ingredients or alloying elements that contribute to the hazard of the mixture or alloy.

(Continued)

Labels

Comparison
The HCS label requirements will now have to be specified. GHS labels have pictograms, as well as specified signal words, and hazard statements. See pictogram table and label element comparison. The HCS provision for no component disclosure on labels is accommodated in the GHS. The use of transport pictograms in nontransport settings is an option, as is the use of a black border for GHS pictograms in domestic settings. See competent authority allowance discussions.

OSHA HCS 29 CFR 1910.1200	GHS
	iii. Where a substance or mixture is supplied exclusively for workplace use, the competent authority may choose to give suppliers discretion to include chemical identities on the SDS, in lieu of including them on labels.
	iv. The competent authority rules for CBI take priority over the rules for product identification. This means that where an ingredient would normally be included on the label, if it meets the competent authority criteria for CBI, its identity does not have to be included on the label.
	e. *Supplier identification*
	1.4.10.5.4 Arrangements for presenting the GHS label elements
	1.4.10.5.4.1 Location of GHS information on the label
	The GHS hazard pictograms, signal word, and hazard statements should be located together on the label. The competent authority may choose to provide a specified layout for the presentation of these and for the presentation of precautionary information, or allow supplier discretion. Specific guidance and examples are provided in the chapters on individual hazard classes.

Workplace Labeling

Comparison

The HCS workplace labeling option is allowed in the GHS. This option is a common practice in many US workplaces. See competent authority allowance discussion.

OSHA HCS 29 CFR 1910.1200	GHS
1910.1200(f) (6)–(7) Workplace Labeling	**1.4.10.5.5.1 Workplace labeling**
(f)(6)	Products falling within the scope of the GHS will carry the GHS label at the point where they are supplied to the workplace, and that label should be maintained on the supplied container in the workplace. The GHS label or label elements should also be used for workplace containers. However, the competent authority can allow employers to use alternative means of giving workers the same information in a different written or displayed format when such a format is more appropriate to the workplace and communicates the information as effectively as the GHS label. For example, label information could be displayed in the work area, rather than on the individual containers. Alternative means of providing workers with the information contained in GHS labels are needed usually where hazardous chemicals are transferred from an original supplier container into a workplace container or system, or where chemicals are produced in a workplace but are not packaged in containers intended for sale or supply. Chemicals that are produced in a workplace may be contained or stored in many different ways such as small samples collected for testing or analysis, piping systems including valves, process or reaction vessels, ore cars, conveyer systems, or freestanding bulk storage of solids. In batch manufacturing processes, one mixing vessel may be used to contain a number of different chemical mixtures.
The employer may use signs, placards, process sheets, batch tickets, operating procedures, or other such written materials in lieu of affixing labels to individual stationary process containers, as long as the alternative method identifies the containers to which it is applicable and conveys the information required by paragraph (f)(5) of this section to be on a label. The written materials shall be readily accessible to the employees in their work area throughout each work shift.	
(f)(7)	
The employer is not required to label portable containers into which hazardous chemicals are transferred from labeled containers, and which are intended only for the immediate use of the employee who performs the transfer. For purposes of this section, drugs that are dispensed by a pharmacy to a health care provider for direct administration to a patient are exempted from labeling.	

(Continued)

Labels

Comparison

The HCS label requirements will now have to be specified. GHS labels have pictograms, as well as specified signal words, and hazard statements. See pictogram table and label element comparison. The HCS provision for no component disclosure on labels is accommodated in the GHS. The use of transport pictograms in nontransport settings is an option, as is the use of a black border for GHS pictograms in domestic settings. See competent authority allowance discussions.

OSHA HCS 29 CFR 1910.1200	GHS
	In many situations, it is impractical to produce a complete GHS label and attach it to the container, due, for example, to container size limitations or lack of access to a process container. Some examples of workplace situations where chemicals may be transferred from supplier containers include containers for laboratory testing or analysis, storage vessels, piping or process reaction systems, or temporary containers where the chemical will be used by one worker within a short time frame.
	Decanted chemicals intended for immediate use could be labeled with the main components and directly refer the user to the supplier label information and SDS.
	All such systems should ensure that there is clear hazard communication. Workers should be trained to understand the specific communication methods used in a workplace. Examples of alternative methods include use of product identifiers together with GHS symbols and other pictograms to describe precautionary measures; use of process flow charts for complex systems to identify chemicals contained in pipes and vessels with links to the appropriate SDS; use of displays with GHS symbols, color, and signal words in piping systems and processing equipment; use of permanent placarding for fixed piping; use of batch tickets or recipes for labeling batch mixing vessels; and use of piping bands with hazard symbols and product identifiers.

Updating Labels

Comparison

The current HCS requirements for updating labels are accommodated by the GHS. Other options could be considered for the purpose of harmonization. See competent authority allowances discussion.

OSHA HCS 29 CFR 1910.1200	GHS
1910.1200(f)(11) Updating Labels (f)(11)	**1.4.7 Updating information**
Chemical manufacturers, importers, distributors, or employers who become newly aware of any significant information regarding the hazards of a chemical shall revise the labels for the chemical within 3 months of becoming aware of the new information. Labels on containers of hazardous chemicals shipped after that time shall contain the new information. If the chemical is not currently produced or imported, the chemical manufacturer, importer, distributor, or employer shall add the information to the label before the chemical is shipped or introduced into the workplace again.	All systems should specify a means of responding in an appropriate and timely manner to new information and updating labels and SDS information accordingly. The following are examples of how this could be achieved.
	1.4.7.2 General guidance on updating of information
	1.4.7.2.1 Suppliers should respond to "new and significant" information they receive about a chemical hazard by updating the label and safety data sheet for that chemical. New and significant information is any information that changes the GHS classification of the substance or mixture and leads to a resulting change in the information provided on the label or any information concerning the chemical and appropriate control measures that may affect the SDS. This could include, for example, new information on the potential adverse chronic health effects of exposure as a result of recently published documentation or test results, even if a change in classification may not yet be triggered.

(*Continued*)

Updating Labels

Comparison

The current HCS requirements for updating labels are accommodated by the GHS. Other options could be considered for the purpose of harmonization. See competent authority allowances discussion.

OSHA HCS 29 CFR 1910.1200	GHS
	1.4.7.2.2 Updating should be carried out promptly on receipt of the information that necessitates the revision. The competent authority may choose to specify a time limit within which the information should be revised. This applies only to labels and SDS for products that are not subject to an approval mechanism such as pesticides. In pesticide labeling systems, where the label is part of the product approval mechanism, suppliers cannot update the supply label on their own initiative. However, when the products are subject to the transport of dangerous goods requirements, the label used should be updated on receipt of the new information, as above.
	1.4.7.2.3 Suppliers should also periodically review the information on which the label and safety data sheet for a substance or mixture are based, even if no new and significant information has been provided to them in respect of that substance or mixture. This will require, for example, a search of chemical hazard databases for new information. The competent authority may choose to specify a time (typically 3–5 years) from the date of original preparation, within which suppliers should review the labels and SDS information.

MSDS/SDS

Comparison

The performance orientation of the HCS MSDS will need to be changed. The GHS requires a 16-section MSDS format with specified sequence and minimum required contents. See separate table for more detailed comparison of MSDS sections/information.

On the basis of requirements in existing systems, there is some discretion in the GHS for determining when an MSDS is required. The hazard pictogram/symbol can be graphically reproduced on the MSDS or the name of the symbol may be provided instead. The level of hazardous components can be given as ranges or concentrations.

OSHA HCS 29 CFR 1910.1200	GHS
1910.1200(g) MSDS	**Table 1.5.2 Minimum information for an SDS**
(g)(2)	1. *Product and company identification*

Each material safety data sheet shall be in English (although the employer may maintain copies in other languages as well) and shall contain at least the following information:	– GHS product identifier
	  – Other means of identification
	  – Recommended use of the chemical and restrictions on use
	  – Supplier's details (including name, address, phone number, etc).
	  – Emergency phone number

(g)(2)(i)	2. *Hazards identification*
The identity used on the label, and, except as provided for in paragraph (i) of this section on trade secrets:	– GHS classification of the substance/mixture and any regional information – GHS label elements, including precautionary statements (Hazard symbols may be provided as a graphical reproduction of the symbols in black and white or the name of the symbol, e.g., flame, skull, and crossbones.)
(g)(2)(i)(A)	– Other hazards that do not result in classification (e.g., dust explosion hazard) or are not covered by the GHS
If the hazardous chemical is a single substance, its chemical and common name(s)	3. *Composition/information on ingredients*
(g)(2)(i)(B)	**Substance**

If the hazardous chemical is a mixture that has been tested as a whole to determine its hazards, the chemical and common name(s) of the ingredients that contribute to these known hazards and the common name(s) of the mixture itself	– Chemical identity
	  – Common name, synonyms, and so on
	  – CAS number, EC number, and so on
	  – Impurities and stabilizing additives that are themselves classified and that contribute to the classification of the substance

(g)(2)(i)(C)	
If the hazardous chemical is a mixture that has not been tested as a whole:	

(Continued)

MSDS/SDS

Comparison

The performance orientation of the HCS MSDS will need to be changed. The GHS requires a 16-section MSDS format with specified sequence and minimum required contents. See separate table for more detailed comparison of MSDS sections/information.

On the basis of requirements in existing systems, there is some discretion in the GHS for determining when an MSDS is required. The hazard pictogram/symbol can be graphically reproduced on the MSDS or the name of the symbol may be provided instead. The level of hazardous components can be given as ranges or concentrations.

OSHA HCS 29 CFR 1910.1200	GHS
(g)(2)(i)(C)(1) The chemical and common name(s) of all ingredients which have been determined to be health hazards, and which comprise 1% or greater of the composition, except that chemicals identified as carcinogens under paragraph (d) of this section shall be listed if the concentrations are 0.1% or greater **(g)(2)(i)(C)(2)** The chemical and common name(s) of all ingredients that have been determined to be health hazards and that comprise less than 1% (0.1% for carcinogens) of the mixture, if there is evidence that the ingredient(s) could be released from the mixture in concentrations that would exceed an established OSHA permissible exposure limit or ACGIH Threshold Limit Value, or could present a health risk to employees **(g)(2)(i)(C)(3)** The chemical and common name(s) of all ingredients that have been determined to present a physical hazard when present in the mixture **(g)(2)(ii)** Physical and chemical characteristics of the hazardous chemical (such as vapor pressure, flash point)	**Mixture** – The chemical identity and concentration or concentration ranges of all ingredients that are hazardous within the meaning of the GHS and are present above their cutoff levels. – Cutoff level for reproductive toxicity, carcinogenicity, and category 1 mutagenicity is $\geq 0.1\%$. – Cutoff level for all other hazard classes is $\geq 1\%$. *Note: For information on ingredients, the competent authority rules for CBI take priority over the rules for product identification.* *4. First-aid measures* – Description of necessary measures, subdivided according to the different routes of exposure, that is, inhalation, skin and eye contact, and ingestion – Most important symptoms/effects, acute and delayed – Indication of immediate medical attention and special treatment needed, if necessary *5. Fire-fighting measures* – Suitable (and unsuitable) extinguishing media – Specific hazards arising from the chemical (e.g., nature of any hazardous combustion products) – Special protective equipment and precautions for firefighters

(g)(2)(iii)
The physical hazards of the hazardous chemical, including the potential for fire, explosion, and reactivity

(g)(2)(iv)
The health hazards of the hazardous chemical, including signs and symptoms of exposure, and any medical conditions that are generally recognized as being aggravated by exposure to the chemical

(g)(2)(v)
The primary route(s) of entry

(g)(2)(vi)
The OSHA permissible exposure limit, ACGIH Threshold Limit Value, and any other exposure limit used or recommended by the chemical manufacturer, importer, or employer preparing the material safety data sheet, where available

(g)(2)(vii)
Whether the hazardous chemical is listed in the National Toxicology Program (NTP) Annual Report on Carcinogens (latest edition) or has been found to be a potential carcinogen in the International Agency for Research on Cancer (IARC) Monographs (latest editions), or by OSHA

(g)(2)(viii)
Any generally applicable precautions for safe handling and use that are known to the chemical manufacturer, importer, or employer preparing the material safety data sheet, including appropriate hygienic practices, protective measures during repair and maintenance of contaminated equipment, and procedures for cleanup of spills and leaks

6. *Accidental release measures*
   – Personal precautions, protective equipment, and emergency procedures
   – Environmental precautions
   – Methods and materials for containment and cleaning up
7. *Handling and storage*
   – Precautions for safe handling
   – Conditions for safe storage, including any incompatibilities
8. *Exposure controls/personal protection*
   – Control parameters, for example, occupational exposure limit values or biological limit values
   – Appropriate engineering controls
   – Individual protection measures, such as personal protective equipment
9. *Physical and chemical properties*
   – Appearance (physical state, color, etc.)
   – Odor
   – Odor threshold
   – pH
   – Melting point/freezing point
   – Initial boiling point and boiling range
   – Flash point
   – Evaporation rate
   – Flammability (solid, gas)
   – Upper/lower flammability or explosive limits
   – Vapor pressure
   – Vapor density
   – Relative density
   – Solubility(ies)
   – Partition coefficient: *n*-octanol/water
   – Auto-ignition temperature decomposition temperature

(Continued)

MSDS/SDS

Comparison

The performance orientation of the HCS MSDS will need to be changed. The GHS requires a 16-section MSDS format with specified sequence and minimum required contents. See separate table for more detailed comparison of MSDS sections/information.

On the basis of requirements in existing systems, there is some discretion in the GHS for determining when an MSDS is required. The hazard pictogram/symbol can be graphically reproduced on the MSDS or the name of the symbol may be provided instead. The level of hazardous components can be given as ranges or concentrations.

OSHA HCS 29 CFR 1910.1200	GHS

(g)(2)(ix)
Any generally applicable control measures that are known to the chemical manufacturer, importer, or employer preparing the material safety data sheet, such as appropriate engineering controls, work practices, or personal protective equipment
	10. *Stability and reactivity—chemical stability*
	  – Possibility of hazardous reactions
	  – Conditions to avoid (e.g., static discharge, shock, or vibration)
	  – Incompatible materials
	  – Hazardous decomposition products

(g)(2)(x) Emergency and first-aid procedures	11. *Toxicological information* Concise but complete and comprehensible description of the various toxicological (health) effects and the available data used to identify those effects, including
(g)(2)(xi) The date of preparation of the material safety data sheet or the last change to it	– Information on the likely routes of exposure (inhalation, ingestion, skin and eye contact)

(g)(2)(xii)
The name, address, and telephone number of the chemical manufacturer, importer, employer, or other responsible party preparing or distributing the material safety data sheet, who can provide additional information on the hazardous chemical and appropriate emergency procedures, if necessary
	– Symptoms related to the physical, chemical, and toxicological characteristics
	  – Delayed and immediate effects and also chronic effects from short- and long-term exposure
	  – Numerical measures of toxicity (such as acute toxicity estimates)

(g)(3)
If no relevant information is found for any given category on the material safety data sheet, the chemical manufacturer, importer, or employer preparing the material safety data sheet shall mark it to indicate that no applicable information was found.

(g)(5) [MSDS Updating]
The chemical manufacturer, importer, or employer preparing the material safety data sheet shall ensure that the information recorded accurately reflects the scientific evidence used in making the hazard determination. If the chemical manufacturer, importer, or employer preparing the material safety data sheet becomes newly aware of any significant information regarding the hazards of a chemical, or ways to protect against the hazards, this new information shall be added to the material safety data sheet within 3 months. If the chemical is not currently being produced or imported, the chemical manufacturer or importer shall add the information to the material safety data sheet before the chemical is introduced into the workplace again.

(g)(10) [MSDS Format]
Material safety data sheets may be kept in any form, including operating procedures, and may be designed to cover groups of hazardous chemicals in a work area where it may be more appropriate to address the hazards of a process rather than individual hazardous chemicals. However, the employer shall ensure that in all cases, the required information is provided for each hazardous chemical and is readily accessible during each work shift to employees when they are in their work area(s).

12. *Ecological information*
 – Ecotoxicity (aquatic and terrestrial, where available)
 – Persistence and degradability
 – Bioaccumulative potential
 – Mobility in soil
 – Other adverse effects
13. *Disposal considerations*
 – Description of waste residues and information on their safe handling and methods of disposal, including any contaminated packaging
14. *Transport information*
 – UN number
 – UN Proper shipping name
 – Transport Hazard class(es)
 – Packing group, if applicable
 – Marine pollutant (yes/no)
 – Special precautions that a user needs to be aware of or needs to comply with in connection with transport or conveyance either within or outside their premises
15. *Regulatory information*
 – Safety, health, and environmental regulations specific for the product in question
16. *Other information*
 – Other information including information on preparation and revision of the SDS

(Continued)

MSDS/SDS

Comparison

The performance orientation of the HCS MSDS will need to be changed. The GHS requires a 16-section MSDS format with specified sequence and minimum required contents. See separate table for more detailed comparison of MSDS sections/information.

On the basis of requirements in existing systems, there is some discretion in the GHS for determining when an MSDS is required. The hazard pictogram/symbol can be graphically reproduced on the MSDS or the name of the symbol may be provided instead. The level of hazardous components can be given as ranges or concentrations.

OSHA HCS 29 CFR 1910.1200	GHS
	1.5.2 Criteria for determining whether an SDS should be produced
	An SDS should be produced for all substances and mixtures that meet the harmonized criteria for physical, health, or environmental hazards under the GHS and for all mixtures that contain substances that meet the criteria for carcinogenic, toxic to reproduction, or target organ systemic toxicity in concentrations exceeding the cutoff limits for SDS specified by the criteria for mixtures (see paragraph 6). The competent authority may choose also to require SDSs for mixtures not meeting the criteria for classification as hazardous but which contain hazardous substances in certain concentrations (see paragraph 6).
	An SDS should be provided on the basis of the following generic cutoff / concentration limits:
	≥1% for acute toxicity, skin corrosion/irritation, serious damage to eyes / eye irritation, respiratory/skin sensitization, mutagenicity category 2, target organ toxicity (single and repeat) exposures, and hazardous to the environment
	≥0.1% for mutagenicity category 1, carcinogenicity, and reproductive toxicity

As noted in the *Classification of Hazardous Substances and Mixtures* (Chapter 1.2, paragraphs 28 through 31), there may be some cases when the available hazard data may justify classification on the basis of other cutoff limits than the generic ones specified in the health and environment hazard class chapters (Chapters 3.2 through 3.10). When such specific cutoffs are used for classification, they should also apply to the obligation to compile an SDS.

Some competent authorities (CA) may require SDSs to be compiled for mixtures that are not classified for acute toxicity or aquatic toxicity as a result of application of the additivity formula, but which contain acutely toxic substances or substances toxic to the aquatic environment in concentrations equal to or greater than 1%.

In accordance with the building block principle, some competent authorities may choose not to regulate certain categories within a hazard class. In such situations, there would be no obligation to compile an SDS.

(The above four paragraphs are not part of the agreed text on hazard communication including SDSs developed by the ILO Working Group on Hazard Communication but have been provided here as additional guidance on the compiling of an SDS.)

Once it is clear that an SDS is required for a substance or a mixture, then the information required to be included in the SDS should in all cases be provided in accordance with GHS requirements.

MSDS/SDS Component Disclosure

Comparison

The values for component disclosure in mixtures vary by endpoint. Some changes will be required on MSDS component disclosure. The level of hazardous components can be given as ranges or concentrations in the MSDS.

OSHA HCS 29 CFR 1910.1200	GHS
(g)(2)(i) The identity used on the label, and, except as provided for in paragraph (i) of this section on trade secrets:	Table 1.5.2 Minimum information for an SDS3 Composition/information on ingredients Substance
(g)(2)(i)(A) If the hazardous chemical is a single substance, its chemical and common name(s)	• Chemical identity • Common name, synonyms, and so on • CAS number, EC number, and so on • Impurities and stabilizing additives that are themselves classified and that contribute to the classification of the substance
(g)(2)(i)(B) If the hazardous chemical is a mixture that has been tested as a whole to determine its hazards, the chemical and common name(s) of the ingredients that contribute to these known hazards, and the common name(s) of the mixture itself	Mixture • The chemical identity and concentration or concentration ranges of all ingredients that are hazardous within the meaning of the GHS and are present above their cutoff levels *Note: For information on ingredients, the competent authority rules for CBI take priority over the rules for product identification.*
(g)(2)(i)(C) If the hazardous chemical is a mixture that has not been tested as a whole:	**1.5.2 Criteria for determining whether an SDS should be produced** An SDS should be produced for all substances and mixtures that meet the harmonized criteria for physical, health, or environmental hazards under the GHS and for all mixtures that contain substances that meet the criteria for carcinogenic, toxic to reproduction, or target organ systemic toxicity in concentrations exceeding the cutoff limits for SDS specified by the criteria for mixtures (see paragraph 1.5.3.1). The competent authority may choose also to require SDSs for mixtures not meeting the criteria for classification as hazardous but which contain hazardous substances in certain concentrations (see paragraph 1.5.3.1).
(g)(2)(i)(C)(1) The chemical and common name(s) of all ingredients that have been determined to be health hazards and that comprise 1% or greater of the composition, except that chemicals identified as carcinogens under paragraph (d) of this section shall be listed if the concentrations are 0.1% or greater	

1.5.3 General guidance for compiling a safety data sheet

1.5.3.1 Cutoff values/concentration limits

1.5.3.1.1 An SDS should be provided on the basis of the generic cutoff values/concentration limit

$\geq$1% for acute toxicity, skin corrosion/irritation, serious damage to eyes/eye irritation, respiratory/skin sensitization, mutagenicity category 2, target organ toxicity (single and repeat) exposures, and hazardous to the environment.

$\geq$0.1% for mutagenicity category 1, carcinogenicity, and reproductive toxicity (sensitizes).

1.5.3.1.2 As noted in the *Classification of Hazardous Substances and Mixtures* (see 1.3.3.2), there may be some cases when the available hazard data may justify classification on the basis of other cutoff values/concentration limits than the generic ones specified in the health and environment hazard class chapters (Chapters 3.2 through 3.10). When such specific cutoff values are used for classification, they should also apply to the obligation to compile an SDS.

1.5.3.1.3 Some competent authorities (CA) may require SDSs to be compiled for mixtures that are not classified for acute toxicity or aquatic toxicity as a result of application of the additivity formula but that contain acutely toxic substances or substances toxic to the aquatic environment in concentrations equal to or greater than 1%.

1.5.3.1.4 In accordance with the building block principle, some competent authorities may choose not to regulate certain categories within a hazard class. In such situations, there would be no obligation to compile an SDS.

(Paragraphs 1.5.3.1.1 through 1.5.3.1.4 are not part of the agreed text on hazard communication including SDSs developed by the ILO Working Group on Hazard Communication but have been provided here as additional guidance on the compiling of an SDS.)

1.5.3.1.5 Once it is clear that an SDS is required for a substance or a mixture, then the information required to be included in the SDS should in all cases be provided in accordance with GHS requirements.

(g)(2)(i)(C)(2)

The chemical and common name(s) of all ingredients that have been determined to be health hazards and that comprise less than 1% (0.1% for carcinogens) of the mixture, if there is evidence that the ingredient(s) could be released from the mixture in concentrations that would exceed an established OSHA permissible exposure limit or ACGIH Threshold Limit Value, or could present a health risk to employees

(g)(2)(i)(C)(3)

The chemical and common name(s) of all ingredients that have been determined to present a physical hazard when present in the mixture

Updating MSDS/SDS

Comparison

The current HCS requirements for updating MSDS are accommodated by the GHS. Other options could be considered for the purpose of harmonization.

OSHA HCS 29 CFR 1910.1200	GHS
1910.1200(g)(5) updating MSDS (g)(5) [MSDS Updating] The chemical manufacturer, importer, or employer preparing the material safety data sheet shall ensure that the information recorded accurately reflects the scientific evidence used in making the hazard determination. If the chemical manufacturer, importer, or employer preparing the material safety data sheet becomes newly aware of any significant information regarding the hazards of a chemical, or ways to protect against the hazards, this new information shall be added to the material safety data sheet within 3 months. If the chemical is not currently being produced or imported, the chemical manufacturer or importer shall add the information to the material safety data sheet before the chemical is introduced into the workplace again.	**1.4.7 Updating information** All systems should specify a means of responding in an appropriate and timely manner to new information and updating labels and SDS information accordingly. The following are examples of how this could be achieved. 1.4.7.2 General guidance on updating of information 1.4.7.2.1 Suppliers should respond to "new and significant" information they receive about a chemical hazard by updating the label and safety data sheet for that chemical. New and significant information is any information that changes the GHS classification of the substance or mixture and leads to a resulting change in the information provided on the label or any information concerning the chemical and appropriate control measures that may affect the SDS. This could include, for example, new information on the potential adverse chronic health effects of exposure as a result of recently published documentation or test results, even if a change in classification may not yet be triggered.

1.4.7.2.2 Updating should be carried out promptly on receipt of the information that necessitates the revision. The competent authority may choose to specify a time limit within which the information should be revised. This applies only to labels and SDS for products that are not subject to an approval mechanism such as pesticides. In pesticide labeling systems, where the label is part of the product approval mechanism, suppliers cannot update the supply label on their own initiative. However, when the products are subject to the transport of dangerous goods requirements, the label used should be updated on receipt of the new information, as above.

1.4.7.2.3 Suppliers should also periodically review the information on which the label and safety data sheet for a substance or mixture are based, even if no new and significant information has been provided to them in respect of that substance or mixture. This will require, for example, a search of chemical hazard databases for new information. The competent authority may choose to specify a time (typically 3–5 years) from the date of original preparation, within which suppliers should review the labels and SDS information.

Information and Training

Comparison
The GHS has broad general training requirements. The HCS has more detailed training requirements than the GHS.

OSHA HCS 29 CFR 1910.1200	GHS
(h) "Employee information and training." **(h)(1)** Employers shall provide employees with effective information and training on hazardous chemicals in their work area at the time of their initial assignment and whenever a new physical or health hazard the employees have not previously been trained about is introduced into their work area. Information and training may be designed to cover categories of hazards (e.g., flammability, carcinogenicity) or specific chemicals. Chemical-specific information must always be available through labels and material safety data sheets. **(h)(2)** "Information." Employees shall be informed of: **(h)(2)(i)** The requirements of this section **(h)(2)(ii)** Any operations in their work area where hazardous chemicals are present **(h)(2)(iii)** The location and availability of the written hazard communication program, including the required list(s) of hazardous chemicals, and material safety data sheets required by this section	**1.4.9 Training** Training users of hazard information is an integral part of hazard communication. Systems should identify the appropriate education and training for GHS target audiences who are required to interpret label and/or SDS information and to take appropriate action in response to chemical hazards. Training requirements should be appropriate for and commensurate with the nature of the work or exposure. Key target audiences for training include workers, emergency responders, and those involved in the preparation of labels, SDS, and hazard communication strategies as part of risk management systems. Others involved in the transport and supply of hazardous chemicals also require training to varying degrees. In addition, systems should consider strategies required for educating consumers in interpreting label information on products that they use.

(h)(3)

"Training." Employee training shall include at least:

(h)(3)(i)

Methods and observations that may be used to detect the presence or release of a hazardous chemical in the work area (such as monitoring conducted by the employer, continuous monitoring devices, visual appearance or odor of hazardous chemicals when being released, etc.)

(h)(3)(ii)

The physical and health hazards of the chemicals in the work area

(h)(3)(iii)

The measures employees can take to protect themselves from these hazards, including specific procedures the employer has implemented to protect employees from exposure to hazardous chemicals, such as appropriate work practices, emergency procedures, and personal protective equipment to be used

(h)(3)(iv)

The details of the hazard communication program developed by the employer, including an explanation of the labeling system and the material safety data sheet, and how employees can obtain and use the appropriate hazard information

Trade Secrets/CBI

Comparison

The GHS provides CBI principles and guidance. The GHS does not harmonize CBI requirements. The HCS is aligned with the CBI principles in the GHS.

OSHA HCS 29 CFR 1910.1200	GHS
(i) "Trade secrets."	**1.4.8 Confidential business information**
(i)(1) Chemical name and other specific identification of a hazardous chemical, from the material safety data sheet, provided that:	1.4.8.1 Systems adopting the GHS should consider what provisions may be appropriate for the protection of confidential business information (CBI). Such provisions should not compromise the health and safety of workers or consumers, or the protection of the environment. As with other parts of the GHS, the rules of the importing country should apply with respect to CBI claims for imported substances and mixtures.
(i)(1)(i) The claim that the information withheld is a trade secret can be supported	
(i)(1)(ii) Information contained in the material safety data sheet concerning the properties and effects of the hazardous chemical is disclosed	1.4.8.2 Where a system chooses to provide for protection of confidential business information, competent authorities should establish appropriate mechanisms, in accordance with national law and practice, and consider
(i)(1)(iii) The material safety data sheet indicates that the specific chemical identity is being withheld as a trade secret	a. Whether the inclusion of certain chemicals or classes of chemicals in the arrangements is appropriate to the needs of the system
(i)(1)(iv) The specific chemical identity is made available to health professionals, employees, and designated representatives in accordance with the applicable provisions of this paragraph	b. What definition of "confidential business information" should apply, taking account of factors such as the accessibility of the information by competitors, intellectual property rights, and the potential harm disclosure would cause to the employer or supplier's business
	c. Appropriate procedures for the disclosure of confidential business information, where necessary to protect the health and safety of workers or consumers, or to protect the environment, and measures to prevent further disclosure

1.4.8.3 Specific provisions for the protection of confidential business information may differ among systems in accordance with national law and practice. However, they should be consistent with the following general principles:

a. For information otherwise required on labels or safety data sheets, CBI claims should be limited to the names of chemicals and their concentrations in mixtures. All other information should be disclosed on the label or safety data sheet, as required.

b. Where CBI has been withheld, the label or chemical safety data sheet should so indicate.

c. CBI should be disclosed to the competent authority upon request. The competent authority should protect the confidentiality of the information in accordance with applicable law and practice.

d. Where a medical professional determines that a medical emergency exists due to exposure to a hazardous chemical or a chemical mixture, mechanisms should be in place to ensure timely disclosure by the supplier or employer or competent authority of any specific confidential information necessary for treatment. The medical professional should maintain the confidentiality of the information.

e. For nonemergency situations, the supplier or employer should ensure disclosure of confidential information to a safety or health professional providing medical or other safety and health services to exposed workers or consumers, and to workers or workers' representatives. Persons requesting the information should provide specific reasons for the disclosure and should agree to use the information only for the purpose of consumer or worker protection, and to otherwise maintain its confidentiality.

f. Where nondisclosure of CBI is challenged, the competent authority should address such challenges or provide for an alternative process for challenges. The supplier or employer should be responsible for supporting the assertion that the withheld information qualifies for CBI protection.

(*Continued*)

(i)(2)
Where a treating physician or nurse determines that a medical emergency exists and the specific chemical identity of a hazardous chemical is necessary for emergency or first-aid treatment, the chemical manufacturer, importer, or employer shall immediately disclose the specific chemical identity of a trade secret chemical to that treating physician or nurse, regardless of the existence of a written statement of need or a confidentiality agreement. The chemical manufacturer, importer, or employer may require a written statement of need and confidentiality agreement, in accordance with the provisions of paragraphs (i)(3) and (4) of this section, as soon as circumstances permit.

(i)(3)
In nonemergency situations, a chemical manufacturer, importer, or employer shall, upon request, disclose a specific chemical identity, otherwise permitted to be withheld under paragraph (i)(1) of this section, to a health professional (i.e., physician, industrial hygienist, toxicologist, epidemiologist, or occupational health nurse) providing medical or other occupational health services to exposed employee(s), and to employees or designated representatives, if:

(i)(3)(i)
The request is in writing

(i)(3)(ii)
The request describes with reasonable detail one or more of the following occupational health needs for the information:

(i)(3)(ii)(A)
To assess the hazards of the chemicals to which employees will be exposed

(i)(3)(ii)(B)
To conduct or assess sampling of the workplace atmosphere to determine employee exposure levels

Trade Secrets/CBI

Comparison
The GHS provides CBI principles and guidance. The GHS does not harmonize CBI requirements. The HCS is aligned with the CBI principles in the GHS.

OSHA HCS 29 CFR 1910.1200	GHS
(i)(3)(ii)(C) To conduct preassignment or periodic medical surveillance of exposed employees	
(i)(3)(ii)(D) To provide medical treatment to exposed employees	
(i)(3)(ii)(E) To select or assess appropriate personal protective equipment for exposed employees	
(i)(3)(ii)(F) To design or assess engineering controls or other protective measures for exposed employees	
(i)(3)(ii)(G) To conduct studies to determine the health effects of exposure	
(i)(3)(iii) The request explains in detail why the disclosure of the specific chemical identity is essential and that, in lieu thereof, the disclosure of the following information to the health professional, employee, or designated representative would not satisfy the purposes described in paragraph (i)(3)(ii) of this section:	
(i)(3)(iii)(A) The properties and effects of the chemical	

(*Continued*)

(i)(3)(iii)(B)
Measures for controlling workers' exposure to the chemical

(i)(3)(iii)(C)
Methods of monitoring and analyzing worker exposure to the chemical

(i)(3)(iii)(D)
Methods of diagnosing and treating harmful exposures to the chemical

(i)(3)(iv)
The request includes a description of the procedures to be used to maintain the confidentiality of the disclosed information.

(i)(3)(v)
The health professional and the employer or contractor of the services of the health professional (i.e., downstream employer, labor organization, or individual employee), employee, or designated representative agree in a written confidentiality agreement that the health professional, employee, or designated representative will not use the trade secret information for any purpose other than the health need(s) asserted and agree not to release the information under any circumstances other than to OSHA, as provided in paragraph (i)(6) of this section, except as authorized by the terms of the agreement or by the chemical manufacturer, importer, or employer.

(i)(4)
The confidentiality agreement authorized by paragraph (i)(3)(iv) of this section

(i)(4)(i)
May restrict the use of the information to the health purposes indicated in the written statement of need

Trade Secrets/CBI

Comparison

The GHS provides CBI principles and guidance. The GHS does not harmonize CBI requirements. The HCS is aligned with the CBI principles in the GHS.

OSHA HCS 29 CFR 1910.1200	GHS
(i)(4)(ii)	
May provide for appropriate legal remedies in the event of a breach of the agreement, including stipulation of a reasonable pre-estimate of likely damages	
(i)(4)(iii)	
May not include requirements for the posting of a penalty bond	
(i)(5)	
Nothing in this standard is meant to preclude the parties from pursuing non-contractual remedies to the extent permitted by law.	
(i)(6)	
If the health professional, employee, or designated representative receiving the trade secret information decides that there is a need to disclose it to OSHA, the chemical manufacturer, importer, or employer who provided the limitations or conditions upon the disclosure of the requested chemical information as may be appropriate to assure that the occupational health services are provided without an undue risk of harm to the chemical manufacturer, importer, or employer.	

(Continued)

(i)(11)

If a citation for a failure to release specific chemical identity information is contested by the chemical manufacturer, importer, or employer, the matter will be adjudicated before the Occupational Safety and Health Review Commission in accordance with the Act's enforcement scheme and the applicable Commission rules of procedure. In accordance with the Commission rules, when a chemical manufacturer, importer, or employer continues to withhold the information during the contest, the Administrative Law Judge may review the citation and supporting documentation "in camera" or issue appropriate orders to protect the confidentiality of such matters.

(i)(12)

Notwithstanding the existence of a trade secret claim, a chemical manufacturer, importer, or employer shall, upon request, disclose to the Assistant Secretary any information that this section requires the chemical manufacturer, importer, or employer to make available. Where there is a trade secret claim, such claim shall be made no later than at the time the information is provided to the Assistant Secretary so that suitable determinations of trade secret status can be made and the necessary protections can be implemented.

(i)(13)

Nothing in this paragraph shall be construed as requiring the disclosure under any circumstances of process or percentage of mixture information that is a trade secret.

Trade Secrets/CBI

Comparison

The GHS provides CBI principles and guidance. The GHS does not harmonize CBI requirements. The HCS is aligned with the CBI principles in the GHS.

OSHA HCS 29 CFR 1910.1200	GHS
Appendix D Definition of Trade Secret (Mandatory) The following is a reprint of the "Restatement of Torts" section 757, comment b (1939): b. "Definition of trade secret." A trade secret may consist of any formula, pattern, device or compilation of information which is used in one's business, and which gives him an opportunity to obtain an advantage over competitors who do not know or use it. It may be a formula for a chemical compound, a process of manufacturing, treating or preserving materials, a pattern for a machine or other device, or a list of customers. It differs from other secret information in a business (see s759 of the Restatement of Torts which is not included in this Appendix) in that it is not simply information as to single or ephemeral events in the conduct of the business, as, for example, the amount or other terms of a secret bid for a contract or the salary of certain employees, or the security investments made or contemplated, or the date fixed for the announcement of a new policy or for bringing out a new model or the like. A trade secret is a process or device for continuous use in the operations of the business. Generally it relates to the production of goods, as, for example, a machine or formula for the production of an article. It may, however, relate to the sale of goods or to other operations in the business, such as a code for determining discounts, rebates or other concessions in a price list or catalog, or a list of specialized customers, or a method of bookkeeping or other office management. "Secrecy." The subject matter of a trade secret must be secret. Matters of public knowledge or of general knowledge in an industry cannot be appropriated by one as his secret. Matters which are completely disclosed by the goods which one markets cannot be his secret.	

(Continued)

Substantially, a trade secret is known only in the particular business in which it is used. It is not requisite that only the proprietor of the business knows it. He may, without losing his protection, communicate it to employees involved in its use. He may likewise communicate it to others pledged to secrecy. Others may also know of it independently, as, for example, when they have discovered the process or formula by independent invention and are keeping it secret. Nevertheless, a substantial element of secrecy must exist, so that, except by the use of improper means, there would be difficulty in acquiring the information. An exact definition of a trade secret is not possible. Some factors to be considered in determining whether given information is one's trade secret are: (1) the extent to which the information is known outside of his business; (2) the extent to which it is known by employees and others involved in his business; (3) the extent of measures taken by him to guard the secrecy of the information; (4) the value of the information to him and his competitors; (5) the amount of effort or money expended by him in developing the information; (6) the ease or difficulty with which the information could be properly acquired or duplicated by others.

"Novelty and prior art." A trade secret may be a device or process which is patentable; but it need not be that. It may be a device or process which is clearly anticipated in the prior art or one which is merely a mechanical improvement that a good mechanic can make. Novelty and invention are not requisite for a trade secret as they are for patentability. These requirements are essential to patentability because a patent protects against unlicensed use of the patented device or process even by one who discovers it properly through independent research. The patent monopoly is a reward to the inventor. But such is not the case with a trade secret. Its protection is not based on a policy of rewarding or otherwise encouraging the development of secret processes or devices.

Trade Secrets/CBI

Comparison
The GHS provides CBI principles and guidance. The GHS does not harmonize CBI requirements. The HCS is aligned with the CBI principles in the GHS.

OSHA HCS 29 CFR 1910.1200	GHS
The protection is merely against breach of faith and reprehensible means of learning another's secret. For this limited protection, it is not appropriate to require also the kind of novelty and invention which is a requisite of patentability. The nature of the secret is, however, an important factor in determining the kind of relief that is appropriate against one who is subject to liability under the rule stated in this section. Thus, if the secret consists of a device or process which is a novel invention, one who acquires the secret wrongfully is ordinarily enjoined from further use of it and is required to account for the profits derived from his past use. If, on the other hand, the secret consists of mechanical improvements that a good mechanic can make without resort to the secret, the wrongdoer's liability may be limited to damages, and an injunction against future use of the improvements made with the aid of the secret may be inappropriate.	

Multiple Hazards/Precedence

Comparison

The HCS will need to consider label requirements for multiple hazards with prescribed pictograms and statements. Are all elements required? Is there a precedence? See competent authority allowances discussions.

OSHA HCS 29 CFR 1910.1200	GHS
Multiple hazards CPL2-2.38D, Appendix A	1.4.10.5.3 Multiple hazards and precedence of hazard information
The label is intended to be an immediate visual reminder of the hazards of a chemical. It is not necessary, however, that every hazard presented by a chemical be listed on the label. Manufacturers, importers, and distributors will have to assess the evidence regarding the product's hazards and must consider exposures under normal circumstances of use or foreseeable emergencies when evaluating what hazards to put on the label. This is not to say that only acute hazards are to be listed on the label or that well-substantiated hazards should be left off the label because they appear on the data sheet.	The following arrangements apply where a substance or mixture presents more than one GHS hazard. It is without prejudice to the building block principle described in the *Purpose, Scope and Application* (Chapter 1.1). Therefore, where a system does not provide information on the label for a particular hazard, the application of the arrangements should be modified accordingly.
	1.4.10.5.3.1 Precedence for the allocation of symbols
	For substances and mixtures covered by the UN Recommendations on the Transport of Dangerous Goods, Model Regulations, the precedence of symbols for physical hazards should follow the rules of the UN Model Regulations. In workplace situations, the competent authority may require all symbols for physical hazards to be used. For health hazards, the following principles of precedence apply:
	(a) If the skull and crossbones applies, the exclamation mark should not appear.
	(b) If the corrosive symbol applies, the exclamation mark should not appear where it is used for skin or eye irritation.
	(c) If the new health hazard symbol appears for respiratory sensitization, the exclamation mark should not appear where it is used for skin sensitization or for skin or eye irritation.
	1.4.10.5.3.2 Precedence for allocation of signal words
	If the signal word "Danger" applies, the signal word "Warning" should not appear.
	1.4.10.5.3.3 Precedence for allocation of hazard statements
	All assigned hazard statements should appear on the label. The competent authority may choose to specify the order in which they appear.

(Continued)

Multiple Hazards/Precedence

Comparison

The HCS will need to consider label requirements for multiple hazards with prescribed pictograms and statements. Are all elements required? Is there a precedence? See competent authority allowances discussions.

OSHA HCS 29 CFR 1910.1200	GHS
	1.4.10.5.4 Arrangements for presenting the GHS label elements
	1.4.10.5.4.1 Location of GHS information on the label
	The GHS hazard pictograms, signal word, and hazard statements should be located together on the label. The competent authority may choose to provide a specified layout for the presentation of these and for the presentation of precautionary information, or allow supplier discretion. Specific guidance and examples are provided in the chapters on individual hazard classes.
	There have been some concerns about how the label elements should appear on different packagings. Specific examples are provided in Annex 6.
	1.4.10.5.4.2 Supplemental information
	The competent authority has the discretion to allow the use of supplemental information subject to the parameters outlined in 1.4.6.3. The competent authority may choose to specify where this information should appear on the label or allow supplier discretion. In either approach, the placement of supplemental information should not impede identification of GHS information.

Hazard Determinations/Classification Provisions

Comparison

The GHS and HCS hazard determination/classification processes. As classification is more involved in the GHS, additional guidance could be useful. The GHS includes weight of evidence in the hazard determination. The HCS has a one positive study threshold. The GHS provides for the one positive study issue within the individual endpoints. In vitro studies are treated differently. Substances not bioavailable or inextricably bound are addressed. Professional/expert judgment is included. Human experience is taken into account. In the HCS, negative findings and data that refute findings of hazard are allowed. The HCS does not address animal welfare.

The GHS addresses the import concept of previously classified substances. Existing data should be accepted when classifying substances under the GHS.

OSHA HCS 29 CFR 1910.1200	GHS
Appendix B Hazard determination (mandatory)	1.3.2.4 Available data, test methods, and test data quality
The quality of a hazard communication program is largely dependent upon the adequacy and accuracy of the hazard determination. The hazard determination requirement of this standard is performance oriented. Chemical manufacturers, importers, and employers evaluating chemicals are not required to follow any specific methods for determining hazards, but they must be able to demonstrate that they have adequately ascertained the hazards of the chemicals produced or imported in accordance with the criteria set forth in this Appendix.	1.3.2.4.1 The GHS itself does not include requirements for testing substances or mixtures. Therefore, there is no requirement under the GHS to generate test data for any hazard class. It is recognized that some parts of regulatory systems do require data to be generated (e.g., pesticides), but these requirements are not related specifically to the GHS. The criteria established for classifying a mixture will allow the use of available data for the mixture itself and/or similar mixtures and/or data for ingredients of the mixture.
... For purposes of this standard, the following criteria shall be used in making hazard determinations that meet the requirements of this standard.	1.3.2.4.2 The classification of a chemical substance or mixture depends both on the criteria and on the reliability of the test methods underpinning the criteria. In some cases, the classification is determined by a pass or fail of a specific test (e.g., the ready biodegradation test for substances or ingredients of mixtures), while in other cases, interpretations are made from dose/response curves and observations during testing. In all cases, the test conditions need to be standardized so that the results are reproducible with a given chemical substance and the standardized test yields "valid" data for defining the hazard class of concern. In this context, validation is the process by which the reliability and the relevance of a procedure are established for a particular purpose.
1. "Carcinogenicity." As described in paragraph (d)(4) of this section and Appendix A of this section, a determination by the National Toxicology Program, the International Agency for Research on Cancer, or OSHA that a chemical is a carcinogen or potential carcinogen will be considered conclusive evidence for purposes of this section. In addition, however, all available scientific data on carcinogenicity must be evaluated in accordance with the provisions of this Appendix and the requirements of the rule.	
...	

(Continued)

Hazard Determinations/Classification Provisions

Comparison

The GHS and HCS hazard determination/classification are self-classification processes. As classification is more involved in the GHS, additional guidance could be useful. The GHS includes weight of evidence in the hazard determination. The HCS has a one positive study threshold. The GHS provides for the one positive study issue within the individual endpoints. In vitro studies are treated differently. Substances not bioavailable or inextricably bound are addressed. Professional/expert judgment is included. Human experience is taken into account. In the HCS, negative findings and data that refute findings of hazard are allowed. The HCS does not address animal welfare.

The GHS addresses the import concept of previously classified substances. Existing data should be accepted when classifying substances under the GHS.

OSHA HCS 29 CFR 1910.1200	GHS
4. "Adequacy and reporting of data." The results of any studies that are designed and conducted according to established scientific principles and that report statistically significant conclusions regarding the health effects of a chemical shall be a sufficient basis for a hazard determination and reported on any material safety data sheet. In vitro studies alone generally do not form the basis for a definitive finding of hazard under the HCS since they have a positive or negative result rather than a statistically significant finding.	1.3.2.4.3 Tests that determine hazardous properties, which are conducted according to internationally recognized scientific principles, can be used for purposes of a hazard determination for health and environmental hazards. The GHS criteria for determining health and environmental hazards are test method neutral, allowing different approaches as long as they are scientifically sound and validated according to international procedures and criteria already referred to in existing systems for the hazard of concern and produce mutually acceptable data. Test methods for determining physical hazards are generally more clear-cut and are specified in the GHS.
The chemical manufacturer, importer, or employer may also report the results of other scientifically valid studies that tend to refute the findings of hazard.	1.3.2.4.4 Both positive and negative results are assembled together in the weight of evidence determination. However, a single positive study performed according to good scientific principles and with statistically and biologically significant positive results may justify classification.
	1.3.2.4.5 *Evidence from humans*
"Human data." Where available, epidemiological studies and case reports of adverse health effects shall be considered in the evaluation.	For classification purposes, reliable epidemiological data and experience on the effects of chemicals on humans (e.g., occupational data, data from accident databases) should be taken into account for the evaluation of human health hazards of a chemical. Testing on humans solely for hazard identification purposes is generally not acceptable.

Not addressed in HCS.

"Animal data." Human evidence of health effects in exposed populations is generally not available for the majority of chemicals produced or used in the workplace. Therefore, the available results of toxicological testing in animal populations shall be used to predict the health effects that may be experienced by exposed workers. In particular, the definitions of certain acute hazards refer to specific animal testing results (see Appendix A).

Hazard evaluation is a process that relies heavily on the professional judgment of the evaluator, particularly in the area of chronic hazards. The performance orientation of the hazard determination does not diminish the duty of the chemical manufacturer, importer, or employer to conduct a thorough evaluation, examining all relevant data and producing a scientifically defensible evaluation.

1.3.2.4.6 *Animal welfare*

The welfare of experimental animals is a concern. This ethical concern includes not only the alleviation of stress and suffering but also, in some countries, the use and consumption of test animals. Where possible and appropriate, tests and experiments that do not require the use of live animals are preferred to those using sentient live experimental animals. To that end, for certain hazards (skin and eye irritation/corrosion or serious damage), testing schemes starting with nonanimal observations/ measurements are included as part of the classification system. For other hazards, such as acute toxicity, alternative animal tests, using fewer animals or causing less suffering, are internationally accepted and should be preferred to the conventional LD_{50} test.

1.3.2.4.7 *Expert judgment*

The approach to classifying mixtures includes the application of expert judgment in a number of areas in order to ensure existing information can be used for as many mixtures as possible to provide protection for human health and the environment. Expert judgment may also be required in interpreting data for hazard classification of substances, especially where weight of evidence determinations are needed.

(Continued)

Hazard Determinations/Classification Provisions

Comparison

The GHS and HCS hazard determination/classification are self-classification processes. As classification is more involved in the GHS, additional guidance could be useful. The GHS includes weight of evidence in the hazard determination. The HCS has a one positive study threshold. The GHS provides for the one positive study issue within the individual endpoints. In vitro studies are treated differently. Substances not bioavailable or inextricably bound are addressed. Professional/expert judgment is included. Human experience is taken into account. In the HCS, negative findings and data that refute findings of hazard are allowed. The HCS does not address animal welfare.

The GHS addresses the import concept of previously classified substances. Existing data should be accepted when classifying substances under the GHS.

OSHA HCS 29 CFR 1910.1200	GHS
Hazard evaluation is a process that relies heavily on the professional judgment of the evaluator, particularly in the area of chronic hazards. The performance orientation of the hazard determination does not diminish the duty of the chemical manufacturer, importer, or employer to conduct a thorough evaluation, examining all relevant data and producing a scientifically defensible evaluation.	*1.3.2.4.8 Previously classified chemicals* One of the general principles established by the IOMC-CG-HCCS states that test data already generated for the classification of chemicals under the existing systems should be accepted when classifying these chemicals under the harmonized system, thereby avoiding duplicative testing and the unnecessary use of test animals. This policy has important implications in those cases where the criteria in the GHS are different from those in an existing system. In some cases, it may be difficult to determine the quality of existing data from older studies. In such cases, expert judgment will be needed. *1.3.2.4.9 Substances/mixtures posing special problems* The effect of a substance or mixture on biological and environmental systems is influenced, among other factors, by the physicochemical properties of the substance or mixture and/or ingredients of the mixture and the way in which ingredient substances are biologically available. Some groups of substances may present special problems in this respect, for example, some polymers and metals. A substance or mixture need not be classified when it can be shown by conclusive experimental data from internationally acceptable test methods that the substance or mixture is not biologically available. Similarly, bioavailability data on ingredients of a mixture should be used where appropriate in conjunction with the harmonized classification criteria when classifying mixtures.

CPL2-2.38D Hazard Determination

Decomposition products that are produced during the normal use of the product or in foreseeable emergencies (e.g., plastics that are injection molded, diesel fuel emissions) are covered. An employer may rely upon the hazard determination performed by the chemical manufacturer. Normally, the chemical manufacturer possesses knowledge of hazardous intermediates, by-products, and decomposition products that can be emitted by their product.

Any substance that is inextricably bound in a product is not covered under the HCS. For example, a hazard determination for a product containing crystalline silica may reveal that it is bound in a rubber elastomer and under normal conditions of use or during foreseeable emergencies cannot become airborne and, therefore, cannot present an inhalation hazard. In such a situation, the crystalline silica need not be indicated as a hazardous ingredient since it cannot result in employee exposure.

"Adequacy and reporting of data." The results of any studies that are designed and conducted according to established scientific principles and that report statistically significant conclusions regarding the health effects of a chemical shall be a sufficient basis for a hazard determination and reported on any material safety data sheet. In vitro studies alone generally do not form the basis for a definitive finding of hazard under the HCS since they have a positive or negative result rather than a statistically significant finding.

The chemical manufacturer, importer, or employer may also report the results of other scientifically valid studies that tend to refute the findings of hazard.

1.3.2.4.10 *Weight of evidence*

1.3.2.4.10.1 For some hazard classes, classification results directly when the data satisfy the criteria. For others, classification of a substance or a mixture is made on the basis of the total weight of evidence. This means that all available information bearing on the determination of toxicity is considered together, including the results of valid in vitro tests, relevant animal data, and human experience such as epidemiological and clinical studies and well-documented case reports and observations.

1.3.2.4.10.2 The quality and consistency of the data are important. Evaluation of substances or mixtures related to the material being classified should be included, as should site of action and mechanism or mode of action study results. Both positive and negative results are assembled together in a single weight of evidence determination.

(Continued)

Hazard Determinations/Classification Provisions

Comparison

The GHS and HCS hazard determination/classification are self-classification processes. As classification is more involved in the GHS, additional guidance could be useful. The GHS includes weight of evidence in the hazard determination. The HCS has a one positive study threshold. The GHS provides for the one positive study issue within the individual endpoints. In vitro studies are treated differently. Substances not bioavailable or inextricably bound are addressed. Professional/expert judgment is included. Human experience is taken into account. In the HCS, negative findings and data that refute findings of hazard are allowed. The HCS does not address animal welfare.

The GHS addresses the import concept of previously classified substances. Existing data should be accepted when classifying substances under the GHS.

OSHA HCS 29 CFR 1910.1200	GHS
	1.3.2.4.10.3 Positive effects that are consistent with the criteria for classification in each chapter, whether seen in humans or animals, will normally justify classification. Where evidence is available from both sources and there is a conflict between the findings, the quality and reliability of the evidence from both sources must be assessed in order to resolve the question of classification. Generally, data of good quality and reliability in humans will have precedence over other data. However, even well-designed and conducted epidemiological studies may lack sufficient numbers of subjects to detect relatively rare but still significant effects, or to assess potentially confounding factors. Positive results from well-conducted animal studies are not necessarily negated by the lack of positive human experience but require an assessment of the robustness and quality of both the human and animal data relative to the expected frequency of occurrence of effects and the impact of potentially confounding factors.
	1.3.2.4.10.4 Route of exposure, mechanistic information, and metabolism studies are pertinent to determining the relevance of an effect in humans. When such information raises doubt about relevance in humans, a lower classification may be warranted. When it is clear that the mechanism or mode of action is not relevant to humans, the substance or mixture should not be classified.

Comparison of Health Hazards

General Comments

The GHS has several health hazard endpoints (e.g., mutagenicity and target organ systemic toxicity) that do not exactly correspond to the HCS hazards. In general, the major difference between the HCS and the GHS is untested mixtures. OSHA has a single 1% cutoff value for all health hazards, except carcinogens at 0.1%. These cutoff values require labels, MSDSs, and disclosure of hazardous components. In the GHS, cutoff values for mixtures vary by endpoint. The GHS cutoff values for labeling, MSDSs, and disclosure can be different. The GHS acute toxicity and irritant hazard determinations for mixtures have more steps.

For substances previously classified under the HCS, existing data should be accepted when these substances are classified under the GHS.

Acute Toxicity

Comparison

Five GHS categories have been included in the GHS Acute Toxicity scheme from which the appropriate elements relevant to means of transport, consumer, worker, and environment protection can be selected. The HCS has two Acute Toxicity hazard categories whose cutoff values do not exactly correspond to the GHS cutoffs. The untested mixture hazard determination is different in the HCS and GHS. The GHS Acute Toxicity hazard determination for mixtures is involved. Acute Toxicity is a common data set.

OSHA HCS 29 CFR 1910.1200	GHS
Criteria	**Criteria**
Substances and mixtures are assigned to one of two acute toxicity hazards on the basis of LD_{50}/LC_{50}:	Substances of this hazard class are assigned to one of five toxicity categories on the basis of LD_{50} (oral, dermal) or LC_{50} (inhalation):
"*Highly toxic.*" A chemical falling within any of the following categories:	*Category 1*
	$LD_{50} < 5$ mg/kg bodyweight (oral)
(a) A chemical that has a median lethal dose (LD_{50}) of 50 mg or less per kilogram of body weight when administered orally to albino rats weighing between 200 and 300 g each.	$LD_{50} < 50$ mg/kg bodyweight (skin/dermal)
	$LC_{50} < 100$ ppm (gas)
	$LC_{50} < 0.5$ (mg/L) (vapor)
(b) A chemical that has a median lethal dose (LD_{50}) of 200 mg or less per kilogram of body weight when administered by continuous contact for 24 h (or less if death occurs within 24 h) with the bare skin of albino rabbits weighing between 2 and 3 kg each.	$LC_{50} < 0.05$ (mg/L) (dust, mist)
	Category 2
	$LD_{50} > 5$ and <50 mg/kg bodyweight (oral)
	$LD_{50} > 50$ and < 200 mg/kg bodyweight (skin/dermal)
	$LD_{50} > 100$ and < 500 ppm gas)
(c) A chemical that has a median lethal concentration (LC_{50}) in air of 200 parts per million by volume or less of gas or vapor, or 2 mg/L or less of mist, fume, or dust, when administered by continuous inhalation for 1 h (or less if death occurs within 1 h) to albino rats weighing between 200 and 300 g each.	$LD_{50} > 0.5$ and < 2.0 (mg/L) (vapor)
	$LC_{50} > 0.05$ and < 0.5 (mg/L) (dust, mist)
	Category 3
	$LD_{50} > 50$ and < 300 mg/kg bodyweight (oral)
	$LD_{50} > 200$ and <1000 mg/kg bodyweight (skin/dermal)
	$LC_{50} > 500$ and < 2500 ppm (gas)
"*Toxic.*" A chemical falling within any of the following categories:	$LC_{50} > 2.0$ and < 10.0 (mg/L) (vapor)
	$LC_{50} > 0.5$ and < 1.0 (mg/L) (dust, mist)
	Category 4
(a) A chemical that has a median lethal dose (LD_{50}) of more than 50 mg/kg but not more than 500 mg/kg of body weight when administered orally to albino rats weighing between 200 and 300 g each.	LD_{50} between 300 and less than 2000 mg/kg bodyweight (oral)
	LD_{50} between 1000 and less than 2000 mg/kg bodyweight (skin/dermal)
	LC_{50} between 2500 and less than 5000 ppm (gas)

(b) A chemical that has a median lethal dose (LD$_{50}$) of more than 200 mg/kg but not more than 1000 mg/kg of body weight when administered by continuous contact for 24 h (or less if death occurs within 24 h) with the bare skin of albino rabbits weighing between 2 and 3 kg each.

(c) A chemical that has a median lethal concentration (LC$_{50}$) in air of more than 200 parts per million but not more than 2000 parts per million by volume of gas or vapor, or more than 2 mg/L but not more than 20 mg/L of mist, fume, or dust, when administered by continuous inhalation for 1 h (or less if death occurs within 1 h) to albino rats weighing between 200 and 300 g each.

For mixtures:

Untested mixtures are assumed to present the same health hazards as components present at ≥1% (by weight or volume).

LC$_{50}$ between 10.0 and less than 20.0 (mg/L) (vapor)
LC$_{50}$ between 1.0 and less than 5.0 (mg/L) (dust, mist)

Category 5

LD$_{50}$ between 2000 and 5000 (oral or skin/dermal)

For gases, vapors, dusts, and mists, LC$_{50}$ in the equivalent range of the oral and dermal LD$_{50}$ (i.e., between 2000 and 5000 mg/kg bodyweight). See also the additional criteria:

- Indication of significant effect in humans
- Any mortality at Category 4
- Significant clinical signs at Category 4
- Indication from other studies.

For mixtures:

Conversion values for range tests or hazard categories are in Table 3.1.2.

"Relevant components" are those present at ≥1% (w/w for solids, liquids, dusts, mists, and vapors and v/v for gases), unless there is a reason to suspect that an ingredient present at <1% is still relevant.

- Apply bridging principles
- If bridging principles do not apply
- If data available for all components or data available to estimate and unknown components ≤10% apply

$$\frac{100}{ATE_{mix}} = \sum_n \frac{C_i}{ATE_i}$$

where

C_i = concentration of ingredient i

n ingredients and i is running from 1 to n

ATE_i = acute toxicity estimate of ingredient i.

- When unknown components >10%, apply

$$\frac{100 - \left(\sum C_{unknown} \text{ if } >10\%\right)}{ATE_{mix}} = \sum_n \frac{C_i}{ATE_i}$$

where

C_i = concentration of ingredient i

n ingredients and i is running from 1 to n

ATE_i = acute toxicity estimate of ingredient i.

- When unknown components >10%, apply

Skin Corrosion

Comparison

A single harmonized GHS Corrosion category is provided. For authorities wanting more than one designation for corrosivity, up to three subcategories are provided within the Corrosive category. Only some authorities will use the subcategories in Corrosive category 1. The HCS has one Corrosion hazard category. The untested mixture hazard determination is different in the HCS and GHS.

OSHA HCS 29 CFR 1910.1200	GHS
Definition/Criteria	**Definition**
"*Corrosive.*" A chemical that causes visible destruction of, or irreversible alterations in, living tissue by chemical action at the site of contact. For example, a chemical is considered to be corrosive if, when tested on the intact skin of albino rabbits by the method described by the US Department of Transportation in Appendix A to 49 CFR part 173, it destroys or changes irreversibly the structure of the tissue at the site of contact following an exposure period of 4 h. This term shall not refer to action on inanimate surfaces. For mixtures: Untested mixtures are assumed to present the same health hazards as components present at ≥1% (by weight or volume).	Skin corrosion means the production of irreversible damage to the skin following the application of a test substance for up to 4 h. **Criteria** Substances and mixtures of this hazard class are assigned to a single harmonized corrosion category. Category 1 1. *For substances and tested mixtures:* • Human experience showing irreversible damage to the skin • Structure/activity or structure/property relationship to a substance or mixture already classified as corrosive • pH extremes of <2 and >11.5 including acid/alkali reserve capacity • Positive results in a valid and accepted in vitro skin corrosion test • Animal experience or test data that indicate that the substance/mixture causes irreversible damage to the skin following exposure of up to 4 h (see Table 3.2.1) 2. *If data for a mixture are not available,* use bridging principles in 3.2.3.2. 3. *If bridging principles do not apply* (a) For mixtures where substances can be added: Classify as corrosive if the sum of the concentrations of corrosive substances in the mixture is >5% (for substances with additivity) (b) For mixtures where substances cannot be added: ≥1% corrosive substance. See 3.2.3.4. For those authorities wanting more than one designation for corrosivity, up to three subcategories are provided within the corrosive category: Subcategory 1A: ≤3 min exposure and ≤1 h observation Subcategory 1B: >3 min ≤1 h exposure and ≤14 days observation Subcategory 1C: >1 h and ≤4 h exposure and ≤14 days observation

Skin Irritation

Comparison

A single harmonized GHS skin irritant category is provided. An additional mild irritant category is available for authorities that want to have more than one skin irritation hazard category. The HCS has one skin irritation hazard category. The untested mixture hazard determination is quite different in the HCS and GHS. The GHS irritant hazard determination for mixtures is involved. Irritation data is a common data set.

OSHA HCS 29 CFR 1910.1200	GHS
Definition/Criteria	**Definition**
"Irritant." A chemical that is not corrosive but that causes a reversible inflammatory effect on living tissue by chemical action at the site of contact. A chemical is a skin irritant if, when tested on the intact skin of albino rabbits by the methods of 16 CFR 1500.41 for 4 h exposure or by other appropriate techniques, it results in an empirical score of five or more.	Skin irritation means the production of reversible damage to the skin following the application of a test substance for up to 4 h.
	Criteria
For mixtures:	Substances and mixtures of this hazard class are assigned to a single irritant category.
Untested mixtures are assumed to present the same health hazards as components present at ≥1% (by weight or volume).	Category 2
	1. *For substances and tested mixtures*
	• Human experience or data showing reversible damage to the skin following exposure of up to 4 h
	• Structure/activity or structure/property relationship to a substance or mixture already classified as an irritant
	• Positive results in a valid and accepted in vitro skin irritation test
	• Animal experience or test data that indicate that the substance/mixture causes reversible damage to the skin following exposure of up to 4 h, mean value of $>2.3 < 4.0$ for erythema/eschar or for edema, or inflammation that persists to the end of the observation period, in two of three tested animals (Table 3.2.2).
	2. *If data for a mixture are not available, use bridging principles in 3.2.3.2.*
	3. *If bridging principles do not apply, classify as an irritant if*
	(a) For mixtures where substances can be added, the sum of concentrations of corrosive substance in the mixture is $>1\%$ but $<5\%$; the sum of the concentrations of irritant substances (Cat 2) is $>10\%$; or the sum of ($10\times$ the concentrations of corrosive ingredients) + (the concentrations of irritant ingredients(Cat 2)) is $>10\%$

(Continued)

Skin Irritation

Comparison

A single harmonized GHS skin irritant category is provided. An additional mild irritant category is available for authorities that want to have more than one skin irritant category. The HCS has one skin irritation hazard category. The untested mixture hazard determination is quite different in the HCS and GHS. The GHS irritant hazard determination for mixtures is involved. Irritation data is a common data set.

OSHA HCS 29 CFR 1910.1200	GHS
	(b) For mixtures where substances cannot be added: ≥3% an irritant substance (Cat 2) (see 3.2.3.4)
	• For those authorities wanting more than one designation for skin irritation, an additional mild irritant category is provided:
	Category 3
	1. *For substances and tested mixtures*
	• Animal experience or test data that indicate that the substance/mixture causes reversible damage to the skin following exposure of up to 4 h, mean value of >1.5 <2.3 for erythema/eschar in two of three tested animals (see Table 3.2.2)
	2. *If data for a mixture are not available* use bridging principles in 3.2.3.2.
	3. *If bridging principles do not apply,* classify as mild irritant if
	• For mixtures where substances can be added, the sum of the concentrations of irritant substances(Cat 2) in the mixture is >1% but <10%
	• The sum of the concentrations of mild irritant substances (Cat 3) is >10%
	• The sum of (10× the concentrations of corrosive substances) + (the concentrations of irritant substances (Cat 2)) is >1% but <10%
	• The sum of (10× the concentrations of corrosive substances) + (the concentrations of irritant substances (Cat 2)) + (the concentrations of mild irritant substances (Cat 3)) is >10%

Serious Eye Damage

Comparison

A single harmonized GHS hazard category is provided for substances that have the potential to seriously damage the eyes. In the GHS, skin corrosives are considered to have serious eye damage. The HCS does not have a quantitative definition. The untested mixture hazard determination is different in the HCS and GHS.

OSHA HCS 29 CFR 1910.1200	GHS
Definition/Criteria	**Definition**
"Corrosive." A chemical that causes visible destruction of, or irreversible alterations in, living tissue by chemical action at the site of contact. *"Irritant."* A chemical that is not corrosive but that causes a reversible inflammatory effect on living tissue by chemical action at the site of contact. A chemical is an eye irritant if so determined under the procedure listed in 16 CFR 1500.42 or other appropriate techniques. For mixtures: Untested mixtures are assumed to present the same health hazards as components present at ≥1% (by weight or volume).	Serious eye damage means the production of tissue damage in the eye, or serious physical decay of vision, following application of a test substance to the anterior surface of the eye, which is not fully reversible within 21 days of application. **Criteria** Substances and mixtures of this hazard class are assigned to a single harmonized hazard category. Category 1—Irreversible Effects 1. *For substances and tested mixtures* • Classification as corrosive to skin • Human experience or data showing damage to the eye, which is not fully reversible within 21 days • Structure/activity or structure/property relationship to a substance or mixture already classified as corrosive • pH extremes of <2 and >11.5 including buffering capacity • Positive results in a valid and accepted in vitro test to assess serious damage to eyes • Animal experience or test data that the substance or mixture produces either (1) in at least one animal, effects on the cornea, iris, or conjunctiva that are not expected to reverse or have not reversed; or (2) in at least two of three tested animals, a positive response of corneal opacity >3 and/or iritis >1.5 (see Table 3.3.1) 2. *If data for a mixture are not available,* use bridging principles in 3.3.3.2. 3. *If bridging principles do not apply* (a) For mixtures where substances can be added: Classify as Category 1 if the sum of the concentrations of substances classified as corrosive to the skin and/or eye Category 1 substances in the mixture is >3% (b) For mixtures where substances cannot be added: >1%. See 3.3.3.3.4.

Eye Irritation

Comparison

A single GHS eye irritant category (Cat 2A) is provided. An additional mild eye irritant category (Cat 2B) is available for authorities that want to have more than one eye irritant category. The HCS has one eye irritation hazard category. In the GHS, skin irritants can be considered to be eye irritants. The untested mixture hazard determination is different in the HCS and GHS. The GHS eye irritant hazard determination for mixtures is involved. Irritation data is a common data set.

OSHA HCS 29 CFR 1910.1200	GHS

Definition/Criteria
 	
"Irritant." A chemical that is not corrosive but that causes a reversible inflammatory effect on living tissue by chemical action at the site of contact. A chemical is an eye irritant if so determined under the procedure listed in 16 CFR 1500.42 or other appropriate techniques.
 	
For mixtures:
 	
Untested mixtures are assumed to present the same health hazards as components present at ≥1% (by weight or volume).
	Definition
	
	Eye irritation means the production of changes in the eye following the application of test substance to the anterior surface of the eye, which are fully reversible within 21 days of application.
	
	Criteria
	
	Substances of this hazard class are assigned to a single harmonized hazard category or for authorities wanting more than one designation for eye irritation, one of two subcategories depending on whether effects are reversible in 21 or 7 days.
	
	For those authorities wanting more than one designation for eye irritation, two subcategories are provided within the eye irritation category:
	
	Category 2A—Irritant
	  1. *Substances and tested mixtures*
	    • Classification as (severe) skin irritant
	    • Human experience or data showing production of changes in the eye that are fully reversible within 21 days

- Structure/activity or structure/property relationship to a substance or mixture already classified as an eye irritant
- Positive results in a valid and accepted in vitro eye irritation test
- Animal experience or test data that indicate that the substance/mixture produces a positive response in at least two of three tested animals of corneal opacity >1, iritis >1, or conjunctival edema (chemosis) >2 (Table 3.3.2).

2. *If data for a mixture are not available*, use bridging principles in 3.3.3.2.

3. *If bridging does not apply*, classify as an irritant (2A) if

   (a) For mixtures where substances can be added: the sum of the concentrations of skin and/or eye Category 1 substances in the mixture is >1% but <3%; the sum of the concentrations of eye irritant substances (Cat 2/2A) is >10%; or the sum of (10× the concentrations of skin and/or eye Category 1 substances) + (the concentrations of eye irritants(2A/2B)) is >10% skin Cat 1 + eye Cat 1 is >1% but <3%; or (10× eye Category 1) + eye Cat 2A/2B is >10%

   (b) For mixtures where substances cannot be added: the sum of the concentrations of eye irritant ingredients (Cat 2) is >3% (see 3.3.3.4)

Category 2B—Mild Irritant

An eye irritant is considered mildly irritating to eyes (Category 2B) when the effects listed above (Cat 2A) are fully reversible within 7 days of observation.

Respiratory Sensitizer

Comparison
The untested mixture hazard determination may be different in the HCS and GHS. OSHA will need to determine how to implement the mixture cutoff values.

OSHA HCS 29 CFR 1910.1200	GHS
Definition/Criteria	**Definition**
"Sensitizer." A chemical that causes a substantial proportion of exposed people or animals to develop an allergic reaction in normal tissue after repeated exposure to the chemical.	Respiratory sensitizer means a substance that induces hypersensitivity of the airways following inhalation of the substance.
For mixtures:	**Criteria**
Untested mixtures are assumed to present the same health hazards as components present at ≥1% (by weight or volume).	Substances and mixtures of this hazard class are assigned to one hazard category.
	1. *For substances and tested mixture*
	– If there is human evidence that the individual substance induces specific respiratory hypersensitivity
	– Where there are positive results from an appropriate animal test
	2. *If the mixture meets the criteria set forth in the "Bridging Principles"* through one of the following:
	(a) Dilution
	(b) Batching
	(c) Substantially similar mixture
	3. *If bridging principles do not apply:* Mixtures containing ≥0.1% or ≥1.0% (solid/liquid) (0.2% gas) of such a substance. See Notes, Table 3.4.1, Chapter 3.4.

Skin Sensitizer

Comparison

The untested mixture hazard determination may be different in the HCS and GHS. OSHA will need to determine how to implement the mixture cutoff values.

OSHA HCS 29 CFR 1910.1200	GHS
Definition/Criteria	**Definition**
"Sensitizer." A chemical that causes a substantial proportion of exposed people or animals to develop an allergic reaction in normal tissue after repeated exposure to the chemical.	Skin sensitizer means a substance that will induce an allergic response following skin contact. The definition for "skin sensitizer" is equivalent to "contact sensitizer."
For mixtures:	**Criteria**
Untested mixtures are assumed to present the same health hazards as components present at ≥1% (by weight or volume).	Substances and mixtures of this hazard class are assigned to one hazard category.
	1. *For substances and tested mixture*
	– If there is evidence in humans that the individual substance can induce sensitization by skin contact in a substantial number of persons
	– Where there are positive results from an appropriate animal test
	2. *If the mixture meets the criteria set forth in the* "Bridging Principles" through one of the following:
	(a) Dilution
	(b) Batching
	(c) Substantially similar mixture
	3. *If bridging principles do not apply:* Mixtures containing ≥0.1% or ≥1.0% of such a substance. See Notes, Table 3.4.1, Chapter 3.4.

Mutagenicity

Comparison

Since they have a positive or negative result rather than a statistically significant finding, the HCS does not generally consider in vitro studies alone to form the basis for a definitive finding of hazard. The HCS addresses chromosomal damage under reproductive toxins. The HCS allows test data on mixtures to be used for all hazard classes. The untested mixture hazard determination may be different in the HCS and GHS.

The GHS provides for two hazard categories to accommodate the weight of evidence. Since this hazard class is not an exact match with the HCS, consideration could be given as to how to implement this hazard class/category.

OSHA HCS 29 CFR 1910.1200	GHS
Definition/Criteria	**Definition**
Chemicals that affect the reproductive capabilities including chromosomal damage (mutations) and effects on fetuses (teratogenesis)	Mutagen means an agent giving rise to an increased occurrence of mutations in populations of cells and/or organisms.
In vitro studies alone generally do not form the basis for a definitive finding of hazard under the HCS since they have a positive or negative result rather than a statistically significant finding.	**Criteria**
	Substances and mixtures of this hazard class are assigned to one of two hazard categories.
	Category 1 has two subcategories.
	Category 1
For mixtures:	Subcategory 1A: Known to induce heritable mutations in the germ cells of humans
Untested mixtures are assumed to present the same health hazards as components present at ≥1% (by weight or volume).	• Positive evidence from human epidemiological studies
	Subcategory 1B: Regarded as if it induces heritable mutations in the germ cells of humans
	• Positive in vivo heritable germ cell mutagenicity tests in mammals
	• Positive in vivo somatic cell mutagenicity tests in mammals and evidence for potential germ cells mutations
	• Tests showing mutagenic effects in human germ cells without demonstration of transmission to progeny
	See criteria in 3.5.2. Mixtures containing ≥0.1% of such a substance.
	Category 2
	Causes concern for man owing to the possibility that it may induce heritable mutations in the germ cells of humans
	• Evidence in mammals and/or in vitro
	• In vivo somatic cell mutagenicity tests, in mammals
	• In vivo somatic cell genotoxicity with in vitro mutagenicity assays
	See criteria in 3.5.2. Mixtures containing ≥1.0% of such a substance.

Carcinogenicity

Comparison

The HCS allows test data on mixtures to be used for all hazard classes. The GHS allows the HCS provision for inclusion on safety data sheets of positive results in any carcinogenicity study performed according to good scientific principles with statistically significant results. The HCS will need to determine the implementation of the GHS mixture cutoff values.

OSHA HCS 29 CFR 1910.1200	GHS
Definition/Criteria	**Definition**
"Carcinogen." A chemical is considered to be a carcinogen if	Carcinogen means a chemical substance or a mixture of chemical substances that induce cancer or increase its incidence.
(a) It has been evaluated by the International Agency for Research on Cancer (IARC) and found to be a carcinogen or potential carcinogen	**Criteria**
	Substances and mixtures of this hazard class are assigned to one of two hazard categories.
	Category 1 has two subcategories.
(b) It is listed as a carcinogen or potential carcinogen in the Annual Report on Carcinogens published by the National Toxicology Program (NTP) (latest edition)	Category 1: Known or presumed human carcinogen
	Category 1A: Known human carcinogen
	• Based on human evidence
	Category 1B: Presumed human carcinogen
	• Strength of evidence with additional considerations
(c) It is regulated by OSHA as a carcinogen OSHA has issued guidance on how IARC classifications apply to the HCS. Similar guidance for the GHS categories would be useful.	• Evidence of animal carcinogenicity (presumed human carcinogen)
	• On a case-by-case basis, limited evidence of carcinogenicity in humans together with limited evidence of carcinogenicity in animals.
	Including mixtures containing ≥0.1% of such a substance.
	Category 2: Suspected human carcinogen
For mixtures:	• Evidence from human and/or animal studies
Untested mixtures are assumed to present a carcinogenic hazard if components present at ≥0.1% are considered to be a carcinogen.	• Strength of evidence together with additional considerations
	Including mixtures containing more than ≥0.1% or ≥1.0% of such a substance. (See Notes 1 and 2 of Table 3.6.1 of Chapter 3.6.)

Reproductive Toxicity

Comparison

The HCS allows test data on mixtures to be used for all hazard classes. The untested mixture hazard determination is different in the HCS and GHS. The HCS will need to determine the implementation of the GHS mixture cutoff values.

OSHA HCS 29 CFR 1910.1200	GHS
Definition/Criteria	**Criteria**
"Reproductive toxins." Chemicals that affect the reproductive capabilities including chromosomal damage (mutations) and effects on fetuses (teratogenesis)	Effects on reproductive ability or capacity and on development are considered separate issues.
Signs and symptoms: Birth defects; sterility	Substances and mixtures of this hazard class are assigned to one of two hazard categories. Category 1 has two subcategories.
Chemicals: Lead; DBCP	Category 1: Known or presumed human reproductive/developmental toxicants
For mixtures:	Category 1A: Known human reproductive or developmental toxicants
Untested mixtures are assumed to present the same health hazards as components present at ≥1% (by weight or volume).	• Human evidence
	Category 1B: Presumed human reproductive/developmental toxicants
	• Evidence from experimental animals
	See criteria in 3.7.2.2.1 to 3.7.2.6.0 of Chapter 3.7. Mixtures containing ≥0.1% or ≥0.3% of such a substance. See Notes 1 and 2 of Table 3.7.1, Chapter 3.7.
	Category 2: Suspected human reproductive/developmental toxicants
	• Evidence from humans or experimental animals, possibly supplemented with other information.
	See criteria in 3.7.2.2.1 to 3.7.2.6.0 of Chapter 3.7. Mixtures containing ≥0.1% or ≥3.0% of such a substance. See Notes 3 and 4 of Table 3.7.1, Chapter 3.7.

Effects on or via Lactation

Comparison
This special category is not specifically addressed in the HCS.

OSHA HCS 29 CFR 1910.1200	GHS
Definition	**Criteria**
Criteria	Effects on or via lactation are assigned to a separate single category.
	Special Category
	Substances that cause concern for the health of breastfed children
	• Absorption, metabolism, distribution, and excretion studies indicating the likelihood that the substance is present in potentially toxic levels in breast milk
	• Evidence from one or two generation studies of adverse effect in the offspring due to transfer in the milk or adverse effect on the quality of the milk
	• Human evidence indicating a hazard to babies during the lactation period
	See criteria in 3.7.2.2.1 to 3.7.2.6.0 and 3.7.3.4 of Chapter 3.7.

Target Organ Systemic Toxicity after Single Exposure

Comparison

The HCS does not distinguish between single and repeat exposure for target organ effects. The GHS provides guidance on dose/concentration value ranges. The untested mixture hazard determination is different in the HCS and GHS. The HCS will need to determine the implementation of the GHS mixture cutoff values.

OSHA HCS 29 CFR 1910.1200	GHS
Definition/Criteria	**Criteria**
"Target organ effects."	All significant health effects that can impair function, both reversible and
The following is a target organ categorization of effects that may occur,	irreversible, immediate and/or delayed are included in the nonlethal
including examples of signs and symptoms and chemicals that have	target organ/systemic toxicity class.
been found to cause such effects. These examples are presented to	Substances and mixtures of this hazard class are assigned to one of two
illustrate the range and diversity of effects and hazards found in the	hazard categories based on evidence for substance or mixture (including
workplace, and the broad scope employers must consider in this area,	bridging).
but are not intended to be all-inclusive.	Category 1: Significant toxicity in humans
a. Hepatotoxins: Chemicals that produce liver damage	Adverse effect on specific organ/systems:
Signs and symptoms: Jaundice; liver enlargement	• Human cases or epidemiological studies
Chemicals: Carbon tetrachloride; nitrosamines	• Animal studies with severe effects at low dose
b. Nephrotoxins: Chemicals that produce kidney damage	• Guidance values in Table 3.8.1 as part of weight of evidence
Signs and symptoms: Edema; proteinuria	evaluation
Chemicals: Halogenated hydrocarbons; uranium	Mixture that lacks sufficient data but contains Category 1 ingredient at a
c. Neurotoxins: Chemicals that produce their primary toxic effects on	concentration of >1% to <10% for some authorities and contains
the nervous system	Category 1 ingredient at a concentration ≥10% for all authorities (see
	Notes of Table 3.8.2)

Signs and symptoms: Narcosis; behavioral changes; decrease in motor functions

Chemicals: Mercury; carbon disulfide

d. Agents that act on the blood or hematopoietic system: They decrease hemoglobin function and deprive the body tissues of oxygen

Signs and symptoms: Cyanosis; loss of consciousness

Chemicals: Carbon monoxide; cyanides

e. Agents that damage the lung; Chemicals that irritate or damage pulmonary tissue

Signs and symptoms: Cough; tightness in chest; shortness of breath

Chemicals: Silica; asbestos

f. Reproductive toxins: Chemicals that affect the reproductive capabilities including chromosomal damage (mutations) and effects on fetuses (teratogenesis)

Signs and symptoms: Birth defects; sterility

Chemicals: Lead; DBCP

g. Cutaneous hazards: Chemicals that affect the dermal layer of the body

Signs and symptoms: Defatting of the skin; rashes; irritation

Chemicals: Ketones; chlorinated compounds

h. Eye hazards: Chemicals that affect the eye or visual capacity

Signs and symptoms: Conjunctivitis; corneal damage

Chemicals: Organic solvents; acids

For mixtures:

Untested mixtures are assumed to present the same health hazards as components present at ≥1%(by weight or volume).

Category 2: Harmful to humans

Adverse effect on specific organ/systems:

• Animal studies with significant toxicity at moderate dose, considering weight of evidence and guidance values in Table 3.8.1

Mixture that lacks sufficient data, but contains Category 1 ingredient: >1% but <10% for some authorities and ≥10% for all authorities (see Notes of Table 3.8.2), and/or contains Category 2 ingredient: >1% but <10% for some authorities and ≥10% for all authorities (see Notes of Table 3.8.2)

Target Organ Systemic Toxicity after Repeat Exposure

Comparison

The HCS does not distinguish between single and repeat exposure for target organ effects. The GHS provides guidance on dose/concentration value ranges. The untested mixture hazard determination may be different in the HCS and GHS. The HCS will need to determine the implementation of the GHS mixture cutoff values.

OSHA HCS 29 CFR 1910.1200	GHS
Definition/Criteria	**Criteria**
"Target organ effects."	All significant health effects that can impair function, both reversible and irreversible, immediate and/or delayed are included in the nonlethal target organ/systemic toxicity class.
The following is a target organ categorization of effects that may occur, including examples of signs and symptoms and chemicals that have been found to cause such effects. These examples are presented to illustrate the range and diversity of effects and hazards found in the workplace and the broad scope employers must consider in this area, but are not intended to be all-inclusive.	Substances and mixtures of this hazard class are assigned to one of two hazard categories based on evidence for substance or mixture (including bridging).
a. Hepatotoxins: Chemicals that produce liver damage	Category 1: Significant toxicity in humans
Signs and symptoms: Jaundice; liver enlargement	Adverse effect on specific organ/systems:
Chemicals: Carbon tetrachloride; nitrosamines	• Human cases or epidemiological studies
b. Nephrotoxins: Chemicals that produce kidney damage	• Animal studies with severe effects at low dose
Signs and symptoms: Edema; proteinuria	• Guidance values in Table 3.9.1, as part of weight of evidence evaluation
Chemicals: Halogenated hydrocarbons; uranium	Mixture that lacks sufficient data but contains Category 1 ingredient at a concentration of >1% to <10% for some authorities, and contains Category 1 ingredient at a concentration ≥10.0% for all authorities (see Notes of Table 3.9.3)
c. Neurotoxins: Chemicals that produce their primary toxic effects on the nervous system	
Signs and symptoms: Narcosis; behavioral changes; decrease in motor functions	

Chemicals: Mercury; carbon disulfide

d. Agents that act on the blood or hematopoietic system: They decrease hemoglobin function and deprive the body tissues of oxygen

Signs and symptoms: Cyanosis; loss of consciousness

Chemicals: Carbon monoxide; cyanides

e. Agents that damage the lung; Chemicals that irritate or damage pulmonary tissue

Signs and symptoms: Cough; tightness in chest; shortness of breath

Chemicals: Silica; asbestos

f. Reproductive toxins: Chemicals that affect the reproductive capabilities including chromosomal damage (mutations) and effects on fetuses (teratogenesis)

Signs and symptoms: Birth defects; sterility

Chemicals: Lead; DBCP

g. Cutaneous hazards: Chemicals that affect the dermal layer of the body

Signs and symptoms: Defatting of the skin; rashes; irritation

Chemicals: Ketones; chlorinated compounds

h. Eye hazards: Chemicals that affect the eye or visual capacity

Signs and symptoms: Conjunctivitis; corneal damage

Chemicals: Organic solvents; acids

For mixtures:

Untested mixtures are assumed to present the same health hazards as components present at ≥1% (by weight or volume).

Category 2: Harmful to humans

Adverse effect on specific organ/systems:

• Animal studies with significant toxicity at moderate dose, considering weight of evidence and guidance values in Table 3.9.1

Mixture that lacks sufficient data, but contains Category 1 ingredient: >1% but <10% for some authorities and ≥10% for all authorities (see Note 3 of Table 3.9.3), and/or contains Category 2 ingredient: >1% but <10% for some authorities and ≥10% for all authorities (see Notes of Table 3.9.3)

Comparison of Physical Hazards

Comparison of OSHA HCS and GHS Criteria

Explosives

Comparison

The HCS has only one hazard category for Explosives and the GHS has six hazard categories. The value of multiple hazard categories for the workplace should be addressed. The HCS does not require testing or specify test methods.

HCS Criteria	GHS Criteria
Test Method	**Test Method**
HCS has no test method.	UN Manual of Tests and Criteria Part I Test Series 2 to 7
Definition	Recommended tests for explosives (including articles)
"Explosive" means a chemical that causes a sudden, almost instantaneous release of pressure, gas, and heat when subjected to sudden shock, pressure, or high temperature.	2(a) UN Gap test
	2(b) Koenen test
	2(c) Time/pressure test
	3(a)(ii) BAM Fallhammer
	3(b)(i) BAM Friction apparatus
	3(c) Thermal stability test at 75°C
	3(d) Small-scale burning test
	4(a) Thermal stability test for unpackaged articles and packaged articles
	4(b)(i) Steel tube drop test for liquids
	4(b)(ii) Twelve-meter drop test for unpackaged articles, packaged articles, and packaged substances
	5(a) Cap sensitivity test
	5(b)(ii) USA DDT test
	5(c) External fire test for Division 1.5
	6(a) Single package test
	6(b) Stack test
	6(c) External fire (bonfire) test
	7(a) EIDS cap test
	7(b) EIDS gap test
	7(c)(ii) Friability test
	7(d)(i) EIDS bullet impact test

(Continued)

Explosives

Comparison

The HCS has only one hazard category for Explosives and the GHS has six hazard categories. The value of multiple hazard categories for the workplace should be addressed. The HCS does not require testing or specify test methods.

HCS Criteria	GHS Criteria
	7(e) EIDS external fire test
	7(f) EIDS slow cook-off test
	7(g) 1.6 Article external fire test
	7(h) 1.6 Article slow cook-off test
	7(j) 1.6 Article bullet impact-off test
	7(k) 1.6 Article stack test
	Definition
	Explosive substance means a solid or liquid substance (or mixture of substances) that is in itself capable by chemical reaction of producing gas at such a temperature and pressure and at such a speed as to cause damage to the surroundings. Pyrotechnic substances are included even when they do not evolve gases.
	Criteria
	Substances, mixtures, and articles of this class are assigned to one of six divisions, 1.1 to 1.6, depending on the type of hazard they present: Division 1.1: Substances and articles that have a mass explosion hazard (a mass explosion is one that affects almost the entire load virtually instantaneously)

Division 1.2: Substances and articles that have a projection hazard but not a mass explosion hazard

Division 1.3: Substances and articles that have a fire hazard and either a minor blast hazard or a minor projection hazard or both, but not a mass explosion hazard:

(i) Combustion of which gives rise to considerable radiant heat

(ii) Which burn one after another, producing minor blast or projection effects or both

Division 1.4: Substances and articles that present no significant hazard: Substances and articles that present only a small hazard in the event of ignition or initiation. The effects are largely confined to the package and no projection of fragments of appreciable size or range is to be expected. An external fire shall not cause a virtually instantaneous explosion of almost the entire contents of the package.

Division 1.5: Very insensitive substances that have a mass explosion hazard: substances that have a mass explosion hazard but are so insensitive that there is very little probability of initiation or of transition from burning to detonation under normal conditions

Division 1.6: Extremely insensitive articles that do not have a mass explosion hazard: articles that contain only extremely insensitive detonating substances and that demonstrate a negligible probability of accidental initiation or propagation

NOTE: Substances that are too unstable for allocation to the above divisions are also to be classified as explosive.

Flammable Gases

Comparison

The HCS has one hazard category for flammable gases. The GHS has two hazard categories. The HCS does not require testing and does not specify test methods.

HCS Criteria	GHS Criteria
Test Method	**Test Method**
HCS has no test method.	ISO 10156:1996
Definition	**Definition**
"Gas, flammable" means: (A) A gas that, at ambient temperature and pressure, forms a flammable mixture with air at a concentration of 13% by volume or less	Flammable gas means a gas having a flammable range with air at 20°C and a standard pressure of 101.3 kPa.
(B) A gas that, at ambient temperature and pressure, forms a range of flammable mixtures with air wider than 12% by volume, regardless of the lower limit	**Criteria**
	Substance sand mixtures of this hazard class are assigned to one of two hazard categories on the basis of the outcome of the test or calculation method:
	1. Gases, which at 20°C and a standard pressure of 101.3kPa:
	(a) are ignitable when in a mixture of 13% or less by volume in air; or
	2. Have a flammable range with air of at least 12 percentage points regardless of the lower flammable limit
	3. Gases, other than those of category 1, which, at 20°C and a standard pressure of 101.3kPa, have a flammable range while mixed in air.

Flammable Aerosols

Comparison

The HCS has one hazard class/category for flammable aerosols. The GHS has two hazard categories. The HCS does not require testing.

HCS Criteria	GHS Criteria
Test Method	**Test Method**
16 CFR 1500.45	GHS Document Annex 11
Definition	**Definition**
"Aerosol, flammable" means an aerosol that, when tested by the method described in 16 CFR 1500.45, yields a flame projection exceeding 18 inches at full valve opening, or a flashback (a flame extending back to the valve) at any degree of valve opening.	*Aerosols* means any nonrefillable receptacles made of metal, glass, or plastics and containing a gas compressed, liquefied, or dissolved under pressure, with or without a liquid, paste, or powder, and fitted with a release device allowing the contents to be ejected as solid or liquid particles in suspension in a gas, as a foam, paste, or powder or in a liquid state or in a gaseous state. Aerosol includes aerosol dispensers.
	Criteria
	Substances and mixtures of this hazard class are assigned to one of two hazard categories on the basis of their components, that is, flammable liquids (see GHS Chapter 2.6), flammable gases (see GHS Chapter 2.2), flammable solids (see GHS Chapter 2.7), and, if applicable, the results of the foam test (for foam aerosols) and of the ignition distance test and enclosed space test (for spray aerosols):
	Note: Flammable components do not cover pyrophoric, self-heating, or water-reactive substances because such components are never used as aerosol contents.
	The chemical heat of combustion (DHc), in kilojoules per gram (kJ/g), is the product of the theoretical heat of combustion (D Hcomb), and a combustion efficiency, usually less than 1.0 (a typical combustion efficiency is 0.95 or 95%).
	For a composite aerosol formulation, the chemical heat of combustion is the summation of the weighted heats of combustion for the individual components, as follows:
	$DH_c \text{ (product)} = S[I\% \times DH_{c(i)}]$
	where DH_c = chemical heat of combustion (kJ/g)
	$I\%$ = weight fraction of component I in the product
	$DH_{c(i)}$ = chemical heat of combustion of component I (kJ/g).
	The chemical heats of combustion can be found in literature, calculated, or determined by tests (see ASTM D 240, ISO/FDIS 13943:1999 (E/F) 86.1 to 86.3 and NFPA 30B).

Oxidizing Gases

Comparison

The HCS covers oxidizers as a class of chemicals with one category. The GHS covers oxidizers by physical state with one hazard category for gases. The HCS does not require testing or specify test methods.

HCS Criteria	GHS Criteria
Test Method HCS has no test method. **Definition** "Oxidizer" means a chemical other than a blasting agent or explosive as defined in 1910.109(a) that initiates or promotes combustion in other materials, thereby causing fire either of itself or through the release of oxygen or other gases.	**Test Method** ISO 10156:1996 **Definition** Oxidizing gas means any gas that may, generally by providing oxygen, cause or contribute to the combustion of other material more than air does. **Criteria** Substances and mixtures of this hazard class are assigned to a single hazard category on the basis that, generally by providing oxygen, they cause or contribute to the combustion of other material more than air does.

Gases under Pressure

Comparison

The HCS has one hazard class/category for compressed gases. The GHS uses physical state as a basis for four groups. The HCS does not require testing or specify test methods.

HCS Criteria	GHS Criteria
Test Method HCS has no test method. **Definition** "Compressed gas" means i. A gas or mixture of gases having, in a container, an absolute pressure exceeding 40 psi at 70°F (21.1°C) ii. A gas or mixture of gases having, in a container, an absolute pressure exceeding 104 psi at 130°F (54.4°C) regardless of the pressure at 70°F (21.1°C) iii. A liquid having a vapor pressure exceeding 40 psi at 100°F (37.8°C) as determined by ASTM D-323-72	**Test Method** For this group of gases, the following information is required to be known: • The vapor pressure at 50°C • The physical state at 20°C at standard ambient pressure • The critical temperature Data can be found in literature, calculated, or determined by testing. Most pure gases are already classified in the UN Model Regulations. **Criteria** Gases are classified, according to their physical state when packaged, into one of four groups as follows: *Compressed gases* A gas that when packaged under pressure is entirely gaseous at −50°C; including all gases with a critical temperature <−50°C *Liquefied gases* A gas that when packaged under pressure is partially liquid at temperatures above −50°C. A distinction is made between i. High-pressure liquefied gas: a gas with a critical temperature between −50°C and +65°C ii. Low-pressure liquefied gas: a gas with a critical temperature above +65°C *Refrigerated liquefied gases* A gas that when packaged is made partially liquid because of its low temperature *Dissolved gases* A gas that when packaged under pressure is dissolved in a liquid phase solvent Note: The critical temperature is the temperature above which a pure gas cannot be liquefied, regardless of the degree of compression.

Flammable Liquids

Comparison

The HCS has two hazard categories for flammable liquids that cover the same flash point range as the four GHS categories. The HCS does not require testing.

HCS Criteria	GHS Criteria
Test Method Flash point: i. Tagliabue Closed Tester (See American National Standard Method of Test for Flash Point by Tag Closed Tester, Z11.24-1979 (ASTM D 56-79)) for liquids with a viscosity of less than 45 Saybolt Universal Seconds (SUS) at 100°F (37.8°C) that do not contain suspended solids and do not have a tendency to form a surface film under test ii. Pensky–Martens Closed Tester (see American National Standard Method of Test for Flash Point by Pensky–Martens Closed Tester, Z11.7-1979 (ASTM D 93-79)) for liquids with a viscosity equal to or greater than 45 SUS at 100°F (37.8°C), or that contain suspended solids, or that have a tendency to form a surface film under test iii. Setaflash Closed Tester (see American National Standard Method of Test for Flash Point by Setaflash Closed Tester (ASTM D3278-78)) **Definition** "Liquid, flammable" means any liquid having a flash point below 100°F (37.8°C), except any mixture having components with flash points of 100°F (37.8°C) or higher, the total of which make up 99% or more of the total volume of the mixture.	**Test Method** Flash point is determined by closed cup methods as provided in GHS Chapter 2.5, paragraph 11. Initial boiling point is also required. **Definition** Flammable liquid means a liquid having a flash point of not more than 93°C. **Criteria** Substances and mixtures of this hazard class are assigned to one of four hazard categories on the basis of the flash point and boiling point: 1. Flash point <23°C and initial boiling point <35°C 2. Flash point < 23°C and initial boiling point >35°C 3. Flash point >23°C and <60°C 4. Flash point > 60°C and <93°C Note 1: Gas oils, diesel, and light heating oils in the flash point range of 55°C to 75°C may be regarded as a special group for some regulatory purposes. Note 2: Liquids with a flash point of more than 35°C may be regarded as nonflammable liquids for some regulatory purposes (e.g., transport) if negative results have been obtained in the sustained combustibility test L.2 of the UN Manual of Tests and Criteria Part III. Note 3: Viscous flammable liquids such as paints, enamels, lacquers, varnishes, adhesives, and polishes may be regarded as a special group for some regulatory purposes (e.g., transport). The classification or the decision to consider these liquids as nonflammable may be determined by the pertinent regulation or competent authority.

Flammable Solids

Comparison

The HCS has one hazard class/category for flammable solids. The GHS has two hazard categories. The HCS does not require testing.

HCS Criteria	GHS Criteria
Test Method	**Test Method**
16 CFR 1500.44	UN Manual of Tests and Criteria Part III Test N.1
Definition	**Definition**
"Solid, flammable" means a solid, other than a blasting agent or explosive as defined in 1910.109(a), that is liable to cause fire through friction, absorption of moisture, spontaneous chemical change, or retained heat from manufacturing or processing, or which can be ignited readily and when ignited burns so vigorously and persistently as to create a serious hazard. A chemical shall be considered to be a flammable solid if, when tested by the method described in 16 CFR 1500.44, it ignites and burns with a self-sustained flame at a rate greater than one-tenth of an inch per second along its major axis.	Flammable solid means a solid that is readily combustible or may cause or contribute to fire through friction.
	Criteria
	Substances and mixtures of this hazard class are assigned to one of two hazard categories on the basis of the outcome of the test:
	1. Burning rate test:
	Substances other than metal powders:
	– Wetted zone does not stop fire
	– Burning time <45 s or burning rate >2.2 mm/s
	Metal powders:
	– Burning time ≤5 min
	$r > 2$ Burning rate test:
	Substances other than metal powders:
	– Wetted zone stops the fire for at least 4 min
	– Burning time <45 s or burning rate >2.2 mm/s
	Metal powders:
	– Burning time >5 min and <10 min

Self-Reactive Substances

Comparison

The HCS has only one hazard category for self-reactive (unstable/reactive) substances. The GHS has seven hazard categories. The value of multiple hazard categories for the workplace should be addressed. The HCS does not require testing or specify test methods.

HCS Criteria	GHS Criteria
Test Method	**Test Method**
HCS has no test method.	UN Manual of Tests and Criteria Part II Test Series A to H
Definition	Recommended tests for Self-Reactive Substance
"Unstable (reactive)" means a chemical that in the pure state, or as produced or transported, will vigorously polymerize, decompose, or condense or will become self-reactive under conditions of shock, pressure, or temperature.	A.6 UN detonation test
	B.1 Detonation test in package
	C.1 Time/pressure test
	C.2 Deflagration test
	D.1 Deflagration test in the package
	E.1 Koenen test
	E.2 Dutch pressure vessel test
	F.4 Modified Trauzl test
	G.1 Thermal explosion test in package
	H.1 United States SADT test (for packages)
	H.2 Adiabatic storage test (for packages, etc.)
	H.4 Heat accumulation storage test (for packages, IBCs, and small tanks)
	Preliminary Safety Assessment Tests
	– Falling weight test for impact sensitivity
	– Friction or impacted friction test for friction sensitivity
	– Test to assess thermal stability and the exothermic decomposition energy
	– Test to assess the effect of ignition
	Definition
	Self-reactive substance means a thermally unstable liquid or solid substance liable to undergo a strongly exothermic decomposition even without participation of oxygen (air). This definition excludes substances or mixtures classified under the GHS as explosive, organic peroxides or as oxidizing.

Criteria

Substances and mixtures of this hazard class are assigned to one of the seven "types" A to G on the basis of the outcome of the tests.

Any self-reactive substance should be considered for classification in this class unless

(a) They are explosives, according to the GHS criteria of Chapter 2.1
(b) They are oxidizing substances, according to the GHS criteria of Chapter 2.13 or 2.14
(c) They are organic peroxides, according to the GHS criteria of Chapter 2.15
(d) Their heat of decomposition is <300 J/g
(e) Their self-accelerating decomposition temperature (SADT) is >75°C for a 50-kg package

Self-reactive substances are classified in one of the seven categories of "types A to G" for this class, according to the following principles:

(A) Any self-reactive substance that can detonate or deflagrate rapidly, as packaged, will be defined as self-reactive substance **Type A**.

(B) Any self-reactive substance possessing explosive properties and which, as packaged, neither detonates nor deflagrates rapidly, but is liable to undergo a thermal explosion in that package will be defined as self-reactive substance **Type B**.

(C) Any self-reactive substance possessing explosive properties when the substance as packaged cannot detonate or deflagrate rapidly or undergo a thermal explosion will be defined as self-reactive substance **Type C**.

(D) Any self-reactive substance that in laboratory testing
   i. Detonates partially, does not deflagrate rapidly, and shows no violent effect when heated under confinement;
   ii. Does not detonate at all, deflagrates slowly, and shows no violent effect when heated under confinement;
   iii. Does not detonate or deflagrate at all and shows a medium effect when heated under confinement;
   will be defined self-reactive substance **Type D**.

(E) Any self-reactive substance that, in laboratory testing, neither detonates nor deflagrates at all and shows low or no effect when heated under confinement will be defined as self-reactive substance **Type E**.

(Continued)

Self-Reactive Substances

Comparison

The HCS has only one hazard category for self-reactive (unstable/reactive) substances. The GHS has seven hazard categories. The value of multiple hazard categories for the workplace should be addressed. The HCS does not require testing or specify test methods.

HCS Criteria	GHS Criteria
	(F) Any self-reactive substance that, in laboratory testing, neither detonates in the cavitated state nor deflagrates at all and shows only a low or no effect when heated under confinement as well as low or no explosive power will be defined self-reactive substance **Type F.**
	(G) Any self-reactive substance that, in laboratory testing, neither detonates in the cavitated state nor deflagrates at all and shows no effect when heated under confinement nor any explosive power, provided that it is thermally stable (self-accelerating decomposition temperature is 60°C to 75°C for a 50-kg package), and, for liquid mixtures, a diluent having a boiling point not less than 150°C is used for desensitization will be defined as self-reactive substance **Type G.** If the mixture is not thermally stable or a diluent having a boiling point less than 150°C is used for desensitization, the mixture shall be defined self-reactive substance **Type F.**

Pyrophoric Liquids

Comparison

The HCS has one hazard class/category for pyrophorics. The GHS covers pyrophorics by physical state with one hazard category for pyrophoric liquids. The HCS does not require testing or specify test methods.

HCS Criteria	GHS Criteria
Test Method	**Test Method**
HCS has no test method.	UN Manual of Tests and Criteria Part III Test N.3
Definition	**Definition**
"Pyrophoric" means a chemical that will ignite spontaneously in air at a temperature of 130°F (54.4°C) or below.	Pyrophoric liquid means a liquid that, even in small quantities, is liable of igniting within 5 min after coming into contact with air.
	Criteria
	Substances and mixtures of this hazard class are assigned to a single hazard category on the basis of the outcome of the test:
	The liquid ignites within 5 min when added to an inert carrier and exposed to air, or it ignites or chars a filter paper on contact with air within 5 min.

Pyrophoric Solids

Comparison

The HCS has one hazard class/category for pyrophorics. The GHS covers pyrophorics by physical state with one hazard category for pyrophoric solids. The HCS does not require testing or specify test methods.

HCS Criteria	GHS Criteria
Test Method HCS has no test method. **Definition** "Pyrophoric" means a chemical that will ignite spontaneously in air at a temperature of 130°F (54.4°C) or below.	**Test Method** UN Manual of Tests and Criteria Part III Test N.2 **Definition** Pyrophoric solid means a solid that, even in small quantities, is liable of igniting within 5 min after coming into contact with air. **Criteria** Substances and mixtures of this hazard class are assigned to a single hazard category on the basis of the outcome of the test: The solid ignites within 5 min of coming into contact with air.

Self-Heating Substances

Comparison

The HCS does not have this exact hazard class. According to the GHS building blocks, OSHA will need to determine if this hazard class will be in the HCS. The HCS does not require testing or specify test methods.

HCS Criteria	GHS Criteria
Test Method	**Test Method**
HCS has no test method.	UN Manual of Tests and Criteria Part III Test N.4
Definition	**Definition**
"Unstable (reactive)" means a chemical that in the pure state, or as produced or transported, will vigorously polymerize, decompose, condense, or will become self-reactive under conditions of shock, pressure or temperature.	Self-heating substance means a solid or liquid substance, other than a pyrophoric substance, that, by reaction with air and without energy supply, is liable to self-heat; this substance differs from a pyrophoric substance in that it will ignite only when in large amounts (kilograms) and after long periods of time (hours or days).
Or	**Criteria**
"Pyrophoric" means a chemical that will ignite spontaneously in air at a temperature of 130°F (54.4°C) or below.	Substances and mixtures of this hazard class are assigned to one of two hazard categories on the basis of the outcome of the test:
	1. A positive result is obtained in a test using a 25-mm sample cube at 140°C.
	2. (a) A positive result is obtained in a test using a 100-mm sample cube at 140°C and a negative result is obtained in a test using a 25-mm cube sample at 140°C and the substance is to be packed in packages with a volume of more than 3 m³.
	(b) A positive result is obtained in a test using a 100-mm sample cube at 140°C and a negative result is obtained in a test using a 25-mm cube sample at 140°C; a positive result is obtained in a test using a 100-mm cube sample at 120°C and the substance is to be packed in packages with a volume of more than 450 L.
	(c) A positive result is obtained in a test using a 100-mm sample cube at 140°C and a negative result is obtained in a test using a 25-mm cube sample at 140°C and a positive result is obtained in a test using a 100-mm cube sample at 100°C.

Substances that on Contact with Water Emit Flammable Gases

Comparison

The HCS has one hazard class/category for water reactive. The GHS has three hazard categories. The HCS does not require testing or specify test methods.

HCS Criteria	GHS Criteria
Test Method HCS has no test method. **Definition** "Water-reactive" means a chemical that reacts with water to release a gas that is either flammable or presents a health hazard.	**Test Method** UN Manual of Tests and Criteria Part III Test N.5 **Definition** *Substances that, on contact with water, emit flammable gases* means a solid or liquid substance or mixture that, by interaction with water, is liable to become spontaneously flammable or to give off flammable gases in dangerous quantities. **Criteria** Substances and mixtures of this hazard class are assigned to one of three hazard categories on the basis of the outcome of the test: 1. Any substance that reacts vigorously with water at ambient temperatures and demonstrates generally a tendency for the gas produced to ignite spontaneously or that reacts readily with water at ambient temperatures such that the rate of evolution of flammable gas is equal to or greater than 10 L/kg of substance over any 1 min. 2. Any substance that reacts readily with water at ambient temperatures such that the maximum rate of evolution of flammable gas is equal to or greater than 20 L/kg of substance per hour and that does not meet the criteria for category 1. 3. Any substance that reacts slowly with water at ambient temperatures such that the maximum rate of evolution of flammable gas is equal to or greater than 1 L/kg of substance per hour and that does not meet the criteria for categories 1 and 2.

Oxidizing Liquids

Comparison

The HCS covers Oxidizers as a class of chemicals with one category. The GHS covers oxidizers by physical state with three hazard categories for liquids. The HCS does not require testing or specify test methods.

HCS Criteria	GHS Criteria
Test Method HCS has no test method. **Definition** "Oxidizer" means a chemical other than a blasting agent or explosive as defined in 1910.109(a) that initiates or promotes combustion in other materials, thereby causing fire either of itself or through the release of oxygen or other gases.	**Test Method** UN Manual of Tests and Criteria Part III Test O.2 **Definition** Oxidizing liquid means a liquid that, while in itself not necessarily combustible, may, generally by yielding oxygen, cause, or contribute to, the combustion of other material. **Criteria** Substances and mixtures of this hazard class are assigned to one of three hazard categories on the basis of the outcome of the test: 1. Any substance that, in the 1:1 mixture, by mass, of substance and cellulose tested, spontaneously ignites; or the mean pressure rise time of a 1:1 mixture, by mass, of substance and cellulose is less than that of a 1:1 mixture, by mass, of 50% perchloric acid and cellulose 2. Any substance that, in the 1:1 mixture, by mass, of substance and cellulose tested, exhibits a mean pressure rise time less than or equal to the mean pressure rise time of a 1:1 mixture, by mass, of 40% aqueous sodium chlorate solution and cellulose, and the criteria for category 1 are not met 3. Any substance that, in the 1:1 mixture, by mass, of substance and cellulose tested, exhibits a mean pressure rise time less than or equal to the mean pressure rise time of a 1:1 mixture, by mass, of 65% aqueous nitric acid and cellulose, and the criteria for category 1 and 2 are not met

Oxidizing Solids

Comparison

The HCS covers oxidizers as a class of chemicals with one category. The GHS covers oxidizers by physical state with three hazard categories for solids. The HCS does not require testing or specify test methods.

HCS Criteria	GHS Criteria
Test Method	**Test Method**
HCS has no test method.	UN Manual of Tests and Criteria Part III Test O.1
Definition	**Definition**
"Oxidizer" means a chemical other than a blasting agent or explosive as defined in 1910.109(a) that initiates or promotes combustion in other materials, thereby causing fire either of itself or through the release of oxygen or other gases.	Oxidizing solid means a solid that, while in itself not necessarily combustible, may, generally by yielding oxygen, cause, or contribute to, the combustion of other material.
	Criteria
	Substances and mixtures of this hazard class are assigned to one of three hazard categories on the basis of the outcome of the test:
	1. Any substance that, in the 4:1 or 1:1 sample-to-cellulose ratio (by mass) tested, exhibits a mean burning time less than the mean burning time of a 3:2 mixture, by mass, of potassium bromate and cellulose
	2. Any substance that, in the 4:1 or 1:1 sample-to-cellulose ratio (by mass) tested, exhibits a mean burning time equal to or less than the mean burning time of a 2:3 mixture (by mass) of potassium bromate and cellulose, and the criteria for category 1 are not met
	3. Any substance that, in the 4:1 or 1:1 sample-to-cellulose ratio (by mass) tested, exhibits a mean burning time equal to or less than the mean burning time of a 3:7 mixture (by mass) of potassium bromate and cellulose, and the criteria for categories 1 and 2 are not met

Organic Peroxides

Comparison

The HCS covers organic peroxides as a class of chemicals with one category. The GHS has seven hazard categories for organic peroxides. The value of multiple hazard categories for the workplace should be addressed. The HCS does not require testing or specify test methods.

HCS Criteria	GHS Criteria
Test Method	**Test Method**
HCS has no test method.	UN Manual of Tests and Criteria Part II Test Series A to H
Definition	Recommended tests for Organic Peroxides
"Organic peroxide" means an organic compound	A.6 UN detonation test
that contains the bivalent -O-O- structure and	B.1 Detonation test in package
which may be considered to be a structural	C.1 Time/pressure test
derivative of hydrogen peroxide where one or	C.2 Deflagration test
both of the hydrogen atoms have been replaced	D.1 Deflagration test in the package
by an organic radical.	E.1 Koenen test
	E.2 Dutch pressure vessel test
	F.4 Modified Trauzl test
	G.1 Thermal explosion test in package
	H.1 United States SADT test (for packages)
	H.2 Adiabatic storage test (for packages, etc.)
	H.4 Heat accumulation storage test (for packages, IBCs, and small tanks)
	Preliminary Safety Assessment Tests
	– Falling weight test for impact sensitivity
	– Friction or impacted friction test for friction sensitivity
	– Test to assess thermal stability and the exothermic decomposition energy
	– Test to assess the effect of ignition
	Definition
	Organic peroxide means a liquid or solid organic substance which contains the bivalent -O-O- structure and may be considered a derivative of hydrogen peroxide, where one or both of the hydrogen atoms have been replaced by organic radicals. The term also includes organic peroxide formulation (mixtures).

(Continued)

Organic Peroxides

Comparison

The HCS covers organic peroxides as a class of chemicals with one category. The GHS has seven hazard categories for organic peroxides. The value of multiple hazard categories for the workplace should be addressed. The HCS does not require testing or specify test methods.

HCS Criteria	GHS Criteria
	Criteria Substances and mixtures of this hazard class are assigned to one of the seven "types" A to G on the basis of the outcome of the tests: Any organic peroxide is considered for classification in this class, unless it contains (a) Not more than 1.0% available oxygen from the organic peroxides when containing not more than 1.0% hydrogen peroxide (b) Not more than 0.5% available oxygen from the organic peroxides when containing more than 1.0% but not more than 7.0% hydrogen peroxide Note: The available oxygen content (%) of an organic peroxide mixture is given by the formula: $$16 \times S\,(n_i \times c_i / m_i)$$ where n_i = number of oxygen groups per molecule of organic peroxide i c_i = concentration (mass%) of organic peroxide i m_i = molecular mass of organic peroxide i Organic peroxides are classified in one of the seven hazard categories of "types A to G" for this class, according to the following principles: (A) Any organic peroxide mixture that can detonate or deflagrate rapidly, as packaged, will be defined as organic peroxide **Type A**. (B) Any organic peroxide mixture possessing explosive properties and which, as packaged, neither detonates nor deflagrates rapidly, but is liable to undergo a thermal explosion in that package will be defined as organic peroxide **Type B**. (C) Any organic peroxide mixture possessing explosive properties when the substance as packaged cannot detonate or deflagrate rapidly or undergo a thermal explosion will be defined as organic peroxide **Type C**.

(D) Any organic peroxide mixture that in laboratory testing

  i. Detonates partially, does not deflagrate rapidly, and shows no violent effect when heated under confinement;

  ii. Does not detonate at all, deflagrates slowly, and shows no violent effect when heated under confinement;

  iii. Does not detonate or deflagrate at all and shows a medium effect when heated under confinement;

  will be defined as organic peroxide **Type D**.

(E) Any organic peroxide mixture that, in laboratory testing, neither detonates nor deflagrates at all and shows low or no effect when heated under confinement will be defined as organic peroxide **Type E**.

(F) Any organic peroxide mixture that, in laboratory testing, neither detonates in the cavitated state nor deflagrates at all and shows only a low or no effect when heated under confinement as well as low or no explosive power will be defined as organic peroxide **Type F**.

(G) Any organic peroxide mixture that, in laboratory testing, neither detonates in the cavitated state nor deflagrates at all and shows no effect when heated under confinement nor any explosive power, provided that it is thermally stable (self-accelerating decomposition temperature is 60°C or higher for a 50-kg package), and, for liquid mixtures, a diluent having a boiling point of not less than 150°C is used for desensitization, will be defined as organic peroxide **Type G**. If the mixture is not thermally stable or a diluent having a boiling point less than 150°C is used for desensitization, the mixture shall be defined as organic peroxide **Type F**.

Note 1: Type G has no hazard communication elements assigned but should be considered for properties belonging to other hazard classes.

Note 2: Types A to G may not be necessary for all systems.

Substances Corrosive to Metal

Comparison

The HCS does not regulate substances corrosive to metal. According to the GHS building block approach, OSHA needs to determine if the HCS will cover this hazard class in the future. The HCS does not have required testing or specified test methods.

HCS Criteria	GHS Criteria
Test Method	**Test Method:**
HCS has no test method or criteria/definition.	Steel: ISO9328 (II): 1991 - Steel type P235
Definition	Aluminum: ASTM G31-72 (1990)—non-clad types 7075-T6 or AZ5GU-T66
HCS does not cover this classification. HCS Appendix A states: This term	**Definition**
shall not refer to action on inanimate surfaces.	*Corrosive to metal* means a substance or a mixture that by chemical action
	will materially damage, or even destroy, metals.
	Criteria
	Corrosion rate on steel or aluminum surfaces exceeding 6.25 mm/year at
	a test temperature of 55°C.
	A substance that is corrosive to metal is classified in a single hazard
	category for this class on the basis of the test.

Comparison of Label Elements

General Comments

- The HCS label requirements are totally performance oriented.
- The GHS labeling requirements are specified: signal words, hazard statements, and pictograms.
- The use of pictograms is a significant change for US workplace labeling.
- If the HCS retains NTP/OSHA/IARC carcinogens, guidance on labeling could be a useful tool.
- US liability concerns are a label consideration.

As part of the HCS–GHS label comparison, a table of GHS pictograms and several examples of ANSI Z129.1 Hazardous Industrial Chemicals–Precautionary Labeling and GHS labels are provided.

Label Elements

Comparison

For each GHS hazard class, detailed GHS label elements (symbol, signal word, hazard statement) have been assigned to the hazard categories (subcategories). Hazard categories reflect the harmonized classification criteria. A summary of the allocation of label elements is provided in GHS Annex 1. There are special arrangements, which apply to the use of certain mixture concentrations in the GHS.

The performance-oriented nature of the OSHA HCS is particularly relevant in the labeling provisions. Implementation of specified GHS elements for all hazard classes and categories will be a major change to the HCS. The HCS requires the "identity" on labels. See the competent authority allowances discussion on product identifier.

HCS	GHS
(f)(1)(i) Identity of the hazardous chemical(s) "Identity" means any chemical or common name that is indicated on the material safety data sheet (MSDS) for the chemical. The identity used shall permit cross-references to be made among the required list of hazardous chemicals, the label, and the MSDS. "Chemical name" means the scientific designation of a chemical in accordance with the nomenclature system developed by the International Union of Pure and Applied Chemistry (IUPAC) or the Chemical Abstracts Service (CAS) rules of nomenclature, or a name that will clearly identify the chemical for the purpose of conducting a hazard evaluation. "Common name" means any designation or identification such as code name, code number, trade name, brand name, or generic name used to identify a chemical other than by its chemical name.	**Product identifier** (i) A product identifier should be used on a GHS label and it should match the product identifier used on the SDS. Where a substance or mixture is covered by the UN Model Regulations on the Transport of Dangerous Goods, the UN proper shipping name should also be used on the package. (ii) The label for a substance should include the chemical identity of the substance. For mixtures or alloys, the label should include the chemical identities of all ingredients or alloying elements that contribute to acute toxicity, skin corrosion or serious eye damage, germ cell mutagenicity, carcinogenicity, reproductive toxicity, skin or respiratory sensitization, or Target Organ Systemic Toxicity (TOST), when these hazards appear on the label. Alternatively, the competent authority may require the inclusion of all ingredients or alloying elements that contribute to the hazard of the mixture or alloy. (iii) Where a substance or mixture is supplied exclusively for workplace use, the competent authority may choose to give suppliers discretion to include chemical identities on the SDS, in lieu of including them on labels. (iv) The competent authority rules for CBI take priority over the rules for product identification. This means that where an ingredient would normally be included on the label, if it meets the competent authority criteria for CBI, its identity does not have to be included on the label. **Signal word** A signal word means a word used to indicate the relative level of severity of hazard and alert the reader to a potential hazard on the label. The signal words used in the GHS are "Danger" and "Warning." "Danger" is used for the more severe hazard categories (i.e., the main hazard categories 1 and 2), while "Warning" is used for the less severe. The tables for each hazard class detail the signal words that have been assigned to each of the hazard categories of the GHS.

(f)(1)(ii) Appropriate hazard warnings

"Hazard warning" means any words, pictures, symbols, or combination thereof appearing on a label or other appropriate form of warning that convey the specific physical and health hazard(s), including target organ effects, of the chemical(s) in the container(s).

(f)(1)(iii) Name and address of the chemical

manufacturer, importer, or other responsible party

Hazard statement

A hazard statement means a phrase assigned to a hazard class and category that describes the nature of the hazards of a hazardous product, including, where appropriate, the degree of hazard. The tables of label elements for each hazard class detail the hazard statements that have been assigned to each of the hazard categories of the GHS.

Hazard pictogram

A pictogram means a graphical composition that includes a symbol plus other graphic elements, such as a border, background pattern, or color, that is intended to convey specific information. There are nine symbols used in the pictograms: flame, flame over circle, exploding bomb, corrosion, gas cylinder, skull and crossbones, exclamation point, environment, and new health hazard. See table.

All hazard pictograms used in the GHS should be in the shape of a square set at a point.

Pictograms prescribed by the GHS but not the UN Recommendations on the Transport of Dangerous Goods, Model Regulations, should have a black symbol on a white background with a red frame sufficiently wide to be clearly visible. However, when such a pictogram appears on a label for a package that will not be exported, the competent authority may choose to give suppliers and employers discretion to use a black border. In addition, competent authorities may allow the use of UN Recommendations on the Transport of Dangerous Goods, Model Regulations pictograms in other use settings where the package is not covered by the Model Regulations.

Pictograms prescribed by the UN Recommendations on the Transport of Dangerous Goods, Model Regulations, will use a background and symbol color as specified by those regulations.

Precautionary statements and pictograms

A precautionary statement means a phrase (and/or pictogram) that describes recommended measures that should be taken to minimize or prevent adverse effects resulting from exposure to a hazardous product, or improper storage or handling of a hazardous product. The GHS label should include appropriate precautionary information, the choice of which is with the labeler or the competent authority. Annex 3 contains examples of precautionary statements, which can be used, and also examples of precautionary pictograms, which can be used.

Supplier identification

The name, address, and telephone number of the manufacturer or supplier of the substance or mixture should be provided on the label.

GHS and Transport Pictograms

General Comments

- The GHS pictograms are provided to assist in evaluating the GHS label elements.
- The transport pictograms are included to show the variation in background and color.

GHS Pictograms and Hazard Classes

- Explosives
- Self-reactives
- Organic peroxides

- Flammables
- Self-reactives
- Pyrophorics
- Self-heating
- Emits flammable gas

- Oxidizers
- Organic peroxides

- Gases under pressure

- Acute toxicity

- Acute toxicity
- Skin irritation
- Eye irritation
- Skin sensitizers

- Carcinogens
- Respiratory sensitizers
- Reproductive toxicity
- Target organ toxicity
- Germ cell mutagens

- Eye corrosion
- Skin corrosion
- Corrosive to metal

- Aquatic toxicity

Physical and Environmental Hazard Symbols

Hazard	Transport Symbols			
Explosive				
Flammability: liquid, solid, gas, pyrophoric, emit flammable gas				
Oxidizer Organic peroxide				
Gases under pressure				
Corrosive to metals				
Environmental				

Label Examples

The GHS label requirements are a significant change from performance-oriented HCS labels. Examples can be useful visual illustrations to understand what is being described verbally. Several label examples are included in this comparison. The ANSI Z 129.1, Hazardous Industrial Chemicals—Precautionary Labeling, label format has been used for the OSHA HCS label format.

Example 1: Label for small workplace container (10 L) packaged inside an outer shipping container—for workplace audience

Hazards (liquid): flammable liquid, flash point = 120°F; oral LD_{50} = 275 mg/kg

Example 2:

Hazards (liquid): moderate skin and eye irritant, possible cancer hazard by inhalation

Example 3: large container (200-L drum) for transport, emergency response, and workplace audiences

Hazards (liquid mixture): classified under GHS as toxic to reproduction, category 1B and flammable liquid, category 3. UN RTDG classification is flammable liquid—UN 1263.

Competent authorities may choose not to require disclosure of ingredient identities on the label of products intended only for workplace use.

EXAMPLE 1

GHS Label

ToxiFlam

Danger!
Toxic if swallowed
Flammable liquid and vapor
Contains: XYZ
Do not taste or swallow. Get medical attention. Do not take internally. Wash thoroughly after handling. Keep away from heat, sparks, and flame. Keep container closed. Use only with adequate ventilation.
FIRST AID
If swallowed, induce vomiting immediately, as directed by medical personnel. Never give anything by mouth to an unconscious person.
See Material Safety Data Sheet for further details regarding safe use of this product.
Company name, Address, Phone number

HAZARDS (Liquid): flammable liquid, flash point = 120°F; oral LD_{50} = 275 mg/kg

EXAMPLE 1

ANSI/OSHA (HCS) Label

ToxiFlam
WARNING!
HARMFUL IF SWALLOWED
FLAMMABLE LIQUID AND VAPOR
Do not taste or swallow. Get medical attention. Do not take internally. Wash thoroughly after handling. Keep away from heat, sparks, and flame. Keep container closed. Use only with adequate ventilation.
FIRST AID
If swallowed, induce vomiting immediately, as directed by medical personnel. Never give anything by mouth to an unconscious person.
IN CASE OF FIRE, use water fog, CO_2, or alcohol foam. Water may be ineffective. Flash point = 120°F
Residue vapor may explode or ignite on ignition; do not cut, drill, grind, or weld on or near this container.
See Material Safety Data Sheet for further details regarding safe use of this product.
Company name, Address, Phone number

HAZARDS (Liquid): flammable liquid, flash point = 120°F; oral LD_{50} = 275 mg/kg

EXAMPLE 2

GHS Label

My Product
Warning!
Cause Skin And Eye Irritation
Suspected of causing cancer by inhalation
Contains: XYZ
Do not breathe vapors or mist. Use only with adequate ventilation. Avoid contact with eyes, skin, and clothing. Wash thoroughly after handling
FIRST AID
EYES: Immediately flush eyes with plenty of water for at least 15 min. Get medical attention.
SKIN: In case of contact, immediately flush skin with plenty of water. Remove contaminated clothing and shoes. Wash clothing before reuse.
Get medical attention if irritation develops and persists.
Company name, Address, Phone number

HAZARDS (Liquid): moderate skin and eye irritant, possible cancer hazard by inhalation

EXAMPLE 2

ANSI/OSHA (HCS) Label

My Product
CAUTION!
MAY CAUSE SKIN AND EYE IRRITATION
ATTENTION! POSSIBLE CANCER
HAZARD—CONTAINS MATERIAL THAT
MAY CAUSE CANCER BASED ON
ANIMAL DATA
Do not breathe vapors or mist.
Use only with adequate ventilation.
Avoid contact with eyes, skin, and clothing.
Wash thoroughly after handling.
FIRST AID
EYES: Immediately flush eyes with plenty of
water for at least 15 min. Get medical
attention.
SKIN: In case of contact, immediately flush
skin with plenty of water. Remove
contaminated clothing and shoes. Wash
clothing before reuse. Get medical attention if
irritation develops and persists.

INHALED: If inhaled, remove to fresh air. If
not breathing, give artificial respiration. If
breathing is difficult, give oxygen. Call a
physician.
INGESTION: If swallowed, DO NOT induce
vomiting unless directed by medical
personnel. Never give anything by mouth to
an unconscious person. Get medical
attention.
SPILL OR LEAK
Take steps to contain liquid and avoid runoff
to waterways and sewer.
FIRE
In case of fire, use water spray, foam, dry
chemical, or CO_2.
HANDLING AND STORAGE
Keep away from strong acids and oxidizers.
Do not apply air pressure, puncture, or weld
on or near this container.
For additional information, read Material
Safety Data Sheet for this product.
24-hour emergency phone number
Company name
Address
Phone number

HAZARDS (Liquid): moderate skin and eye irritant, possible cancer hazard by inhalation

GHS Label for large container (200-L drum) for transport, emergency response, and workplace audiences

---

ZZZ Red Paint   UN 1263 Paint
Danger!
May damage fertility or the unborn child
Highly flammable liquid and vapor
**Contains lead pigments and cellosolve acetate

Keep away from heat and ignition sources. Keep away from food and drink. Avoid contact with skin and eyes and inhalation of vapor. Wash hands thoroughly after use and before eating

FIRST AID

For skin contact, remove contaminated clothing and wash affected area thoroughly with water. If irritation develops, seek medical attention. For eye contact, immediately flush eyes with flowing water for at least 15 min and seek medical attention.

GHS Example PLC, Leeds, England. Telephone 44 999 999 9999

** Competent authorities may choose not to require disclosure of ingredient identities on the label of products intended only for workplace use.

Comparison of MSDS Elements

General Comments

The GHS MSDS has more extensive information requirements than the HCS, for example, transportation classifications and other pertinent product regulatory requirements. In general, medium and large companies are using a 16-section MSDS format to comply with ANSI and ISO MSDS standards, EU directive, and other global MSDS requirements. The GHS 16-section MSDS has information requirements that smaller companies may not currently be providing on MSDSs. However, information such as transportation classifications and other product regulatory information should be available.

Comparison

The MSDS information required by the HCS can be provided in any format. OSHA Form 174 is optional. The GHS SDS requires 16 sections with a specified sequence and minimum specified information. The GHS SDS has minimum information requirements. The few HCS elements that are not specifically covered could be added: medical conditions aggravated and protective measures during maintenance and repair. The component disclosure requirements are different in the HCS and GHS. A key difference is the GHS requirement for information that OSHA would not usually regulate: transport information, ecological information, and the other pertinent regulatory information in Section 15.

GHS MSDS Sections	GHS SDS	OSHA HCS	OSHA Form 174
1. Product and company identification	• GHS product identifier • Other means of identification • Recommended use of the chemical and restrictions on use • Supplier's details (including name, address, phone number, etc.) • Emergency phone number	• Product identity same as on label • Name address and telephone number of the manufacturer, distributor, employer, or other responsible party	• Identity (as used on label and list) • Manufacturer's name • Address (number, street, city, state and zip code) • Emergency telephone number • Telephone number for information • Date prepared • Signature of preparer
2. Hazards identification	• GHS classification of the substance/mixture and any regional information • GHS label elements, including precautionary statements (hazard symbols may be provided as a graphical reproduction of the symbols in black and white or the name of the symbol, e.g., flame, skull and crossbones) • Other hazards that do not result in classification (e.g., dust explosion hazard) or are not covered by the GHS	• Health hazards including acute and chronic effects, listing target organs or systems • Signs and symptoms of exposure • Conditions generally recognized as aggravated by exposure • Primary routes of exposure • If listed as a carcinogen by OSHA, IARC, or NTP • See Sections 5, 9, and 10 for physical hazards	• Route(s) of entry • Inhalation • Skin • Ingestion • Health hazards (acute and chronic) • Carcinogenicity • NTP • IARC • OSHA • Signs and symptoms of exposure • Medical conditions generally aggravated by exposure

3. Composition/information on ingredients	• Substance–chemical identity • Common name, synonyms, and so on • CAS number, EC number, and so on • Impurities and stabilizing additives that are themselves classified and that contribute to the classification of the substance • Mixture • The chemical identity and concentration or concentration ranges of all ingredients that are hazardous within the meaning of the GHS and are present above their cutoff levels. • Note: For information on ingredients, the competent authority rules for CBI take priority over the rules for product identification	• Chemical and common name of ingredients contributing to known hazards • For untested mixtures, the chemical and common name of ingredients at 1% or more that present a health hazard and those that present a physical hazard in the mixture • Ingredients at 0.1% or greater, if listed carcinogens	• Hazardous components (specific chemical identity, common name(s) • OSHA PEL • ACGIHTLV • Other recommended limits • % (optional)
4. First-aid measures	• Description of necessary measures, subdivided according to the different routes of exposure, that is, inhalation, skin and eye contact, and ingestion • Most important symptoms/effects, acute and delayed. • Indication of immediate medical attention and special treatment needed, if necessary	• Emergency and first-aid procedures	• Emergency and first-aid procedures

(Continued)

MSDS Comparison by Section

Comparison

The MSDS information required by the HCS can be provided in any format. OSHA Form 174 is optional. The GHS SDS requires 16 sections with a specified sequence and minimum specified information. The GHS SDS has minimum information requirements. The few HCS elements that are not specifically covered could be added: medical conditions aggravated and protective measures during maintenance and repair. The component disclosure requirements are different in the HCS and GHS. A key difference is the GHS requirement for information that OSHA would not usually regulate: transport information, ecological information, and the other pertinent regulatory information in Section 15.

GHS MSDS Sections	GHS SDS	OSHA HCS	OSHA Form 174
5. Firefighting measures	• Suitable (and unsuitable) extinguishing media • Specific hazards arising from the chemical (e.g., nature of any hazardous combustion products) • Special protective equipment and precautions for firefighters	• Flammable property information such as flash point • Physical hazards • Generally applicable control measures	• Flash point (METHOD USED) • Flammable limits • LEL, UEL • Extinguishing media • Special firefighting procedures • Unusual fire and explosion hazards
6. Accidental release measures	• Personal precautions, protective equipment, and emergency procedures • Environmental precautions • Methods and materials for containment and cleaning up	• Procedures for cleanup of spills and leaks	• Steps to be taken in case material is released or spilled
7. Handling and storage	• Precautions for safe handling • Conditions for safe storage, including any incompatibilities	• Precautions for safe handling and use, including appropriate hygienic practices	• Precautions to be taken in handling and storing • Other precautions

8. Exposure controls/personal protection

- Control parameters, for example, occupational exposure limit values or biological limit values
- Appropriate engineering controls
- Individual protection measures, such as personal protective equipment

- General applicable control measures
- Appropriate engineering controls and work practices
- Protective measures during maintenance and repair
- Personal protective equipment
- Permissible exposure levels, threshold limit values, listed by OSHA, ACGIH, or established company limits

- Respiratory protection (specify type)
- Ventilation
- Local exhaust
- Special
- Mechanical (general)
- Other
- Protective gloves
- Eye protection
- Other protective clothing or equipment
- Work/hygienic practices

9. Physical and chemical properties

- Appearance (physical state, color, etc.)
- Odor
- Odor threshold
- pH
- Melting point/freezing point
- Initial boiling point and boiling range
- Flash point:
- Evaporation rate
- Flammability (solid, gas)
- Upper/lower flammability or explosive limits
- Vapor pressure
- Vapor density
- Relative density
- Solubility(ies)
- Partition coefficient: n-octanol/water:
- Auto-ignition temperature
- Decomposition temperature

- Physical and chemical characteristics of hazardous chemicals such as vapor pressure and density

- Boiling point
- Specific gravity ($H_2O = 1$)
- Vapor pressure (mm Hg)
- Melting point
- Vapor density (air = 1)
- Evaporation rate (butyl acetate = 1)
- Solubility in water
- Appearance and odor

(Continued)

MSDS Comparison by Section

Comparison

The MSDS information required by the HCS can be provided in any format. OSHA Form 174 is optional. The GHS SDS requires 16 sections with a specified sequence and minimum specified information. The GHS SDS has minimum information requirements. The few HCS elements that are not specifically covered could be added: medical conditions aggravated and protective measures during maintenance and repair. The component disclosure requirements are different in the HCS and GHS. A key difference is the GHS requirement for information that OSHA would not usually regulate: transport information, ecological information, and the other pertinent regulatory information in Section 15.

GHS MSDS Sections	GHS SDS	OSHA HCS	OSHA Form 174
10. Stability and reactivity	• Chemical stability • Possibility of hazardous reactions • Conditions to avoid (e.g., static discharge, shock or vibration) • Incompatible materials • Hazardous decomposition products	• Organic peroxides, pyrophoric, unstable (reactive), or water-reactive hazards • Physical hazards, including reactivity and hazardous polymerization	• Stability (unstable/stable) • Conditions to avoid • Incompatibility (materials to avoid) • Hazardous decomposition or by-products • Hazardous polymerization • Conditions to avoid • See Section 2
11. Toxicological information	• Concise but complete and comprehensible description of the various toxicological (health) effects and the available data used to identify those effects, including: • Information on the likely routes of exposure (inhalation, ingestion, skin and eye contact) • Symptoms related to the physical, chemical, and toxicological characteristics • Delayed and immediate effects and also chronic effects from short- and long-term exposure • Numerical measures of toxicity (such as acute toxicity estimates)	• See also Section 2 • Health hazards including acute and chronic effects, listing target organs or systems • Signs and symptoms of exposure • Primary routes of exposure • If listed as a carcinogen by OSHA, IARC, or NTP	

12. Ecological information	• Ecotoxicity (aquatic and terrestrial, where available) • Persistence and degradability • Bioaccumulative potential • Mobility in soil • Other adverse effects	• No present requirements
13. Disposal considerations	• Description of waste residues and information on their safe handling and methods of disposal, including any contaminated packaging	• See Section 7 • No present requirements • Waste disposal method
14. Transport information	• UN number • UN Proper shipping name • Transport hazard class(es) • Packing group, if applicable • Marine pollutant (Y/N) • Special precautions that a user needs to be aware of or needs to comply with in connection with transport or conveyance either within or outside their premises	• No present requirements
15. Regulatory information	• Safety, health, and environmental regulations specific for the product in question	• No present requirements
16. Other information	• Other information including information on preparation and revision of the SDS	• Date of preparation of MSDS or date of last change • See Section 1

Comparison of MSDS Elements

Comparison

The GHS SDS has minimum information requirements. The few HCS elements that are not specifically covered could be added: medical conditions aggravated and protective measures during maintenance and repair. The component disclosure requirements are different in the HCS and GHS. A key MSDS difference is the GHS requirement for information that OSHA does not regulate: transport information, ecological information, and the other pertinent regulatory information in Section 15.

OSHA HCS	GHS
• Identity	• GHS product identifier
	• Other means of identification
	• Recommended use of the chemical and restrictions onuses
• Chemical and common names of substances	**Substance**
	• Chemical identity
	• Common name, synonyms, and so on
	• CAS number, EC number, and so on
	• Impurities and stabilizing additives that are themselves classified and that contribute to the classification of the substance.
	Note: For information on ingredients, the competent authority rules for CBI take priority over the rules for product identification.
Mixture	**Mixture**
• If the hazardous chemical is a mixture that has been tested as a whole to determine its hazards, the chemical and common name(s)of the ingredients that contribute to these known hazards, and the common name(s) of the mixture itself; or,	• The chemical identity and concentration or concentration ranges of all ingredients that are hazardous within the meaning of the GHS and are present above their cutoff levels. See individual health hazards for cutoff values.
• If the hazardous chemical is an admixture that has not been tested as a whole	*Note: For information on ingredients, the competent authority rules for CBI take priority over the rules for product identification.*
– The chemical and common name(s) of all ingredients that are health hazards and are present at ≥1%, except carcinogens, shall be listed at ≥0.1%	

- The chemical and common name(s) of all ingredients that are health hazards and are present at <1% (<0.1% for carcinogens), if there is evidence that the ingredient(s) could be released from the mixture in concentrations that would exceed an established OSHA PEL or ACGIH TLV, or could present a health risk to employees
- The chemical and common name(s) of all ingredients that have been determined to present a physical hazard when present in the mixture

- Chemical and physical characteristics (vapor pressure, flash point, etc.)

  - Appearance (physical state, color, etc.)
  - Odor
  - Odor threshold
  - pH
  - Melting point/freezing point
  - Initial boiling point and boiling range
  - Flash point
  - Evaporation rate
  - Flammability (solid, gas)
  - Upper/lower flammability or explosive limits
  - Vapor pressure
  - Vapor density
  - Relative density
  - Solubility(ies)
  - Partition coefficient: *n*-octanol/water
  - Auto-ignition temperature
  - Decomposition temperature

- Physical hazards (fire, explosion, reactivity)

  - GHS classification of the substance/mixture and any national or regional information
  - Chemical stability
  - Possibility of hazardous reactions
  - Conditions to avoid (e.g., static discharge, shock or vibration)
  - Incompatible materials
  - Hazardous decomposition products

(*Continued*)

Comparison of MSDS Elements

Comparison

The GHS SDS has minimum information requirements. The few HCS elements that are not specifically covered could be added: medical conditions aggravated and protective measures during maintenance and repair. The component disclosure requirements are different in the HCS and GHS. A key MSDS difference is the GHS requirement for information that OSHA does not regulate: transport information, ecological information, and the other pertinent regulatory information in Section 15.

OSHA HCS	GHS
• OSHA, IARC, NTP Carcinogen	
• Health hazards	• GHS classification of the substance/mixture and any national or regional information
	• Concise but complete and comprehensible description of the various toxicological (health) effects and the available data used to identify those effects, including
	• Delayed and immediate effects and also chronic effects from short- and long-term exposure
	• Numerical measures of toxicity (such as acute toxicity estimates)
	• Available data used to identify those toxicological (health) effects
• Signs and symptoms of exposure	• Most important symptoms/effects, acute and delayed
	• Symptoms related to the physical, chemical, and toxicological characteristics
• Medical conditions aggravated	
• Primary route of entry	• Information on the likely routes of exposure (inhalation, ingestion, skin and eye contact)
• First-aid procedures	• Description of necessary first-aid measures, subdivided according to the different routes of exposure, that is, inhalation, skin and eye contact, and ingestion
	• Indication of immediate medical attention and special treatment needed, if necessary
• Protective measures during maintenance	

- Precautions for safe handling

- Hygienic practices

- Personal protective equipment

- OSHA PEL, ACGIH TLV, other limits

- General applicable control measures
- Engineering controls
- Cleanup procedures for spills or leaks

- Date of preparation/changes to MSDS
- Name/address/phone for additional information

- Emergency procedures

- Description of waste residues and information on their safe handling and methods of disposal, including the disposal of any contaminated packaging
- Other hazards that do not result in classification (e.g., dust explosion hazard) or are not covered by the GHS
- Environmental precautions
- Personal precautions, protective equipment, and emergency procedures
- Individual protection measures, such as personal protective equipment
- Individual protection measures, such as personal protective equipment
- Control parameters, for example, occupational exposure limit values or biological limit values
- Control parameters
- Appropriate engineering controls
- Personal precautions, protective equipment, and emergency procedures
- Methods and materials for containment and cleaning up
- Information on preparation and revision of the SDS
- Supplier's details (including name, address, phone number, etc.)
- GHS label elements, including precautionary statements (hazard symbols may be provided as a graphical reproduction of the symbols in black and white or the name of the symbol, e.g., flame, skull and crossbones)
- Suitable (and unsuitable) extinguishing media
- Specific hazards arising from the chemical (e.g., nature of any hazardous combustion products)
- Special protective equipment and precautions for firefighters
- Emergency phone number

(Continued)

Comparison of MSDS Elements

Comparison

The GHS SDS has minimum information requirements. The few HCS elements that are not specifically covered could be added: medical conditions aggravated and protective measures during maintenance and repair. The component disclosure requirements are different in the HCS and GHS. A key MSDS difference is the GHS requirement for information that OSHA does not regulate: transport information, ecological information, and the other pertinent regulatory information in Section 15.

OSHA HCS	GHS
	• Ecotoxicity (aquatic and terrestrial, where available)
	• Persistence and degradability
	• Bioaccumulative potential
	• Mobility in soil
	• Other adverse effects
	• UN number
	• UN Proper shipping name
	• Transport hazard class(es)
	• Packing group, if applicable
	• Marine pollutant (yes/no)
	• Special precautions that a user needs to be aware of or needs to comply with in connection with transport or conveyance either within or outside their premises
	• Safety, health, and environmental regulations specific for the product in question
	• Other information including information on preparation and revision of the SDS

3

Math Review

For most of us in the occupational safety and health profession, it has been several years since we've sat in a classroom and had to solve problems handed out by an instructor, who makes it look so easy while we're in the classroom. However, what the instructor may not remember is that we all come from a wide variety of backgrounds or may not remember the basics required to completely understand what we're doing. The professional safety practitioner requires a solid background in mathematics to perform adequately. During the course of this book, there are numerous equations and applications that require a basic understanding of mathematics. It is for this reason that I've included this math review chapter. This chapter is not intended to provide a complete understanding of mathematics, but rather to serve only as a review. Some may find it very basic, while others will appreciate the review.

Order of Operations

One of the first items that we must understand is that there is an order to performing any mathematical equation. Included below is an age-old acronym and saying that will help you remember the order of operations.

PEMDAS: "Please Excuse My Dear Aunt Sally"

Simply put, PEMDAS identifies the order of operations listed below:

1. Parentheses
2. Exponents
3. Multiplication
4. Division
5. Addition
6. Subtraction

For example, in the following equation there are five operations required.

$4 + [-1(-2-1)]^2$	
$= 4 + [-1(-3)]^2$	Solve everything inside the parentheses, including multiplication and subtraction.
$= 4 + [3]^2$	Solve everything inside the parentheses (brackets) and then solve the exponents.
$= 4 + 9$	Finally, perform the addition and/or subtraction.
$= 13$	Final result.

Basic Rules of Positive and Negative Numbers

Addition/Subtraction

$(+) + (+) = (+)$	Positive plus a Positive will equal a Positive.
$(+) + (-) = (+$ or $-)$	Positive plus a Negative will equal a Positive or Negative.
$(-) + (-) = (-)$	Negative plus a Negative will equal a Negative.

Multiplication/Division

$(+) \times (+) = (+)$	Positive times a Positive will equal a Positive.
$(+) \times (-) = (-)$	Positive times a Negative will equal a Negative.
$(-) \times (-) = (+)$	Negative times a Negative will equal a Positive.

Understanding Exponents

An exponent can be defined as a quantity representing the power to which some other quantity is raised. Exponents do not have to be numbers or constants; they can be variables. They are often positive *whole numbers*, but they can be *negative numbers*, *fractional numbers*, *irrational numbers*, or *complex numbers*.

A number with an exponent is written as follows and contains two parts (the Base and the Exponent):

$$\underset{\uparrow}{\underline{10}}{}^{6} \nwarrow \text{Exponent}$$

Base

This represents the following:

$$10^6 = 10 \times 10 \times 10 \times 10 \times 10 \times 10 = 1{,}000{,}000.$$

When solving for negative exponents, it will be written as

$$10^{-6}.$$

This represents the following:

$$10^{-6} = \frac{1}{10 \times 10 \times 10 \times 10 \times 10 \times 10}.$$
$$= 0.000001$$

Scientific Notation

Scientific notation (standard form) is a mathematical way to express values or numbers too large or too small to be easily written in a standard decimal notation.

For example 392,000 can be written in scientific notation as follows:

$$3.92 \times 10^5.$$

It can also be written as 3.92 E 05, where E is representative of "×10."

Another example is the number 0.00432. Written in scientific notation, it would look like this:

$$4.32 \times 10^{-3} \text{ OR } 4.32 \text{ E}{-03}.$$

Multiplication and Division Using Scientific Notation

Example

$$\frac{(3.21 \times 10^5)(4.88 \times 10^{-6})}{(5.9 \times 10^{-4})}$$
$$= \frac{3.21 \times 4.88}{5.9}$$
$$= \mathbf{2.65}$$

For now, disregard the exponents and multiply and divide the coefficients.

Now, we will solve for the exponents. The exponents in the *numerator* (top portion) are 5 and −6, while the exponent in the *denominator* (bottom portion) is −4. Therefore, we would ADD the exponents in the numerator (5 + (−6)) and SUBTRACT the exponent in the denominator (−4), as follows:

$$5 + (-6) - (-4)$$
$$= 3$$

We have now solved this equation, with the result being

$$2.65 \times 10^3.$$

Engineering Notation

During the course of your professional career, you will encounter the use of engineering notation. Engineering notation is a version of scientific notation in which the power of 10 must be a multiple of three. For example, 10^6 can be written as 1000^2. Instead of writing powers of 10, the *System International (SI)* uses prefixes to describe numbers. An example of this is 10^{-1} or 0.01, which is described as deci in the SI system. By way of example, 0.01 (or 10^{-1}) liters of a substance is described as a *deciliter* or *dl*. Table 3.1 describes the most often encountered engineering notation terms in the safety and health profession.

TABLE 3.1

Common Engineering Notation Symbols and Prefixes

10^n	Prefix	Symbol	Decimal
10^9	giga	G	1,000,000,000
10^6	mega	M	1,000,000
10^3	kilo	k	1000
10^2	hecto	h	100
10^1	deca	d	10
10^0	None	None	1
10^{-1}	deci	d	0.1
10^{-2}	centi	c	0.01
10^{-3}	milli	m	0.001
10^{-6}	micro	μ	0.000001
10^{-9}	nano	n	0.000000001
10^{-12}	pico	p	0.000000000001

Absolute Values

The *absolute value* of any number is the value of the corresponding arithmetic number that is positive. Absolute values are represented mathematically by two enclosed bars.

For example: $|-3| = 3$.

Absolute values play a particular importance when solving ventilation equations.

Logarithms

In working with logarithms, remember that there are three basic rules. These rules are identified below:

- $\log_b(mn) = \log_b(m) + \log_b(n)$
- $\log_b(m/n) = \log_b(m) - \log_b(n)$
- $\log_b(m^n) = n \times \log_b(m)$

Example 1:

$$\log(4 \times 5) = \log 4 + \log 5$$
$$\log(4 \times 5) = 0.601 + 0.699$$
$$\log(4 \times 5) = 1.299.$$

Example 2:

$$\log \frac{4}{5} = \log 4 - \log 5$$
$$\log 0.8 = 0.602 - 0.6989$$
$$\log 0.8 = -0.0969.$$

Example 3:

$$\log 4^5 = 5 \times \log 4$$
$$\log 4^5 = 5 \times 0.602$$
$$\log 4^5 = 3.01.$$

Formula or Equation Transpositions

In any equation, the purpose is to determine the value of an unknown variable. The key step in solving for the unknown is to isolate the unknown variable to one side of the equation. Fair and equal treatment to both sides of the equation is required. In other words, do to one side as you do to the other side.

For example, solving for x in the following equation, we must first isolate x on one side by itself.

$6x + 5 = 29$		
	$= 6x + 5 - 5 = 29 - 5$	Subtract 5 from both sides.
	$= 6x = 24$	Now we can isolate x by dividing both sides by 6.
	$= \dfrac{6x}{6} = \dfrac{24}{6}$	
	$= x = 4$	The value of $x = 4$.

To illustrate transposition further, the following two examples, using basic safety calculations, are offered.

In the first example, we will use the ventilation equation $Q = VA$,

where

Q = volumetric rate (cubic feet per minute or cfm)
V = velocity of air (feet per minute or fpm)
A = cross-sectional area of duct (SF or ft²)

NOTE: It is extremely important to pay particular attention to the units of measurement when solving safety equations. Many times, the questions on the Associate Safety Professional (ASP) or Certified Safety Professional (CSP) exams are written in different units to test your detailed knowledge and understanding. For example, the question may give you a duct measurement of 5 inches wide and 8 inches long. You will need to convert into feet in order to obtain the correct result.

Solve for A in the following case, given that the total volume of air flow is 3225 cfm and the air flow is 350 fpm. $Q = 3225$ cfm; $V = 350$ fpm; and $A = ?$

$3225 \text{ cfm} = (350 \text{ fpm})(A),$	where A is the unknown.
$= \dfrac{3225 \text{ cfm}}{350 \text{ fpm}} = \dfrac{(350 \text{ fpm})(A)}{(350 \text{ fpm})}$	Isolate A by itself by dividing both sides by 350 fpm.
$= 9.21 \text{ sf or ft}^2.$	Remember that A is represented in square feet.

A second example of transposition using a basic electrical equation is to solve for the amperage, given that there are 110 volts (*V*) and the resistance (*R*) is 8 ohms (Ω). The basic equation is

$$V = IR,$$

where
    V = volts
    I = current (measured in amperes)
    R = resistance (measured in ohms or Ω)

Inserting the known values into the equation, we are left with

110 V = (I)(8 Ω)	*I* being the unknown variable.
$= \dfrac{110\,V}{8\,\Omega} = \dfrac{(I)(8\,\Omega)}{(8\,\Omega)}$	Isolate *I* by itself by dividing both sides by 8 Ω.
= 13.75 A.	

Factorials

Factorials are quite simple to understand. They are products with an exclamation point, such as 6!. It simply means to multiply from 1 to the product in the factors. For example, 6! is solved as follows: $6! = 1 \times 2 \times 3 \times 4 \times 5 \times 6 = 720$. Perform ordinary mathematical functions using factorials as you would any other function. For example:

$$\frac{4!}{6!}$$
$$= \frac{1 \times 2 \times 3 \times 4}{1 \times 2 \times 3 \times 4 \times 5 \times 6}$$
$$= \frac{24}{720}$$
$$= 0.33.$$

NOTE: For many reasons, 0! is always 1. You should commit this to memory when working with factorials.

Common Geometric Equations

Throughout your career and specifically on the ASP and CSP exams, you will have the need to determine square feet, cubic feet, or other dimensions of various geometric spaces or containers. It is amazing to me that over the years, when needed, some of the basic equations have simply vanished from my mind for the moment and I've had to go to a handy reference source to review the equation. Therefore, I've included several of the basic equations to help you in your review. Many questions on the certification exams will be asked of you on the premise that you already know how to solve for the unknown and will not directly provide this information in the scenario. For example, a question may ask you to calculate the required overflow protection (dike area), given that you have two tanks measuring 4 feet in diameter and 10 feet long. In order to calculate the required capacity of the dike area, you must first determine the maximum amount of liquid potentially stored in the tanks. Then, you must calculate the perimeter and height of the proposed dike.

Area	
Square or rectangle	L (length) $\times$ W (width)
Circle	πr^2, where π is 3.14159
Triangle	$\frac{1}{2} b$ (base) $\times$ h (height)

NOTE: Area of an object will always result in sq. ft., sq. inches, or other unit of measurement.

Distance	
Circle (also called circumference)	πd (diameter) or $2\pi r$ (radius)
Square or rectangle	$2L + 2W$

Volume	
Square or rectangle	$L \times W \times H$
Cylinder	$\pi r^2 \times h$
Sphere	$4/3\pi r^3$
Cone	$\dfrac{\pi h r^2}{3}$

Pythagorean Equation

The *Pythagorean equation* is extremely useful in solving many problems encountered in the safety and health profession. It is used to determine the lengths of any side of a triangle and can be used further to identify the calculated degree of any angle. Fortunately, for the exam candidate, this equation is provided on the exam reference sheet, so you will not have to commit this one to memory.

The *Pythagorean equation* is as follows:

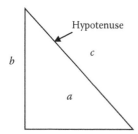

$a^2 + b^2 = c^2$

Note: The hypotenuse of any right triangle is always the longest side.

A practical safety example using the *Pythagorean equation* is a ladder leaning against a building. At the base, the ladder is 4 feet from the wall. It touches the wall at the top at a height of 12 feet. How long is the ladder? Inserting the known values into the equation, we have the following:

$$a = 4 \text{ feet, } b = 12 \text{ feet, and } c = ?$$

Therefore,

$4^2 + 12^2 = c^2$	
$= 16 + 144 = c^2$	There are no parentheses, so we solve for the exponents.
$= 160 = c^2$	c^2 is isolated, but we now need to solve for c by obtaining the square root of both sides of the equation.
$= \sqrt{160} = \sqrt{c^2}$	
$= 12.65 \text{ ft} = c$	The length of the ladder is 12.65 feet.

Basic Trigonometric Functions

Trigonometry can help solve a lot of the problems that you may see on the certification exam or when encountering "real world" problems. Therefore, it is necessary to at least understand the basics of sine, cosine, tangent, arcsine, arccosine, and arccotangent.

Sine

The *sine* (*sin*) of an angle is equal to the opposite (length) divided by the hypotenuse (length). For example:

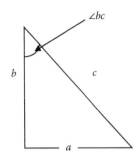

∠bc represents the angle (∠) created by the intersection of sides *b* and *c*.

Given the following information, solve for ∠bc:

a = 3 ft

b = 4 ft

c = 5 ft (the hypotenuse)

Using *sine (sin)*, we input the following data into the equation:

$\sin \angle bc = \dfrac{3}{5}$	Remember that the sine is equal to the opposite side divided by the hypotenuse.
$\sin \angle bc = 0.6$	The sin of ∠bc is 0.6. However, this does not provide us with the angle. To obtain the angle we need to determine the arcsine.

Note: The arcsine is the inverse function of sine.

Now that we have determined that the sin ∠bc = 0.6, we use the inverse function (arcsine) to calculate the angle. It is usually written as arcsin or sin⁻¹. By way of example, we take the sin ∠bc above, which is 0.6 and obtain the arcsin or sin⁻¹. We do this by inputting 0.6 into the calculator and hitting the inverse or 2nd key for sin, giving us an angle of 36.87° and rounding off to 37°.

Cosine

The *cosine (cos)* of an angle is determined by dividing the adjacent side (length) by the hypotenuse (length). Given the same data as discussed under the Sine section, we can determine the angle by using the cosine function, as follows:

$\cos \angle bc = \dfrac{4}{5}$	Remember that the cosine is equal to the adjacent side divided by the hypotenuse.
$= \cos \angle bc = 0.8$	The sin of ∠bc is 0.8. However, this does not provide us with the angle. To obtain the angle, we need to determine the arccosine.

Now that we have determined that the cos ∠*bc* = 0.8, we use the inverse function (arccosine) to calculate the angle. It is usually written as arccos or cos⁻¹. By way of example, we take the cos ∠*bc* above, which is 0.8 and obtain the arccos or cos⁻¹. We do this by inputting 0.8 into the calculator and hitting the inverse or 2nd key for cos, giving us an angle of 36.87° and rounding off to 37°.

Tangent

The *tangent* (*tan*) of an angle is determined by dividing the opposite side (length) by the adjacent (length). Given the same data as discussed under the Sine section, we can determine the angle by using the tangent function, as follows:

$\tan \angle bc = \dfrac{3}{4}$	Remember that the tangent is equal to the opposite side divided by the adjacent side.
$= \tan \angle bc = 0.75$	The tan of ∠*bc* is 0.75. However, this does not provide us with the angle. To obtain the angle, we need to determine the arctangent.

Now that we have determined that the tan ∠*bc* = 0.8, we use the inverse function (arctangent) to calculate the angle. It is usually written as arctan or tan⁻¹. By way of example, we take the tan ∠*bc* above, which is 0.75 and obtain the arctan or tan⁻¹. We do this by inputting 0.75 into the calculator and hitting the inverse or 2nd key for tan, giving us an angle of 36.87° and rounding off to 37°.

Arcsine, Arccosine, and Arctangent

As mentioned previously, the *arcsine*, *arccosine*, and *arctangent* are all inverse functions of sine, cosine, and tangent, respectively. Mathematically, they can be written as follows:

Arcsine: arcsin = sin⁻¹

Arccosine: arccos = cos⁻¹

Arctangent: arctan = tan⁻¹

As a practical safety application, we have a ramp that rises 4 feet in a 10-feet run. What is the angle of the ramp?

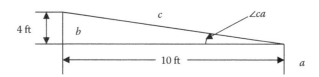

Since we know the lengths of the adjacent side and opposite sides (remember that c in this case is the hypotenuse), we will solve for the tangent of $\angle ca$.

$$\tan \angle ca = \frac{4 \ (\text{opposite})}{10 \ (\text{adjacent})} = 0.4$$

or written another way:

$$\angle ca = \tan^{-1} = 0.4.$$

Now that we have determined that the tan $\angle ca = 0.4$, we use the inverse function (arctangent) to calculate the angle. It is usually written as arctan or $\tan^{-1}$. By way of example, we take the tan $\angle ca$ above, which is 0.4 and obtain the arctan or $\tan^{-1}$. We do this by inputting 0.4 into the calculator and hitting the inverse or 2nd key for tan, giving us an angle of 21.8° and rounding off to 22°. Therefore, the ramp is on a 22° incline.

Another example calculating sling loads is as follows:

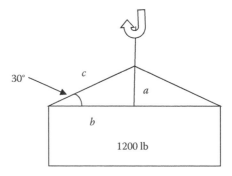

In the example above, we have an object weighing 1200 lb supported by two slings at 30° angles. We divide the weight by the number of slings (2), which is 600 lb. Now we have the weight at side a (the opposite). We also know that the angle is 30°. By using the sine function, we can determine the weight at each leg of the sling.

$$\sin 30° = \frac{600}{c}$$ Isolate *c* by itself by multiplying both sides by *c*.

$$= \sin 30°(c) = \frac{(600)(c)}{c}$$ Now divide both sides by sin 30°.

$$= \sin 30°(c) = 600$$ Now divide both sides by sin 30°.

$$= \frac{\sin 30°(c)}{\sin 30°} = \frac{600}{\sin 30°}$$

$$= c = \frac{600}{\sin 30°}$$ Now use the sin function key and divide 600 by 0.5.

$$= c = \frac{600}{0.5}$$

$$= c = 1200$$ The weight on each sling is 1200 lb.

Quadratic Equation

Solving for multiple unknowns can often be difficult. It is for this reason that the safety professional must know how to calculate them using the *quadratic equation*. The quadratic equation uses the *numerical coefficients* from the following: $ax^2 + bx + c = 0$. We are simply setting each factor to 0. The value of *x* is given by the following (*quadratic equation*):

$$x_1, x_2 = \frac{-b \pm \sqrt{b^2 - 4ac}}{2a}.$$

Using $2x^2 - 4x - 3$, we can solve for *x* using the quadratic equation. The coefficients are as follows: $a = (2)$, $b = (-4)$, and $c = (-3)$.
We insert the coefficients into the quadratic equation as follows:

$$x_1, x_2 = \frac{-(-4) \pm \sqrt{(-4)^2 - 4(2)(-3)}}{2(2)}$$ Remember the order of operations and begin solving the equation. We will solve for *x*.

$$x_1, x_2 = \frac{4 \pm \sqrt{16 + 24}}{4}$$

$$x_1, x_2 = \frac{4 \pm \sqrt{40}}{4}$$

$$x_1, x_2 = \frac{4 \pm 6.32}{4}$$ Complete the math and then solve for *x*.

$$x_1 = \frac{4 \pm 6.32}{4} = \frac{10.32}{4} = 2.58 \qquad x_1 = 2.58. \text{ Now we will solve for } x_2$$

$$x_2 = \frac{4 - 6.32}{4} = \frac{-2.32}{4} = -0.58 \qquad x_2 = -0.58$$

Calculator

One of the single most important items that you are allowed to bring with you when taking the exam is your calculator. It is recommended that you bring two in case the battery in one is used up. In the latest version of the BCSP (Board of Certified Safety Professionals) guidelines, you are authorized to bring any of the following types of calculators:

- Casio models (FX-115, FX-250, FX-260, FX-300)
- Hewlett-Packard models (HP 9, HP10, HP12, HP30)
- Texas Instruments models (TI30, TI-34, TI-35, TI36)

My personal preference is the TI30, since I am very familiar with the model. It is recommended that regardless of the calculator you use, you should spend a good deal of time practicing with it before taking the exam. This will help in saving valuable test-taking time and will also allow you to know the specific functions and capabilities of the calculator.

Summary

This chapter has been included to provide a basic review of math required to perform more complex equations throughout the course of this book. Hopefully, it has laid the groundwork for preparing you to move forward. If, however, you are having difficulty at this point, it may be wise to consider taking a more in-depth course in algebra and trigonometry at the local community college before attempting to take the examination. With a good understanding of the information provided in this chapter, you should be able to solve any of the equations or problems listed throughout the book, which will show how to work each and every equation listed on the CSP exam reference sheet.

4

Particulates and Gases

Every safety professional requires more than a basic understanding of particulates and gases, as they are a common hazard in today's work environment. In order to understand particulates and gases, you must have a good working knowledge of basic chemistry. Therefore, I will begin this chapter by going through a basic review of chemistry. As in Chapter 3, this review is not intended to teach an entire semester course on chemistry, but rather, to serve as a general review.

Periodic Table of the Elements

We begin our discussion of basic chemistry by reviewing the *periodic table of the elements*. The periodic table of the elements (or Periodic Table) displays all the known chemical elements in a tabular form. Currently, there are 118 individual elements listed on the periodic table. The key elements of the Periodic Table are shown in the diagram below:

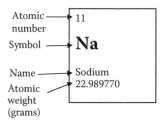

An explanation of each of these terms is listed as follows:

Atomic Number

The *atomic number* is the number of protons found in the nucleus of an atom.

Atomic Mass

The *atomic mass* or weight of an atom includes protons, neutrons, and electrons. This weight is measured in grams. The atomic weight of any element is the average weight (of the natural isotopes) compared to carbon. The mass in grams of any element that is equal to that element's atomic weight is called one *gram-molecular weight*, or one *mole (6.02 × 10²³ atoms = Avogadro's number)*.

NOTE: When preparing for the Associate Safety Professional (ASP)/Certified Safety Professional (CSP) examination, you must know the atomic weights of the more common elements. The more common elements and their atomic weights are as follows: sodium (Na) = 23, hydrogen (H) = 1, carbon (C) = 12, nitrogen (N) = 14, and oxygen (O) = 16. These common elements and their weights should be committed to memory.

Atoms

An atom is a basic unit of matter, which contains a nucleus. The *nucleus* contains positively charged *protons* and neutrally charged *neutrons*. Surrounding the nucleus is a cloud of negatively charged *electrons*. These electrons are bound to the nucleus by means of an electromagnetic force. The number of protons in an atom is always equal to the atomic number. The number of neutrons in an atom is equal to the atomic weight minus the atomic number. The number of electrons in an atom is dependent on the charge of the atom. For example, a neutrally charged atom has a number of electrons equal to the number of protons. For those atoms having a positive or negative charge, the number of electrons will change. In some cases, electrons are lost, which creates a charge to the atom. These charged atoms are referred to as *ions*. In the case of ions, you would subtract the charge from the atomic number. In the case of a positively charged ion, you would have fewer electrons than protons.

Electron shells are considered to be orbits around the nucleus. The number of electron shells present in an atom is determined by the number of electrons present. Each shell can contain a limited number of electrons. One way to think of shells is in terms of levels or distances away from the nucleus. These electron shells are labeled as 1, 2, 3, 4,… or k, l, m, n, o,…. The following is a list of shells (levels) and their maximum number of electrons in that shell:

Shell Level	Maximum No. of Electrons
1	2
2	8
3	16
4	32
5	64

For example, sodium (Na) has an atomic number of 11, as identified on the periodic table. It is a neutrally charged atom and therefore will also have

11 electrons. There will be 2 electrons in the first shell, 8 electrons in the second shell and 1 electron in the third (or outer) shell. The electrons in the outermost shell are referred to as *valence electrons*. Shells with a complete number of valence electrons are considered to be inert. Shells without a complete number or one or two valence electrons tend to be more reactive. The illustration below will show the various shells and their respective numbers:

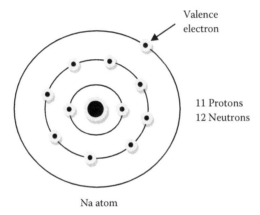

Na atom

Chemical Bonding

The two types of chemical bonding are ionic bonding and covalent bonding. *Ionic bonds* are formed when two atoms exchange electrons, as in the bonding of sodium (Na) and chlorine (Cl) to form table salt (NaCl). One electron from the sodium atom is lost to the chlorine atom, forming NaCl to stabilize the molecule, as illustrated below:

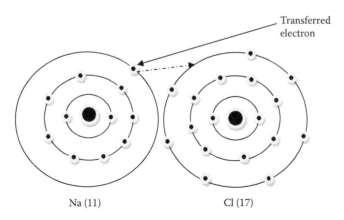

Na (11)   Cl (17)

Covalent bonds are those formed when electrons are shared between two atoms in a molecule, generally between atoms of the nonmetals. These atoms gain electronic stability by sharing electrons so that they look like atoms of the noble gas nearest to them in atomic number. Covalent bonding is illustrated below:

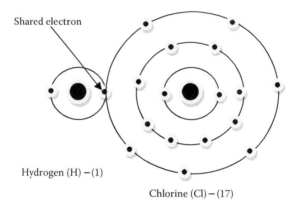

Shared electron

Hydrogen (H) – (1)

Chlorine (Cl) – (17)

Moles

As described above, a *mole* is an *SI* unit, which describes the atomic weight in terms of grams per mole. When we weigh 1 mol of a substance on a balance, this is called a *molar mass* and has the unit grams per mole. This idea is critical as it will be used throughout this chapter and in the course of your professional career. An example of this is as follows:

NaCl or sodium chloride. Sodium has an atomic weight of 22.99 g and chlorine has an atomic weight of 35.453 g for a combined weight of 58.44 g, which is equivalent to 58.44 gram-molecular weight or 58.44 g/mol. Remember Avogadro's number of 6.02×10^{23}. One mole of NaCl has 6.02×10^{23} atoms.

Molecules and Compounds

A *molecule* is a unit of matter formed by the chemical combination of two or more atoms, usually from different elements, but can be the same element. Chlorine (Cl) combines with hydrogen to form hydrochloric acid (or HCl). This is an example of atoms combining by chemical bonding to form a molecule. A *compound* is a substance composed of all the same kinds of molecules. For example, water is a compound composed entirely of water molecules (H_2O).

Mixtures

Compounds should not be confused with mixtures. Mixtures are not the result of chemical bonding; rather, they consist of separate elements or compounds that are mixed together. *Mixtures* include solutions, when a solute (e.g., sugar) is dissolved in a solvent (water). Like all mixtures, each component of a solution retains its own chemical properties.

Chemical Formulas

Chemical formulas are a shorthand way of expressing the composition of molecules and the way they react with one another. Hydrochloric acid, the chemical combination of hydrogen and chlorine, is abbreviated as HCl. This is the chemical formula for hydrochloric acid. In this case, there is one atom of hydrogen and one atom of chlorine in every molecule of HCl. If there is more than one identical atom in a molecule, the number is shown as a subscript after the symbol for that atom, as in the chemical formula for sulfuric acid (H_2SO_4). When writing a chemical formula, one should remember the "Law of Conservation," which states that matter can be neither created nor destroyed in a chemical reaction. Therefore, when writing chemical equations, the same number of atoms on one side of the equation must equal those on the opposite side of the equation. For example, in the case of carbon dioxide, there is one carbon atom that combines with two oxygen atoms. This is illustrated as

$$C \text{ (carbon)} + O_2 = CO_2.$$

Written as a chemical equation:

$$C + O_2 \xrightarrow{\text{yields}} CO_2.$$

You can see from the illustration above that there remain one carbon and two oxygen atoms on each side of the equation. The illustration below shows the basic components of a chemical equation:

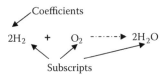

In this equation, the coefficients are multiplied by the subscript to give you a total. As you can see, there are 4 hydrogen atoms and 2 oxygen atoms on each side of the equation.

NOTE: When there is no subscript behind a symbol, it is assumed to be 1.

Atomic Weight of Compounds

To determine the atomic weight of compounds, one can simply add the atomic weight of the individual atoms, remembering to multiply the individual weights by the number of atoms of each present in the compound. For example, let's take hydrogen sulfate (H_2SO_4) and determine the atomic weight. As you can see, there are 2 hydrogen atoms, 1 sulfur atom, and 4 oxygen atoms in this compound. The individual weights are as follows: hydrogen (H), 1; sulfur (S), 32.066; and oxygen (O), 15.9994.

> Hydrogen: 2 atoms × 1 (atomic weight)   = 2
> Sulfur: 1 atom × 32.066 (atomic weight)   = 32.066
> Oxygen: 4 atoms × 15.9994 (atomic weight) = <u>63.9976</u>
> 98.0636 g

The total compound weight is 98.0636 g. Therefore, 1 mol of H_2SO_4 would be 98.0636 g.

Percentage of Element in a Compound (by Weight)

As a safety professional, you often have the need to determine the percentage of an element in a compound. To calculate the percentage of the element in a compound, we use the following method, as in the case of H_2SO_4:

We know that a mole of H_2SO_4 contains 98.0636 g. Let's determine the individual element percentage.

> Hydrogen: 2 g/98.0636 g × 100   = 2.04%
> Sulfur: 32.066 g/98.0636 g × 100   = 32.70%
> Oxygen: 63.9976 g/98.0636 g × 100 = <u>65.26%</u>
> 100%

Acids, Bases, and pH's

In determining whether or not a solution is an acid or base, we must determine its *pH*. The pH is calculated as follows:

$$pH = -\log|H^+|.$$

For example, you have a solution that contains 10^{-2} ions per liter. What is the pH of this solution?

> $pH = -\log|10^{-2}|$
> $pH = -\log|0.01|$
> $pH = -(-2)$
> $pH = 2.$

The pH is a scalar measurement (0–14) that determines whether a chemical or solution is an acid, base, or neutral. A neutral solution is considered to be 7. Any pH below 7 is considered an acid, while a pH above 7 is considered a base. This is illustrated graphically below:

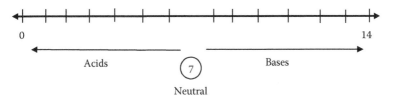

Gas Laws

There are four basic gas laws that need to be understood. They are Boyle's law, Charles' law, the ideal gas law, and the combined gas law.

Boyle's Law

Boyle's law states that at constant temperature, a fixed mass of gas occupies a volume that is inversely proportional to the pressure exerted upon it. Boyle's law written mathematically is

$$P_1V_1 = P_2V_2,$$

where P = pressure exerted and V = volume of gas.

Example

A 3.7-L (liter) volume of gas exerts a pressure of 760 mm Hg. What would be the volume of the gas if the pressure is decreased to 630 mm Hg?

$$(3.7 \text{ L})(760 \text{ mm Hg}) = (630 \text{ mm Hg})(V_2)$$

$$\frac{(3.7 \text{ L})(760 \text{ mm Hg})}{630 \text{ mm Hg}} = \frac{(630 \text{ mm Hg})(V_2)}{630 \text{ mm Hg}}$$

$$4.46 \text{ L} = V_2.$$

As you can see, when the pressure is reduced, the volume increases, according to Boyle's law.

Charles' Law

Charles' law states that at constant pressure, the volume occupied by a fixed mass of gas is directly proportional to the absolute temperature. Written mathematically, Charles' law is shown below:

$$\frac{V_1}{T_1} = \frac{V_2}{T_2},$$

where V = volume of gas and T = temperature of gas (Kelvin temperature scale).

Example

A container contains a volume of 4.2 L of gas at a temperature of 32°C. What is the volume of gas if the temperature is increased to 37°C?

Remember that the temperature is measured in Kelvin; therefore, we must convert degrees Celsius to Kelvin. We do this as follows:

$$t_K = t_{°C} + 273,$$

where t_K = temperature in Kelvin and $t_{°C}$ = temperature in Celsius.

$$t_K = 32°C + 273.$$
$$t_K = 305$$

The temperature 32°C is equal to 305 K, and the temperature 37°C is equal to 310 K.

We can now insert our known variables into the equation for Charles' law, as follows:

$$\frac{4.2\ \text{L}}{305\ \text{K}} = \frac{V_2}{310\ \text{K}}$$

$$(310\ \text{K})\frac{4.2\ \text{L}}{310\ \text{K}} = \frac{V_2}{310\ \text{K}}(310\ \text{K})$$

$$4.27\ \text{L} = V_2.$$

Ideal Gas Law

An ideal gas can be defined as a gas, where all internal energy is in kinetic energy form and any change causes a change in temperature. The ideal gas law is written as follows:

$$PV = nRT,$$

where

P = absolute pressure (total pressure exerted on a system [also written as psia]). Absolute pressure is equal to the gauge pressure plus the atmospheric pressure (unit of atmosphere or atm)

V = volume (liters)

T = temperature (Kelvin)

n = number of molecules (moles)

R = universal gas constant (8.3145 J/mol K)

Before we solve an example problem using the equation for the ideal gas law, it is first necessary to discuss some of the finer points in this equation. As you can see, these variables must be in specific units for the equation to be valid.

Conversion Factors for Converting Pressure Units to Atmospheric Pressure Units

To convert pressures to atmosphere pressures, use the following conversion factors:

1 atm = 14.6959488 pounds per square inch (psi)

1 atm = 29.9246899 inches of mercury (in Hg)

1 atm = 760 mm mercury (mm Hg)

1 atm = 760 torr (torr)

1 atm = 101,325 pascals (Pa)

1 atm = 101.325 kilopascals (kPa)

1 atm = 1.01325 bar (bar)

Conversion Factors for Converting Units of Volume to Liters

To convert units of volume to liters, use the following conversion factors:

1 L = 1000 milliliters (mL)

1 L = 1000 cubic centimeters (cm^3)

1 L = 1 cubic decimeter (dm^3)

1 L = 0.001 cubic meters (m^3)

1 L = 0.264172051 US gallons (gal)

1 L = 1.0566882 US quarts (qts)

Converting Grams to Moles

To convert from grams to moles, divide the number of grams by the molar mass (MM). For example, 1 mol of HCN is equal to 27 g. Three moles would equal 81 g. Thus, to determine the number of moles of HCN, divide 81 g by 27 g to convert to 3 mol.

Example

What is the volume of 2 gram-mole of gas at an absolute pressure of 28.3 psia and a temperature of 42°C?

We must first convert the temperature to Kelvin. This is done as follows:

$$t_K = 42°C + 273 = 315 \text{ K}.$$

Now, insert the known data into the ideal gas law equation, as follows:

$$28.3 \text{ psia}(V) = 2 \text{ g-mol}(1.206)(315 \text{ K})$$

$$\frac{28.3 \text{ psia}(V)}{28.3 \text{ psia}} = \frac{759.78}{28.3 \text{ psia}}$$

$$V = 26.85 \text{ L}.$$

NOTE: 1 g-mol of ideal gas occupies 24.04 L (usually rounded to 24 L) or 0.02404 m³.

Universal Gas Constant

The universal gas constant or R value in the ideal gas law is dependent upon the units used for pressure, temperature, volume, and number of moles. The table below is given to you on the Board of Certified Safety Professionals (BCSP) examination reference sheet.[1]

| | | | Value Gas Constant | | | | |
| | | | Absolute Pressure | | | | |
Volume	Temperature	Moles	atm	psi	mm Hg	in Hg	ft. H₂O
ft³	K	g	0.00290	0.0426	2.20	0.00867	0.0982
		lb	1.31	19.31	999.0	39.3	44.6
	°R	g	0.00161	0.2366	1.22	0.0482	0.0546
		lb	0.730	10.73	555.0	21.8	24.8
L	K	g	0.08205	1.206	62.4	2.45	2.78
		lb	37.2	547.0	28,300.0	1113.0	1262.0
	°R	g	0.0456	0.670	34.6	1.36	1.55
		lb	20.7	304.0	15,700.0	619.0	701.0

For example, using the above chart provided on the BCSP examination reference sheet, if the volume (V), were in liters, temperature in Kelvin, moles in grams, and the pressure in psi, the universal gas constant or R value would be 1.206. Insert this value into your equation for the universal gas constant.

Combined Gas Law

The combined gas law combines both Charles' law and Boyle's law and is written mathematically as follows:

$$\frac{P_1 V_1}{T_1} = \frac{P_2 V_2}{T_2},$$

where

P = absolute pressure (atm)
V = volume of gas (liters)
T = temperature of gas (Kelvin temperature scale)

Example

A gas has a volume of 3.9 L, a pressure of 1.3 atm, and a temperature of 295 K. If the pressure remains constant, but the gas is heated to 308 K, what is the new volume of the gas?

$$\frac{1.3 \text{ atm}(3.9 \text{ L})}{295 \text{ K}} = \frac{1.3 \text{ atm}(V_2)}{308 \text{ K}}$$

$$\frac{308 \text{ K}(1.3 \text{ atm})(3.9 \text{ L})}{295 \text{ K}} = \frac{(1.3 \text{ atm})(V_2)308 \text{ K}}{308 \text{ K}}$$

$$\frac{1561.56 \text{ atm-L}}{295 \text{ K}(1.3 \text{ atm})} = \frac{1.3 \text{ atm}(V_2)}{1.3 \text{ atm}}$$

$$4.1 \text{ L} = V_2.$$

Concentrations of Vapors, Gases, and Particulates

As a way of introduction into this section, I would first like to provide some definitions that you will find helpful in understanding this section.

Vapors: A gaseous form of a substance that is normally a solid or liquid at room temperature.

Gases: Substances that completely occupy a space and can be converted to a liquid or solid by increasing or decreasing temperature.

Vapor pressure: The pressure that a vapor at equilibrium with a pure liquid at a given temperature exerts to the surrounding atmosphere.

Particulate: Fine solid or liquid particles, such as dust, fog, mist, smoke, or spray.

Dust: Solid particles generated by mechanical action (crushing, grinding, etc.). Size ranges are usually from 0.1 to 30 μm.

Fume: Airborne solid particles formed by condensation of vapor (i.e., welding fumes). Size ranges from 0.001 to 1.0 μm.

Mist: Suspended liquid droplets generated by condensation or atomization (fogs are formed by condensation). Size ranges from 0.01 to 10 μm.

Fibers: Particulate with an aspect ratio (length to width) of 3:1.

Standard Temperature and Pressure

Throughout your professional career and specifically on the ASP/CSP exam, you encounter the term *STP* (*Standard Temperature and Pressure*). It is critical that you understand its meaning, as it will drastically change the outcome of your calculations, if performed incorrectly. For the purposes of discussing concentrations of gases and vapors, STP is the standard temperature and pressure of 25°C and 1 atm. This standard is provided on the BCSP examination reference sheet. When discussing physical sciences, STP is equal to 0°C and 1 atm, and when discussing ventilation, STP is equal to 70°F and 1 atm. These last two are also listed on the BCSP examination reference sheet.

Standards and Regulations

There are two standards that we need to reference before proceeding. The first standard is the *Permissible Exposure Limit or PEL*, which is established by the US Occupational Safety and Health Administration (OSHA), for each contaminant regulated. The PEL establishes the maximum exposure concentration to a contaminant that an employee can be exposed to in an 8-h workday. The OSHA PEL is the only standard that is enforceable by law. The second standard to be discussed is the *Threshold Limit Value or TLV*, which is established by the American Conference of Governmental Industrial Hygienist (ACGIH). The TLV establishes the exposure time for various contaminants in an 8-h workday. The majority of contaminants having a TLV are consistent with the requirements of the PEL. However, for some contaminants listed, the exposure concentrations are more stringent than the PEL. This is primarily attributed to the fact that ACGIH does not consider the economic impact in establishing the standard. The TLV is not enforceable under the law. When faced with a conflict in the standards, it is strongly recommended that the safety professional adhere to the more stringent standard. This is in keeping with the philosophy of holding the safety and health of the employee paramount in any decision.

Time-Weighted Average

A *time-weighted average or TWA* is the concentration that an employee is exposed to for a specified period. The PEL and TLV are both established for a TWA of 8 h. The average person should have no adverse health effects to the exposure at these concentrations. The TWA is based on, or rather, compared to the OSHA PEL or the ACGIH TLV. Each contaminant has its own unique PEL and can be found in 29 CFR 1910.1000 (Table Z-1). The TLV can be found on the material safety data sheet (MSDS) or various other sources.

To calculate a TWA, we use the following equation:

$$TWA = \frac{(C_1 T_1) + (C_2 T_2) + \ldots (C_n T_n)}{\text{total hours worked}},$$

where C = concentration (ppm or mg/m³) and T = time of exposure (minutes or hours).

Example

An employee is exposed to carbon dioxide (CO_2) concentrations of 5300 ppm for 2 h, 3500 ppm for 3 h, and 5125 ppm for 3 h. What is the TWA?

$$TWA = \frac{(5200 \text{ ppm})(2 \text{ h}) + (3500 \text{ ppm})(3 \text{ h}) + (5125 \text{ ppm})(3 \text{ h})}{8 \text{ h}}$$

$$TWA = \frac{(10,600 \text{ ppm}) + (10,500 \text{ ppm}) + (15,375 \text{ ppm})}{8 \text{ h}}$$

$$TWA = \frac{36,475 \text{ ppm}}{8 \text{ h}}$$

$$TWA = 4559.38 \text{ ppm}.$$

On the basis of this calculation, the employee was exposed to a TWA concentration of 4559.38 ppm in an 8-h period. Now we look in Table Z-1 to obtain the PEL for CO_2, which is 5000 ppm. Therefore, 4559.38 ppm is less than the PEL. This employee has not exceeded the standard.

Calculating PELs/TLVs for Periods Greater than 8 h

Many places of employment do not work a standard 8-h workday. Sometimes they work a 10- or 12-h work shift. How does an 8-h PEL/TLV compare to a 12-h shift concentration? To calculate a PEL/TLV for a period greater than 8 h, you would use the following equation:

$$PEL_{n \text{ hours}} = \frac{PEL \text{ (or TLV)} \times 8 \text{ h}}{(\text{actual shift worked in hours})}.$$

Example

OSHA has established a PEL of 5.0 mg/m³ for respirable dust. What would be the modified PEL for an employee working a 12-h shift?

$$PEL_{12\,hour} = \frac{5.0\frac{mg}{m^3} \times 8\,h}{12\,h}$$

$$PEL_{12\,hour} = \frac{40\,mg/m^3}{12\,h}$$

$$PEL_{12\,hour} = 3.33\,mg/m^3.$$

The established 12-hour PEL for respirable dust would be 3.33 mg/m³.

Gaseous Mixtures

Mixtures are simply the mixing of two or more chemicals. Sometimes, employees are exposed to several contaminants at the same time. How do we determine if the employee has been overexposed? To determine exposure for gaseous mixtures, we use the following equation:

$$TLV \text{ (or PEL)}_{m\text{-gaseous}} = \frac{C_1}{TLV_1} + \frac{C_2}{TLV_2} + \dots \frac{C_n}{TLV_n}.$$

If the result for this equation is greater than 1, then the mixture TLV (gaseous) has been exceeded.

Example

Given the following information, determine whether an employee's exposure to a gaseous mixture has been exceeded. Chemical A has a TLV of 200 ppm and a measured concentration of 150 ppm; Chemical B has a TLV of 300 ppm and a measured concentration of 100 ppm; and Chemical C has a TLV of 500 ppm and a measured concentration of 100 ppm.

$$TLV_{m\text{-gaseous}} = \frac{150\,ppm}{200\,ppm} + \frac{100\,ppm}{300\,ppm} + \frac{100\,ppm}{500\,ppm}$$

$$TLV_{m\text{-gaseous}} = 0.75\,ppm + 0.333\,ppm + 0.2\,ppm$$

$$TLV_{m\text{-gaseous}} = 1.28\,ppm.$$

The result is 1.28, which is greater than 1; therefore, this employee was overexposed to the gaseous mixture.

Liquid Mixtures

Just as in gaseous mixtures, it is necessary for the safety professional to determine an employee's exposure to liquid mixtures. To do this, we use the following equation:

$$TLV_{m\text{-liquid}} = \frac{1}{\left(\dfrac{f_1}{TLV_1} + \dfrac{f_2}{TLV_2} + \cdots \dfrac{f_n}{TLV_n}\right)},$$

where f = fraction of chemical (weight percent of liquid mixture) and TLV = threshold limit value of the chemical.

Example

A liquid mixture contains 40% (by weight) heptanes, 20% methyl chloroform, and the balance is perchloroethylene. Assuming that all of this liquid evaporates and that atmospheric composition is similar to the original liquid, calculate the TLV exposure for the gaseous mixture generated from this liquid mixture. The TLVs are as follows: heptane = 1600 mg/m³, methyl chloroform = 1900 mg/m³, and perchloroethylene = 335 mg/m³.

$$TLV_{m\text{-liquid}} = \frac{1}{\left(\dfrac{0.4}{1600 \text{ mg/m}^3} + \dfrac{0.2}{1900 \text{ mg/m}^3} + \dfrac{0.4}{335 \text{ mg/m}^3}\right)}$$

$$TLV_{m\text{-liquid}} = \frac{1}{0.0015 \text{ mg/m}^3}$$

$$TLV_{m\text{-liquid}} = 645.46 \text{ mg/m}^3$$

645.46 mg/m³ is the TLV or PEL for this liquid mixture. This is the maximum concentration that an employee can be exposed to for this liquid mixture in an 8-h period.

Percentage of TLV Mixture

To determine the percentage of the TLV/PEL, use the following equation:

$$\%TLV_{mixture} = \left(\frac{f_1}{TLV_1} + \frac{f_2}{TLV_2} + \cdots \frac{f_n}{TLV_n}\right),$$

where f = measured concentration of contaminant and TLV = established TLV of contaminant.

Example

An employee is exposed to 425 mg/m³ of heptane, 270 mg/m³ of methyl chloroform, and 175 mg/m³ of perchloroethylene. What is the percentage of the TLV that the employee has been exposed to for this mixture, using 645.46 mg/m³ as the TLV?

$$\%TLV_{mixture} = \left(\frac{425 \text{ mg/m}^3}{1600 \text{ mg/m}^3} + \frac{270 \text{ mg/m}^3}{1900 \text{ mg/m}^3} + \frac{175 \text{ mg/m}^3}{335 \text{ mg/m}^3} \right) 100$$

$$\%TLV_{mixture} = (0.27 + 0.14 + 0.52)100$$

$$\%TLV_{mixture} = 93\%.$$

This employee has been exposed to 93% of the established TLV for this mixture or approximately 600.28 mg/m³.

Calculating PEL/TLV for Silica

One contaminant, silica (SiO_2), does not have a specific PEL listed in Table Z-1. In order to determine the PEL/TLV for silica, you must use the following equation, which is not listed on the BCSP examination reference sheet, but may be asked on the examination:

$$PEL_{silica} = \frac{10 \text{ mg/m}^3}{\%SiO_2 + 2},$$

where $\%SiO_2$ = the percentage of silica in the batch.

Example

A company uses a batch powder that contains 4% silica. When using this material, what would be the PEL or TLV?

$$PEL_{silica} = \frac{10 \text{ mg/m}^3}{\%SiO_2 + 2}$$

$$= \frac{10 \text{ mg/m}^3}{4 + 2}$$

$$= 1.67 \text{ mg/m}^3.$$

Converting mg/m³ to ppm

To convert mg/m³ to ppm, use the following equation:

$$ppm = \frac{mg/m^3 \times 24.45}{MW},$$

where
    ppm = parts per million
    mg/m^3 = measured mg/m^3 of the contaminant
    MW = molecular weight of contaminant
    24.45 = constant = 1 g-mol

Example

Convert 22 mg/m^3 of H_2S (hydrogen sulfide) to ppm. The MW of hydrogen (H) is 1 and the MW of sulfur (S) is 32.

Converting ppm to mg/m³

To convert from ppm to mg/m^3, use the following equation:

$$mg/m^3 = \frac{(ppm)(MW)}{24.45},$$

where
    ppm = parts per million
    mg/m^3 = measured mg/m^3 of the contaminant
    MW = molecular weight of contaminant
    24.45 = constant = 1 g-mol

Example

Convert 16.3 ppm of hydrogen sulfide to mg/m^3. MW: H = 1; S = 32.

$$mg/m^3 = \frac{(16.3 \text{ ppm})(33 \text{ g/mol})}{14.45 \text{ g/mol}}$$

$$mg/m^3 = \frac{537.9 \text{ ppm}}{24.45}$$

$$mg/m^3 = 22.$$

Lower Flammability Limit of Mixtures

The flammability range for any chemical is the range of concentrations that exist between the lower flammability limit (LFL) and the upper flammability limit (UFL). The range varies from chemical to chemical. This information can be conveniently located on the material safety data sheet (MSDS) or other sources. For example, if an individual chemical has an LFL of 1.7% to

8.1%, ignition only occurs if the concentration is between this range. To calculate the LFL of a mixture, use the following equation:

$$LFL_{mixture} = \frac{1}{\left(\dfrac{f_1}{LFL_1} + \dfrac{f_2}{LFL_2} + \ldots \dfrac{f_n}{LFL_n}\right)},$$

where f = fraction of chemical in the mixture and LFL = lower flammability limit.

Example

A liquid mixture contains 40% heptanes, 20% methyl chloroform, and the balance is perchloroethylene. What is the LFL of the mixture? LFLs: heptane = 1.2%, methyl chloroform = 1.7%, and perchloroethylene = 1.1%.

$$LFL_{mixture} = \frac{1}{\left(\dfrac{0.4}{1.2} + \dfrac{0.2}{1.7} + \dfrac{0.4}{1.1}\right)}$$

$$LFL_{mixture} = \frac{1}{(0.333 + 0.118 + 0.364)}$$

$$LFL_{mixture} = \frac{1}{0.815}$$

$$LFL_{mixture} = 1.23\%.$$

Referenced Equations

Boyle's Law

$$P_1V_1 = P_2V_2,$$

where P = pressure exerted and V = volume of gas.

Charles' Law

$$\frac{V_1}{T_1} = \frac{V_2}{T_2},$$

where V = volume of gas and T = temperature of gas (Kelvin temperature scale).

Ideal Gas Law

$$PV = nRT,$$

where
   $P =$ absolute pressure (total pressure exerted on a system [also written as psia]). Absolute pressure is equal to the gauge pressure plus the atmospheric pressure (unit of atmosphere or atm)
   $V =$ volume (liters)
   $T =$ Temperature (Kelvin)
   $n =$ number of molecules (moles)
   $R =$ universal gas constant (8.3145 J/mol K)

Combined Gas Law

$$\frac{P_1V_1}{T_1} = \frac{P_2V_2}{T_2},$$

where
   $P =$ absolute pressure (atm)
   $V =$ volume of gas (liters)
   $T =$ temperature of gas (Kelvin temperature scale)

Time-Weighted Average

$$TWA = \frac{(C_1T_1) + (C_2T_2) + ...(C_nT_n)}{\text{total hours worked}},$$

where $C =$ concentration (ppm or mg/m³) and $T =$ time of exposure (minutes or hours).

Calculating PELs/TLVs for Periods Greater than 8 h

Gaseous Mixtures

$$\text{TLV (or PEL)}_{\text{m-gaseous}} = \frac{C_1}{TLV_1} + \frac{C_2}{TLV_2} + ... \frac{C_n}{TLV_n}.$$

Liquid Mixtures

$$\text{TLV}_{\text{m-liquid}} = \frac{1}{\left(\dfrac{f_1}{\text{TLV}_1} + \dfrac{f_2}{\text{TLV}_2} + \cdots \dfrac{f_n}{\text{TLV}_n} \right)},$$

where f = fraction of chemical (weight percent of liquid mixture) and TLV = threshold limit value of the chemical.

Percentage of TLV for Mixtures

$$\%\text{TLV}_{\text{mixture}} = \left(\dfrac{f_1}{\text{TLV}_1} + \dfrac{f_2}{\text{TLV}_2} + \cdots \dfrac{f_n}{\text{TLV}_n} \right),$$

where f = fraction of chemical (weight percent of liquid mixture) and TLV = threshold limit value of the chemical.

Calculating PEL/TLV for Silica

$$\text{PEL}_{\text{silica}} = \frac{10 \text{ mg/m}^3}{\%\text{SiO}_2 + 2},$$

where $\%\text{SiO}_2$ = the percentage of silica in the batch.

Converting mg/m³ to ppm

$$\text{ppm} = \frac{\text{mg/m}^3 \times 24.45}{\text{MW}},$$

where
  ppm = parts per million
  mg/m³ = measured mg/m³ of the contaminant
  MW = molecular weight of contaminant
  24.45 = constant = 1 g-mol

Converting ppm to mg/m³

To convert from ppm to mg/m³, use the following equation:

$$mg/m^3 = \frac{(ppm)(MW)}{24.45},$$

where
ppm = parts per million
mg/m^3 = measured mg/m^3 of the contaminant
MW = molecular weight of contaminant
24.45 = constant = 1 g-mol

LFLs of Mixtures

$$LFL_{mixture} = \frac{1}{\left(\dfrac{f_1}{LFL_1} + \dfrac{f_2}{LFL_2} + \dots \dfrac{f_n}{LFL_n}\right)},$$

where f = fraction of chemical in the mixture and LFL = lower flammability limit.

Key Information to Remember on Particulates and Gases

1. The atomic number is the number of protons in a nucleus.
2. The atomic mass or weight of an atom includes protons, neutrons, and electrons.
3. Avogadro's number is 6.02×10^{23} atoms.
4. One mole is equal to 6.02×10^{23} atoms.
5. Atomic weights of the common elements are as follows: sodium (Na) = 23, hydrogen (H) = 1, carbon (C) = 12, nitrogen (N) = 14, oxygen (O) = 16, and sulfur (S) = 32.
6. Valence electrons are those electrons in the outer shell of an atom. A complete number in a shell creates a more stabilized atom.
7. Ionic bonding is the transfer of an electron from one atom to another.
8. Covalent bonding is the sharing of electrons between two atoms.

9. The law of conservation should be kept in mind when writing chemical formulas. Energy or mass cannot be created or destroyed, but it can change its form.

10. The universal gas constant value is dependent upon the units of pressure, volume, temperature, and moles.

11. In industrial hygiene equations, the STP is equal to 25°C and 1 atm.

12. Ventilation STP is equal to 70°F and 1 atm.

13. Air density STP = 0.075 lb/ft^3 at 70°F and 1 atm.

14. Physical science STP = 0°C and 1 atm.

Reference

1. Board of Certified Safety Professionals, *Examination Reference Sheet, Gas Constant Value*.

5

Toxicology

Toxicology is the study of the adverse effects of chemicals on living organisms. It is critical for the practicing safety professional, at all levels, to have a solid understanding of basic toxicology (Wikipedia[1]). Hemlock, opium, and toxic metals, such as antimony, copper, and lead, were described as early as 1500 BC. The Greek physician Dioscorides (ca. AD 50–90) attempted to classify poisons into three categories, plant poison, animal poison, and mineral poisons. In the 15th century, physicians recognized that certain occupations were exposed to toxic substances. Workers involved in goldsmithing, mining, printing, and pottery making were exposed to lead and mercury, which was documented in *On the Miner's Sickness and Other Diseases of Miners* (Paracelsus, 1567) and *Discourse on the Diseases of Workers* (Ramazzini, 1700). Paracelsus is often referred to as the father of toxicology.

Since the days of Parcelsus and Ramazzini, great advances have taken place in the field of toxicology. The saying "The Dose Makes the Poison" is very true. Everything can be poisonous to humans and animals, given enough dosage. For example, water is a requirement for humans to survive. However, if someone were to drink enough water, it could result in water intoxication or death. This chapter will cover the basics of toxicology, the major diseases, and associated toxins.

Definitions

Before getting into the chapter and our review of toxicology, it is necessary to have a basic understanding of the terms and definitions.

Toxicology: The science dealing with the study of poisons and their effects.

Toxicity: The degree to which a substance can harm an exposed organism.

Hazard: The potential for a substance to cause harm (toxicity plus exposure).

Risk: The quantitative measurement or estimate of a hazard.

Poison: A substance that is harmful to an organism.

Xenobiotic: A substance that is found in an organism but that is not normally produced or expected to be present in it.

Dose–response theory: Increasing the dose increases the effect.

Tolerance: Increased ability to withstand exposure.

Chronic exposure: Long-term exposure (days, weeks, months, and years).

Acute exposure: Short-term exposure (minutes, hours, and days).

NOEL: No observed effect level.

NOAEL: No observed adverse effect level.

LOEL: Lowest observed effect level.

LOAEL: Lowest observed adverse effect level.

Latency period: The period between exposure and the onset of symptoms.

LD_{50}: Lethal dose. It is the amount of material (normally liquid or solid) that kills 50% of the laboratory animals.

LC_{50}: Lethal concentration. It is the amount of material (normally airborne concentrations) that kills 50% of laboratory animals.

IDLH: Immediately dangerous to life and health.

Routes of Entry

When discussing poisons and toxins and their effects on humans or animals, one of the first things to consider is their method of entry into the body. In this section, we will discuss inhalation, ingestion, absorption, and percutaneous and intravenous routes of entry.

Inhalation

Entry through inhalation is obviously respiratory in nature. Therefore, it is necessary to review the basics of the respiratory system in terms of its anatomy and physiology. The respiratory system consists of the nose, nasal passages, mouth, epiglottis, pharynx, larynx, trachea, bronchi, and alveoli. Air enters the body through the nose or mouth and travels through to the alveoli (see Figure 5.1[2]). The body's natural defense against foreign bodies include mucus in the linings of the nose and sinus cavities and cilia in the nose. Particles can be trapped and expelled through coughing by the mucus. Furthermore, particles can also be captured by the cilia. Once the air (and the gaseous makeup of air) enters the alveoli (air sacs) that lie near blood vessels, the molecules of air enter the bloodstream through permutation. Oxygen

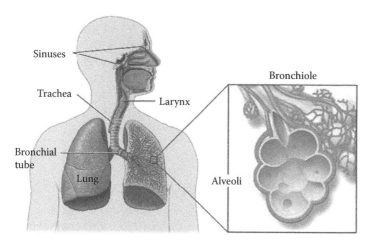

FIGURE 5.1
Diagram of the respiratory system. (From Nucleus Medical Art, 2009. Si555509999 Respiratory System, http://64.143.176.9/library/healthguide/en-us/support/topic.asp?hwid=ug2841.) (Illustration Copyright 2009 Nucleus Medical Art, all rights reserved. http://www.nucleusinc .com.)

enters the bloodstream and carbon dioxide is expelled through the respiratory system as outgoing air. This process is known as *respiration*. In a normal relaxed state, humans breathe in and out 10–14 times per minute, with each breath lasting 4–6 s. In a minute, 4.3–5.7 L of air is breathed in. As a person's activity increases, so does their respiratory rate, which can be up to one breath per second and a total intake of up to 120 L.[1] This volume is referred to as the *tidal volume*.

Ingestion

Ingestion of toxic or other substances is through the mouth. When swallowing a substance, the substance enters the digestive system and ultimately into the bloodstream through the stomach or intestinal tract. A chemical does not normally enter through this route as quickly as in inhalation. However, through this route, the substance can travel throughout the system, just as in other routes of entry.

Absorption

A substance can be absorbed through the skin and enter the bloodstream and travel throughout the body. The skin acts as a natural barrier against foreign substances. A foreign substance will enter the bloodstream through the skin more rapidly when the skin has lacerations or abrasions.

Percutaneous and Intravenous Injections

Percutaneous and intravenous injections are primarily medical procedures that invade the skin intentionally in order to administer a regulated dose of medicine. However, on occasion, an accidental injection such as a needlestick may occur.

Dose–Response Relationship

The *dose–response relationship*, or *exposure–response relationship*, describes the change in effect on an organism caused by differing levels of exposure (or doses) to a stressor (usually a chemical) after a certain exposure time (see Figure 5.2).

In other words, the higher the dose, the greater the response. A lethal dose (50%) is demonstrated in Figure 5.3 (lethal dose 50%).

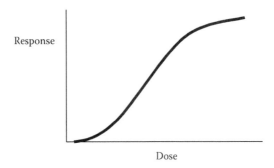

FIGURE 5.2
Dose–response curve.

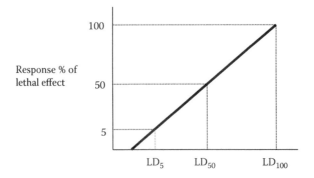

FIGURE 5.3
Graphic representation of lethal dose/concentration.

Exposures to Chemical Combinations

Any number of combinations of chemicals is possible in today's work environments. Various chemicals react differently with other chemicals and also have varying effects on humans and animals in these varying combinations. These effects can be described as additive, antagonistic, synergistic, or potentiating. Each of these is described in the following.

Additive: The combined effect of the chemicals is equal to the sum of each chemical acting independently. To use a mathematical example, $2 + 2 = 4$.

Antagonistic: When combined, the two chemicals interfere with each other. Example: $3 + (-2) = 1$.

Synergistic: The combined effect of two chemicals is much greater than the sum of the effect of each agent acting independently. Example: $2 + 2 = 10$.

Potentiating: One substance does not have a toxic effect on a certain organ system; however, when it combines with another chemical, it makes the combination of the two greater. An example is $2 + 0 = 10$.

Other classifications of chemicals or toxins include carcinogens, co-carcinogens, epigenetic, genotoxic, mutagen, clastogen, and teratogen, which are described in the following.

Carcinogen: Any substance or agent known to cause cancer. Carcinogens do not adhere to the dose–response curve.

Co-carcinogen: These agents, when applied immediately prior to or with a genotoxic carcinogen, enhance the oncogenic (cancerous) effect of the agent.

Epigenetic: Changes in phenotype (appearance) or gene expression caused by mechanisms other than changes in the underlying DNA sequence.

Genotoxic: These materials are known to be potentially mutagenic and carcinogenic in nature. They act directly by altering the DNA.

Mutagen: A physical or chemical agent that changes the genetic material (usually DNA) of an organism and thus increases the frequency of mutagens above the natural background level.

Teratogens: Any agent that can disturb the development of an embryo or fetus.

Stages of Cancer

There are four stages of cancer. They are as follows:

- Initiation
- Latency period
- Promotion
- Progression

Initiation

Initiation can be described as the first step in cancer development. Initiators, if not already reactive with DNA, become altered via drug-metabolizing enzymes in the body, which makes them available to cause changes in the DNA, or mutations. These initiators are often specific to the particular tissue types or species. The effects of initiators are irreversible, until its death. Any daughter cell reproduced from an initiator also carries the mutations.

Latency Period

As with any chemical, the latency period can be described as the period from exposure to onset of symptoms. In cancer development, the latency period is the period where the mutations lie dormant.

Promotion

Once a mutation occurs as a result of the initiator, it is susceptible to the effects of promoters. The compounds promote the proliferation of the cell, giving rise to a large number of daughter cells containing the mutation created by the initiator.

Progression

The term *progression* refers to the transformation of a benign tumor to a neoplasm and to malignancy.

Types of Poisons

Poisons can be described as hepatoxin, nephrotoxin, or neurotoxin. *Hepatoxins* are those chemicals or agents having an adverse effect on the liver or blood. *Nephrotoxins* are those chemicals or agents having an adverse effect on the kidneys, and *neurotoxins* are those agents having an adverse effect on the nervous system.

Ames Testing

Ames testing is a procedure to determine whether or not a chemical is a mutagen. There is a high correlation between the mutagenic trait of a chemical and its ability to cause cancer. The bacterium used in the test is a strain of *Salmonella typhimurium* that carries a defective (mutant) gene, making it unable to synthesize the amino acid histidine (His) from the ingredients in its culture medium.

Cohort Study

A cohort study is a study in which subjects who presently have a certain condition or receive a particular treatment are followed over time and compared with another group who are not affected by the condition under investigation. For research purposes, a cohort is any group of individuals who are linked in some way or who have experienced the same significant life event within a given period.

Advantages

One of the advantages of cohort analysis is that the study design does not require strict random assignment of subjects, which is, in many cases, unethical or improbable.

Disadvantages

One of the most difficult tasks in cohort studies is to assess whether associations between cohort and dependent variables derived from the studies are of a causal nature or not. Cohort studies are subject to the influence of factors over which the investigators most often do not have full control, and that findings from these studies are more open to threats to validity than those of studies with an experimental research design.

Case–Control Study

A study that compares two groups of people: those with the disease or condition under study and a very similar group of people who do not have the disease or condition (controls). Researchers study the medical and lifestyle histories of the people in each group to learn what factors may be associated with the disease or condition.

Cross-Sectional Study

Cross-sectional studies form a class of research methods that involve observation of some subset of a population of items all at the same time, in which groups can be compared at different ages with respect to independent variables, such as IQ and memory.

Common Occupational Diseases and Disorders Caused by a Contributing Agent or Substance

An *occupational disease* refers to those illnesses caused by exposures in the workplace. The list of diseases or illnesses that follow is common in today's work environments. The list is not all-inclusive but is provided as a review and is likely to be seen on the Associate Safety Professional/Certified Safety Professional examinations.

Asbestosis and Asbestos-Related Illnesses

Asbestos is a naturally occurring mineral used for high-temperature areas, decorative treatments, and binding of materials. The inhalation of the fibers can result in asbestos, cancer, and mesothelioma.[3]

Asbestosis is respiratory disease, caused by the inhalation of asbestos fibers. The fibers, when inhaled, reach the alveoli, where they create a "scarring" of the lung tissue. As you recall in the section Inhalation earlier, gas exchanges between the respiratory and circulatory system are made in the alveoli. The alveoli become scarred and prohibit the normal exchange of oxygen and carbon dioxide. The severity of the disease depends entirely on how long the person was exposed to asbestos and in what concentrations. There is a strong correlation between asbestos exposure and smoking to the development of lung cancer. There are no known cures for asbestos. However, stopping the exposure to asbestos may slow or stop the progression. The latency period for asbestosis is 10 to 20 years. The latency period for asbestos-related cancer is 20 to 30 years. *Mesothelioma* is an asbestos-related cancer of the lining of the lung (pleura) or the lining of the abdominal cavity (peritoneum). Mesothelioma is a very rare form of cancer. Mesothelioma is incurable, but several treatments are available including surgery, chemotherapy, and radiation.

The Occupational Safety and Health Administration (OSHA) permissible exposure limit (PEL) for asbestos is 0.1 fibers/cm^3. The method of sampling for asbestos uses a 0.8-micron mixed-cellulose ester filter (MCEF) and phase contrast microscopy. When testing for airborne concentrations in schools (K5–Grade 12), the Asbestos Hazard Emergency Response Act applies and requires that clearance monitoring be performed using 0.45-micron MCEF cassettes, with a 2-inch cowl and transmission electron microscopy. The OSHA standard for asbestos is 29 CFR 1910.1001.

Brucellosis

Brucellosis, which is also called Bang's disease, Gibraltar fever, Malta fever, Maltese fever, Mediterranean fever, rock fever, or undulant fever, is a highly contagious zoonosis caused by ingestion of unsterilized milk or meat from infected animals or close contact with their secretions. Most cases are caused by occupational exposure to infected animals or the ingestion of unpasteurized dairy products. The latency period for brucellosis is 3 to 60 days.[4]

Benzene-Related Illnesses

Benzene is an aromatic hydrocarbon that is produced by the burning of natural products. It is a component of products derived from coal and petroleum and is found in gasoline and other fuels. Benzene is used in the manufacture of plastics, detergents, pesticides, and other chemicals. Research has shown benzene to be a carcinogen (cancer causing). With exposures from less than 5 years to more than 30 years, individuals have developed, and died from, leukemia. Long-term exposure may affect bone marrow and blood production. Long-term exposure to high levels of benzene in the air can cause leukemia, particularly acute myelogenous leukemia. Benzene has been determined by the US Department of Health and Human Services to be a known carcinogen. Short-term exposure to high levels of benzene can cause drowsiness, dizziness, unconsciousness, and death.[5]

The OSHA 8-h PEL is 10 ppm, with a ceiling limit of 25 ppm for a 10-min period. The maximum peak concentration for an 8-h shift is 50 ppm. The OSHA standard for benzene is 29 CFR 1910.1028.

Byssinosis

Byssinosis is a disease of the lungs as a result of breathing in cotton dust or dusts from other vegetable fibers such as flax, hemp, or sisal while at work. Byssinosis is most common in people who work in the textile industry. Those who are sensitive to the dust can have an asthma-like condition after being exposed. In those with asthma, being exposed to the dust makes breathing more difficult, but in byssinosis, the symptoms usually go away by the end of the workweek. After long periods of exposure, symptoms can continue throughout the week without improving.

Methods of prevention in the United States have reduced the number of cases, but byssinosis is still common in developing countries. Smoking increases the risk for this disease. Being exposed to the dust many times can lead to chronic lung disease and shortness of breath or wheezing.[6]

The OSHA PEL for cotton dust exposure is 1.0 mg/m³. The OSHA standard is 29 CFR 1910.1043.

Arsenic-Related Illnesses

Chronic exposure to arsenic in the occupational setting, skin lesions, and peripheral neuropathy are the most common adverse effects. Patchy hyperpigmentation is the classic skin lesion of chronic arsenic exposure. Other adverse effects include anemia, leukopenia, thrombocytopenia, eosinophilia, and liver injury. Arsenic exposure occurs when workers sand or burn this wood. Arsenic is used as an alloy in lead-acid batteries. Inorganic arsenic is no longer used in agriculture in the United States. Organic arsenic pesticides (cacodylic acid, disodium methyl arsenate, and monosodium methyl arsenate) are used on cotton.[7] The high-risk occupations at greatest risk of arsenic poisoning include the following:

- Applying arsenic preservatives to wood
- Manufacturing of pesticides containing arsenic
- Sawing or sanding arsenic-treated wood
- Smelting or casting lead
- Smelting or refining of zinc or copper

The current OSHA PEL for arsenic is 10 μg/m³ (8 h time-weighted average [TWA]). The National Institute of Occupational Safety and Health (NIOSH) 15-min limit for arsenic (airborne concentration) is 2 μg/m³.

Berylliosis and Beryllium-Related Illnesses

Berylliosis is a lung disease resulting from the inhalation of beryllium. All compounds of beryllium should be considered potentially harmful, if inhaled, even in minute quantities. The current OSHA PEL for beryllium exposure is 2.0 μg/m³ (8 h TWA). It has a ceiling limit of 5 μg/m³ not to be exceeded for more than 30 min and never to exceed a limit of 25 μg/m³. Beryllium is used in lightweight metals, especially in valves in the aeronautical and space industries. Chronic beryllium disease (CBD) may occur among people who are exposed to the dust or fumes from beryllium metal, metal oxides, alloys, ceramics, or salts. It occurs when people inhale beryllium in these forms. CBD usually has a very slow onset, and even very small amounts of exposure to beryllium can cause the disease in some people. Acute beryllium disease usually has a quick onset and has symptoms that resemble those of pneumonia or bronchitis. The acute form of the disease is believed to occur as a result of exposures well above the current PEL. This form of beryllium disease is now rare.

Recent data suggest that exposures to beryllium even at levels below the 2-μg/m^3 PEL may have caused CBD in some workers. Therefore, employers should consider providing their beryllium-exposed workers with air-purifying respirators equipped with 100-series filters (either N-, P-, or R-type) or, where appropriate, powered air-purifying respirators equipped with HEPA filters, particularly in areas where material containing beryllium can become airborne.[8]

Copper-Related Illnesses

The primary methods of entry into the body for occupational copper exposure are through inhalation or absorption through the eyes or skin. Short-term exposure to copper fumes causes irritation of the eyes, nose, and throat, and a flu-like illness called *metal fume fever* can result. Metal fume fever symptoms include fever, muscle aches, nausea, chills, dry throat, cough, and weakness. It may also include a metallic or sweet taste in the mouth of the affected employee. Long-term exposure to copper fumes owing to repeated exposure may cause the skin and hair to change color.[9]

The OSHA PEL for copper fume is 0.1 mg/m^3 (8 h TWA). The OSHA PEL for copper dust and mist is 1.0 mg/m^3. Initial medical screening for employees potentially exposed to copper should include a thorough history of chronic respiratory diseases and *Wilson's disease* (hepatolenticular degeneration).

Cadmium-Related Illnesses

Acute exposure to cadmium can cause delayed pulmonary edema and acute renal failure after inhalation of high concentrations of the fume. Chronic exposure to cadmium dust and fume is toxic primarily to the kidneys with secondary effects on the bones (osteomalacia). Biological monitoring of cadmium-exposed workers is mandated by OSHA to prevent chronic renal disease. Studies have shown the latency period to be approximately 10 years.

High-risk occupational exposures are due to the following[10]:

- Brazing using cadmium-based solder
- Machining or welding on cadmium-alloyed or cadmium-plated steel
- Making cadmium-containing products
- Plating metal with cadmium
- Reclaiming scrap metals containing lead, cadmium, beryllium, and mercury
- Removing cadmium coatings
- Smelting or casting lead
- Smelting or refining zinc and copper

The OSHA PEL for an 8-h shift is 5 µg/m^3. The *action level* established by OSHA is 2.5 µg/m^3. The OSHA standard for cadmium is 29 CFR 1910.1027.

Chromium-Related Illnesses

Chronic exposure to chromium in the occupational environment most often results in contact dermatitis and ulcerations of the skin and nasal mucosa. Hexavalent chromium is the most toxic form of chromium. Inhalation of chromium dust, mist, or fumes can produce acute bronchoconstriction, probably through a direct irritant mechanism. Industrial uses of hexavalent chromium compounds include chromate pigments in dyes, paints, inks, and plastics; chromates added as anticorrosive agents to paints, primers, and other surface coatings; and chromic acid electroplated onto metal parts to provide a decorative or protective coating. Hexavalent chromium can also be formed when performing "hot work" such as welding on stainless steel or melting chromium metal. In these situations, the chromium is not originally hexavalent, but the high temperatures involved in the process result in oxidation that converts the chromium to a hexavalent state.[11]

High-risk occupational exposures to chromium are due to the following:

- Welding stainless steel
- Heating or machining chromium alloys
- Mining or crushing chromium ores
- Plating metal with chromium
- Producing chromium alloys or chromate pigments
- Removing chromate-containing paints by abrasive blasting
- Spraying and sanding chromate-containing paints
- Using chromates or dichromates in printing
- Using chromates or dichromates in tanning
- Working as cement floorer with exposure to chromate salts

The OSHA PEL for hexavalent chromium is 5 µg/m^3 for an 8-h TWA.

Coal Dust–Related Illnesses

Coal workers' pneumoconiosis (also known as black lung disease) is a lung disease caused by inhaling coal mine dust. Although some miners never develop the disease, others may develop the early signs after less than 10 years of mining experience. According to recent studies by the NIOSH, about 1 of every 20 miners participating in our program has x-ray evidence of some pneumoconiosis.

In its early stages, called simple pneumoconiosis, the disease may not prevent an employee from working or carrying on most normal activities. In some miners, the disease never becomes more severe. In other miners, the disease progresses from simple to complicated pneumoconiosis, a condition also called progressive massive fibrosis (PMF).

Unfortunately, there is no cure for the damage that the dust has already done to the lungs. However, preventing coal workers' pneumoconiosis is among the highest priorities in protecting the health of the coal miner.[12]

The current OSHA PEL for the respirable fraction of coal dust (less than 5% silica) is 2.4 mg/m^3 TWA concentration.

Cobalt-Related Illnesses

Acute Exposure

Acute exposure to cobalt metal, dust, and fume is characterized by irritation of the eyes and, to a lesser extent, irritation of the skin. In sensitized individuals, exposure causes an asthma-like attack, with wheezing, bronchospasm, and dyspnea. Ingestion of cobalt may cause nausea, vomiting, diarrhea, and a sensation of hotness.

Chronic Exposure

Chronic exposure to cobalt metal, dust, or fume may cause respiratory or dermatologic signs and symptoms. After skin sensitization, contact with cobalt causes eruptions of dermatitis in creases and on frictional surfaces of the arms, legs, and neck. After sensitization of the respiratory system, cobalt exposure causes an obstructive lung disease with wheezing, cough, and shortness of breath. Chronic respiratory exposure results in reduced lung function, increased fibrotic changes on chest x-ray, production of scanty mucoid sputum, and shortness of breath. Chronic cobalt poisoning may cause polycythemia, hyperplasia of the bone marrow and thyroid gland, pericardial effusion, and damage to the alpha cells of the pancreas. High-risk occupational exposures to cobalt are as follows[13]:

- Machining cobalt alloys
- Manufacturing cemented carbide materials or tools
- Using cemented carbide materials or tools

The current OSHA PEL for cobalt metal, dust, and fume (as Co) is 0.1 mg/m^3 of air as an 8-h TWA concentration (29 CFR 1910.1000, Table Z-1).

Formaldehyde-Related Illnesses

Studies indicate that formaldehyde is a potential human carcinogen. Formaldehyde is a colorless, strong-smelling gas. Commonly used as a preservative in medical laboratories and mortuaries, formaldehyde is also found in other products such as chemicals, particle board, household products, glues, permanent press fabrics, paper product coatings, fiberboard, and plywood. It is also widely used as an industrial fungicide, germicide, and disinfectant.

Workers can inhale formaldehyde as a gas or vapor or absorb it through the skin as a liquid. They can be exposed during the treatment of textiles and the production of resins. Besides health care professionals and medical laboratory technicians, groups at potentially high risk include mortuary employees as well as teachers and students who handle biological specimens preserved with formaldehyde or formalin.[14]

The PEL for formaldehyde in all workplaces (including general industry, construction, and maritime, but not in agriculture) covered by the OSH Act is 0.75 ppm measured as an 8-h TWA. The standard includes a 2-ppm short-term exposure limit (i.e., maximum exposure allowed during a 15-min period). The "action level" is 0.5 ppm measured over 8 h. The OSHA standard for formaldehyde is 29 CFR 1910.1048.

Lead-Related Illnesses

Workers are exposed to lead by breathing in lead dust or fumes from work activities, or particles of lead may be swallowed from eating, drinking, or smoking in lead-contaminated areas. Workers may also take home lead dust on their skin, hair, work clothes, and boots. When lead enters the body, it can be stored in bones and tissues for a very long time. This means that lead can build up in the body and cause harm in the long term. Early effects of lead poisoning may be hard to detect. However, employer training and awareness programs can help workers recognize symptoms and seek prompt medical attention.[15]

Health effects related to occupational exposure to lead are as follows:

- Brain disorders
- Anemia
- Nerve disorders
- High blood pressure
- Kidney disorders
- Reproductive disorders
- Decreased red blood cells
- Slowed reflexes

The OSHA 8-h PEL is 50 $\mu g/m^3$. For those exposed to air concentrations at or above the action level of 30 $\mu g/m^3$ for more than 30 days per year, OSHA mandates periodic determination of blood lead levels. If a blood lead level is found to be greater than 40 $\mu g/dL$, the worker must be notified in writing and provided with a medical examination. If a worker's one-time blood lead level reaches 60 µg/dL (or averages 50 µg/dL or more on three or more tests), the employer is obligated to remove the employee from excessive exposure, with maintenance of seniority and pay, until the employee's blood lead level falls below 40 µg/dL. The OSHA standard for occupational exposure to lead is 29 CFR 1910.1025. OSHA's lead standard can be located in 29 CFR 1910.1025.

Manganese-Related Illnesses

Workers exposed to Mn have higher average urine and blood levels than unexposed workers, but these tests will not predict disease severity in a particular case. Similarly, abnormalities in the globus pallidus of the brain by magnetic resonance imaging (MRI) study reflect recent exposure to Mn but are not diagnostic of disease. Neuropsychological testing can identify workers with advanced disease. Symptoms first appear after a chronic exposure of at least several months in duration:

1. *Prodromal Phase*: Evidence of cognitive dysfunction and emotional disturbance begin to occur prior to severe motor and neurological dysfunction; symptoms may include fatigue, anorexia, muscle pain, nervousness, irritability, violent outbursts, insomnia, decreased libido, and labile affect; workers may also report headache, hypersomnia, spasms, weariness of the legs, and arthralgias.

2. *Intermediate Phase*: Compulsive uncontrollable laughter or crying, clumsiness of movement, exaggeration of reflexes in lower limbs, speech disorders, visual hallucinations, excessive sweating, excessive salivation, and confusion.

3. *Established Phase*: Muscular weakness, difficulty in walking, stiffness, impaired speech, mask-like face, increased muscular tone, slow and shuffling gait, micrographia, and resting tremors; the syndrome is indistinguishable from Parkinson's disease; Parkinsonism caused by manganese has the following that distinguish it from Parkinson's disease: psychiatric symptoms early in the disease, cock walk, tendency to fall backward, less frequent resting tremor, more frequent dystonia, lack of response to dopaminomimetics, abnormal MRI showing manganese in the pallidum and caudate nucleus, and normal fluorodopa positron emission tomography scan. Characteristic findings of manganese poisoning are poor steadiness of hands, difficulty doing rapid alternating movements, muscular rigidity, and postural instability. Urine and plasma Mn levels are ineffective biomarkers. Blood Mn "may reflect average elevated daily air Mn levels when exposures are on-going and relatively consistent."[16]

High-risk occupational exposures to manganese are as follows:

- Machining manganese alloys
- Mining or crushing manganese ores
- Using ferromanganese in alloy production
- Using manganese-containing welding rods

The OSHA 8-h PEL for manganese (dust) is 5 mg/m^3.

Mercury-Related Illnesses

Mercury can cause peripheral neuropathy and neuropsychiatric disorders after chronic exposure. Inhalation of elemental mercury vapor is the most common exposure route leading to occupational mercury poisoning. The key to preventing chronic mercury poisoning is to reduce spills and to clean up ones that occur. Other forms of mercury poisoning follow ingestion of inorganic mercury and organic mercury compounds.

High-risk occupational exposures to mercury include the following[17]:

- Extracting mercury ore
- Fabricating measuring devices containing mercury
- Manufacturing and repairing mercury fluorescent lights
- Manufacturing or using mercury dental amalgams
- Reclaiming scrap metal containing lead, cadmium, beryllium, or mercury
- Using mercury to extract gold
- Working in the mercury cell room in a chloralkali plant

The OSHA PEL for an 8-h TWA is 0.01 mg/m^3, with a ceiling limit of 0.04 mg/m^3.

Pneumoconiosis

Pneumoconiosis is a group of interstitial lung diseases caused by the inhalation of certain dusts and the lung tissue's reaction to the dust. The principal cause of the pneumoconioses is workplace exposure. The primary pneumoconioses are asbestosis, silicosis, and coal workers' pneumoconiosis. As their names imply, they are caused by inhalation of asbestos fibers, silica dust, and coal mine dust. We have already discussed asbestos-related illnesses. Therefore, we will discuss the remaining two in the following.[18]

Silica-Related Illnesses

As mentioned, silicosis is a pneumoconiosis illness. Complicated silicosis refers to the development of PMF in which pulmonary nodules coalesce into larger conglomerations. PMF has been observed to develop in a small percentage of patients with simple chronic silicosis and a large percentage of patients with accelerated silicosis. Accelerated silicosis is similar to simple silicosis, but the disease develops after a heavier exposure over a shorter period. The latency is 2 to 5 years for accelerated silicosis, versus greater than 10 years for simple silicosis. Accelerated silicosis follows a course of increasing dyspnea. Silicosis is associated with an increased risk of autoimmune diseases.[19]

High-risk occupational exposures to silica are as follows:

- Blasting, drilling, removing, or crushing rock, concrete, or brick
- Grinding or cutting tiles, stones, concrete, bricks, or terrazzo
- Loading or dumping dusty rock, stone, or sand
- Making products from silica powder/stone or other fibrogenic minerals
- Use abrasives containing silica or silicon carbide
- Use handheld saw or grinder to remove brick mortar

The OSHA 8-h PEL is calculated as follows:

$$PEL_{silica} = \frac{10 \text{ mg/m}^3}{\%SiO_2 + 2},$$

where $\%SiO_2$ = the percentage of silica in the batch.

Zinc-Related Illnesses

Zinc is a metal that is normally found in small amounts in nature. It is used in many commercial industries and can be released into the environment during mining and smelting (metal processing) activities. People living near smelters or industries using zinc could be exposed to higher levels of zinc by drinking water, breathing air, and touching soil that contains the metal. Exposure to high levels of zinc over long periods may cause adverse health effects.

A short-term illness called *metal fume fever* can result if workers breathe very high levels of zinc dust or fumes. This condition, which usually lasts from 24 to 48 h, causes chills, fever, excessive sweating, and weakness. Long-term effects of breathing zinc dust or fumes are not known.[20]

The current OSHA PEL for zinc oxide is 15 mg/m^3 of air for total dust, and 5 mg/m^3 for the respirable fraction as an 8-h TWA concentration (29 CFR 1910.1000, Table Z-1).

Aluminum-Related Illnesses

Aluminum dust is an eye and respiratory tract irritant in humans. Soluble aluminum salts are irritants when inhaled as aerosols. Although inhalation of aluminum powder of particle size 1.2 um, given over 10- or 20-min periods several times weekly, resulted in no adverse health effects among thousands of workers over several years, several other studies report x-ray evidence of pulmonary fibrosis. Some patients on long-term hemodialysis develop speech disorders, dementia, or convulsions. This syndrome is associated with increased concentration of aluminum in serum, brain, muscle, and bone. There is some evidence that Alzheimer's disease may be linked to aluminum content in the body. Analysis of the aluminum content in the brains of persons dying from Alzheimer's have shown increased levels, although brain aluminum levels vary greatly. A second correlating factor is that neurofibrillary tangles (NFTs) have been identified in both aluminum encephalopathy and in Alzheimer's disease. However, it has been shown that the NFTs produced by the two conditions are structurally and chemically different and that NFTs are present in several other neurological disorders. It appears that the aluminum content of the brain is less an issue relating to exposure to aluminum than an issue of a blood–brain barrier defect or compromise of some kind.[21]

High-risk occupational exposures to aluminum are as follows:

- The processing and transportation of aluminum
- Melting or soldering of electrical transmission lines
- Used in the construction, manufacturing, explosives, petrochemical, and paper industries
- Used in desalinization, cryogenic technology, permanent magnets, and as a substitute for copper
- Used in testing for gold, arsenic, and mercury
- Used in sugar refining, alloying metals, as a chemical intermediate, and in containers for fissionable reactor fuels

The current OSHA PEL for aluminum is 15 mg/m^3 of air for total dust and 5 mg/m^3 for the respirable fraction, as an 8-h TWA concentration (29 CFR 1910.1000, Table Z-1).

Antimony-Related Illnesses

Antimony is a silvery-white metal that is found in the earth's crust. Antimony ores are mined and then mixed with other metals to form antimony alloys or combined with oxygen to form antimony oxide.

Little antimony is currently mined in the United States. It is brought into this country from other countries for processing. However, there are companies in the United States that produce antimony as a by-product of smelting lead and other metals.

Antimony isn't used alone because it breaks easily, but when mixed into alloys, it is used in lead storage batteries, solder, sheet and pipe metal, bearings, castings, and pewter. Antimony oxide is added to textiles and plastics to prevent them from catching fire. It is also used in paints, ceramics, and fireworks, and as enamels for plastics, metal, and glass.

Exposure to antimony at high levels can result in a variety of adverse health effects. Breathing high levels for a long time can irritate your eyes and lungs and can cause heart and lung problems, stomach pain, diarrhea, vomiting, and stomach ulcers.

Ingesting large doses of antimony can cause vomiting. We don't know what other effects may be caused by ingesting it. Long-term animal studies have reported liver damage and blood changes when animals ingested antimony. Antimony can irritate the skin if it is left on it. Antimony can have beneficial effects when used for medical reasons. It has been used as a medicine to treat people infected with parasites.[22]

The OSHA has set an occupational exposure limit of 0.5 milligrams of antimony per cubic meter of air (0.5 mg/m^3) for an 8-h workday, 40-h workweek.

Exposure to antimony occurs in the workplace or from skin contact with soil at hazardous waste sites. Breathing high levels of antimony for a long time can irritate the eyes and lungs and can cause problems with the lungs, heart, and stomach.

Dust-Related Illnesses

Excessive dust in the workplace can be highly dangerous on a number of levels. First, although rare, a cloud of concentrated dust is potentially combustible and can, therefore, cause explosions so it's important that companies keep their working environments as relatively dust free as possible, so they can avoid such potential catastrophes. However, the most common problem associated with dust in the workplace arises from dust-related illnesses, which have been one of the major occupational diseases identified.

Common Environments for Contracting Dust-Related Illnesses

All workplaces need to carry out cleaning duties and pay particular regard to hygiene issues and, for the most part, in places such as an office for example, dust should not present too much of a problem. However, there are many industries that need to be especially vigilant. Here is a list of some of the more common working environments where excess dust can create a real problem.

- Mines and quarries—dust from coal, flint and silica
- Construction sites—dust from cement and asbestos
- Farming and agriculture—dust from grain

- Carpentry and joinery—dust from wood
- Bakeries and mills—dust from flour
- Textiles—dust from materials like leather

Workers can suffer from a variety of illnesses and medical conditions as a result of working in dust-filled environments. Depending on the nature of the work, some of these ailments can become more serious than others. The range of dust-related industries and conditions encompass eye and nose damage, rashes, and other conditions, asthma, silicosis, asbestosis, mesothelioma, and lung cancer related to asbestos. Pneumoconiosis, which is the name given to diseases such as those caused by the likes of asbestosis and silicosis, is a broad term that describes any condition that affects the lungs causing inflammation or scarring of the lung tissue. One of the major worries is that it can often take several decades for a person to develop any symptoms of pneumoconiosis, which can manifest itself in things like excess coughing, breathing difficulties, and even weight loss.

The current OSHA 8-h PEL for PONR (Particulates Otherwise Not Regulated) respirable fraction is 5.0 mg/m^3 and 15 mg/m^3 for total particulates.

Thallium-Related Illnesses

Pure thallium is a bluish-white metal that is found in trace amounts in the earth's crust. In the past, thallium was obtained as a by-product from smelting other metals; however, it has not been produced in the United States since 1984. Currently, all the thallium is obtained from imports and from thallium reserves.

In its pure form, thallium is odorless and tasteless. It can also be found combined with other substances such as bromine, chlorine, fluorine, and iodine. When it's combined, it appears colorless-to-white or yellow.

Thallium is used mostly in manufacturing electronic devices, switches, and closures, primarily for the semiconductor industry. It also has limited use in the manufacture of special glass and for certain medical procedures.

The current OSHA 8-h PEL for thallium is 0.1 mg/m^3. Thallium is also designated by OSHA as a skin hazard.[23]

Pesticide-Related Illnesses

The organophosphate and *N*-methyl carbamate insecticides cause accumulation of acetylcholine at nerve endings by poisoning the acetylcholinesterase enzyme. In carbamate poisoning, the inhibition of the enzyme is rapidly reversible, and the workers are often improved by the time of arrival at the clinic or emergency room. Organophosphates can irreversibly bind to the enzyme so that normal enzyme activity can only be restored after the cells synthesize new acetylcholinesterase. This process takes up to 60 days. The primary route of occupational exposure is through the skin. Some of the health effects associated with pesticides are diarrhea, urination, miosis, bradycardia, emesis, lacrimation and salivation, secretions, and sweating. The primary route of occupational exposure is through the skin.[24]

High-risk occupational exposures to pesticides are as follows:

- Application of organophosphates or work in fields after an application
- Work with toxic chemicals that could be spilled or released

The current OSHA 8-h TWA PELs for pesticides are chemical specific and are available on the MSDS or in 29 CFR 1910.1000 Table Z-1.

Key Information to Remember on Toxicology

1. Ames testing is a procedure to determine whether or not a chemical is a mutagen.
2. Aluminum used in bauxite ore can cause lung cancer, emphysema, and pneumoconiosis.
3. Asbestos used as insulation and numerous other products can cause asbestosis, lung cancer, and mesothelioma.
4. Arsenic from abrasive blasting can cause lung cancer or hemoglobinuria.
5. Benzene can cause granulocytic leukemia or myelogenous leukemia.
6. Beryllium from ore processing can cause CBD or berylliosis.
7. Cadmium from abrasive blasting can cause renal damage.
8. Creosote coal tar from wood preservatives causes lung cancer.
9. Chromium exposure can lead to the development of lung cancer.
10. Cobalt causes hard metal disease or lung cancer.
11. Nickel causes lung and nasopharynx cancer.

References

1. Plog, B. A., 2001. *Fundamentals of Industrial Hygiene*, 5th Edition, National Safety Council, Itasca, IL, p. 45.
2. Nucleus Medical Art, 2009. *Si555509999 Respiratory System*. Available at http://64.143.176.9/library/healthguide/en-us/support/topic.asp?hwid=ug2841.

3. National Institute of Health, 2010. Available at http://hazmap.nlm.nih.gov/cgi-bin/hazmap_generic?tbl=TblDiseases&id=1.
4. National Institute of Health, 2010. Available at http://hazmap.nlm.nih.gov/cgi-bin/hazmap_generic?tbl=TblDiseases&id=31.
5. National Institute of Health, 2010. Available at http://hazmap.nlm.nih.gov/cgi-bin/hazmap_generic?tbl=TblAgents&id=5.
6. National Institute of Health, 2010. Available at http://hazmap.nlm.nih.gov/cgi-bin/hazmap_generic?tbl=TblDiseases&id=255.
7. National Institute of Health, 2010. Available at http://hazmap.nlm.nih.gov/cgi-bin/hazmap_generic?tbl=TblDiseases&id=36.
8. National Institute of Health, 2010. Available at http://hazmap.nlm.nih.gov/cgi-bin/hazmap_generic?tbl=TblDiseases&id=26.
9. National Institute of Health, 2010. Available at http://hazmap.nlm.nih.gov/cgi-bin/hazmap_generic?tbl=TblAgents&id=374.
10. National Institute of Health, 2010. Available at http://hazmap.nlm.nih.gov/cgi-bin/hazmap_generic?tbl=TblDiseases&id=6.
11. National Institute of Health, 2010. Available at http://hazmap.nlm.nih.gov/cgi-bin/hazmap_generic?tbl=TblDiseases&id=17.
12. National Institute of Health, 2010. Available at http://hazmap.nlm.nih.gov/cgi-bin/hazmap_generic?tbl=TblDiseases&id=117.
13. National Institute of Health, 2010. Available at http://hazmap.nlm.nih.gov/cgi-bin/hazmap_generic?tbl=TblAgents&id=37.
14. National Institute of Health, 2010. Available at http://hazmap.nlm.nih.gov/cgi-bin/hazmap_generic?tbl=TblAgents&id=271.
15. National Institute of Health, 2010. Available at http://hazmap.nlm.nih.gov/cgi-bin/hazmap_generic?tbl=TblAgents&id=10.
16. National Institute of Health, 2010. Available at http://hazmap.nlm.nih.gov/cgi-bin/hazmap_generic?tbl=TblAgents&id=39.
17. National Institute of Health, 2010. Available at http://hazmap.nlm.nih.gov/cgi-bin/hazmap_generic?tbl=TblDiseases&id=13.
18. National Institute of Health, 2010. Available at http://hazmap.nlm.nih.gov/cgi-bin/hazmap_generic?tbl=TblDiseases&id=552.
19. National Institute of Health, 2010. Available at http://hazmap.nlm.nih.gov/cgi-bin/hazmap_generic?tbl=TblDiseases&id=260.
20. National Institute of Health, 2010. Available at http://hazmap.nlm.nih.gov/cgi-bin/hazmap_generic?tbl=TblDiseases&id=69.
21. National Institute of Health, 2010. Available at http://hazmap.nlm.nih.gov/cgi-bin/hazmap_generic?tbl=TblAgents&id=287.
22. National Institute of Health, 2010. Available at http://hazmap.nlm.nih.gov/cgi-bin/hazmap_generic?tbl=TblAgents&id=58.
23. National Institute of Health, 2010. Available at http://hazmap.nlm.nih.gov/cgi-bin/hazmap_generic?tbl=TblAgents&id=65.
24. National Institute of Health, 2010. Available at http://hazmap.nlm.nih.gov/cgi-bin/hazmap_generic?tbl=TblDiseases&id=66.

6

Industrial Hygiene Air Sampling

Industrial hygiene is the science and art dedicated to the *anticipation, recognition, evaluation,* and *control* of workplace hazards that may cause worker injuries or illnesses. Today's industrial hygienist is a person trained and educated in physics, biology, chemistry, safety, engineering, or environmental sciences. In addition, a Certified Industrial Hygienist (CIH) is someone with the requisite education, who has applied to the American Board of Industrial Hygiene (ABIH) and who has successfully completed an examination. The industrial hygienist must have a broad spectrum of knowledge in safety, ventilation, illumination, radiation, biological sciences, noise, ergonomics, thermal stressors, and a wide variety of other topics. Each of these topics will be discussed in detail within their own individual chapters. The industrial hygienist, while concerned with physical safety, focuses on the prevention of worker illnesses caused by workplace exposures to various chemicals, substances, and elements.

Anticipation of Hazards

The industrial hygienist must be capable of anticipating workplace hazards. Before ever arriving at a job site or facility, the industrial hygienist can perform preliminary research to determine the type of work and potential exposures that may be present. This can be performed by asking and answering the following questions:

- What type of facility or site is it?
- What types of operations or processes are being performed?
- What hazardous materials are present or likely to be present?
- Are facility schematics or drawings available for review?
- What standard operating procedures are available?
- Are MSDSs available for review?

Recognition of Hazards

This phase of the assessment process is conducted primarily through walkthrough inspections by the industrial hygienist. This process is more qualitative than quantitative in nature. During the walkthrough inspection, the industrial hygienist will begin to identify the specific processes and operations and the potential hazards that each may pose to workers. Once the walkthrough inspection has been completed, the industrial hygienist will develop a strategy to further evaluate the extent each hazard poses.

Evaluation of Hazards

The evaluation process may require the use of specialized sampling equipment to actually quantify the workers' exposure to the specific hazards. Depending on the type of hazard, representative samples will be collected using various methodologies that will be discussed in this chapter. Once samples have been analyzed by an accredited laboratory, then the industrial hygienist will formulate a plan to control the hazards.

Control of Hazards

Once the plan to control the hazards has been established, the industrial hygienist will develop a plan of action to eliminate, minimize, or mitigate the hazards, which may include engineering controls, administrative controls, or personal protective equipment. Through the elimination or mitigation of a hazard, the risk to the employee can be drastically reduced.

Definitions

Vapors: A gaseous form of a substance that is normally a solid or liquid at room temperature.

Gases: Substances that completely occupy a space and can be converted to a liquid or solid by increasing or decreasing temperature.

Vapor pressure: The pressure that a vapor at equilibrium with a pure liquid at a given temperature exerts to the surrounding atmosphere.

Particulate: Fine solid or liquid particles, such as dust, fog, mist, smoke, or spray.

Dust: Solid particles generated by mechanical action (crushing, grinding, etc.). Size ranges are usually from 0.1 to 30 μm.

Fume: Airborne solid particles formed by condensation of vapor (i.e., welding fumes). Size ranges from 0.001 to 1.0 μm.

Mist: Suspended liquid droplets generated by condensation or atomization (fogs are formed by condensation). Size ranges from 0.01 to 10 μm.

Fibers: Particulate with an aspect ratio (length to width) of 3:1.

Air Sampling

There is a wide variety of industrial hygiene sampling equipment available for air sampling. There is such a variety that each and every one cannot be discussed. In this chapter, we will discuss the type and applications of the most commonly used equipment and media.

There are several reasons why air sampling and testing are conducted. They may include the following:

- To determine compliance with regulations
- To assess worker exposure to determine if personal protective equipment being used is adequate
- To monitor implemented control measures
- To evaluate contaminant emissions
- Documentation for legal reasons

The types of air sampling that can be performed include the following:

- *Grab Sampling*: It can be performed using direct-reading instruments or a known volume of air collected in a container, such as a Tedlar Bag, and sent to a laboratory.
- *Personal Sampling*: An employee wears a sampling device, such as a pump, which draws air across a filter media for sample collection, which is sent to a laboratory. This is the preferred method of sampling when evaluating worker exposure.

- *Area Sampling*: It uses the same type of media as in personal sampling; however, the sampling device and media are stationary inside a room.
- *Integrated Sampling*: Integrated sampling is collecting one or more personal air samples in order to estimate the workers' 8-h time-weighted average exposure.
- *Direct Reading*: Direct-reading instruments are calibrated for specific contaminants. Air samples are drawn into the direct-reading instrument and analyzed by several methods, including, but not limited to, fiber optics, photo ionization detection, or flame ionization detection methods.[1]

Sampling Methodology

Sampling methodologies are most frequently determined by the National Institute of Occupational Safety and Health (*NIOSH Manual of Analytical Methods*), which is available in hard copy or can be located at http://www.cdc .gov/niosh/docs/2003-154/. Another source to use when making preliminary plans to conduct air sampling is the analytical laboratory itself. The selection of the laboratory is a critical step in the sampling process. It is highly recommended that you use an American Industrial Hygiene Association (AIHA)–accredited laboratory. By using an AIHA-accredited laboratory, you are assured of the laboratory quality assurance/quality control and ultimately the validity of the sample analysis. The AIHA laboratory director can also serve as a great resource when attempting to determine the methodology of choice for your particular work environment. One other source to refer to regarding sampling methodologies is the Occupational Safety and Health Administration (OSHA) Index of Analytical Methods available at the following Internet address: http://www.osha.gov/dts/sltc/methods/toc.html.

Equipment Selection

The methodology chosen for the particular sampling dictates the type of equipment used in the collection of air samples. The equipment of choice is based on the methodology and the contaminant that you are sampling for. Is the contaminant a gas, vapor, mist, fume, or particulate? Are you sampling for respirable or total portions of the sample? These are questions that need to be determined prior to sampling. We will now discuss the types of sampling equipment used in air sampling.

Air Sampling Pumps

Air sampling pumps are one of the most common types of equipment used to collect both personal and area samples. They can be classified as either low-volume or high-volume sampling pumps, or personal or area sampling pumps, respectively. Low-volume air sampling pumps can range from 100 to 5000 cc/m (0.1–5 lpm), whereas most high-volume air sampling pumps have an adjustable flow rate of 5000 to 30,000 cc/m (5–30 lpm). Low-volume pumps are typically used to collect personal samples, while high-volume air sampling pumps are used to collect area samples. Both types of air sampling pumps draw volumes at known flow rates through a filter/media through means of a pump. Some of the air sampling pumps used in industrial hygiene sampling have a "critical orifice," which is a hole of specific measurement drilled into a plate to create a constant flow rate.

Piston and Bellows Air Pumps

A piston-type pump uses a handheld pump similar to the operation of a bicycle pump to draw a known volume of air across an absorbent tube designed for specific contaminants. A bellows-type pump is used for the same purposes as the piston-type pump. Both piston and bellows pumps are used primarily for screening purposes during walkthrough surveys of a facility to make an initial assessment of potential exposures.

Direct-Reading Instruments

Direct-reading instruments are valuable tools for detecting and measuring worker exposure to gases, vapors, aerosols, and fine particulates suspended in air. These instruments permit real-time or near real-time measurements, and their use is specifically required by some OSHA standards. There are many types of instruments available, each of which is designed for a specific monitoring purpose. Proper operation of direct-reading instruments is essential to ensure that accurate information is obtained when evaluating air contaminants.[1]

One type of direct-reading instrument is the multi-gas meter used in confined space air sampling prior to and during entry. The multi-gas meter uses various sensors installed inside the meter and may include oxygen, sulfur dioxide, carbon monoxide, carbon dioxide, hydrogen sulfide, and combustible gases. It is imperative that these meters be zeroed and calibrated prior to each use to insure the accuracy of the meter. The manufacturer's operating instructions should be read, understood, and followed. Operators of multi-gas meters should be thoroughly trained. All too often, operators have a 5 or 10-min instruction period and are expected to understand the operation and interpretation of data. As a safety professional, part of your responsibility is to insure that adequate training is performed.

NOTE: When working in a potentially explosive environment, insure that all sampling equipment is classified by the manufacturer as "intrinsically safe" prior to using.

Cyclones

On occasion, when sampling, various apparatus are to be used in conjunction with the air sampling pumps. To be discussed in this section are cyclones. Cyclones are devices that discard larger particles. They are used in conjunction with sampling filters and cassettes and air sampling pumps. An example of their use is when collecting samples for the respirable fraction of airborne particulates. The cyclone will discard particulates larger than 4.0 μm in diameter and deposit the respirable fraction onto the sampling media. A diagram of a cyclone configuration is shown in Figure 6.1.

Sampling Media

Just as in the varied type of equipment, there are also varied types of sampling media used in air sampling. We will discuss some of the more common sample media in this section.

Filters

Again, the specific filter used for sampling will depend on the sampling methodology chosen and the particular air contaminant that you are sampling. There is a wide variety of filters to select from, including the following:

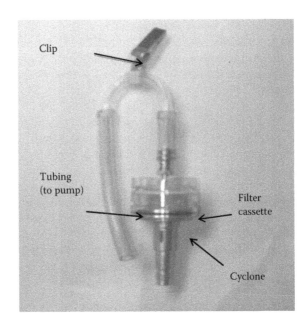

FIGURE 6.1
Cyclone-filter cassette configuration.

- Polyvinyl chloride (PVC) filters
- Mixed cellulose ester filters (MCEF)
- Glass filters
- Matched-weight filters
- Tare-weighted filters

Filters are used primarily to sample for particulates, such as total and respirable particulates, metals, lead, zinc, and so on. An example of a filter and cassette is shown in Figure 6.2.[2]

Sorbent Tubes

Sorbent tubes are glass-encapsulated tubes with various media inside. They are used primarily for sampling of gases and vapors. When using a sorbent tube to test for gases and vapors, take precautionary measures to prevent cuts when breaking the tips of the tube. A diagram of a sorbent tube is shown in Figure 6.3.[2]

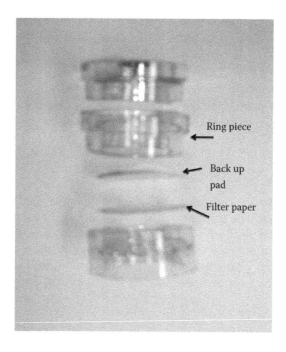

FIGURE 6.2
Example of a filter and cassette. (From Occupational Safety and Health Administration, 2008. *"OSHA Technical Manual—Personal Sampling for Air Contaminants,"* http://www.osha.gov/dts /osta/otm/otm_ii/otm_ii_1.htm.)

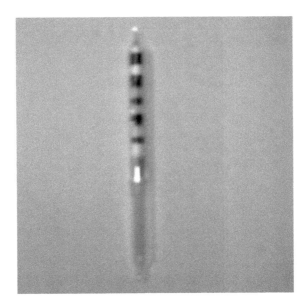

FIGURE 6.3
Photo of a sorbent tube. (From Occupational Safety and Health Administration, 2008. *"OSHA Technical Manual—Personal Sampling for Air Contaminants,"* http://www.osha.gov/dts/osta/otm/otm_ii/otm_ii_1.htm.)

Sample Collection Bags or Canisters

Sample collection bags or canisters are used to take grab samples of gases or vapors. Either positive or negative pressure can be used to fill the bags or canisters with the air from the sample location. The bag or canister is then sealed and sent to the laboratory for analysis.

Passive Samplers

Passive samplers are commercially available sample media worn by the worker as a badge. The collection process is controlled by a physical process such as diffusion through static air or permeation through a membrane, without using an air sampling pump. The badge is worn by an employee for a documented time (on and off times recorded) and then sealed and sent to a laboratory for analysis. A photo of a passive sampler is shown in Figure 6.4.[2]

Sampling Pump Calibration

One of the most important issues involving industrial hygiene air sampling is sample validity and reproducibility. Therefore, air sampling pumps must be pre- and post-sample calibrated using a primary standard or a secondary

FIGURE 6.4
Passive sampler. (From Occupational Safety and Health Administration, 2008. "*OSHA Technical Manual—Personal Sampling for Air Contaminants,*" http://www.osha.gov/dts/osta/otm/otm_ii /otm_ii_1.htm.)

standard that is traceable to a primary standard. Examples of primary and secondary standards used in industrial hygiene sampling equipment calibration include the following:

- Primary Standards
  - Spirometer
  - Bubble burette
  - Electronic soap bubble flow meter
- Secondary Standards
  - Wet test gas meter
  - Dry gas meter
  - Rotameters

In general, the procedure for calibrating air sampling pumps is listed below:

1. Insure that the air sampling pump batteries are fully charged prior to calibrating.
2. Connect the inlet port of the air sampling pump to the filter media, as shown in Figure 6.5, via the tubing.

3. Connect the filter media, as shown in Figure 6.5, to the calibration device.

4. Inspect all connections.

5. Turn the air sampling pump to the "ON" position.

6. Wet the inside of the electronic bubble meter. *Note: This may take several trial runs to fully wet the inside of the meter.*

7. Once the inside of the meter is completely wet, such that the soap bubble is free flowing, take a measurement by depressing the button on the electronic bubble meter.

8. If the measurement on the meter is not where you would like the flow rate to be, make the necessary adjustments on the air sampling pump. Repeat this step until your flow is as close to the desired rate as possible.

9. Take a minimum of three measurements, which should be ±2% of each other. Once three measurements within the ±2% range have been achieved, average the three readings. This is now your calibrated flow rate. Record this flow rate on the calibration sheet and the air sampling work sheet.

10. Disconnect the tubing from the filter media and electronic bubble meter. Your pump is now calibrated and ready for use. *Note: The filter media used is only for calibration purposes and should not be used for sampling purposes.*

11. This procedure should be completed prior to sampling and after sampling has been completed to insure that the flow rate is fairly constant throughout the sampling process. The post-sampling calibrated

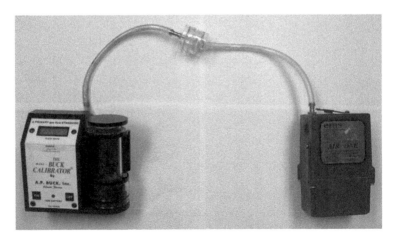

FIGURE 6.5
Example of air sampling pump calibration configuration.

flow should be within ±5% of the pre-calibrated flow rate. If the post-sampling flow rate is outside ±5% of the pre-sampling flow rate, samples must be discarded.

Determining Minimum and Maximum Sample Volumes

Prior to collecting samples, it is necessary to determine the minimum and maximum volumes of air to collect in order to have a statistically valid sample. Before explaining how to calculate the minimum and maximum sample volumes, it is first necessary to define some terms. *Limit of detection* is the lowest level that can be determined to be statistically different from a blank sample. *Limit of quantification* is the concentration level above which quantitative results may be obtained with a certain degree of confidence. The limit of detection and limit of quantification can be obtained from the analytical procedure or by consulting the analytical laboratory prior to sampling. *Target concentration* is an estimate of the airborne concentrations of the contaminant being tested. The use of previous data, direct-reading instruments, or professional judgment can be used to estimate the contaminant concentration. The *upper measurement limit* is the useful limit (milligrams of analyte per sample) of the analytical instrument (Dinardi,[3] p. 180). The minimum sample volume can be calculated using the following equation:

$$V_{min} = \frac{LOQ \times 10^3}{C_t},$$

where:
V_{min} = minimum sample volume (L)
LOQ = limit of quantification (mg)
C_t = contaminant target concentration (mg/m³)

For example, using NIOSH Analytical Method 2555, you would like to collect valid samples for acetone. Given a range of 2.7–295 µg and a target concentration of 420 ppm (1016.4 mg/m³), what is the minimum sample volume required?

$$V_{min} = \frac{0.0027 \, mg \times 10^3 \left(\dfrac{L}{m^3} \right)}{1016.4 \, mg/m^3}$$

$$= \frac{2.7 \, L}{1016.4}$$

$$= 0.003 \, L.$$

To calculate the maximum sampling volume, use the following equation:

$$V_{max} = \frac{W_{max} \times 10^3}{C_t},$$

where:
V_{max} = maximum sample volume (L)
W_{max} = maximum analyte per sample (mg)
C_t = contaminant target concentration

Given the same information above and a W_{max} of 3 mg, what is the maximum sample volume to collect a statistically valid sample and to avoid sample breakthrough?

$$V_{max} = \frac{3.0 \, mg \times 10^3 \, L/m^3}{1016.4 \, mg/m^3}$$

$$= \frac{3000 \, L}{1016.4}$$

$$= 2.95 \text{ or } 3.0 \, L.$$

Determining the Minimum Number of Samples to Collect

One of the most commonly asked question regarding air sampling is "How many samples do I need to collect to insure compliance with exposure limits?" Table 6.1 illustrates a 90% confidence that one worker will be in the top 10% of the exposure within the group.

The Sampling Process

Once you have determined the contaminant you are sampling for, the methodology to use, the equipment to use, the appropriate media on which to collect the sample, and the number of samples to collect, you are ready to begin sampling. To collect air samples, perform the following steps:

TABLE 6.1

Table of Minimum Number of Samples to Collect

Employee Group Size (N)	1	2	3	4	5	6	7	8	9	10	11–12	13–14	15–17	18–20	21–24	25–29	30–37	38–49	50	[a]
Min. no. of samples collected	1	2	3	4	5	6	7	7	8	9	10	11	12	13	14	15	16	17	18	22

[a] Exposure in highest 10% of N.

1. Assemble the sampling equipment.

2. Perform pre-sampling calibration in accordance with the previously mentioned procedures.

3. Record the calibration flow rates on the air sampling work sheet (see Figure 6.6).

Air Sampling Worksheet

U. S. Department of Labor
Occupational Safety and Health Administration

1. Reporting ID 5555555	2. Inspection Number 123456789	3. Sampling Number 497330105	
4. Establishment Name J & X Casting		5. Sampling Date: 06–14–07	6. Shipping Date: 06–15–07
7. Person Performing Sampling (Signature) Signature		8. Print Last Name RIMA	9. CSHO ID Z1234
10. Employee (Name, Address, Telephone Number): (123) 456-7859		14. Exposure Information	a. Number: 2 / b. Duration: 3.5 Yrs/ en person
B.J. Albrecht, 950 Lego Road Pixar City, CA 99999		c. Frequency: 6 hr./day	
11. Job Title: Brass Squeeze Molder Machine Operator – 12 years	12. Occupation Code	15. Weather Conditions: Indoors	16. Photo(s): Y
13. PPE (Type and effectiveness): Safety glasses and ear plugs, no respirator worn		17. Pump Checks and Adjustments: 7:30 – ok, 8:30 – ok, 9:30 – ok, 10:30 – ok, 11:30 – ok, 12:30 – ok, 1:30 – ok, 2:30 – ok	

18. Job Description, Operation, Work Location(s), Ventilation, and Controls

Operates brass squeeze molding machine. Fills and compacts sand into mold. Finished molds placed on pouring lines. There are fans but no exhaust ventilation.

Cont'd

19. Pump Number: 10337		Sampling Data					
20. Lab Sample Number							
21. Sample Submission Number	ER300						
22. Sample Type	P						
23. Sample Media	MCEF						
24. Filter/Tube Number	ER300						
25. Time On/Off	6:30am	1:00pm					
	12:30pm	2:48pm					
26. Total Time (in minutes)	360	108					
27. Flow Rate ☒ l/min ☐ cc/min	2.13	2.13					
28. Volume (in liters)	766.8	230	= 996.8 Total volume				
29. Net Sample Weight (in mg)							
30. Analyze Samples for:	31. Indicate Which Samples to Induce In TWA, Ceiling, etc. Calculations						
Welding Fume (Lead & Cadmium)	T						

32. Interferences and IH Comments to Lab	33. Supporting Samples a. Blanks: ER302 b. Bulks	34. Chain of Custody a. Seals Intact? b. Rec'd in Lab c. Rec'd by Anal. d. Ana. Completed e. Calc Checked f. Supr OK'd	Initials Y N	Date

Case File Page ___ of ___

OSHA-91A (Rev. 1/84)

FIGURE 6.6
Air sampling work sheet (OSHA Technical Manual) sample chain-of-custody.

4. Determine which employees are to be sampled.

5. Inform the employees to be sampled as to the purpose of the sampling and how it will affect them (i.e., wearing of a personal pump, perform work in normal fashion, avoid intentionally contaminating the sample, etc.).

6. Label the samples with an appropriate numbering system prior to placing them on the employees.

7. Record the individual's name, location of sample (area and operation description), and employee identification number or last four digits of the social security number on the air sampling work sheet.

8. Place the pump securely on the employee with a belt or strap.

9. Run the sampling tube with the media on the end of it to the employee's breathing zone (i.e., in the region below the nasal passages or chest region). Take precautions when placing the tubing not to restrict the employee's movements.

10. When the employee is about to begin work or enter a contaminated area, turn the pump on and record the time on the air sampling work sheet.

11. Once the sampling is completed, remove the sample and seal openings to the media. Record the time off on the air sampling work sheet.

12. Review the information provided on the air sampling work sheet for accuracy and thoroughness.

13. Package samples in accordance with instructions from the laboratory and prepare for shipping.

14. Retain copies of the air sampling work sheet and copies of the shipping documents, which will serve as a chain-of-custody to the laboratory.

Most personal sampling required by OSHA can be performed by trained technicians. However, some air sampling may require the services of a CIH. To locate a CIH, contact the ABIH for a list of practicing CIHs.

One of the most vital considerations in any sampling is that the information, methodology and results may all be "fair game" in any legal proceedings. Therefore, a sample chain-of-custody that tracks the sample from point of collection to the laboratory is required. This chain-of-custody ensures a level of confidence that the analytical results reported are true and accurate and are the results from the actual samples received and that no tampering after collection has been done to the sample.

The person collecting the sample will initiate the chain-of-custody and each person that the sample is passed to will sign that they received the sample and that they passed it along to the next person in the chain, until

Date Shipped	# of samples	Sample Type/Media		Project Name or Number	
Purchase Order No.		Contact		Telephone Number	
Turnaround Time ☐ Same Day ☐ 2 Day ☐ 1 Day ☐ STD		Special Instructions and/or Unusual Conditions:		☐ Fax results: ( )___-_____ ☐ Email results – Email _____	
For Lab Use Only:		Sample # or Sample Area	Sample Date	Sample Volume (liters)	Analysis Requested

CHAIN OF CUSTODY RECORD

SAMPLES HAVE BEEN SEALED FOR TRANSPORT AND DELIVERED TO LABORATORY VIA:		Sign here to initiate chain of custody	
Carrier		Date	

Date/Time	Condition of Sample	Samples Received by:	Samples Released by:

FIGURE 6.7
Sample chain-of-custody form.

the sample has reached the laboratory. Once the analysis of the sample has been completed, the original, completed copy of the chain-of-custody will be returned to the client for their records. Figure 6.7 is a representation of the information included on a typical chain-of-custody form.

Industrial Hygiene Sampling and Record-Keeping Procedures

If you are a safety professional responsible for either collecting industrial hygiene samples or contracting the sampling to an outside industrial hygiene firm, record keeping is of the utmost importance. Furthermore, if you are employed by any sizeable company, keeping up with the required sampling and conducting qualitative and quantitative industrial hygiene assessment, the task can be overwhelming. Therefore, I strongly recommend some type of industrial hygiene software to not only maintain accurate records for the sampling and sample results. Following is a list and short discussion on a few commercially available software programs. Some of these I have personal experience with, while others I'm only familiar with through an Internet search and strictly based on information provided from their websites. By listing or not listing the software here is by no means to be considered a recommendation or lack of a recommendation. Each program listed or not listed has its own uses, advantages, and disadvantages, and therefore, you should conduct your own research to determine the best fit for your particular organization.

Before beginning your research into industrial hygiene software, and most definitely before purchasing, you should consider the following issues:

- *Budgetary constraints*: The majority of the programs listed here can be very expensive, ranging from $30,000 to $125,000 depending on the various modules that are needed.

- *Expected performance*: It is extremely important to determine what your specific needs are when purchasing the industrial hygiene software. Are you looking just for a program to record the samples that you collected and analyzed? However, if you have hundreds of employees that require medical surveillance, hearing conservation audiograms, various types of samples, you may need a more sophisticated software package. Do you want to record training inside of the software package? These are just a few of the questions that you must ask yourself before beginning and purchasing the software.

- *Security*: This is a critical piece of the puzzle. I would strongly recommend that you get the Information Technology, Human Resources, and even the Legal staff involved early on in the process. Some of the information being input into the system may be covered either under the Health Insurance Portability and Accountability Act (HIPAA) Privacy Rule or even under a collective bargaining agreement (CBA). Simple information, such as birth dates, may be considered sensitive information, but may be required to accurately populate the software. For example, the frequency of chest x-rays for asbestos workers varies with age and most of the software programs are triggered on birth dates. The information input into the system must be adequately secure, as many of them are cloud-based packages.

Spiramid

Headquartered in Chantilly, Virginia, Spiramid, LLC, was formed in 2001 to specialize in the Environmental, Health, and Safety (EHS) Software. Some of the major features of the Spiramid software are as follows:

- Qualitative Exposure Assessment
- IH Sample Data Management
- Chemical and Physical Stressor Data Management
- Equipment Management
- Fit Testing Records
- SEG Management

Spiramid can be purchased as a core group of modules with additional add-on modules. One of the features of this software is the automatic uploading using LIMs application, which allows the laboratory to directly input analytical results into your software (http://spiramid.com/products /industrial-hygiene-software/).[4]

Medgate

Medgate is headquartered in Ontario, Canada, specializing in EHS software. Based on information from their website at http://www.medgate.com/software /industrial-hygiene/,[5] the major features of their software are as follows:

- Equipment Management
- Monitoring
- Respirator Fit Testing
- Survey
- Laboratory Requisition
- Qualitative Exposure Monitoring
- SEG Management

ProcessMAP

I, personally, am not very familiar with this software solution, but a review of their website demonstrates that a module for industrial hygiene is part of a larger EHS program. You can visit their website at http://www.processmap .com/ehs-software/integrated-ehs-management-system/industrial-hygiene -management for a more thorough review.[6]

Key Information to Remember on
Industrial Hygiene Air Sampling

1. Industrial hygiene is the science and art dedicated to the anticipation, recognition, evaluation, and control of workplace hazards that may cause worker injuries and illnesses.

2. The reasons for conducting air sampling are to determine compliance with regulations, to assess worker exposures to determine if PPE is adequate, to monitor implemented control measures, to evaluate contaminant emissions, and to provide documentation.

3. Grab sampling is collecting a known volume of air in a container for laboratory analysis or by a direct-reading instrument.

4. Personal sampling involves an employee wearing an air sampler while he or she performs his or her normal work routine.

5. Area sampling uses the same type of pump and media as in personal sampling, but the sampling device is stationary throughout the sampling period.

6. Integrated sampling involves collecting one or more samples and then combining them to estimate the workers' 8-h time-weighted average exposure.

7. When analyzing samples, it is highly recommended that you use an AIHA-accredited laboratory.

8. Before sampling, determine whether the contaminant is a gas, vapor, mist, fume, or particulate.

9. Filter media are used primarily to sample for particulates, such as total and respirable particulates, metals, lead, zinc, and so on.

10. Sorbent tubes are used for sampling gases and vapors.

11. Air sampling pumps and direct-reading instruments must be pre- and post-sample calibrated using a primary standard or a secondary standard traceable to a primary standard.

12. If the post-sampling flow rate is outside the ±5% of the pre-sampling flow rate, samples must be discarded.

13. Limit of detection is the lowest level that can be determined to be statistically different from a blank sample.

14. Limit of quantification is the concentration level above which quantitative results may be obtained with a certain degree of confidence.

15. Target concentration is an estimate of the airborne concentrations of the contaminant being tested.

16. When working in a potentially explosive environment, insure that all sampling equipment is classified by the manufacturer as "intrinsically safe" prior to using.

References

1. Occupational Safety and Health Administration (OSHA), 2008. *Direct Reading Instruments*. Available at http://www.osha.gov/sltc/directreadinginstruments/index.html.
2. Occupational Safety and Health Administration, 2008. *OSHA Technical Manual—Personal Sampling for Air Contaminants*. Available at http://www.osha.gov/dts/osta/otm/otm_ii/otm_ii_1.htm.
3. Dinardi, S. R. (Ed.), 2003. *The Occupational Environment: Its Evaluation, Control, and Management*, American Industrial Hygiene Association Press, Fairfax, VA, p. 180.
4. http://spiramid.com/products/industrial-hygiene-software/.
5. http://www.medgate.com/software/industrial-hygiene/.
6. http://www.processmap.com/ehs-software/integrated-ehs-management-system/industrial-hygiene-management.

7

Ventilation

In order to reduce or eliminate any hazard from the work area, there are three methods that the safety professional must adhere to. These methods include *Engineering Control, Administrative Control,* and *Personal Protective Equipment.* As we all know, the best (permanent) method is to engineer any hazard out of the workplace. One way to reduce or eliminate the concentration of hazardous materials from the workplace is through ventilation. Today's safety professional must have more than just a general knowledge of ventilation and ventilation systems. He or she must have an in-depth knowledge. Furthermore, the Associate Safety Professional (ASP)/Certified Safety Professional (CSP) exam candidate will be expected to answer numerous questions regarding ventilation on the examination. It is for this reason that this chapter on ventilation has been included. Using the Board of Certified Safety Professionals (BCSP) exam reference sheet[1], this chapter will walk through each equation and explain, in detail, the use and purpose of the equation and provide examples and solutions to various scenarios.

Purpose for Using Ventilation

The primary purposes for using ventilation are to (a) maintain adequate oxygen supply, (b) control hazardous concentrations of chemicals, (c) remove odors, (d) control temperature and humidity, and (e) remove contaminants at the source, before they enter the workplace. By understanding how to use and control ventilation, the safety professional can successfully eliminate or greatly reduce any concentration of contaminant in the air.

Types and Selection of Ventilation

Ventilation is basically divided into three categories, *general, dilution,* and *local* ventilation.

General Ventilation

General ventilation is primarily used for comfort, such as temperature, humidity, and odor control, such as the air conditioning and heating system in your home.

Dilution Ventilation

Dilution ventilation is a system designed to dilute contaminants by mixing fresh air with contaminated air. Dilution ventilation is used primarily in health and fire protection. The components of a dilution system include the source of air exhaust, source of air supply, a duct system, and a method to filter and temper incoming air.

Dilution ventilation is used to control the following:

- Contaminants of moderate toxicity
- A large number of sources
- Intermittent exposures
- Where emission sources are well distributed

Local (Exhaust) Ventilation

Local (exhaust) ventilation systems are designed to control contaminants at the source before mixing with breathing air occurs. When selecting which type of ventilation system to design and install (especially in the industrial environment), use the following criteria to make the selection.

Local (exhaust) ventilation is used to control the following:

- Highly toxic substances
- Single-source emissions
- Direct worker exposures

NOTE: Energy is required to move air. Air is constantly in turbulence moving at 25 feet per minute.

General Concepts of Ventilation Notes

1. Air movement results from difference in pressure.
2. Difference in pressure can be attained by heating or by mechanical means.
3. A temperature gradient contributes to the ventilation.

Principles of Air Movement

Calculating for Volumetric Air Flow

In order to begin to understand ventilation, it is necessary to understand the principles of air movement. Air flow rates are calculated by multiplying the velocity by the cross-sectional area of the hood or duct in which the air flows. This is represented mathematically as

$$Q = V \times A,$$

where
  Q = volumetric flow rate (expressed in cfm [cubic feet per minute])
  V = velocity of the air (expressed in fpm [feet per minute])
  A = cross-sectional area of duct (expressed in sf [square feet])

An example of this would be to calculate the volumetric flow rate of a duct, given the following information: The duct is a round duct measuring 8 inches in diameter and the measured flow rate is 378 fpm. The first step would be to calculate the cross-sectional area of the duct. *(Note: Remember to convert the unit of measurement into feet.)*
We determine that the radius (1/2 the diameter) is equal to 4 inches or 0.33 feet. Using the equation for the area of a circle, the duct cross section is 0.35 sf. Insert these data into the equation as follows:

$$Q = 378 \text{ fpm} \times 0.35 \text{ sf}$$

$$Q = 132.3 \text{ or } 132 \text{ cfm}$$

Calculating Static Pressure, Velocity Pressure, and Total Pressure

Air moves from an area of higher pressure to those of a lower pressure. Exhaust fans work by pushing air upstream of the fan, thus lowering the pressure on the downstream side. This pressure differential causes air to flow into the hood or duct. This pressure created by the fan is referred to as *static pressure*. Static pressure on the downstream side of the fan is positive and negative on the upstream side. *Velocity pressure* is the pressure in the direction of flow necessary to cause the air at rest to flow at a given velocity. The *total pressure* on a ventilation system equals the sum of static and velocity pressure. Total pressure is written mathematically as

$$TP = SP + VP,$$

where
  TP = total pressure (in inches, water gauge)
  SP = static pressure (in inches, water gauge)
  VP = velocity pressure (in inches, water gauge)

Calculating Velocity of Air

Another relationship to understand is the relationship between velocity and velocity pressure. Airflow velocity is important in various ways. Airflow velocity is used to capture contaminants and overcome cross-drafts, transportation of contaminants through the duct, and balancing of "losses" (as described previously in this chapter) in the system and in the discharge of the contaminant from the stack. This is described as the "magnitude" of the system, which is a function of the velocity pressure. This relationship is written mathematically as

$$V = 4005\sqrt{VP},$$

where
  V = velocity (fpm)
  VP = velocity pressure (H_2O)

Example

Given a velocity pressure of 0.2", determine the velocity of air.

$$V = 4005\sqrt{0.2}$$
$$V = 4005(0.45)$$
$$V = 1802 \text{ fpm.}$$

Contaminant Generation

When using dilution ventilation, it is important to understand that the flow rate of fresh air is determined by (a) contaminant generation, (b) proper mixing, and (c) target final concentration. The methods used apply to uniform rates of generation and low to moderate toxicity. The accumulation of contaminants is equal to the generation minus the removal. The following equation is used to determine the concentration buildup that will occur over a given period:

$$\ln\left(\frac{G - Q'C_2}{G - Q'C_1}\right) = \frac{Q'(t_2 - t_1)}{V},$$

where
  $G =$ rate of generation of contaminant (cfm)
  $Q' = (Q/K)$
  $K =$ design distribution constant (a constant factor [1–10])
  $C =$ concentration at a given time (parts per million [ppm])
  $V =$ volume of room or enclosure (ft³)
  $Q =$ flow rate (cfm)
  $t_2 - t_1 =$ time interval or Δt

Example

Acetone is being generated under the following conditions: $G = 1.4$ cfm, $Q' = 2500$ cfm, $V = 75,000$ ft³, $C_1 = 0$, and $K = 2$. How long will it take (in minutes) before the concentration reaches 175 ppm?

The question is straightforward and asks that we calculate for time, which is the unknown. We can use the above equation but must first manipulate it in order to obtain the desired outcome (time). We therefore set up the equation as follows:

$$\Delta t = -\frac{V}{Q'}\left[\ln\left(\frac{G - Q'C_2}{G}\right)\right].$$

We can now insert the data from our example into the equation and solve for time, as follows:

$$\Delta t = -\frac{75,000 \text{ ft}^3}{2500 \text{ cfm}}\left[\ln\left(\frac{1.4 \text{ cfm} - 2500 \text{ cfm}(0.000175)}{1.4 \text{ cfm}}\right)\right]$$

$$\Delta t = -30\left[\ln\left(\frac{0.9625}{1.4}\right)\right]$$

$$\Delta t = -30\,[\ln\,0.6875]$$

$$\Delta t = -30\,(-0.3747)$$

$$\Delta t = 11.24 \text{ min}.$$

NOTE: In the above equation, it was necessary to convert the concentration of 175 ppm to a decimal number. We did this by taking 175 and dividing by 10^6, with the result being 0.000175.

Simplified, it will take approximately 11.24 min for an area with zero concentration of acetone, and a volume of 75,000 ft³, generating 1.4 cfm of acetone, and a Q' of 2500 cfm to reach a concentration of 175 ppm.

Example

Another example problem may ask you to calculate the concentration after an established timeframe. For example, given the same data as listed in the previous problem, what would be the concentration of acetone after 60 min?

Again, we must manipulate the original equation in order to obtain the requested information (concentration). We can use the original equation and manipulate it as follows:

$$C_2 = \frac{G\left[1-e^{\left(-\frac{Q'\Delta t}{V}\right)}\right]}{Q'} \times 10^6.$$

Now, we can insert the known data from the previous problem, as follows:

$$C_2 = \frac{1.4 \text{ cfm}\left[1-e^{\left(-\frac{2500 \text{ cfm } (60)}{75,000 \text{ ft}^3}\right)}\right]}{2500 \text{ cfm}} \times 10^6$$

$$C_2 = \frac{1.4 \text{ cfm } [1-0.1353]}{2500} \times 10^6$$

$$C_2 = \frac{1.4 \text{ cfm } (0.8647)}{2500 \text{ cfm}} \times 10^6$$

$$C_2 = 0.000484 \times 10^6$$

$$C_2 = 484.2 \text{ ppm}.$$

Given the criteria for the equation in the previous problem, you can see that after 60 min, the concentration of acetone will be 484.2 ppm.

Calculating Purge Rates

Assuming the air in a room is contaminated, but the generation of new contaminant has ceased, how long will it take to reduce the air concentration to an acceptable limit? We utilize the following equation to determine the time required to reduce the concentration:

$$\ln\left(\frac{C_2}{C_1}\right) = -\frac{Q'}{V}(t_2 - t_1),$$

where
C_1 = the measured concentration
C_2 = the desired concentration
$Q' = (Q/K)$
K = design distribution constant (a constant factor [1–10])
Q = flow rate (cfm)
$t_2 - t_1$ = time interval or Δt

Example

Using the information provided in the acetone problem above, how much time would be required (assuming all sources of generation have ceased) to reduce the contamination to 50 ppm?

$$\ln\left(\frac{0.00005 \text{ ppm}}{0.000175 \text{ ppm}}\right) = -\frac{2500 \text{ cfm}}{75,000 \text{ ft}^2}(t_2 - t_1)$$

$$\ln 0.28 = -0.03333 (\Delta t)$$

$$\frac{-1.2729}{-0.03333} = \frac{-0.03333 (\Delta t)}{-0.03333}$$

$$38.2 \text{ min} = \Delta t.$$

Given the information above, it will take 38.2 min to reduce the concentration of acetone from 175 to 50 ppm. (*Note: Remember to convert ppm to a decimal number.*)

Example

In another case, given the same information, you may be asked to determine the concentration after a given period. In order to calculate this, you will need to manipulate the equation as follows:

$$C_2 = C_1 e^{\left(-\frac{Q'\Delta t}{V}\right)} \times 10^6$$

Now, we can insert the known data to solve for the unknown, as follows:

$$C_2 = (0.000175 \text{ ppm})e^{\left(-\frac{(2500 \text{ cfm})(60 \text{ min})}{75,000 \text{ ft}^3}\right)} \times 10^6$$
$$C_2 = (0.000175 \text{ ppm})e^{(-2)} \times 10^6$$
$$C_2 = (0.000175 \text{ ppm})(0.1353) \times 10^6$$
$$C_2 = 23.7 \text{ ppm}.$$

Given the information above, the concentration of acetone after 60 min, with no additional generation of contaminant, will be 23.7 ppm.

Steady-State Concentration

The objective of dilution ventilation is to maintain a steady-state concentration, which is less than the threshold limit value (TLV) or permissible exposure limit (PEL) for the contaminant of concern. We can calculate this using the following equation:

$$Q' = \frac{G}{C},$$

where

Q' = the effective rate of ventilation corrected for incomplete mixing, cfm ($Q' = Q/K$)

K = design distribution constant to allow for incomplete mixing of contaminant air (1–10)

C = concentration of gas or vapor at time t, ppm

For example, given that $Q' = 2500$ cfm and $G = 1.2$ cfm, what is the steady-state concentration of the gas or vapor?

Inserting the known data into the equation, we have the following:

$$2500 \text{ cfm} = \frac{1.2 \text{ cfm}}{C}$$

$$\frac{2500 \text{ cfm}}{1.2 \text{ cfm}} = C$$

$$480 \text{ ppm} = C.$$

Calculating Rate of Generation for Liquid Solvents

In order to calculate the rate of generation for liquid solvents, we utilize the following equation, which also accounts for incomplete mixing of air:

$$Q = \frac{403 \times 10^6 \times SG \times ER \times K}{MW \times C},$$

where

Q = actual ventilation rate (cfm)

SG = specific gravity of volatile liquid

ER = evaporation rate of liquid (pints/minute)

K = design distribution constant to allow for incomplete mixing of contaminant air (1–10)

MW = molecular weight of liquid
C = desired concentration of gas or vapor at time t (ppm). *Note: Normally the TLV or PEL.*

Example

Methyl ethyl ketone (MEK) is evaporating from a container at a rate of 1.2 pints every 60 min. Determine the actual ventilation rate given the following information: PEL = 200 ppm, SG = 0.81, MW = 72.11, and assume that $K = 4$.

Insert the known data into the equation as follows:

$$Q = \frac{403 \times 10^6 \times 0.81 \times \dfrac{1.2 \text{ pints}}{60 \text{ minutes}} \times 4}{72.11 \times 200 \text{ ppm}}$$

$$Q = \frac{26,114,400}{14,422}$$

$$Q = 1810.7 \text{ cfm}$$

Calculating Vapor or Gaseous Concentrations

In order to calculate the vapor or gaseous concentrations, we use the following equation:

$$C = \frac{P_v \times 10^6}{P_b},$$

where
C = concentration (ppm)
P_v = pressure of chemical (mm Hg)
P_b = barometric pressure (mm Hg)

Example

Given a chemical with a pressure of 350 mm Hg, what would be the concentration at a barometric pressure of 760 mm Hg?

Insert the data into the equation as follows:

$$C_{ppm} = \frac{350 \text{ mm Hg} \times 10^6}{760 \text{ mm Hg}}$$

$$C_{ppm} = \frac{3.5 \times 10^8 \text{ mm Hg}}{760 \text{ mm Hg}}$$

$$C_{ppm} = 460,526.3 \text{ ppm}.$$

While not specifically listed on the BCSP exam reference sheet, a useful equation for calculating concentration of vapors as a percentage is

$$C_{percentage} = \frac{P_v \times 100}{P_b}.$$

Given a chemical with a vapor pressure of 350 mm Hg and a barometric pressure of 760 mm Hg, what is the concentration percentage?

$$C_{percentage} = \frac{350 \text{ mm Hg} \times 100}{760 \text{ mm Hg}}$$

$$C_{percentage} = 46.05\%.$$

Calculating Room Air Changes

Determining the number of air changes in a room is done by utilizing the following equation, which is NOT listed on the BCSP examination reference sheet:

$$N_{changes} = \frac{60Q}{V_r},$$

where
  Q = actual ventilation rate (cfm)
  V_r = volume of room (cubic feet or ft³)
  $N_{changes}$ = number of room air changes

Example

Given that $Q = 2500$ cfm and $V_r = 75,000$ ft³, calculate the number of room air changes that is occurring.

$$N_{changes} = \frac{60 \text{ min}(2500 \text{ cfm})}{75,000 \text{ ft}^3}$$

$$N_{changes} = 2.$$

In this example, we calculate the air to change 2 times every 60 min or 2 air changes per hour.

Calculating Concentration of a Contaminant with Dilution Ventilation

Now we will calculate the concentration of a contaminant, using air changes per hour. To do this, we will utilize the following equation:

$$C = \frac{G}{Q'}\left(1 - e^{\left(-\frac{Nt}{60}\right)}\right) \times 10^6,$$

where
   C = concentration at a given time (ppm)
   G = rate of generation of contaminant (cfm)
   $Q' = (Q/K)$
   K = design distribution constant (a constant factor [1–10])
   Q = flow rate (cfm)
   t = time (in hours)

NOTE: The equation listed on the BCSP examination reference sheet does not multiply the product of the equation by 10^6. Therefore, it will be necessary for you to remember this correction.

Example

As the CSP, you are asked to determine the concentration of a contaminant in a storage area having a ventilation system that provides 7 air changes per 1 h. The volumetric flow rate is 2500 cfm, and the contaminant is being generated at a rate of 1.2 cfm. What is the concentration in parts per million?

$$C = \frac{1.2}{2500 \text{ cfm}}\left(1 - e^{\left(-\frac{7 \times 1}{60}\right)}\right) \times 10^6$$
$$C = 0.00048(1 - 0.89) \times 10^6$$
$$C = 52.8 \text{ ppm.}$$

Local Exhaust Ventilation

As we mentioned previously in this chapter, local (exhaust) ventilation systems are designed to control contaminants at the source before mixing with breathing air occurs. They are used primarily to control the concentrations at the source of highly toxic contaminants. A local exhaust ventilation system consists of the following parts: (a) hood, (b) duct, (c) air-cleaning device, (d) fan, and (e) stack. The proper design of hoods is necessary to capture the contaminant at the source of release. The term *capture velocity* is defined as the minimum velocity of hood-induced air necessary to capture the contaminant.

Canopy Hood

Canopy hoods are used where hot gases and vapors are encountered and workers do not work directly over the source of emissions. The major drawback to using canopy hoods is the potential for the worker to place himself or herself between the contaminant and the exhaust stream.

Down Draft Hood

Down draft ventilation is used where heavier than air contaminants exist, which are not being propelled away from the source of contaminant release. This type of ventilation system draws the air downward and away from the worker's breathing zone.

Enclosure Hood

An example of an enclosure hood is the laboratory hood. This hood actually encloses the contaminant source and the air is forced in an opposite direction (upward, downward, or backward away from the worker).

Receiving Hood

Receiving hood exhaust systems are used at the point of contaminant generation. An example of this system is a welding fume exhaust system that can be placed very close to the point of contaminant generation.

Openings

There are basically two types of hood openings. Openings can be either flanged or plain. Flanged openings have some type of "lip" on them (see Figure 7.1). Flanges are sometimes designed to create a certain desired airflow.

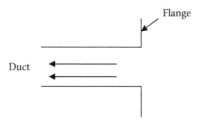

FIGURE 7.1
Flanged opening.

Duct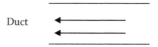

FIGURE 7.2
Plain opening.

Plain openings do not have a flange and therefore air movement is directly into the duct. See Figure 7.2 for a diagram of a plain opening.

Calculating Hood Entry Losses

All hoods, regardless of the type, have significant energy loss, which must be accounted for. The hood entry loss represents the energy necessary to overcome the losses caused by air moving through and into the duct. Before discussing hood entry loss calculations, it is first necessary to show how to calculate the coefficient of entry loss, which is the square root of the ratio of duct velocity pressure to hood status suction. This is written mathematically as follows:

$$C_e = \sqrt{\frac{VP}{SP_h}},$$

where
C_e = coefficient of entry loss
VP = velocity pressure of the duct ("wg)
SP_h = static pressure of the hood ("wg)

Theoretically, if there are no losses, then SP_h would equal to $VP + C_e = 1.0$. However, as we have previously mentioned, all hoods have some loss.

Example

Calculate the coefficient of entry loss given an exhaust system VP of 1.25 "wg and a static pressure of the hood of 1.82 "wg.

$$C_e = \sqrt{\frac{1.25}{1.82}} = 0.829.$$

Understanding how to calculate for the coefficient of entry loss, we can now discuss the equation for calculating the hood entry loss, which is described mathematically as follows:

$$h_e = \frac{\left(1 - C_e^2\right) VP}{C_e^2},$$

where
  h_e = hood entry loss ("wg)
  C_e = coefficient of entry loss
  VP = velocity pressure of duct ("wg)

Example

A local exhaust system has a VP pressure of 1.37 "wg and a C_e of 0.829. What is the hood entry loss of this system?

$$h_e = \frac{(1-0.829^2)1.37 \text{ "wg}}{0.829^2}$$

$$h_e = \frac{(0.3128)1.37}{0.687}$$

$$h_e = 0.624 \text{ "wg.}$$

Now that we have discussed the relationship between hood entry loss, velocity pressure, and the coefficient of entry loss, we can calculate the static pressure of the hood, using the following equation:

$$SP_h = VP + h_e,$$

where
  SP_h = static pressure of the hood ("wg)
  h_e = overall hood entry loss ("wg)
  VP = duct velocity pressure ("wg)

NOTE: When calculating for SP_h, it is understood that the static pressure of the hood is always positive. A more accurate way of writing the equation would be as follows:

$$|SP_h| = VP + h_e.$$

Example

As the safety professional, you are called upon to calculate the static pressure of the hood with the following information: VP = 0.2 "wg and h_e = 0.1 "wg.
  Insert the data into the equation as follows:

$$|SP_h| = 0.2 \text{ "wg} + 0.1 \text{ "wg}$$

$$|SP_h| = 0.3 \text{ "wg.}$$

Calculating Airflow Velocity

Calculating airflow velocity in a local exhaust ventilation system requires that we account for the coefficient of entry loss. We can calculate the airflow velocity and account for the coefficient of entry loss by using the following equation:

$$V = 4005C_e\sqrt{SP_h},$$

where
 V = velocity of air (fpm)
 C_e = coefficient of entry loss
 SP_h = static pressure of the hood ("wg)

Example

Given that $C_e = 0.829$ and SP $= 0.3$ "wg, what is the airflow velocity of the system?

$$V = 4005(0.829)\sqrt{0.3}$$
$$V = 3320(0.55)$$
$$V = 1826 \text{ fpm.}$$

Calculating Capture Velocity for Plain Opening Hood

As mentioned in Calculating Hood Entry Losses, suction sources have contours and other imperfections that create hood entry losses, which further have an impact on capture velocities. Therefore, it is necessary to discuss the relationship between distance (from the source), the air flow, and the capture velocity. We can calculate the capture velocity using the following equation:

$$V = \frac{Q}{10X^2 + A},$$

where
 V = velocity (fpm)
 Q = flow rate (cfm)
 X = source distance from hood opening (ft) (*Note: The equation is only accurate for a limited distance of 1.5 times the diameter of a round duct or the side of a rectangle or square duct.*)
 A = area (square feet [ft²])

Example

Calculate the capture velocity of a round duct measuring 10 inches in diameter with the contaminant source being 1.2 feet from the duct opening. The flow rate is 600 cfm.

The first step would be to determine the area of the duct, using the circle area equation.

$$A_{circle} = \pi r^2$$
$$A_{circle} = 3.14 \times 0.417^2$$
$$A_{circle} = 3.14(0.1736)$$
$$A_{circle} = 0.55 \text{ ft}^2.$$

Next, we insert the known data into the equation as follows:

$$V = \frac{600 \text{ cfm}}{10(1.2 \text{ ft})^2(0.55 \text{ ft}^2)}$$
$$V = \frac{600}{7.92}$$
$$V = 75.76 \text{ fpm}.$$

Therefore, the capture velocity under this given scenario is 75.76 fpm.

Ducts

Exhaust ducts are used to convey the contaminated air from the hood to the air cleaner or stack. When air moves through exhaust ducts, a certain amount of friction loss occurs because of the friction of the air stream with the duct. The selection of exhaust duct size is based on minimizing friction loss while maintaining an adequate transport velocity to keep particulate matter from settling out. When two or more ducts branch off from a single exhaust source, a special problem is encountered. After determining the amount of airflow required in each branch of the system, the designer must assure that the exhaust volume is properly proportioned between the two branches. This is done in one of two ways: (1) balancing the two branches by the proper sizing of ducts and fittings to assure proper distribution, or (2) blast gates, which are slide gates that can be pushed into the duct to partially block the airflow to lower the amount of air entering that branch.

One equation that is not listed on the BCSP examination reference sheet is

$$V_1 A_1 = V_2 A_2,$$

where
    V = velocity of air (fpm)
    A = cross-sectional area (ft²)

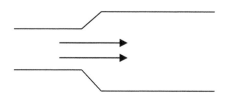

Fans

Fans generate the airflow volume (Q) of the system against airflow resistance presented by the system. In other words, they create the SP differential needed to cause the desired airflow. There are two basic types of fans: (1) axial and (2) centrifugal. Axial fans create airflow when the air enters and leaves along the axis of rotation. Centrifugal fans create airflow as the air enters along the axis of rotation and leaves perpendicularly (accelerates centrifugally) through the blades.

Calculating Static Pressure of the Fan (SP_h)

In order to calculate the static pressure of the fan (SP_h), we use the following equation:

$$SP_{fan} = SP_{out} - SP_{in} - VP_{in}.$$

Example

Calculate the static pressure of the fan, given the following information: $SP_{out} = 1.8$ "wg, $SP_{in} = 0.3$ "wg, and $VP_{in} = 0.4$ "wg.

$$SP_{fan} = 1.8 - 0.3 - 0.4$$
$$SP_{fan} = 1.1 \text{ "wg.}$$

Air-Cleaning Devices

Air cleaners or air pollution devices remove the contaminant from the air stream to protect the community, protect the fan, recover materials, and enable circulation.

Ventilation Measurement Equipment

There are several different types of ventilation measuring equipment. The most common types include pitot tubes, rotating vane anemometers, and thermal anemometers.

Pitot Tubes

Pitot tubes are probes that are inserted into the duct system and connected to a manometer. Pitot tubes are used to measure various pressures within the system.

Rotating Vane Anemometers

Rotating vane anemometers are used to measure airflow through large supply and exhaust systems. It is recommended that the size of the rotating vane anemometer should not exceed 5% of the cross-sectional area of the duct.

Thermal Anemometers

Thermal anemometers are primarily digital instruments that measure the heat removed by an air stream as it passes over a probe, which allows for calibration to the velocity of the air stream at a given density. The probe can be inserted directly into the air stream. This is a useful instrument but the user should understand that errors can be created by the movement and location of the probe. It should be perpendicular to the air stream and maintained in that position until such time as the measurement is stabilized. Furthermore, the user should understand that the probe is extremely fragile; thus, precautions should be in place to protect the probe from damage.

Key Information to Remember on Ventilation

1. Pitot tubes usage is limited to velocities at or below 600–800 fpm.
2. Blast gates are used to balance the airflow in ducts of different sizes.
3. *Capture velocity* is defined as the minimum velocity of hood-induced air necessary to capture the contaminant.
4. When calculating the static pressure of the hood, remember that the SP_h is always positive.
5. Static pressure on the downstream side of the fan is positive and negative on the upstream side.
6. Backward curved fan blades are the most efficient.

7. Centrifugal fans are the best for local exhaust ventilation systems.

8. The equation for calculating capture velocities of plain hood openings is only accurate for a limited distance of 1.5 times the diameter of a round duct or the side of a rectangle or square duct.

Reference

1. Board of Certified Safety Professionals, *Examination Reference Sheet, Ventilation*. Available at http://www.eng.auburn.edu/ise/courses/insy3020/BCSP%20 Formula%20Sheet.pdf.

8

Noise and Hearing Conservation Program

One of the most important and more common areas of concern for the occupational safety and health professional is noise and noise exposure. This chapter is dedicated to providing the exam candidate and the experienced professional with a basic understanding of the requirements of the hearing conservation program of the Occupational Safety and Health Administration (OSHA), along with how to correctly use the equations to determine potential noise exposures. Every equation provided on the Certified Safety Professional (CSP) exam reference sheet will be discussed in detail and example problems for each will be provided.

OSHA's Hearing Conservation Program

The OSHA standard for the hearing conservation program is *29 CFR 1910.95*. This standard covers a wide variety of industries and covers an even wider variety of topics. The first thing that we will discuss is *Who is included in the Hearing Conservation Program?* It may seem easy enough to include all employees in your company into the program. However, the hearing conservation program can become very expensive, especially in the current economic hardships that some companies find themselves. OSHA requires that an employer implement a continuing hearing conservation program whenever an employee's noise exposure equals or exceeds an 8-h time-weighted average (TWA) sound level of 85 decibels (dB) on the A-scale (slow response), or equivalently a dose of 50%.[1]

Who Is Included in the Hearing Conservation Program?

As mentioned above, any employee whose noise exposure equals or exceeds an 8-h TWA sound level of 85 dB on the A-scale (slow response), or equivalently a dose of 50%, must be included in the hearing conservation program (29 CFR 1910.95). According to the OSHA standard, this noise exposure is not reduced by any noise attenuation devices (earplugs, ear muffs, or bands) and is not to be used in reducing the noise exposure for purposes of avoiding including an employee into the hearing conservation program.

Monitoring

When information indicates that any employee's noise exposure may equal or exceed an 8-h TWA, the employer shall develop and implement a monitoring program. The sampling strategy should be designed to identify employees for inclusion into the program. For high-mobility employees, representative sampling may be used. All continuous, intermittent, and impulsive sound levels from 80 to 139 dB shall be integrated into the measurements. Instruments used to measure employee noise exposure shall be calibrated to ensure measurement accuracy. Monitoring shall be repeated whenever a change in production, process, equipment, or controls occurs.[1]

Sound Measuring Instruments

There are several types of instruments that can be used to measure an employee's sound level exposure. The *Type 1 sound level meter* is used for precision measurements in the field and has an accuracy of ±1 dB, while the *Type 2 sound level meter* is used for general purposes in the field and has an accuracy of ±2 dB. The Type 2 sound level meter is the minimum measuring device allowed for determining an employee's noise level exposure under the OSHA Hearing Conservation standard. Sound level meters can be used to spot check employee exposures, identify potential noise sources needing further evaluation, and assist in the feasibility of engineering controls.

One of the more useful sound measuring devices is the *octave band analyzer*. An octave band analyzer is a sound level meter that measures noise levels at various frequencies. It is useful in helping to analyze sources of noise, based on frequencies, determining specific noise attenuation devices to recommend, and dividing noise into frequency components. Most octave band analyzers provide readings for 31.5, 63, 125, 250, 500, 1000, 2000, 4000, 8000, and 16,000 Hz.

The final sound measuring device that we will discuss is the *noise dosimeter*. Noise dosimeters are basically a sound level meter, but they are worn by the employee for a period (usually 8 h). The noise dosimeter computes not only the overall noise exposure but also the *dose*. Once completed, the information provided in the noise dosimeter can be printed out and placed in an employee's file.[1]

Employee Notification

OSHA requires that any employee exposed at or above an 8-h TWA of 85 dbA shall be notified of the results of the monitoring. In addition, OSHA also requires that the affected employee or their representative be provided the opportunity to observe any noise measurements conducted.

Audiometric Testing

Under the Hearing Conservation standard, OSHA requires that audiometric testing be performed on any employee included into the program. The testing must be conducted by an audiologist, otolaryngologist, or technician certified by Council for Accreditation in Occupational Hearing Conservation (CAOHC). The baseline audiogram must be conducted within 6 months of first exposure greater than 85 dBA, with an exception being made for those companies that utilize the services of a mobile test van, which extends the requirement to a maximum of 1 year. Audiometric testing to establish the baseline audiogram shall be preceded by 14 h without exposure to excessive noise. The annual audiogram is conducted annually after the baseline. A revised baseline can be done, but only by a certified audiologist, otolaryngologist, or licensed physician. Audiometric tests must be pure tone, air conduction tests in the 500-, 1000-, 2000-, 3000-, 4000-, and 6000-Hz range; 8000 Hz is also recommended. Audiometers must meet the requirements of ANSI S3.6-1969.[1]

Standard Threshold Shift

The *Standard Threshold Shift* (*STS*) is a change in hearing threshold relative to the baseline audiogram of an *average* of 10 dB or more at 2000, 3000, and 4000 Hz in either ear. If an employee indicates an STS, this is recorded on the OSHA 300 log as a recordable injury. In determining whether an STS has occurred, allowance may be made for the contribution of aging (*presbycusis—natural hearing loss caused by the aging process*), using the Calculation and Application of Age Correction to Audiograms (defined in 29 CFR 1910.95 [appendix F]).

Training Program

Any employee entered into the hearing conservation program must be trained initially within 30 days of the exposure and repeated annually thereafter, as long as they are in the program. This training must include, as a minimum, the following:

- Effects of noise on hearing
- Purpose of hearing protectors (advantages, disadvantages, attenuation of various types, and instructions on selection, fitting, use, and care)
- Purpose of audiometric testing[1]

Record Keeping

Audiometric test record keeping requires the following information to be identified:

- Name and job classification of the employee
- Date of the audiogram
- Examiner's name

- Date of the last acoustic calibration
- Employee's most recent noise exposure assessment

Noise exposure measurement records are required every 2 years. Audiometric testing must be maintained in an employee's records for the duration of the employment.[1]

Noise-Related Definitions

In order to totally understand some of the items discussed in this chapter, it is necessary to understand the basic terms and their meaning.

Noise

Noise, simply defined, is unwanted sound. Sound travels in the form of waves, similar to dropping a rock in the middle of a still body of water. The waves ripple outward from the source. Noise can be harmful to humans, by creating a hearing loss, if uncontrolled.

Continuous Noise

Continuous noise is a sound that has unbroken sound waves of one or more different frequencies.

Intermittent Noise

Intermittent noise is a sound that has broken sound waves or is noncontinuous.

Sound

Sound is *oscillations* in pressure above and below the ambient atmospheric pressure, generated by a vibrating surface or turbulent fluid flow, causing high- and low-pressure areas to be formed, which propagate away from the source. Mechanical vibrations are transmitted through an elastic medium. A sensation is produced by stimulation of organs of hearing by vibration (AIHA,[2] p. 79).

Frequency (*f*)

Frequency is the number of complete cycles of a sound wave in 1 s. They are measured in Hertz (Hz). The normal range of hearing for human beings is 20–20,000 Hz.

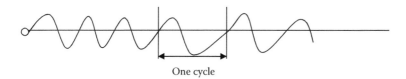

One cycle

Period

A *period* is the time required for one cycle. It is the reciprocal of frequency.

Speed or Velocity of Sound (*c*)

The speed or velocity of sound (*c*) is dependent upon the medium density and compressibility. To provide examples of the speed of sound, we look at the following:

- Speed of sound in air = 332 m/s
- Speed of sound in water = 1500 m/s
- Speed of sound in steel = 6100 m/s

Speed is equal to the wavelength × frequency (cycles per second). $c = \lambda f$.

Wavelength (λ)

A wavelength is the distance traveled during one pressure cycle (Figure 8.1).

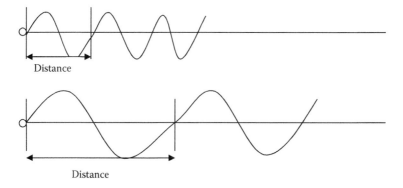

Distance

Distance

FIGURE 8.1
Diagram of a wavelength.

Anatomy and Physiology of the Ear

There are three major parts of the ear: the outer, middle, and inner ear. The outer ear is designed to collect sound vibrations and consists of the *pinna* (ear lobe), which gathers sound waves and directs it toward the *external auditory canal* and onto the *tympanic membrane* (eardrum). The middle ear transmits mechanical vibrations from the air into vibrations in fluid. The middle ear consists of the *ossicle* (the *incus, stapes,* and *malleus* [the three smallest bones in the human body]); the *Eustachian tube,* which equalizes pressure between the middle and outer ear; the *oval window;* and the *round window.* The inner ear (also referred to as the inner-transduction) changes mechanical waves in a liquid to chemical impulses sent to the brain. The inner ear consists of the *cochlea* (*snail shell–shaped organ*) and the *organ of Corti,* which is the essential receptor end organ for hearing and contains the hair cells.

Types of Hearing Loss

There are basically five types of hearing loss as a result of noise-induced exposure. They are listed as follows:

Conductive—It occurs in the outer and middle ear. The loss of "loudness" only occurs. Conductive hearing loss can occur as a result of built-up ear wax, physical obstruction of the Eustachian tube, or a perforation of the tympanic membrane.

Sensorineural—It occurs in the inner ear. As you recall, the inner ear contains the *organ of Corti,* where the "hair cells" are located. Damage to these organs is primarily irreversible. Damage as a result of hazardous or excessive noise exposure tends to occur first at the higher frequencies.

Mixed—Hearing loss can occur as a combination of both conductive and sensorineural.

Central nervous system (*CNS*)—Hearing loss related to the CNS occurs between the inner ear and the brain and may have many causes not related to noise exposure.

Psychogenic—It occurs in the mind. Psychogenic hearing loss may occur as a result of some type of emotional trauma.

Major Causes of Hearing Loss

- Obstruction and disease
- Acoustic trauma
- Presbycusis/sociocusis
- Noise induced

NOTE: Presbycusis is hearing loss as a result of the aging process, whereas sociocusis is hearing loss as a result of continuous noise exposure not related to an occupation. For example, listening to loud music continuously would be classified as a sociocusis-related hearing loss.

OSHA Permissible Noise Exposures

In accordance with the OSHA Hearing standard (29 CFR 1910.95(b)(2)), an employee can be subjected to the following noise exposures for the specific times mentioned (Table 8.1).

It is strongly recommended that you commit Table 8.1 to memory, as there may be questions on the examination that are related to these exposure durations. In the following, we will discuss how to calculate these times.

TABLE 8.1

Permissible Noise Exposures

Exposure Times (h)	Sound Level (dbA)
8	90
6	92
4	95
3	97
2	100
1 1/2	102
1	105
1/2	110
1/4	115

Source: United States Department of Labor, Occupational Safety and Health Administration, 29 CFR 1910.95.

Entities of Noise

There are three entities of noise/sound. They are *sound power* (L_w), *sound intensity* (L_I), and *sound pressure* (L_p). Power does not change; it is constant. Intensity decreases with an increase in distance according to the inverse square law.

Sound Power (L_w)

The sound power of a source is expressed in terms of its sound power level and is written as (L_w or PWL). *Note: Commit to memory the sound power reference level, which is 10^{-12} W.*

Sound Pressure (L_p)

The amplitude of sound pressure disturbance can be related to the displacement amplitude of the vibrating sound source. Pressure is expressed as force per unit area. The preferred unit of pressure is the pascal (Pa) or 1 N/m². *Note: Commit to memory the sound pressure reference level, which is 0.00002 N/m².*

Sound Intensity (L_I)

The sound intensity level is similar to the sound pressure level, except that intensity is a vector quantity, having both magnitude and direction (DiNardi,[2] p. 436).

Noise Calculations

As mentioned in the preface, the primary purpose of this book is to assist the exam candidate in preparing to take and successfully pass the Associate Safety Professional (ASP) and CSP exams. In the following pages, we will discuss the equations provided by the Board of Certified Safety Professionals (BCSP) on the exam reference sheet, by explaining the meaning of the symbols and providing examples of their usage and examples of how to solve the equation, given a specific scenario.

Calculating Permissible Noise Exposures

As mentioned earlier in OSHA Permissible Noise Exposures, there exists an equation to help you calculate permissible exposures at various noise levels.

This equation is useful whenever a noise exposure is not listed in Table 8.1. The equation is as follows:

$$T = \frac{8}{2^{\left[\left(\frac{L-90}{5}\right)\right]}}.$$

For example, an employee is exposed to 80 dBA for 4 h, what is the permissible exposure time?

$$T = \frac{8}{2^{\left[\left(\frac{80-90}{5}\right)\right]}}$$

$$T = \frac{8}{2^{-2}}$$

$$T = \frac{8}{0.25}$$

$$T = 32 \text{ h}.$$

In other words, using this equation, the employee could be exposed to 80 dBA for up to 32 h.

NOTE: The American Board of Industrial Hygiene subtracts 85 from the noise exposure in the denominator exponent.

Calculating Noise Dosage

Noise dosage is the amount of noise that an individual has been exposed to for a measured period. To calculate noise dosage, we use the following equation provided on the ASP/CSP exam reference sheet:

$$D = 100 \left[\sum_{i=1}^{N} \frac{C_i}{T_i} \right],$$

where
  D = dosage (or effective dose)
  C_i = actual exposure time
  T_i = allowed exposure time (from Table 8.1)

Given the following information, calculate the employee's total exposure to noise, expressed as a dose or percentage of PEL: 90 dB for 4 h, 92 dB for 2 h, and 80 dB for 2 h. Using the equation above, we insert the data as follows:

$$D = 100 \left[\sum_{i=1}^{N} \frac{C_i}{T_i} \right]$$

$$D = 100 \left[\sum_{i=1}^{N} \left(\frac{4}{8} \right) + \left(\frac{2}{6} \right) + \left(\frac{2}{32} \right) \right]$$

$$D = 100(0.89)$$

$$D = 89\%.$$

To further explain, using the permissible exposure calculation on Table 8.1, we determine that an employee is permitted to be exposed to 90 dBA for 8 h, 92 dBA for 6 h, and 80 dBA for 32 h.

Another helpful equation to determine employee noise dosage exposure is

$$D = \left(\frac{C_1}{T_1} \right) + \left(\frac{C_2}{T_2} \right) + \dots \left(\frac{C_n}{T_n} \right).$$

Inserting exposures of 90 dBA for 2 h, 92 dBA for 2 h, and 95 for 4 h:

$$D = \left(\frac{2}{8} \right) + \left(\frac{2}{6} \right) + \left(\frac{4}{4} \right)$$

$$D = 0.25 + 0.33 + 1$$

$$D = 1.58.$$

The dosage is greater than 1 and therefore the permissible exposure limit has been exceeded. This equation is not listed on the BCSP exam reference sheet, but can be quite useful in determining if exposures have been exceeded.

Converting Noise Dosage into TWA

In order to convert noise dosimetry readings into a TWA, we utilize the following equation:

$$TWA = 16.61 \log_{10} \left[\frac{D}{100} \right] + 90,$$

where
  TWA = time-weighted average
  D = measured dose

Given the example that an employee has been monitored using a noise dosimeter and the results are that the employee has been exposed to 72% of the permissible exposure. What is the TWA, expressed in a sound pressure level?

$$TWA = 16.61 \log_{10}\left[\frac{72}{100}\right] + 90$$
$$TWA = 16.61 \log_{10}[0.72] + 90$$
$$TWA = (16.61)(-0.14) + 90$$
$$TWA = -2.37 + 90$$
$$TWA = 87.63 \text{ dBA.}$$

On the basis of a noise dosimeter measurement of 72% of TWA, the TWA expressed in decibels would be 87.63 or 88 dBA.

Combining Noise Levels

Many times, especially in industrial environments, there are numerous noise sources present in the work area. An employee may have exposures from various sources with different noise levels at the same. You cannot just add the sound level measurements and obtain an average. Hence, how do we accurately combine the noise levels? There are two methods that will be described. The first method is the decibel addition method, which utilizes Table 8.2.

To use this method, simply take the higher reading and subtract the lower reading and determine the distance. Once this is complete, add the amount to be added (corresponding difference in sound levels) to the higher reading. If combining more than two sources, calculate each combination separately. For example, there are four machines present in a facility. Machine 1 measures 79 dB, Machine 2 measures 85 dB, Machine 3 measures 90 dB, and Machine 4 measures 95 dB. What is the combined noise level?

Using Table 8.2, we perform the following steps:

Step 1: (85 dB Machine 2 (*highest reading*) − 79 dB) = Difference of 6. Go to Table 8.2 and look at the corresponding addition to a difference of 6, which is 1. Therefore, we add 1 to the highest level, which is 85 + 1 = 86.

TABLE 8.2

Decibel Addition Table

Difference between Two Decibel Levels to Be Added (dB)	Amount to Be Added to a Larger Level to Obtain Decibel Sum (dB)
0	3.0
1	2.6
2	2.1
3	1.8
4	1.4
5	1.2
6	1.0
7	0.8
8	0.6
9	0.5
10	0.4
11	0.3
12	0.2

Note: This table is, in essence, a summary of adding logarithms.

Step 2: We take the combination measurement for Step 1 (86 dB) and determine the difference between Machine 3 (90 dB – 86 dB = 4 dB). Again, look at Table 8.2 and determine that based on a difference of 4 dB, you would add 1.4 dB. Now we add 1.4 to the highest of this combination and determine it to be 91.4 dB.

Step 3: Take the combination measurement from Step 2 (91.4 dB) and obtain the difference between this and Machine 4, which is 95 dB. The difference is 3.6 or 4. Go to Table 8.2 and you will see that you should add 1.4 to 95 dB. The final combined noise level is 96.4 dB.

As you can easily see, this method is quite time consuming, as well as cumbersome. Now we will look at the second method for determining a combination of noise levels, by using the following equation:

$$L_{pt} = 10\log\left[\sum_{i=1}^{N} 10^{\left(\frac{L_{pi}}{10}\right)}\right],$$

where
L_{pt} = combined sound pressure level
L_{pi} = individual measured sound pressure level

Using the same data in the previous example (79, 85, 90, and 95), we insert these into the equation.

$$L_{pt} = 10 \log \left[\sum_{i=1}^{N} \left(10^{\frac{79}{10}} \right) + \left(10^{\frac{85}{10}} \right) + \left(10^{\frac{90}{10}} \right) + \left(10^{\frac{95}{10}} \right) \right]$$

$$L_{pt} = 10 \log[4,557,938,250]$$

$$L_{pt} = 10(9.659)$$

$$L_{pt} = 96.59 \text{ dB}.$$

Calculating Sound Levels at Various Distances

From time to time, it may be necessary to calculate sound levels taken at one distance and extrapolate the sound level at another distance. To do this, we use the following equation:

$$dB_1 = dB_0 + 20 \log_{10} \left(\frac{d_0}{d_1} \right),$$

where
  dB_0 = the original sound level measurement
  dB_1 = the calculated sound level measurement at another distance
  d_0 = the original distance where noise measurement was taken
  d_1 = the second distance that you would like to calculate the sound level reading for

Let's say that you have taken a sound level measurement 4 ft from a noise source and that measurement was 92 dBA. You would like to know what that noise level measurement would be at 10 ft. Inputting these data into the equation, we have

$$dB_1 = 92 \text{ dBA} + 20 \log_{10} \left(\frac{4 \text{ ft}}{10 \text{ ft}} \right)$$

$$dB_1 = 92 \text{ dBA} + 20 \log_{10}(0.4)$$

$$dB_1 = 92 \text{ dBA} + 20(-0.398)$$

$$dB_1 = 92 + (-7.96)$$

$$dB_1 = 84 \text{ dB}.$$

If the original measurement of 92 dBA was taken at 4 ft from the source, the sound level measurement at 10 ft should be 84 dBA, based on the results of this calculation.

Calculating Sound Power Levels (L_w)

To calculate the sound power level, we use the following equation:

$$L_w = 10 \log_{10} \frac{W}{W_0},$$

where
   L_w = sound power level
   W = acoustic power in watts
   W_0 = reference acoustic power (10^{-12} W)

For example, you have a source that is generating 0.02 W. What is the sound power level in decibels?

$$L_w = 10 \log_{10} \frac{0.02}{10^{-12}}$$
$$L_w = 10 \log_{10} 2 \times 10^{10}$$
$$L_w = (10)(10.30)$$
$$L_w = 103 \text{ dB.}$$

Calculating Sound Pressure (L_p) Levels (or SPL)

To calculate the sound pressure level, we use the following equation:

$$L_p = 20 \log_{10} \frac{p}{p_0} \text{ dB,}$$

where
   L_p or SPL = sound pressure level (dB)
   p = measured sound pressure level (N/m²)
   p_0 = reference sound pressure level (0.00002 N/m²)

Given a measured sound pressure level of 0.002 N/m², what would the sound pressure level be, expressed in decibels?
Inserting the data into the equation, we have:

$$L_p = 20 \log_{10} \frac{0.002}{0.00002} \text{ dB}$$
$$L_p = 20 \log_{10} 100 \text{ dB}$$
$$L_p = 20(2) \text{ dB}$$
$$L_p = 40 \text{ dB.}$$

Calculating Sound Intensity Levels (L_I)

To calculate sound intensity levels, we utilize the following equation:

$$I = \frac{p^2}{\rho c},$$

NOTE: It must be noted that the equation on the exam reference sheet does not show the RMS sound pressure squared. Therefore, it is necessary for you to remember the correction in this equation if performing sound intensity level calculations.

where
  I = sound intensity (W/m²)
  p = sound pressure level (N/m²)
  ρ = the density of the medium (in air, 1.2 kg/m²)
  c = the speed of sound (in air, it is 344 m/s)

Given a sound pressure level of 0.0796 N/m², calculate the sound intensity level in watts per square meter.

$$I = \frac{(0.0796 \text{ N/m}^2)^2}{(1.2 \text{ kg/m}^3)(344 \text{ m/s})}$$

$$I = \frac{0.006336}{412.8}$$

$$I = 0.0000153 \text{ W/m}^2.$$

You can insert the intensity level of 0.0000153 W/m² into the equation for sound power levels and determine the equivalent decibel level to be approximately 72 dB.

Calculating Room Absorption

It must be noted that the control of noise through bonding surfaces is limited in its effectiveness and can become quite costly owing to the large surface areas present. With this being said, to calculate the noise reduction from bonding surfaces, we can use the following equation:

$$\text{NR} = 10 \log_{10} \frac{(A_2)}{(A_1)},$$

where
  NR = noise reduction (dB)
  A_1 = total number of absorption units (sabins) in the room before treatment
  A_2 = total number of absorption units (sabins) in the room after treatment

NOTE: A sabin is a unit of sound measurement. It measures how well 1 ft^2 of any surface texture in a room is able to absorb sound reflections. A simple method for determining sabins is to multiply the square footage of surface area by the "noise reduction coefficient."

For example, you have a 20 ft × 20 ft room with hardwood flooring and you wish to install carpeting over the flooring in order to reduce sound pressure levels. What would be the overall noise reduction in decibels? (NRC [noise reduction coefficient]: hardwood, 0.15; carpeting, 0.48)

The first step would be to determine the number of square feet present, which is 400 SF. The next step would be to determine the number of sabins present before and after installation of the carpeting. We do this by multiplying 400 SF × the NRC for each of the materials and determine that the number of sabins presents before treatment is 60 and the number of sabins present after treatment is 192. Using this information, we can now solve for the noise reduction expressed in decibels.

$$NR = 10 \log_{10} \frac{192 \text{ sabins}}{60 \text{ sabins}}$$
$$NR = 10 \log_{10} 3.2$$
$$NR = 10(0.505)$$
$$NR = 5.05 \text{ dB.}$$

The reduction in noise, expressed in decibels, by installing carpeting is 5 dB. As you can see, this method of controlling noise is not very effective and could become very expensive in the process.

Calculating Absorption along a Transmission Path

Normal day-to-day duties of the professional safety manager do not require him or her to calculate absorptions. However, for purposes of preparing the candidate for the ASP/CSP exam, we will provide the equation and examples of how to use it. One of the most common examples given to show absorption along a transmission path is the commercial muffler. The equation provided on the exam reference sheet is

$$NR = \frac{12.6 \, P\alpha^{1.4}}{A} \, dB/ft,$$

where
  NR = noise reduction (decibels per foot of length)
   P = perimeter of the duct (in inches)

α = absorption coefficient of the lining material at the frequency of interest

A = cross-sectional area of the duct (in inches)

By way of example, you have a duct perimeter of 5 ft (60 inches) in length with a diameter of 1 ft. The absorption coefficient for the packing material is rated at 0.54. What is the NR in decibels per foot of length?

$$NR = \frac{12.6(60)(0.54^{1.4})}{113}$$

$$NR = \frac{319.06}{113}$$

$$NR = 2.82 \text{ dB/ft.}$$

Key Information to Remember on Noise

1. Fourteen hours away from noise environment is required prior to audiometric testing.
2. OSHA's hearing conservation standard is 29 CFR 1910.95.
3. Any employee with an occupational noise exposure equaling or exceeding 85 dBA or a dose of 50% must be included in the Hearing Conservation Program.
4. The Type 1 sound level meter is the minimum standard and has an accuracy of ±2 dB.
5. Monitoring is required by 29 CFR 1910.95 every 2 years.
6. Octave band analyzers determine sound level readings at various frequencies.
7. STS is a change in hearing relative to the baseline audiogram of an average of 10 dB or more at 2000, 3000, and 4000 Hz in either ear.
8. The reference acoustic power for sound power levels is 10^{-12} W.
9. The reference sound pressure level is 0.00002 N/m^2.
10. A sabin is equal to square footage of surface times the noise reduction coefficient.

References

1. United States Department of Labor, Occupational Safety and Health Administration, *29 CFR 1910.95*.
2. DiNardi, S. (Ed.), 2003. *The Occupational Environment: Its Evaluation, Control and Management*, American Industrial Hygiene Association, Fairfax, VA, p. 436.

9

Biological Hazards

Biological hazards (biohazards) are substances that are biological in nature and pose a threat to the health of living organisms. Sources of biological hazards include bacteria, viruses, insects, plants, birds, animals, and humans. These sources can cause a variety of health effects ranging from skin irritation and allergies to infections (e.g., tuberculosis, AIDS), cancer, and so on. Infections from biological agents can be caused by bacteria, viruses, rickettsia, chlamydia, and fungi. Parasites include protozoa, helminths, and arthropods. Toxic and allergenic substances include bites from animals or "sticks" from plants.

Biological hazards have become a leading cause of concern in the workplace. The Occupational Safety and Health Administration (OSHA) implemented the Blood-Borne Pathogen standard (29 CFR 1910.1030) to reduce the potential for occupational illnesses related to biological hazards. Inside this chapter, we will discuss the more common biological illnesses of occupational concern and the means to prevent injuries and illnesses related to them.

Bacterial Diseases

Anthrax

The etiological agent (causative agent) for anthrax is *Bacillus anthracis*. Most forms of anthrax are lethal. The primary occupations at risk of contracting anthrax include agricultural workers and occupations handling goat hair, wool, and hides and veterinarians. *B. anthracis* is a rod-shaped, gram-positive, aerobic bacterium. The bacterium normally rests in endospore form in the soil and can survive for decades in this state. Animals that graze on the vegetation contaminated with the bacterium ingest the bacterium and become infected. Anthrax can enter the human body through ingestion, inhalation, or cutaneous routes. Normally, infected humans do not infect noninfected humans. However, the clothing and personal items of an infected humans can become contaminated with the spores and cause infections in others through the inhalation, ingestion, or cutaneous contamination from the spores left on these items. Preventive measures include protective, impermeable clothing and equipment; prevention of skin contact, especially open wounds; and the use of high-efficiency respiratory protection. Vaccines are available to aid in preventive measures, but they are only useful in the prevention of anthrax if it is administered well in advance of any exposure. If exposed workers are given the vaccination series, they should receive annual booster injections.

Brucellosis

Brucellosis is an infectious disease caused by the bacteria of the genus *Brucella*. Occupations of interest in regard to brucellosis include meatpacking house employees and inspectors, livestock producers, and marketers. These bacteria are primarily passed among animals, and they cause disease in many different vertebrates. Various *Brucella* species affect sheep, goats, cattle, deer, elk, pigs, dogs, and several other animals. Humans become infected by coming in contact with animals or animal products that are contaminated with these bacteria. In humans, brucellosis can cause a range of symptoms that are similar to the flu and may include fever, sweats, headaches, back pains, and physical weakness. Severe infections of the central nervous systems or lining of the heart may occur. Brucellosis can also cause long-lasting or chronic symptoms that include recurrent fevers, joint pain, and fatigue.

Preventive measures include avoiding the consumption of unpasteurized milk, cheese, or ice cream while traveling. Hunters and animal herdsmen should use rubber gloves when handling viscera of animals. There is no vaccine available for humans.[1]

Leptospirosis

Leptospirosis is a bacterial disease that affects humans and animals. It is caused by bacteria of the genus *Leptospira*. In humans, it causes a wide range of symptoms, and some infected persons may have no symptoms at all. Symptoms of leptospirosis include high fever, severe headache, chills, muscle aches, and vomiting, and may include jaundice (yellow skin and eyes), red eyes, abdominal pain, diarrhea, or a rash. If the disease is not treated, the patient could develop kidney damage, meningitis (inflammation of the membrane around the brain and spinal cord), liver failure, and respiratory distress. In rare cases, death occurs.[2] Occupations of special interest for leptospirosis include farmers, field workers, sugarcane workers, meatpacking house workers, sewer workers, miners, and military personnel.

The risk of acquiring leptospirosis can be greatly reduced by not swimming or wading in water that might be contaminated with animal urine. Protective clothing or footwear should be worn by those exposed to contaminated water or soil because of their job or recreational activities. Leptospirosis is treated with antibiotics, such as doxycycline or penicillin, which should be given early in the course of the disease. Intravenous antibiotics may be required for persons with more severe symptoms. Persons with symptoms suggestive of leptospirosis should contact a health care provider.[2]

Plague

The primary occupations concerned with plague include shepherds, farmers, ranchers, hunters, and geologists. Plague is an infectious disease of animals and humans caused by a bacterium named *Yersinia pestis*. People usually get plague from being bitten by a rodent flea that is carrying the plague bacterium or by handling an infected animal. Millions of people in Europe died from plague in the Middle Ages, when human homes and places of work were inhabited by flea-infested rats. Today, modern antibiotics are effective against plague, but if an infected person is not treated promptly, the disease is likely to cause illness or death.[3]

Attempts to eliminate fleas and wild rodents from the natural environment in plague-infected areas are impractical. However, controlling rodents and their fleas around places where people live, work, and play is very important in preventing human disease. Therefore, preventive measures are directed to home, work, and recreational settings where the risk of acquiring plague is high. A combined approach using the following methods is recommended:

- Environmental sanitation
- Educating the public on ways to prevent plague exposures
- Preventive antibiotic therapy[4]

Tetanus

Tetanus (also known as "lockjaw") is a serious disease that causes painful tightening of the muscles, usually all over the body. It can lead to "locking" of the jaw so the victim cannot open his mouth or swallow. It is a disease of the nervous system caused by *Clostridium tetani* bacteria, a rod-shaped, anaerobic bacterium. It is found as spores in soil or as parasites in the gastrointestinal tract of animals. Tetanus leads to death in approximately 1 in 10 cases.[5] *C. tetani* enters the body through a break in the skin. However, tetanus is not transmitted from person to person. Tetanus can be prevented through the use of an effective vaccine. Occupations at greatest risk of contracting tetanus include those who work around domestic animals and soil.

Tuberculosis

Tuberculosis, also referred to as TB, is a disease caused by a bacterium called *Mycobacterium tuberculosis*. The bacteria usually attack the lungs, but TB bacteria can attack any part of the body such as the kidney, spine, and brain. If not treated properly, TB disease can be fatal. TB disease was once the leading cause of death in the United States. It is spread through the air, when people who have the disease cough, sneeze, or spit. Most infections in humans result in an asymptomatic (without symptoms), latent infection, and approximately

1 in 10 latent infections eventually progresses to active disease, which, if left untreated, kills more than 50% of its victims. Occupations at greatest risk include health care workers, prison employees and inmates, homeless shelter employees, and drug treatment center employees.

All health care settings need an infection-control program designed to ensure prompt detection, airborne precautions, and treatment of persons who have suspected or confirmed TB disease. In order to be effective, the primary emphasis of the TB infection-control program should be on achieving these three goals.[6]

In all health care settings, particularly those in which persons are at high risk for exposure, policies and procedures for TB control should be developed, reviewed periodically, and evaluated for effectiveness to determine the actions necessary to minimize the risk for transmission of TB.[6]

Tularemia

Tularemia is a disease of animals and humans caused by the bacterium *Francisella tularensis*. Rabbits, hares, and rodents are especially susceptible and often die in large numbers during outbreaks. Humans can become infected through several routes, including tick and deer fly bites, skin contact with infected animals, ingestion of contaminated water, or inhalation of contaminated dusts or aerosols. In addition, humans could be exposed as a result of bioterrorism. Symptoms vary depending upon the route of infection. Although tularemia can be life threatening, most infections can be treated successfully with antibiotics. Steps to prevent tularemia include use of insect repellent, wearing gloves when handling sick or dead animals, and not mowing over dead animals. In the United States, naturally occurring infections have been reported from all states except Hawaii.[7] Primary occupations at risk include forestry workers, butchers, and meat plant operators.

Cat Scratch Fever (Cat Scratch Disease)

Cat scratch disease (CSD) is a bacterial disease caused by *Bartonella henselae*. Most people with CSD have been bitten or scratched by a cat and developed a mild infection at the point of injury. Lymph nodes, especially those around the head, neck, and upper limbs, become swollen. Additionally, a person with CSD may experience fever, headache, fatigue, and a poor appetite. Measures to reduce the risk of contracting CSD include the following:

- Avoid "rough play" with cats, especially kittens. This includes any activity that may lead to cat scratches and bites.
- Wash cat bites and scratches immediately and thoroughly with running water and soap.

- Do not allow cats to lick open wounds that you may have.
- Control fleas.
- If you develop an infection (with pus and pronounced swelling) where you were scratched or bitten by a cat or develop symptoms, including fever, headache, swollen lymph nodes, and fatigue, contact your physician.[8]

Occupations at greatest risk include animal laboratory workers, veterinarians, and animal housing employees.

Viral Diseases

Hepatitis A

Hepatitis A is an acute infectious disease of the liver caused by the hepatitis A virus (HAV). Person-to-person transmission through the fecal–oral route (i.e., ingestion of something that has been contaminated with the feces of an infected person) is the primary means of HAV transmission in the United States. Most infections result from close personal contact with an infected household member or sex partner. Common-source outbreaks and sporadic cases can also occur from exposure to fecally contaminated food or water. Uncooked HAV-contaminated foods have been recognized as a source of outbreaks. Cooked foods can also transmit HAV if the temperature during food preparation is inadequate to kill the virus or if food is contaminated after cooking, as occurs in outbreaks associated with infected food handlers. Waterborne outbreaks are infrequent in developed countries with well-maintained sanitation and water supplies.[9] Occupations with the greatest potential exposure include daycare center workers, food preparation workers, and sewer and sanitation workers.

Vaccination with the full, two-dose series of hepatitis A vaccine is the best way to prevent HAV infection. Hepatitis A vaccine has been licensed in the United States for use in persons 12 months of age and older. The vaccine is recommended for persons who are more likely to get HAV infection or are more likely to get seriously ill if they get hepatitis A.[9]

Hepatitis B

Hepatitis B is caused by infection with the hepatitis B virus (HBV). The incubation period from the time of exposure to onset of symptoms is 6 weeks to 6 months. HBV is found in highest concentrations in blood and in lower concentrations in other body fluids (e.g., semen, vaginal secretions, and wound exudates). HBV infection can be self-limited or chronic.

HBV is efficiently transmitted by percutaneous or mucous membrane exposure to infectious blood or body fluids that contain blood. The primary risk factors that have been associated with infection are unprotected sex with an infected partner, birth to an infected mother, unprotected sex with more than one partner, men who have sex with other men (MSM), history of other sexually transmitted diseases (STDs), and illegal injection drug use.

The national strategy of the Centers for Disease Control and Prevention (CDC) to eliminate transmission of HBV infection includes

- Prevention of perinatal infection through routine screening of all pregnant women for HBsAg and immunoprophylaxis of infants born to HBsAg-positive mothers and infants born to mothers with unknown HBsAg status
- Routine infant vaccination
- Vaccination of previously unvaccinated children and adolescents through age 18 years
- Vaccination of previously unvaccinated adults at increased risk for infection

High vaccination coverage rates, with subsequent declines in acute hepatitis B incidence, have been achieved among infants and adolescents. In contrast, vaccination coverage among the majority of high-risk adult groups (e.g., persons with more than one sex partner in the previous 6 months, MSM, and injection drug users) has remained low, and the majority of new infections occur in these high-risk groups. STD clinics and other settings that provide services targeted to high-risk adults are ideal sites in which to provide hepatitis B vaccination to adults at risk for HBV infection. All unvaccinated adults seeking services in these settings should be assumed to be at risk for hepatitis B and should receive hepatitis B vaccination.[10] In addition, those employees working in the health care industry or emergency medicine pose the greatest occupational exposures.

Orf (Sore Mouth Disease)

Sore mouth is caused by a poxvirus (specifically orf virus) and is found all over the world. The scabs of infected animals contain virus, can fall off, remain in the environment, and serve as a source of infection to susceptible animals. A flock can become infected through contaminated bedding, feed, or trucks, or by direct contact with infected animals (e.g., replacements brought onto the operation or at shows).[11]

People can become infected with the virus that causes sore mouth. A person who comes into contact with virus from an infected animal or equipment (such as a harness that has rubbed the animal's sores) can potentially get infected. People often develop sores on their hands. The sore may be painful and can last for 2 months. People do not infect other people. Sores usually heal without scarring.[11]

The virus that causes sore mouth is spread to people by touching infected animals and their equipment. Some animals may or may not have visible sores but may still be able to spread the virus.[11]

The two ways to protect yourself and others include

1. Wearing nonporous (i.e., rubber or latex) gloves when handling sheep or goats, especially when you have an open cut or sore and are handling the animals mouth/muzzle area.
2. Practicing good hand hygiene by washing with clean, warm water and soap for at least 20 s (or using a waterless alcohol-based hand rub when soap is not available and hands are not visibly soiled).[11]

Occupations at greatest risk of exposure include shepherds, stockyard workers, and shearers (sheep and goats).

Rabies

Rabies is a preventable viral disease of mammals most often transmitted through the bite of a rabid animal. The vast majority of rabies cases reported to the CDC each year occur in wild animals like raccoons, skunks, bats, and foxes.[12]

The rabies virus infects the central nervous system, ultimately causing disease in the brain and death. The early symptoms of rabies in people are similar to that of many other illnesses, including fever, headache, and general weakness or discomfort. As the disease progresses, more specific symptoms appear and may include insomnia, anxiety, confusion, slight or partial paralysis, excitation, hallucinations, agitation, hypersalivation (increase in saliva), difficulty swallowing, and hydrophobia (fear of water). Death usually occurs within days of the onset of these symptoms.[12]

All species of mammals are susceptible to rabies virus infection, but only a few species are important as reservoirs for the disease. In the United States, distinct strains of rabies virus have been identified in raccoons, skunks, foxes, and coyotes. Several species of insectivorous bats are also reservoirs for strains of the rabies virus.[12]

Transmission of rabies virus usually begins when infected saliva of a host is passed to an uninfected animal. The most common mode of rabies virus transmission is through the bite and virus-containing saliva of an infected host, although transmission has been rarely documented via other routes such as contamination of mucous membranes (i.e., eyes, nose, and mouth), aerosol transmission, and corneal and organ transplantations.[12]

For people who have never been vaccinated against rabies previously, postexposure anti-rabies vaccination should always include administration of both passive antibody and vaccine.

The combination of human rabies immune globulin and vaccine is recommended for both bite and nonbite exposures, regardless of the interval between exposure and initiation of treatment.[12]

People who have been previously vaccinated or are receiving preexposure vaccination for rabies should receive only vaccine.[12] Occupations at greatest risk include veterinarians, wild animal handlers, cave explorers, and farmers and ranchers.

Rickettsial and Chlamydia Diseases

Psittacosis (Ornithosis)

Psittacosis is a zoonotic infectious disease caused by a bacterium called *Chlamydophila psittaci* and contracted not only from parrots, such as macaws, cockatiels, and budgerigars, but also from pigeons, sparrows, ducks, hens, gulls, and many other species of bird. Infection is acquired by inhaling dried secretions from infected birds. The incubation period is 5 to 19 days. Although all birds are susceptible, pet birds (parrots, parakeets, macaws, and cockatiels) and poultry (turkeys and ducks) are most frequently involved in transmission to humans. Bird owners, pet shop employees, and veterinarians are among the highest occupations at risk. Outbreaks of psittacosis in poultry processing plants have been reported.[13]

Preventive measures include continued education, and when handling potentially infected birds, such as cleaning of the cages, employees should wear the appropriate protective equipment, which includes the wearing of an N95 respirator or a higher level of protection. According to the National Association of State Veterinarians, Inc., surgical masks may not be effective in preventing the transmission of this disease.

Rocky Mountain Spotted Fever

Rocky Mountain spotted fever is the most severe tick-borne rickettsial illness in the United States. Occupations at greatest risk include military personnel, foresters, rangers, ranchers, farmers, trappers, construction workers, and lumber workers. This disease is caused by infection with the bacterial organism *Rickettsia rickettsii*. The organism that causes Rocky Mountain spotted fever is transmitted by the bite of an infected tick. The American dog tick (*Dermacentor variabilis*) and Rocky Mountain wood tick (*Dermacentor andersoni*) are the primary arthropods (vectors) that transmit Rocky Mountain spotted fever bacteria in the United States. The brown dog tick *Rhipicephalus sanguineus* has also been implicated as a vector as well as the tick *Amblyomma cajennense* in countries south of the United States.[14]

It is unreasonable to assume that a person can completely eliminate activities that may result in tick exposure. Therefore, take the following precautions to protect yourself when exposed to natural areas where ticks are present:

- Wear light-colored clothing that allows you to see ticks that are crawling on your clothing. Tuck your pants legs into your socks so that ticks cannot crawl up the inside of your pants legs.

- Apply repellents to discourage tick attachment. Repellents containing permethrin can be sprayed on boots and clothing and will last for several days. Repellents containing DEET (*N,N*-diethyl-*m*-toluamide) can be applied to the skin but will last only a few hours before reapplication is necessary. Use DEET with caution on children.

- Conduct a body check upon return from potentially tick-infested areas by searching your entire body for ticks. Use a handheld or full-length mirror to view all parts of your body. Remove any tick you find on your body.

Check children for ticks, especially in the hair, when returning from potentially tick-infested areas. Ticks may also be carried into the household on clothing and pets and only attach later, so both should be examined carefully to exclude ticks.[15]

Q Fever

Q fever is a zoonotic disease caused by *Coxiella burnetii,* a species of bacteria that is distributed globally. Occupations at greatest risk include veterinarians, dairy farmers, ranchers, stockyard workers, slaughterhouse employees, wool handlers, and rendering plant workers. In 1999, Q fever became a notifiable disease in the United States but reporting is not required in many other countries. Because the disease is underreported, scientists cannot reliably assess how many cases of Q fever have actually occurred worldwide. Many human infections are inapparent.[16]

Cattle, sheep, and goats are the primary reservoirs of *C. burnetii*. Infection has been noted in a wide variety of other animals, including other species of livestock and in domesticated pets. *C. burnetii* does not usually cause clinical disease in these animals, although abortion in goats and sheep has been linked to *C. burnetii* infection. Organisms are excreted in milk, urine, and feces of infected animals. Most importantly, during birthing, the organisms are shed in high numbers within the amniotic fluids and the placenta. The organisms are resistant to heat, drying, and many common disinfectants. These features enable the bacteria to survive for long periods in the environment. Infection of humans usually occurs by inhalation of these organisms from air that contains airborne barnyard dust contaminated by dried placental material, birth fluids, and excreta of infected herd animals. Humans are often very susceptible to the disease, and very few organisms may be required to cause infection.[16]

Ingestion of contaminated milk, followed by regurgitation and inspiration of the contaminated food, is a less common mode of transmission. Other modes of transmission to humans, including tick bites and human-to-human transmission, are rare.[16]

The following measures should be used in the prevention and control of Q fever:

- Educate the public on sources of infection.
- Appropriately dispose of placenta, birth products, fetal membranes, and aborted fetuses at facilities housing sheep and goats.
- Restrict access to barns and laboratories used in housing potentially infected animals.
- Use only pasteurized milk and milk products.
- Use appropriate procedures for bagging, autoclaving, and washing of laboratory clothing.
- Vaccinate (where possible) individuals engaged in research with pregnant sheep or live *C. burnetii*.
- Quarantine imported animals.
- Ensure that holding facilities for sheep should be located away from populated areas. Animals should be routinely tested for antibodies to *C. burnetii,* and measures should be implemented to prevent airflow to other occupied areas.
- Counsel persons at highest risk for developing chronic Q fever, especially persons with preexisting cardiac valvular disease or individuals with vascular grafts.

A vaccine for Q fever has been developed and has successfully protected humans in occupational settings in Australia. However, this vaccine is not commercially available in the United States.[16]

Fungal Diseases

Aspergillosis

Aspergillus is a fungus (or mold) that is very common in the environment. It is found in soil, on plants, and in decaying plant matter. It is also found in household dust, building materials, and even in spices and some food items. There are lots of different types of *Aspergillus*, but the most common ones are *Aspergillus fumigatus* and *Aspergillus flavus*. Some others are *Aspergillus terreus, Aspergillus nidulans,* and *Aspergillus niger.*[17]

Aspergillosis is a disease caused by *Aspergillus*. Farmers and grain workers are the primary occupations of concern regarding aspergillosis. There are many different kinds of aspergillosis. One kind is allergic bronchopulmonary aspergillosis (also called ABPA), a condition where the fungus causes allergic respiratory symptoms, such as wheezing and coughing, but does not actually invade and destroy tissue. Another kind of aspergillosis is invasive aspergillosis, a disease that usually affects people with immune system problems. In this condition, the fungus invades and damages tissues in the body. Invasive aspergillosis most commonly affects the lungs but can also cause infection in many other organs and can spread throughout the body.[17]

Since *Aspergillus* is so common in the environment, most people breathe in *Aspergillus* spores every day. It is probably impossible to completely avoid breathing in some *Aspergillus* spores. For people with healthy immune systems, this does not cause harm, and the immune system is able to get rid of the spores. But for people with compromised immune systems, breathing in *Aspergillus* spores, especially breathing in a lot of spores (such as in a very dusty environment), can lead to infection. Studies have shown that invasive aspergillosis can occur during building renovation or construction. Outbreaks of *Aspergillus* skin infections have been traced to contaminated biomedical devices.[17]

It is almost impossible to avoid all exposure to this fungus. It is present in the environment. However, for persons who are very immunocompromised, some measures that may be helpful include avoidance of dusty environments and activities where dust exposure is likely (such as construction zones), wearing N95 respirators when traveling near dusty environments, and avoidance of activities such as gardening and lawn work. Other air quality improvement measures such as HEPA filtration may be used in health care settings, and prophylactic antifungal medication may, in some circumstances, be prescribed by your doctor.[17]

Candidiasis

Candidiasis is also known as *thrush* and is a fungal infection of the *Candida* species, with *Candida albicans* being the most common. Occupations at greatest risk include dishwashers, bartenders, cooks, bakers, poultry, and packinghouse workers. Candidiasis is usually localized as infections of the skin or mucosal membranes of the oral cavity (thrush), the pharynx or esophagus, the gastrointestinal tract, the urinary bladder, or the genitalia. This disease can be transmitted sexually through intercourse. Prevention includes frequent hand washing and avoidance of hand-to-mouth or hand-to-genitalia contact when handling contaminated items.

Coccidiomycosis

Coccidiomycosis, also known as Valley Fever, is a fungal disease caused by *Coccidioides* species. These organisms live in the soil of semiarid areas. It is endemic in areas such as the southwestern United States, parts of Mexico, and South America. It is a reportable disease in states where the disease is endemic, such as California, New Mexico, Arizona, and Nevada. Of people who live in an endemic region, approximately 10%–50% will have evidence of exposure to *Coccidioides*. Most of the people who get the disease are people who live in or visit places where the fungus is in the soil and who engage in activities that expose them to dust (such as construction, agricultural work, military field training, and archeological exploration). The infection is not spread from person to person or from animals to people. The infectious form of the fungus exists when the fungus grows in the environment. The fungus changes its form when it infects a person, and this form cannot be transmitted from one person to another. Avoidance of dusty environments in endemic regions may help prevent infection. In addition, persons at risk for severe disease should avoid activities that may result in dust exposure, such as digging.[18]

Histoplasmosis

Histoplasmosis is a disease caused by the fungus *Histoplasma capsulatum*. Its symptoms vary greatly, but the disease primarily affects the lungs. Occasionally, other organs are affected. This form of the disease is called disseminated histoplasmosis, and it can be fatal if untreated. Occupations at greatest risk include environmental remediation workers, farmers, poultry workers, and veterinarians. *H. capsulatum* grows in soil and material contaminated with bat or bird droppings. Spores become airborne when contaminated soil is disturbed. Breathing the spores causes infection. The disease is not transmitted from an infected person to someone else.

It is not practical to test or decontaminate most sites that may be contaminated with *H. capsulatum*, but the following precautions can be taken to reduce a person's risk of exposure:

- Avoid areas that may harbor the fungus, for example, accumulations of bird or bat droppings.
- Before starting a job or activity having a risk for exposure to *H. capsulatum*, consult the NIOSH/NCID document *Histoplasmosis: Protecting Workers at Risk*. This document contains information on work practices and personal protective equipment that will reduce the risk of infection. A copy of this document can be obtained by requesting publication no. 2005-109 from the National Institute for Occupational Safety and Health. You can also request additional information by calling 1-800-CDC-INFO.[19]

Biological Safety

The primary objective of any biological safety program is to contain any harmful or potentially harmful biological agents from inside a controlled environment. The purpose of *primary containment* is to protect personnel and the immediate laboratory environment from exposure to infectious agents. The purpose of *secondary containment* is the protection of the environment outside of the laboratory. There are three basic elements to a containment program, including laboratory practice and technique, safety equipment, and facility design. Each of these will be discussed in this chapter.

Laboratory Practice and Technique

Any laboratory working with hazardous or potentially hazardous agents should develop standard operating procedures in place prior to handling any hazardous agents. The procedures and practices should be strictly enforced at all times to prevent contamination or exposure. Employees working with these agents should be thoroughly trained and knowledgeable of these procedures, including emergency action plans. This training should be carefully reviewed and updated periodically to insure that the employees are protected. Whenever there are new procedures or new agents introduced into the laboratory, retraining should be conducted.

Some of the standard laboratory practices included in the operating manual should include the following:

- Limited and controlled access into the laboratory.
- Work areas, particularly horizontal surfaces, should be cleaned and decontaminated daily.
- Contaminated materials, including cleaning materials, should be decontaminated prior to disposal.
- Eating, drinking, smoking, or gum chewing should be strictly PROHIBITED inside the controlled area.
- Frequent hand washing should be enforced.
- Protective clothing should be worn to prevent contamination of street clothes.
- Depending on the agents inside the laboratory, showering of employees inside a decontamination station may be required.
- Training.

Safety Equipment

Specific safety equipment for each laboratory will be different. However, the basic safety equipment includes biological safety cabinets, personal protective equipment, and safety centrifuge cups. Biological safety cabinets are classified as Class I, Class II, or Class III. Class I cabinets are considered partial containment cabinets and have the following features:

- Room air flows through fixed front opening.
- Approximately 8 inches.
- Minimum velocity of 75 linear fpm.
- Prevents aerosols generated in cabinet from escaping to room.
- Not appropriate for experimental systems vulnerable to airborne contamination.

Class II cabinets are laminar flow cabinets and have the following features:

- Protects the worker and research material.
- Curtain of room air entering the grille at forward edge of opening to the work surface.
- Partial recirculation of HEPA-filtered air.
- Downward flow of HEPA-filtered air creates contaminant-free zone.

Class II, Type A
- Fixed work opening.
- Minimum inflow velocity of 75 linear fpm.
- 70% recirculation.
- Minimum vertical velocity of 75 linear fpm.
- Not for flammable solvents, toxic agents, or radioactive materials.

Class II, Type B1
- Vertical sliding sash.
- 100 linear fpm at 8-inch work opening.
- 50 linear fpm downward vertical air velocity.
- 70% of air flowing through work area is exhausted.
- Not recommended for explosive vapors.

Class II, Type B2
- HEPA downflow air is from laboratory or outside air.
- Minimum inflow velocity of 100 fpm.
- 100% exhaust to outside through HEPA; no recirculation within cabinet.
- Used for low- to moderate-risk biological agents, toxic chemicals, and radionuclides.

Class II, Type B3

- HEPA downflow air is from laboratory or outside air.
- Minimum inflow velocity of 100 fpm.
- 70% recirculated air is exhausted to outside through HEPA.
- Used for low- to moderate-risk biological agents, minute or trace amounts of toxic chemicals, and radionuclides.

Class III (gas tight, negative pressure)

- Provides physical barrier between agent and worker.
- Highest degree of worker protection.
- Arm-length rubber gloves and sealed front panel.
- Air drawn into cabinet through HEPA filtration.[20]

Facility Design and Construction

Facility design and construction serves as secondary barriers. That is, the primary purpose is to prevent the escape of hazardous materials or agents outside of the controlled environment. Obviously, the design and construction of the laboratory will depend primarily on the nature and potential hazards of the particular agent.

Biosafety Levels

The level of safety when working with biological agents depends primarily on the potential hazards presented by the specific agent. In this section, we will discuss the four levels of biosafety. When discussing the biosafety levels, keep in mind the three major areas of containment, laboratory practices and techniques, and safety equipment and facility design.

Biosafety Level I

This level represents the basic level of containment and relies on standard microbiological practices. There are no special primary or secondary barriers required, except for hand washing facilities. This level is appropriate for any undergraduate or secondary educational training laboratory.

Biosafety Level II

Practices and techniques for biosafety level II include the use of standard microbiological techniques, training of personnel, collection of baseline serum samples taken and stored, warning signage, and personal protective equipment. Equipment for biosafety level II include Class I or Class II biological safety cabinets. Access to the laboratory should be limited and decontamination of equipment and instruments. This level of biosafety is adequate for most medical laboratories.

Biosafety Level III

With biosafety level III, greater emphasis is placed on primary and secondary barriers. Practices and techniques include performing all manipulations of agents in a biological safety cabinet or other enclosed equipment, such as a gas-tight aerosol generation chamber and strict controlled access to the laboratory environment. The use of biological cabinets (Class I, II, or III) and personal protective equipment should be strictly enforced. This level of safety is adequate for laboratories that perform work on indigenous or exotic agents with a potential for respiratory transmission that may cause serious and potentially lethal infection.

Biosafety Level IV

Biosafety level IV is the highest level of protection. Practices and techniques to prevent infection or escape of hazardous agents include all manipulations being performed in biological cabinets, sealing openings to the laboratory, standard microbiological practices, warning signage, and frequent hand washing. Safety equipment includes biological safety cabinets (Class I, II, or III), and depending on the hazardous agent, special engineering may be required. An air lock between the laboratory and public areas should be installed to prevent the escape of the hazardous agent.

Key Information to Remember on Biological Hazards

1. OSHA's Blood-Borne Pathogen Standard is found in 29 CFR 1910.1030.
2. The etiological agent for anthrax is *B. anthracis*.
3. Plague is an infectious disease of animals and humans caused by a bacterium named *Y. pestis*.
4. Tetanus, also known as "lockjaw," is a disease of the nervous system caused by *C. tetani*.
5. The bacterium *M. tuberculosis* causes tuberculosis.
6. Person-to-person transmission through the fecal–oral route is the primary means of HAV in the United States.

7. Hepatitis B can be transmitted by percutaneous or mucous membrane exposure to infectious blood or body fluids that contain blood.

8. *Aspergillus* is a fungus (mold) that can cause aspergillosis in farmers and grain workers.

9. Three primary preventive measures in biological safety include laboratory practice and techniques, safety equipment, and facility construction and design.

10. There are four levels of biosafety (Levels I, II, III, and IV).

References

1. Center for Disease Control, 2010. Available at http://www.cdc.gov/ncidod/dbmd/diseaseinfo/brucellosis_g.htm.
2. Center for Disease Control, 2010. Available at http://www.cdc.gov/ncidod/dbmd/diseaseinfo/leptospirosis_g.htm.
3. Center for Disease Control, 2010. Available at http://www.cdc.gov/ncidod/dvbid/plague/index.htm.
4. Center for Disease Control, 2010. Available at http://www.cdc.gov/ncidod/dvbid/plague/prevent.htm.
5. Center for Disease Control, 2010. Available at http://www.cdc.gov/vaccines/vpd-vac/tetanus/default.htm.
6. Center for Disease Control, 2010. Available at http://www.cdc.gov/tb/topic/basics/default.htm.
7. Center for Disease Control, 2010. Available at http://www.cdc.gov/Tularemia/.
8. Center for Disease Control, 2010. Available at http://www.cdc.gov/healthypets/diseases/catscratch.htm.
9. Center for Disease Control, 2010. Available at http://www.cdc.gov/hepatitis/HAV/index.htm.
10. Center for Disease Control, 2010. Available at http://www.cdc.gov/hepatitis/HBV/index.htm.
11. Center for Disease Control, 2010. Available at http://www.cdc.gov/ncidod/dvrd/orf_virus/.
12. Center for Disease Control, 2010. Available at http://www.cdc.gov/rabies/.
13. Center for Disease Control, 2010. Available at http://www.cdc.gov/ncidod/dbmd/diseaseinfo/psittacosis_t.htm.
14. Center for Disease Control, 2010. Available at http://www.cdc.gov/ticks/diseases/rocky_mountain_spotted_fever/faq.html.
15. Center for Disease Control, 2010. Available at http://www.cdc.gov/ticks/prevention.html.
16. Center for Disease Control, 2010. Available at http://www.cdc.gov/ncidod/dvrd/qfever/index.htm.
17. Center for Disease Control, 2010. Available at http://www.cdc.gov/nczved/divisions/dfbmd/diseases/aspergillosis/.

18. Center for Disease Control, 2010. Available at http://www.cdc.gov/nczved /divisions/dfbmd/diseases/coccidioidomycosis/.
19. Center for Disease Control, 2010. Available at http://www.cdc.gov/nczved /divisions/dfbmd/diseases/histoplasmosis/.
20. Fleeger, A. and Lillquist, D., 2006. *Industrial Hygiene Reference & Study Guide*, American Industrial Hygiene Association, Fairfax, VA, pp. 160–162.

10

Fire Protection and Prevention

An area of major concern for today's safety professional is the prevention of fires and the protection of personnel and property from the effects of fires. To fully understand the preventive measures to be taken, you must first understand the causes of fires. We begin this chapter by discussing the terminology that will be used.

Definitions

Combustion: Combustion is a chemical reaction that occurs between a fuel and an oxidizing agent that produces energy, usually in the form of heat and light. When a fuel is burned, the carbon reacts with the oxygen and can form either carbon monoxide (CO) or carbon dioxide (CO_2).

Heat of combustion: The amount of heat in calories evolved by the combustion of 1 g weight of a substance.

Combustible liquid: Any liquid having a flash point at or above 140°F and below 200°F.

Flammable liquid: Any liquid having a flash point below 140°F and having a vapor pressure not exceeding 40 psia at 100°F.

Flash point: The minimum temperature at which a liquid gives off vapor within a test vessel in sufficient concentration to form an ignitable mixture with air near the surface of the liquid. The flash point is normally an indication of susceptibility to ignition.

Lower flammability limit: The lower end of the concentration range of a flammable solvent at a given temperature and pressure for which air/vapor mixtures can ignite. The lower flammability limit (LFL) is usually expressed in volume percent.

Upper flammability limit: The maximum concentration of a combustible substance capable of propagating a flame through a homogeneous combustible mixture. The upper flammability limit (UFL) is usually expressed in volume percent.

Transfer of Heat

In this section, we will discuss three primary methods of heat transfer, which includes (1) radiation, (2) convection, and (3) conduction. It is important to understand the process of heat transfer. Heat always moves from a warmer place to a cooler place. Hot objects in a cooler room will eventually cool to room temperature. Cold objects in a warmer room will heat up to room temperature.

Heat Transfer by Radiation

Radiation heat is the amount of thermal radiation between two or more objects (bodies). Radiation travels in wavelengths and does not require that the two objects be in contact, as radiation can travel through a vacuum or space. We can calculate the radiation heat transfer rate by using the following equation:

$$\frac{Q}{A} = \varepsilon \sigma T^4,$$

where

Q/A = heat flux (energy/time area)
A = area of heat transfer
ε = emissivity of radiating body
σ = Stefan–Boltzmann constant
T = absolute temperature

Heat Transfer by Convection

Convection can be defined as the process whereby thermal energy is transferred by movement of a heated fluid such as liquid or air. There can be natural convection and forced convection. The rate of convective heat transfer can be calculated as follows:

$$Q = hA(T_s - T_b),$$

where

$Q =$ convective heat transfer rate
$h =$ heat transfer coefficient
$A =$ surface area of the heat being transferred
$T_s =$ surface temperature
$T_b =$ temperature of the fluid at bulk temperature

Heat Transfer by Conduction

Conduction is the transfer of thermal energy between two objects in contact with each other. The conductive heat transfer rate is calculated as follows:

$$Q = kA\left(\frac{T_1 - T_2}{L}\right),$$

where

Q = conductive heat transfer rate
k = thermal conductivity constant of material
A = surface area
T_1 = temperature on one side of a surface
T_2 = temperature on the opposite side of a surface
L = thickness of a material (distance of one side to the opposite side of a material)

Fire Tetrahedron

There are four components that are necessary to sustain combustion. Without the presence of all four components, fires will not exist. The four components include fuel, oxygen, heat, and a chain reaction. The fire tetrahedron is illustrated in Figure 10.1.

Understanding the fire tetrahedron is important not only in determining how to extinguish fires but also in preventing them from occurring. Fuels

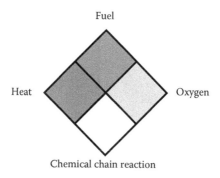

FIGURE 10.1
Fire tetrahedron.

are present everywhere and can include solids, liquids, gases, or vapors. The type of fuel in a fire will determine the type of extinguishing media that must be used.

Classification of Fires

According to the National Fire Protection Association (NFPA), fires can be classified as A, B, C, D, or K.

Class A Fires

Class A fires involve common combustibles such as wood, paper, cloth, rubber, trash, and plastics. They are common in typical commercial and home settings but can occur anywhere these types of materials are found.

Class B Fires

Class B fires involve flammable liquids, gases, solvents, oil, gasoline, paint, lacquers, tars, and other synthetic or oil-based products. Class B fires often spread rapidly and, unless properly secured, can reflash after the flames are extinguished.

Class C Fires

Class C fires involve energized electrical equipment, such as wiring, controls, motors, data processing panels, or appliances. They can be caused by a spark, power surge, or short circuit and typically occur in locations that are difficult to reach and see.

Class D Fires

Class D fires involve combustible metals such as magnesium and sodium. Combustible metal fires are unique industrial hazards that require special dry powder agents to extinguish.

Class K Fires

Class K fires involve combustible cooking media such as oils and grease commonly found in commercial kitchens. The new cooking media formulations used for commercial food preparation require a special wet chemical extinguishing agent that is specially suited for extinguishing and suppressing these extremely hot fires that have the ability to reflash.

Portable Fire Extinguishers

It is important to understand the classification of fires when selecting the appropriate type of fire extinguisher. For example, water should not be used to put out a Class C (electrical) fire, since it presents a shock hazard. Fire extinguishers are labeled in such as manner as to clearly identify the class of fires that they can extinguish. The new labeling system for fire extinguishers uses pictures to indicate what types of fires extinguishers can be used on and red diagonal lines through the types of fires they should not be used on. Figure 10.2 shows the symbols of the new labeling system.

In addition, fire extinguishers also come with number ratings to indicate how large a fire they can contain. For example, Class A extinguishers may have the numeric rating of 1, which would indicate 1 gal of extinguishing material. A higher rating number indicates increased levels of firefighting protection in each progressive extinguisher. The number ratings on Class B or C fire extinguishers indicate how many square feet of coverage the unit will contain. For example, a numeric rating of 5BC indicates a 5 ft² of coverage area for Class B and C fires. A multipurpose fire extinguisher may be rated at (example) 2A–10BC, which means that the extinguisher will cover up to 3000 ft² for Class A fire and 10 ft² for B/C fires. Table 10.1 shows the hazard level, minimum fire extinguisher rating required, maximum travel distance to a fire extinguisher, and the maximum square footage of coverage.[1]

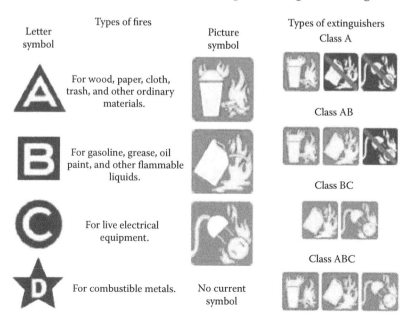

FIGURE 10.2
Fire extinguisher symbols (new labeling system).

TABLE 10.1

Portable Fire Extinguishers and Travel Distances by Hazard

Hazard Level	Minimum Fire Extinguisher Rating	Maximum Travel Distance to a Fire Extinguisher	Maximum Square Footage
Light hazard	2A10B:C	75 ft	Maximum coverage of 3000 ft² per unit of A (2A can cover 6000 ft²)
Ordinary hazard with moderate amounts of Class A combustibles and only minimal amounts of flammable/combustible liquids and gases	2A10B:C	75 ft	Maximum coverage of 1500 ft² per unit of A (2A extinguisher in this case can cover 3000 ft²)
Ordinary hazard with moderate amounts of flammable/combustible liquids and gases	Option 1 2A10B:C	30 ft	Maximum coverage of 1500 ft² per unit of A (2A extinguisher can cover 3000 ft²)
	Option 2 2A20B:C	50 ft	Maximum coverage of 1500 ft² per unit of A (2A extinguisher can cover 3000 ft²)
Extra hazard—all types	Option 1 4A40B:C	30 ft	Maximum coverage of 1000 ft² per unit of A (4A extinguisher can cover 4000 ft²)
	Option 2 4A80B:C	50 ft	Maximum coverage of 1000 ft² per unit of A (4A extinguisher can cover 4000 ft²)
Commercial kitchens	2A1B:C;K	30 ft	Maximum coverage of 1000 ft² per unit of A (2A extinguisher can cover 2000 ft²)

Source: Livermore-Pleasanton Fire Department, Livermore, California, http://www.ci.livermore.ca.us/LPFD/pdfs/fpfireextgl.pdf.

Hydrostatic Testing (Portable Fire Extinguishers)

The Occupational Safety and Health Administration (OSHA) requires that portable fire extinguishers undergo hydrostatic testing at intervals listed in Table 10.2, except under the following circumstances:

- When the unit has been repaired by soldering, welding, brazing, or use of patching compounds
- When the cylinder or shell threads are damaged
- When there is corrosion that has caused pitting, including corrosion under removable name plate assemblies

TABLE 10.2

OSHA's Hydrostatic Test Intervals

Type of Extinguisher	Test Interval (years)
Soda acid (soldered brass shells) (until January 1, 1982)	a
Soda acid (stainless steel shell)	5
Cartridge operated water and/or antifreeze	5
Stored pressure water and/or antifreeze	5
Wetting agent	5
Foam (soldered brass shells) (until January 1, 1982)	a
Foam (stainless steel shell)	5
Aqueous film forming foam (AFFF)	5
Loaded stream	5
Dry chemical with stainless steel	5
Carbon dioxide	5
Dry chemical, stored pressure, with mild steel, brazed brass, or aluminum shells	12
Dry chemical, cartridge or cylinder operated, with mild steel shells	12
Halon 1211	12
Halon 1301	12
Dry powder, cartridge, or cylinder operated with mild steel shells	12

[a] Extinguishers having shells constructed of copper or brass joined by soft solder or rivets shall not be hydrostatically tested and shall be removed from services by January 1, 1982.

- When the extinguisher has been burned in a fire
- When a calcium chloride extinguished agent has been used in a stainless steel shell

Fire Extinguisher Inspections and Service Requirements

An employer is responsible for ensuring that the inspection, maintenance, and testing of all portable fire extinguishers are conducted. Portable fire extinguishers shall be visually inspected monthly. Furthermore, the employer shall ensure that each portable fire extinguisher is serviced annually and tested in accordance with Table 10.2.[2]

Automatic Sprinkler Systems

There are several categories of sprinkler systems that may include dry pipe, wet pipe, deluge, combined dry pipe and preaction, and sprinklers that are designed for limited water supply systems. A fire sprinkler system consists of a water supply that provides adequate pressure and flow rate to a water distribution piping system, which has sprinkler heads attached. The

TABLE 10.3

Sprinkler Head Color Codes

Maximum Ceiling Temperature		
°C	°F	Color Code
38	100	Uncolored
66	150	White
107	225	Blue
150	300	Red
190	375	Green
218	425	Orange
246	475	Orange

sprinkler heads are held closed by either a heat-sensitive glass bulb or a two-part metal link held together with fusible alloy. The sprinkler heads have varying degrees of temperature sensitivities and are color coded. These color codes are identified in Table 10.3.

Dry Pipe Systems

Dry pipe systems are normally installed in areas where there is a potential for freezing, which would render the system basically inoperable when needed. For example, dry pipe systems are normally installed in parking garages, unheated buildings, or outside canopies that may be attached to buildings. According to the NFPA, dry pipe systems cannot be installed unless the range of ambient temperatures reaches below 40°F. In this type of system, water is not present in the pipe until the system is triggered and the water is released into the distribution system.

Wet Pipe Systems

Wet pipe systems are the more frequently installed systems. In this type of system, water is present at all times inside the piping distribution system. Once the sprinkler head is opened, the water in the distribution system will flow out of the distribution system.

Deluge Systems

Deluge systems are primarily installed in areas where there are special hazards where the rapid spread of fire is a major concern. In deluge systems, the heat sensors from the sprinkler heads have been removed by design. There is no water stored in the distribution system. The water is released into the piping distribution system by the activation of a *deluge valve*, which is activated by a fire alarm system. Deluge systems must be manually reset, by resetting the deluge valve.

Preaction Systems

Preaction systems are basically hybrids of wet, dry, and deluge systems. In a preaction system, a water supply valve is added to a dry pipe system. The valve itself is opened by the activation of a fire detection system, such as a fire alarm or smoke detector. Once activated, the water supply valve will send water to the distribution system. This type of system is normally placed in areas where accidental discharge is undesirable. Typical areas include data centers, museums, and art centers.

Water Spray Systems

Water spray systems operate in the same fashion as do deluge systems. However, the piping and discharge nozzle spray patterns are designed to protect a uniquely configured hazard. Such patterns are designed for three-dimensional components or equipment, whereas the deluge systems are designed to cover the floor area of a space.

Foam Water Sprinkler Systems

For special applications, the foam water sprinkler may be used. Foam water sprinkler systems discharge a mixture of water and low expansion foam concentrate, which results in a foam spray from the sprinkler head. These systems are usually used in areas that have high challenge fires, such as flammable liquids and airport hangars.

Fire Hydrants

A fire hydrant is an active fire protection measure that provides a source of water from the municipal water system or other source. Buildings located near fire hydrants may qualify for special insurance rate reductions on the basis of the proximity to the hydrant. Fire hydrants are color coded to indicate their specific water flow rate. Table 10.4 illustrates the color codes and flow rates of various fire hydrants.

TABLE 10.4

Fire Hydrant Color Code System

Fire Hydrant Top Color	Flow Rate (GPM, gallons per minute)
Red	500
Orange	500–1000
Green	1000–1500
Blue	>1500

Fire Detection

There are wide varieties of fire detectors on the market, including heat detectors, rate-compensation detectors, rate-of-rise detectors, pneumatic detectors, smoke detectors, and flame detectors. We will discuss each of these individually in this section.

Heat Detectors

Heat detectors are devices that are installed in fixed locations. These devices respond to changes in the ambient temperature above a predetermined temperature. Once the ambient temperature rises above the predetermined temperature, the alarm is triggered, indicating a potential fire. When selecting a heat detector, it is important to know the maximum normal temperature. For example, installing a 135°F heat detector near a beam of a metal building in hot climates may falsely trigger the alarm.

Rate-of-Rise Heat Detectors

Rate-of-rise heat detectors react to the sudden change or rise in ambient temperature from a normal baseline condition. The sudden temperature increase that matches the predetermined alarm criteria will cause an alarm.

Rate-Compensation Detectors

While the fixed-temperature heat detector will not initiate an alarm until the air temperature near the ceiling exceeds the design operating point, the rate-of-rise detector will, since the effect of flaming fires is to heat the surrounding air rapidly. Rate-of-rise detectors are designed to compensate for the normal changes in ambient temperature that are expected under nonfire conditions.

Smoke Detectors

Smoke detectors are classified according to their operating principles, which include ionization- and photoelectric-type detectors. *Ionization detectors* operate by using a small amount of radioactive material to ionize the air within a sensing chamber inside the detector. The ionization of the air permits the air to conduct electricity between two electrodes within this chamber. As smoke enters the chamber, the smoke particles become ionized and reduce the conductivity of the air between the electrodes. This reduction in conductivity between the electrodes is sensed and will cause the detector to respond.

Photoelectric detectors operate on one of three different principles: light obscuration principle, light scattering principle, or cloud chamber principle.

Light obscuration photoelectric detectors operate by projecting a light beam onto a photosensitive device. As smoke particles enter the chamber between the light source and the photosensitive device, the light intensity diminishes and initiates an alarm at a predetermined level. Light scattering detectors operate with a light source and photosensitive device. In this device, the photosensitive device is not in the light beam path. As smoke particles enter the sensing chamber, light is reflected (scattered) from the smoke particles onto the photosensitive device causing the detector to alarm. The cloud chamber operating principle draws an air sample from the protected area into a high-humidity chamber within the detector. After the air samples have been raised to high humidity, the pressure is slightly lowered. If smoke particles are in the chamber, a cloud will form. The density of this cloud is measured with a photoelectric device. When the "cloud" reaches a predetermined density, the detector will respond with an alarm.[3]

Fire Hydrants

Fire hydrant caps are color coded to indicate the specific flow rates of the hydrants. Table 10.4 illustrates the fire hydrant color and flow rates.

Flammable and Combustible Liquids

As mentioned earlier in Definitions, a flammable liquid is any liquid having a flash point below 140°F and a vapor pressure not exceeding 40 psia at 100°F. A combustible liquid is any liquid having a flash point at or above 140°F and below 200°F. The NFPA has classified flammable liquids into three classes according to their boiling points. These classes are identified in Table 10.5.

When taking the ASP or CSP examination, it is highly recommended that you commit to memory Tables 10.5 and 10.6.

TABLE 10.5

Characteristics of Flammable Liquids by Class

Classes of Flammable Liquids	
Class	Characteristics
IA	Flash point below 73°F; boiling point below 100°F
IB	Flash point below 73°F; boiling point at or above 100°F
IC	Flash point at or above 73°F, but below 100°F

TABLE 10.6

Characteristics of Combustible Liquids by Class

Classes of Combustible Liquids	
Class	Characteristics
II	Liquids having flash points at or above 100°F and below 140°F
IIIA	Liquids having flash points at or above 140°F and below 200°F
IIIB	Liquids having flash points at or above 200°F

The NFPA has also classified combustible liquids into three classes according to their flash points. These classifications are illustrated in Table 10.6.

To further understand flammable and combustible liquids, it is necessary to understand some of the individual properties associated with them. Each of the properties listed below will be referenced on material safety data sheets (MSDSs), along with their assigned values for the substance listed on the MSDS. The following are key properties that will be discussed in more detail.

Flash Point

We have already defined flash point as the minimum temperature at which a liquid gives off vapor within a test vessel in sufficient concentration to form an ignitable mixture with air near the surface of the liquid. The flash point is normally an indication of susceptibility to ignition.

Vapor Pressure

The *vapor pressure* of a liquid is defined as the pressure exerted by the molecules that escapes from the liquid to form a separate vapor phase above the liquid surface. The pressure exerted by the vapor phase is called the vapor or saturation pressure. Vapor or saturation pressure depends on temperature. As the temperature of a liquid or solid increases, its vapor pressure also increases. Conversely, vapor pressure decreases as the temperature decreases.

If fluids consist of more than one component, then components with high vapor pressures are called light components and those with lower vapor pressures are called heavy components.

Fire Point

Fire point is closely related to flash point. However, fire point is defined as the temperature at which a substance will give off a vapor that will burn continuously after ignition. Usually, the fire point is higher than the flash point.

Flammable and Explosive Limit Ranges

Flammability limits, or *explosive limits,* give the proportion of combustible gases in a mixture. Concentrations between the lower and upper limits of the mixture are flammable or explosive. The lower flammable limit (LFL) or the lower explosive limit (LEL) is the leanest mixture that is still flammable or explosive. Conversely, the upper flammable limit (UFL) or upper explosive limit (UEL) is the richest mixture that is still flammable or combustible. These concentrations are given in percentage of air.

Table 10.7 lists some of the more commonly used gases that are used as fuel or in combustion processes along with their LFL and UFL.

Autoignition Temperature

The autoignition temperature, also referred to as kindling point, is the lowest temperature at which a substance will ignite in a normal atmosphere without an external source of ignition from a spark or flame. The temperature at which a chemical will ignite decreases as the pressure increases or oxygen concentration increases.

Specific Gravity

The specific gravity describes the density of a liquid compared to the density of water. Those liquids with a specific gravity of one or less (≤ 1) are lighter than water and those with specific gravities of greater than or equal to one (≥ 1) are heavier than water. The specific gravity is particularly important when addressing chemical spills.

Vapor Density

Like specific gravity, vapor density is the measure of the density of a substance compared to air. Those gases and vapors with densities greater than or equal to one (≥ 1) are heavier than air and will tend to sink to lower levels. Those gases and vapors with densities less than or equal to one (≤ 1) are lighter than air and will tend to float upward. Knowing the vapor density of a substance is useful in determining the location of the ventilation system. For example, those substances with vapor densities of greater than one will sink. Therefore, the ventilation should be located near the floor or lower level of the area.

Evaporation Rate

The evaporation rate is the rate at which a liquid is converted to vapor at a given temperature and pressure. Butylacetate, which has an evaporation rate of 1, is the standard by which most substances are compared to. Therefore, if a substance has an evaporation rate of less than 1, it will evaporate more quickly than butylacetate.

TABLE 10.7

Lower and Upper Flammability/Explosive Limits for
Common Chemicals

Chemical	LFL/LEL	UFL/UEL
Acetaldehyde	4	60
Acetone	2.6	12.8
Acetylene	2.5	81
Ammonia	15	28
Arsine	5.1	78
Benzene	1.35	6.65
n-Butane	1.86	8.41
iso-Butane	1.80	8.44
iso-Butene	1.8	9.0
Butylene	1.98	9.65
Carbon disulfide	1.3	50
Carbon monoxide	12	75
Cyclohexane	1.3	8
Cyclopropane	2.4	10.4
Diethyl ether	1.9	36
Ethane	3	12.4
Ethylene	2.75	28.6
Ethyl alcohol	3.3	19
Ethyl chloride	3.8	15.4
Fuel oil no. 1	0.7	5
Hydrogen	4	75
Isobutane	1.8	9.6
Isopropyl alcohol	2	12
Gasoline	1.4	7.6
Kerosene	0.7	5
Methane	5	15
Methyl alcohol	6.7	36
Methyl chloride	10.7	17.4
Methyl ethyl ketone	1.8	10
Naphthalene	0.9	5.9
n-Heptane	1.0	6.0
n-Hexane	1.25	7.0
n-Pentane	1.65	7.7
Neohexane	1.19	7.58
Neopentane	1.38	7.22
n-Octane	0.95	3.20
iso-Octane	0.79	5.94
n-Pentane	1.4	7.8
iso-Pentane	1.32	9.16

TABLE 10.7 (CONTINUED)

Lower and Upper Flammability/Explosive Limits for
Common Chemicals

Chemical	LFL/LEL	UFL/UEL
Propane	2.1	10.1
Propylene	2.0	11.1
Silane	1.5	98
Styrene	1.1	6.1
Toluene	1.27	6.75
Triptane	1.08	0.69
p-Xylene	1.0	6.0

Note: Limits are for gas and air at 20°C and atmospheric pressure.

Water Solubility

Water solubility, which is also known as aqueous solubility, is the maximum amount of a substance that can dissolve in water at equilibrium at a given temperature and pressure. Water solubility (S_w) values are usually expressed as moles of solute per liter. The term *water solubility* is primarily used in many environmental studies to help determine the fate of chemicals in the environment (http://toxics.usgs.gov/definitions/water_solubility.html).

Boiling Point

The boiling point of a substance is the temperature at which the vapor pressure of the liquid is equal to the environmental pressure surrounding the liquid. It is the temperature point where the liquid is converted into a vapor.

Storage Requirements for Flammable and Combustible Liquids

Tables 10.8[4] and 10.9 show the maximum allowable size of containers for flammable and combustible liquids, respectively.

TABLE 10.8

Maximum Allowable Size of Containers for Flammable Liquids (OSHA)

Container Type	Flammable Liquids		
	Class IA	Class IB	Class IC
Glass or approved plastic	1 pt	1 qt	1 gal
Metal (other than DOT drums)	1 gal	5 gal	5 gal
Safety cans	2 gal	5 gal	5 gal
Metal drums (DOT specifications)	60 gal	60 gal	60 gal
Approved portable tanks	660 gal	660 gal	660 gal

TABLE 10.9

Maximum Allowable Size of Containers for Combustible Liquids (OSHA)

	Combustible Liquids	
Container Type	Class II	Class III
Glass or approved plastic	1 gal	1 gal
Metal (other than DOT drums)	5 gal	5 gal
Safety cans	5 gal	5 gal
Metal drums (DOT specifications)	60 gal	60 gal
Approved portable tanks	660 gal	660 gal

TABLE 10.10

Inside Storage Room Maximum Quantities per Square Foot (OSHA)

Fire Protection Provided[a]	Fire Resistance (h)	Maximum Size (ft²)	Total Allowable Quantities (gal/ft²)
Yes	2	500	10
No	2	500	5
Yes	1	150	4
No	1	150	2

[a] Examples of fire protection systems include sprinkler, water spray, carbon dioxide, or other systems.

Inside Storage Rooms for Flammable or Combustible Liquids

When flammable or combustible liquids are stored in inside rooms, the storage room shall be equipped with either a gravity or a mechanical exhaust ventilation system. This system shall be designed to provide for a complete change of air within the room at least six times per hour. In addition, in each inside storage room, there shall be maintained one clear aisle at least 3 ft wide. Containers over 30 gal capacity will not be stacked one upon the other. Table 10.10 illustrates the maximum storage capacity for any inside storage room.

Key Information to Remember on Fire Protection and Prevention

1. Radiation heat is the amount of thermal radiation between two or more objects (bodies).
2. Convection can be defined as the process whereby thermal energy is transferred by movement of a heated fluid such as liquid or air.

3. Conduction is the transfer of thermal energy between two objects in contact with each other.

4. According to the NFPA, fires can be classified as A, B, C, D, or K.

5. Portable fire extinguishers shall be visually inspected monthly.

6. Portable fire extinguishers shall be serviced annually and hydrostatically tested at 5 or 12 years, depending on their shell.

7. The lower flammability limit or lower explosive limit is the leanest mixture that is still flammable or explosive.

8. The upper flammability limit or upper explosive limit is the richest mixture that is still flammable or combustible.

References

1. Livermore-Pleasanton Fire Department, Livermore, California, 2010. Available at http://www.ci.livermore.ca.us/LPFD/pdfs/fpfireextgl.pdf.
2. U.S. Department of Labor, Occupational Safety and Health Administration, 2010, *29 CFR 1910.157, Table L-1*.
3. Marchetti, L., 2010. PDH Online, *Course M110A, Module 5—Fire Detection Devices*, pp. 1–7. Available at http://www.pdhcenter.com/courses/m110a/Module5.pdf.
4. U.S. Department of Labor, Occupational Safety and Health Administration, 2010, *29 CFR 1910.106, Table H-12*.

11

Thermal Stressors

In today's work environments, there are a wide variety of temperatures, not only outdoor temperatures, but indoors, as well. Thermal stress can be a problem in indoor operations, such as boiler maintenance after a "shutdown," smelting operations, laundries and kitchens, foundries, brick firing and ceramic operations, and work in refrigerated rooms or freezers. Oftentimes, the personal protective equipment used to protect employees from one type of injury can create additional hazards from thermal stress. Thermal stress should be considered when selecting the appropriate personal protective equipment and to determine work/rest cycles.

Thermal stress, whether heat or cold, can increase the risk of accidents in the workplace by causing fatigue. Employees exposed to extreme hot or cold temperatures may have a hard time concentrating on work and ignore safety. The Occupational Safety and Health Administration (OSHA) does not have a special standard for thermal stress. However, since thermal stress is considered a hazard, the employee is protected under OSHA's General Duty Clause, which states that "employers shall provide a workplace free of recognized hazards...."[1]

Heat Stress

Heat stress is of great concern in the workplace, as well as during recreational activities. It is caused by a combination of factors and tends to increase body temperature, heart rate, and sweating.

Sources of Heat Stress

There are four basic sources of heat to the body, which includes (1) radiation, (2) convection, (3) conduction, and (4) metabolic. It is important to understand the process of heat transfer. Heat always moves from a warmer place to a cooler place. Hot objects in a cooler room will eventually cool to room temperature. Cold objects in a warmer room will heat up to room temperature. *Radiation* heat is the amount of thermal radiation between two or more objects (bodies). Radiation travels in wavelengths and does not require that the two objects be in contact, as radiation can travel through a vacuum or space. *Convection* can be defined as the process whereby thermal energy is transferred by movement of a heated fluid such as liquid or air. *Conduction* is the transfer of thermal energy between two objects in contact with each other. *Metabolic* heat is generated from within the body through work. Metabolic heat is the source of greatest concern, since it has potential to overheat the body.

Human Body Reaction to Heat

What actually happens to the human body when exposed to heat? The human body, as with all other organisms, has a system of thermoregulation, which helps maintain a body temperature within a certain range. The survival range for a human body is between 97°F and 100°F with a normal temperature of approximately 98.6°F. When exposed to heat extremes, the body must regulate itself to stay within this temperature range. To do so, the body must get rid of the excess heat, through varying the rate and amount of blood circulation through the skin and the release of sweat from the glands. This is accomplished when the heart begins to pump more blood, the blood vessels expand to accommodate the increased flow, and the capillaries that are located along the upper layer of the skin begin to fill with blood. The blood then circulates closer to the surface of the skin, and excess heat is lost to the cooler environment. The cooled blood returns to the core of the body and picks up more heat and carries it to the surface. This process is repeated until the core temperature reaches a state of homeostasis.

The sweat that appears on the surface of the skin is evaporated by the cooler temperatures outside of the body. The evaporation of sweat cools the skin, thereby eliminating heat from the body. The problem that is encountered in this process is that when environmental temperatures outside of the body approach normal skin temperature, the cooling process becomes more difficult. However, the heart continues to pump blood to the surface, the sweat glands pour liquids containing electrolytes onto the surface of the skin, and the evaporation of the sweat becomes the principal effective means of cooling.

If the sweat is not removed from the skin by evaporation, it will not cool the body. Under conditions of high humidity, the evaporation of sweat from the skin is decreased and the body's efforts to maintain acceptable body temperature may be significantly impaired. With so much blood going to the external surface of the body, less is going to the active muscles, the brain, and other organs; strength declines; and fatigue occurs sooner than normal. Alertness and mental capacity also may be affected. Workers who must perform delicate or detailed work may find their accuracy decrease, and others may find their comprehension and retention of information lowered.[1]

Safety-Related Issues of Heat

In addition to the health-related issues, there are a wide variety of safety issues related to heat and heat stress. Heat in the workplace tends to promote accidents by means of sweaty and slippery palms, dizziness, fatigue, decreased alertness, and possibly the fogging of safety glasses or goggles. Physical discomfort increases irritability, anger, and other emotional states that sometimes cause workers to overlook safety procedures or to divert attention from hazardous tasks. Combined with longer shifts, there is a greater risk of accidents and health problems.

Health-Related Issues of Heat

Extended exposures to heat extremes bring about a wide variety of heat-induced disorders. Managers and supervisors should insure that all workers exposed to potential heat be trained in the signs and symptoms of these disorders, and how to prevent them. The disorders that will be discussed include heat rash, heat cramps, heat syncope, dehydration, heat exhaustion, and heat stroke.

Heat Rash

Heat rash is most likely to occur in hot, humid environments where sweat is not easily removed from the surface of the skin through evaporation and the skin remains wet most of the time. For example, the sweat remains trapped in the clothing of an employee and can possibly cause a rash that is related to heat. The causes of heat rash are prolonged, uninterrupted sweating and poor hygiene. The signs and symptoms of heat rash include skin eruptions, itching red skin, and reduced sweating. The prevention of heat rash includes keeping the skin clean and periodically allowing the skin to dry. If heat rash occurs, clean and dry the skin, change into dry clothing, and reduce heat exposure. Medical attention may include topical ointments to prevent any bacteriological infection.

Heat Cramps

Heat cramps result in individuals who sweat profusely in heat, drink large quantities of water, and do not adequately replace the body salts that are lost through sweating. Heat cramps are painful spasms of the muscles created by the loss of electrolytes. The signs and symptoms of heat cramps include painful muscle cramps and incapacitating pain in the muscle. To prevent heat cramps, workers should drink plenty of fluids and add salt to their diets. Normal salting of food is considered adequate to reduce the possibility of heat cramps. In extremely hot environments, drinking of water is encouraged. However, about every third time a person drinks, it is recommended that they drink some type of sports drink to help replace the lost electrolytes.

Heat Syncope

Heat syncope is fainting as a result of exposure to heat. It is caused by mild overheating with inadequate water or salt. The signs and symptoms of heat syncope include blurred vision, fainting, and normal body temperature. The main cause of heat syncope is pooling of blood in the legs and skin from prolonged static posture and heat exposure. To prevent heat syncope, it is important to flex the leg muscles several times before moving and stand or sit up slowly. Should heat syncope occur, have the affected employee lie on their back in a cool environment and drink water.

Dehydration

Dehydration occurs when the amount of water leaving the body is greater than the amount being taken in. There are no early symptoms of dehydration. However, signs and symptoms in later stages include fatigue, weakness, dry mouth, and the loss of work capacity and increased response time. A rule of thumb when referring to dehydration is that if you are thirsty, then dehydration has already occurred. Major contributing factors to dehydration include excess fluid loss (through sweating or other means), alcohol consumption, and some medications. In order to prevent dehydration, employees are encouraged to drink water frequently and to use normal salt in their diets. Once dehydration has occurred, fluid and salt replacement are required.

Heat Exhaustion

Heat exhaustion is caused by the loss of large amounts of fluid and salts through excessive sweating. Workers suffering from heat exhaustion still sweat but experience extreme fatigue or loss of consciousness. If heat exhaustion is not treated, the illness may advance to heat stroke. The signs and symptoms of heat exhaustion include fatigue, weakness, blurred vision, dizziness, high pulse rate, profuse sweating, low blood pressure, pale face, clammy skin, collapse, nausea and vomiting, headaches, and slightly increased body temperature. The major cause of heat exhaustion is dehydration, low level of acclimation, and low level of physical fitness. To prevent heat exhaustion, an employee should drink water or other fluids (sports drinks) frequently, take normal salt in their diets, take rest breaks to a cool area, and acclimate themselves to the environment. The treatment for heat exhaustion is to have the affected person lie down flat on their back in a cool environment, drink water, cool their skin with a cool mist or wet cloth, and loosen their clothing (especially at the neck, waist, and feet or ankles). Many medical experts conclude that once a person has experienced heat exhaustion, they are more susceptible to a second experience with heat exhaustion. Therefore, it would be helpful to know which persons are more susceptible before beginning a project in a heat environment.

Heat Stroke

Heat stroke is considered a **MEDICAL EMERGENCY**! It occurs when the body's temperature regulation system fails and sweating becomes inadequate or stops entirely. Without treatment, death or brain damage is almost certain. The signs and symptoms of heat stroke include chills; restlessness; irritability; mental confusion; euphoria; red face; disorientation; hot, dry skin; sweating stops; erratic behavior; collapse; shivering; unconsciousness; convulsions; and a core body temperature greater than 104°F. The major cause of heat stroke is excessive workloads in hot environments. To prevent heat stroke, persons should acclimate themselves to the environment, lead a healthy lifestyle, adhere to an appropriate work/rest cycle, drink plenty of fluids, maintain a proper diet, and self-determine the amount and length of heat stress exposure they can take.

Control Methods

As with any hazard associated with the workplace, OSHA requires that engineering controls, administrative controls, and personal protective equipment be used in that order to prevent work-related injuries.

Engineering Controls

To control heat in the work environments, a wide variety of techniques can be utilized. *General ventilation* is used to dilute hot air with cooler air. General ventilation works better in cooler climates than in hot ones. It is usually permanently installed to handle larger areas or entire buildings. When dealing with smaller areas, portable or local exhaust systems may be more effective or practical. *Air treatment or air cooling* differs from ventilation because it reduces the temperature of the air by removing the heat and sometimes the humidity from the air. *Air conditioning* is an effective method for cooling, but it is expensive to install and operate. An alternative to air conditioning is the use of chillers to circulate cool water through heat exchangers over which air from the ventilation system is then passed. *Convection fans* can be set up in hot areas and can increase heat exchange and the rate of evaporation, provided that the air temperature is less than body temperature. *Insulating* hot surfaces or tools that generate or conduct heat can also reduce heat conduction to workers. *Heat shields* can be used to reduce radiant heat when placed between a heat source and workers. Polished surfaces make the best barriers, although special glass or metal mesh surfaces can be used if visibility is a concern. Shields should be located so that they do not interfere with air flow, unless they are being used to reduce convective heating. Mechanization of work procedures can often make it possible to isolate workers from the heat source (air-conditioned booth) and increase productivity by decreasing the rest time needed and metabolic heat produced. *Reduction in air humidity* can be accomplished by removing the water from the air with chillers or mechanical refrigeration. By lowering the humidity in the air, the rate of evaporation of sweat increases.[1]

Administrative and Work Practice Controls

Acclimation or Acclimatization

Acclimation or acclimatization is the process of the body adapting to a change in temperature. For humans, the process of acclimatization can take up to 2 weeks. After acclimation occurs, the body can perform normal activities in the new environment with fewer cardiovascular demands. Since the process of acclimation is gradual, the work demands on a worker should also be gradual. A good rule of thumb for workers exposed to extreme temperatures is to expose them to approximately 20% on each day for 5 days. Good physical conditioning also reduces the cardiovascular demands on people. New employees should be closely monitored for signs and symptoms of heat-related injuries.

Fluid Replacement

While working in hot environments, workers can produce as much as 2–3 gal of sweat. Therefore, the replacement of these fluids should be about equal to the amount of sweat produced. In addition to fluid replacement, workers should also continue to eat normal or near-normal meals, which is often difficult since the appetite in hot environments is greatly decreased. Fluids high in caffeine are not recommended. Extremely cold fluids should also be avoided in hot environments. The amount of fluids to be replaced at various temperatures will be discussed later.

Work/Rest Cycles

Depending on the amount of work being performed and the environmental temperatures, workers in hot environments need more break periods than do unexposed workers do. Rest breaks are used to reduce the metabolic heat produced by the body during work and to aid in cooling. When determining work/rest cycles, the Wet Bulb Globe Temperature (WBGT) Index and the workload must be evaluated. To determine the metabolic rate (time-weighted average), we can use the following equation:

$$\text{Average}_{\text{Metabolic}} = \frac{(M_1)(T_1) + (M_2)(T_2) + \ldots (M_n)(T_n)}{(T_1) + (T_2) + \ldots (T_n)},$$

where
   M = metabolic rate (kcal/min)
   T = time (min)

   Table 11.1 illustrates an assessment of work in relation to body position and movement.
   Table 11.2 illustrates an assessment of work in relation to type of work being performed.

TABLE 11.1

Assessment of Work by Body Position and Movement

Body Position and Movement	kcal/min[a]
Sitting	0.3
Standing	0.6
Walking	2.0–3.0
Walking uphill	Add 0.8 for every meter or yard rise

Source: U.S. Department of Labor, Occupational Safety and Health Administration, *Technical Manual (TED 01-00-015)*, http://www.osha.gov/dts/osta/otm /otm_iii/otm_iii_4.html.

[a] For a "standard" worker of 70 kg (154 lb) body weight and 1.8 m² (19.4 ft²) body surface.

TABLE 11.2

Assessment of Work by Type of Work Being Performed

Type of Work	kcal/min	Range (kcal/min)
Hand Work		
Light	0.4	0.2–1.2
Heavy	0.9	
Work: One Arm		
Light	1.0	0.7–2.5
Heavy	1.7	
Work: Both Arms		
Light	1.5	1.0–3.5
Heavy	2.5	
Work: Whole Body		
Light	3.5	2.5–15.0
Moderate	5.0	
Heavy	7.0	
Very heavy	9.0	

Source: U.S. Department of Labor, Occupational Safety and Health Administration, *Technical Manual (TED 01-00-015)*, http://www .osha.gov/dts/osta/otm/otm_iii/otm_iii_4.html.

Using the equation above, determine the average metabolic rate for an employee who shovels gravel for 1.5 h at a metabolic rate of 7.0 kcal/min, takes a break for 30 min in a sitting position (metabolic rate = 0.3 kcal/min), resumes working on constructing a concrete form for 4 h (metabolic rate = 3.5 kcal/min), and finally pours and finishes the concrete (metabolic rate = 5.0 kcal/min) for 2 h. What is the average metabolic rate?

Average_M

$$= \frac{(7.0 \text{ kcal})(90 \text{ min}) + (0.3 \text{ kcal})(30 \text{ min}) + (3.5 \text{ kcal})(240 \text{ min}) + (5.0 \text{ kcal})(120 \text{ min})}{480 \text{ min}}$$

$$= \frac{630 \text{ kcal} + 9 \text{ kcal} + 840 \text{ kcal} + 600 \text{ kcal}}{480}$$

$$= \frac{2079}{480}$$

$$= 4.33 \text{ kcal/min.}$$

The average metabolic rate for this work is 4.33 kcal/min. Multiplied by 60 min, the metabolic rate would be 259.8 kcal/h. Workload categories are determined by averaging metabolic rates for the tasks and then ranking them as follows:

- Light work: up to 200 kcal/h
- Medium work: 200–350 kcal/h
- Heavy work: 350–500 kcal/h

Therefore, the total work described in the example would be categorized as moderate work. Determining the average metabolic rate is only part of the equation in setting work rest schedules. We must now calculate the WBGT Index, using the appropriate calculation. The average WBGT is calculated as follows:

$$\text{Average}_\text{WBGT} = \frac{(\text{WBGT}_1)(t_1) \pm (\text{WBGT}_2)(t_2) + \ldots + (\text{WBGT}_n)(t_n)}{(t_1) + (t_2) + \ldots (t_n)}.$$

For indoor and outdoor conditions with no solar load, WBGT is calculated as

$$\text{WBGT} = 0.7 \text{ WB} + 0.3 \text{ GT.}$$

For outdoors with a solar load, WBGT is calculated as

$$\text{WBGT} = 0.7 \text{ WB} + 0.2 \text{ GT} + 0.1 \text{ DB,}$$

where
$$\begin{aligned}
\text{WBGT} &= \text{Wet Bulb Globe Temperature Index} \\
\text{WB} &= \text{wet-bulb temperature} \\
\text{DB} &= \text{dry-bulb temperature} \\
\text{GT} &= \text{globe temperature}
\end{aligned}$$

By calculating both the WBGT Index and the average metabolic rate of the employee, we can now establish the work/rest cycle for each group of employees. The work/rest cycle can be determined by the use of Table 11.3 provided by the ACGIH Threshold Limit Values for Chemical Substances and Physical Agents and Biological Indices.

By way of example, the work scenario described previously was performed outdoors as follows:

WBGT of 79° for 90 min, WBGT of 74° for 30 min, WBGT of 86° for 240 min, and 88° for 120 min. What is the average WBGT Index?

TABLE 11.3

Permissible Heat Exposure Threshold Limit Value

	Workload		
Work/Rest Cycle	Light	Moderate	Heavy
Continuous	30.0°C (86°F)	26.7°C (80°F)	25.0°C (77°F)
75% Work, 25% rest, each hour	30.6°C (87°F)	28.0°C (82°F)	25.9°C (78°F)
50% Work, 50% rest, each hour	31.4°C (89°F)	29.4°C (85°F)	27.9°C (82°F)
25% Work, 75% rest, each hour	32.2°C (90°F)	31.1°C (88°F)	30.0°C (86°F)

Source: U.S. Department of Labor, Occupational Safety and Health Administration, *Technical Manual (TED 01-00-015)*, http://www.osha.gov/dts/osta/otm/otm_iii/otm_iii_4.html.

Note: These TLVs are based on the assumption that nearly all acclimatized, fully clothed workers with adequate water and salt intake should be able to function effectively under the given working conditions without exceeding a deep body temperature of 38°C (100.4°F). They are also based on the assumption that the WBGT of the resting place is the same or very close to that of the workplace. Where the WBGT of the work area is different from that of the rest area, a time-weighted average should be used (consult the ACGIH *1992–1993 Threshold Limit Values for Chemical Substances and Physical Agents and Biological Exposure Indices*).

These TLVs apply to physically fit and acclimatized individuals wearing light summer clothing. If heavier clothing that impedes sweat or has a higher insulation value is required, the permissible heat exposure TLVs in Table 11.3 must be reduced by the corrections shown in Table 11.4.

TABLE 11.4

WBGT Correction Factors for Clothing

Clothing Type	Clo[a] Value	WBGT Correction (°C)	WBGT Correction (°F)
Summer, lightweight working clothing	0.6	0	32
Cotton overalls	1.0	−2	28
Winter work clothing	1.4	−4	25
Water barrier, permeable	1.2	−6	21

Source: U.S. Department of Labor, Occupational Safety and Health Administration, *Technical Manual (TED 01-00-015)*, http://www.osha.gov/dts/osta/otm/otm_iii/otm_iii_4.html.

[a] Clo: Insulation value of clothing. One clo = 5.55 kcal/m^2/h of heat exchange by radiation and convection for each degree Celsius difference in temperature between the skin and the adjusted dry-bulb temperature.

Average$_{WBGT}$

$$= \frac{(79°)(90\,min)+(74°)(30\,min)+(86°)(240\,min)+(88°)(120\,min)}{480\,min}$$

$$= \frac{(7110)+(2220)+(20,640)+(10,560)}{480}$$

$$= \frac{40,530}{480}$$

$$= 84.44°F.$$

Using Table 11.3, the appropriate work/rest cycle for moderate work would be to work for 30 min and rest for 30 min each hour.

Personal Protective Equipment

Reflective clothing, which can vary from aprons and jackets to suits that completely enclose the worker from neck to feet, can stop the skin from absorbing radiant heat. However, since most reflective clothing does not allow air exchange through the garment, the reduction of radiant heat must more than offset the corresponding loss in evaporative cooling. Reflective clothing should be worn as loosely as possible. In situations where radiant heat is high, auxiliary cooling systems can be used under the reflective clothing. Auxiliary body cooling can also be used to reduce some of the heat stress. Auxiliary cooling options include ice vests, water-cooled garments, circulating air, and wetted terry-cloth coveralls or cotton suits (effective under low humidity). Respirators and other personal protective equipment increases the stress on a worker, and this stress contributes to overall heat stress. Chemical protection suits will also add to the heat stress problem, as well as protective vests worn by military and police officers. The effects of this protective equipment must be considered when determining workloads and work/rest cycles.[1]

In addition, employees exposed to direct sunlight should protect their exposed skin (neck, face, hands, and arms) using a sunblock/cream of at least 15 SPF. A wetted scarf can also be worn around the neck to aid in cooling.[2]

Training

Training is the key to good work practices. An effective heat stress training program should include at least the following elements:

- Knowledge of the hazards of heat stress
- Recognition of predisposing factors, danger signs, and symptoms
- Awareness of first-aid procedures for, and the potential effects of heat-related disorders

- Employee responsibilities in avoiding heat stress
- Use of personal protective clothing and equipment
- Physical fitness to prevent heat-related disorders
- Fluid replacement
- Dangers of drugs and alcohol in hot work environments
- Purpose and coverage of environmental and medical surveillance programs

Prevention of Heat Stress Injuries

Heat-related disorders can be prevented through various means. As you have seen in this chapter, predisposing factors to heat-related disorders may be age, fairness of skin, health, medication, caffeine and alcohol consumption, and general physical conditioning. Therefore, prevention of heat-related injuries begins with knowing these factors. Additional measures of prevention include the following:

- Learning the signs and symptoms of heat-related disorders
- Training workers about heat-related illnesses
- Performing the heaviest work in the coolest part of the day
- Acclimation of employees
- Using the buddy system (working in pairs)
- Adequate fluid replacement (one cup of water every 15–20 min)
- Wear light, loose-fitting, breathable clothing
- Adhere to the work/rest cycle
- Avoid caffeine and alcoholic beverages
- Wear appropriate personal protective equipment

Cold Stress

As we discussed in the Heat Stress section of this chapter, heat (and particularly body heat) can be lost through four different means: (a) radiation, (b) conduction, (c) convection, and (d) evaporation. Through radiation, body heat is lost to the colder environment, since heat will always travel to the cooler area. Body heat may be lost through the direct contact with another cooler object. Through convection, molecules against the surface of a warm body are moved away and replaced by cold ones. For example, wind chill may play an important factor in body heat by removing heat and replacing it with cold against the skin. In evaporation, heat and fluid are lost to the environment from sweating and respiration. A decreased fluid level makes the body more susceptible to hypothermia and other cold injuries.

In the Heat Stress section of this chapter, we also discussed how heat is gained through the normal functioning of the body to regulate its internal temperature. When exposed to cold extremes, the body's first reaction is to conserve body heat by reducing blood circulation through the skin, which makes the skin an insulating layer. The next reaction when confronted with cold extremes is that the body begins to shiver in order to increase the rate of metabolism. Shivering is actually a good sign that cold stress is significant and hypothermia may be present. Behavioral responses to cold include increasing clothing insulation, increasing activity, and seeking warmth.

Safety Problems Related to Cold

Just as in heat stress, when employees are uncomfortable (stress), they do not pay as much attention to their jobs or tasks at hand. Therefore, accidents are more likely to occur from inattention. Cold stress will also cause employees to become more rapidly fatigued.

Cold-Related Injuries and Illnesses

Cold-related injuries can be divided into two types of injuries that include nonfreezing and freezing injuries. Nonfreezing injuries can include chilblains, immersion injuries, and mild, moderate, and severe hypothermia. A freezing injury includes frostbite of varying degrees. Obviously, nonfreezing injuries occur when the environmental temperatures are above 32°F, while freezing injuries occur when temperatures fall below freezing. Each of these injuries will be discussed individually.

Chilblains

Chilblains, also known as *pernio* or *perniosis*, are itchy and tender red or purple bumps that occur as a reaction to cold temperatures. Chilblains result from the shutting down of the blood vessels in cold conditions. They occur several hours after exposure to the cold in temperate, humid climates and are sometimes aggravated by exposure to the sun. The small arteries and veins in the skin constrict from the cold and rewarming results in leakage of blood into the tissues and swelling of the skin. Contributing factors to chilblains include the following:

- Familial tendency
- Peripheral vascular disease owing to diabetes, smoking, or hyperlipidemia
- Low body weight or poor nutrition

- Hormonal changes (chilblains actually improve during pregnancy)
- Connective tissue disease
- Bone marrow disorders

There is very little treatment for chilblains. Topical steroid creams applied to the skin may relieve the itching. Antibiotic ointments or oral antibiotics may help prevent secondary infection as a result of scratching. Chilblain preventive measures may include the following:

- Insulated and heated homes and workplaces
- Warm clothing especially gloves, thick woolen socks, and comfortable protective footwear
- Avoid medicines that might constrict blood vessels, including caffeine, decongestants, and diet aids
- Exercise vigorously before going outside
- Wear cotton-lined waterproof gloves for wet work
- Apply sunscreen to exposed skin even on dull days
- Refrain from smoking
- Maintain physical conditioning

Immersion Injuries (Trench Foot)

Immersion injures result from prolonged exposure to cold weather, usually 12 h or longer at temperatures of 50°F–70°F (10°C–21°C) or for shorter periods at or near 32°F (0°C). Trench foot is an immersion injury seen in trench warfare or other environments where mobility is limited and dry boots and socks are unobtainable. The injured part is cold, swollen, waxy-white, with cyanotic burgundy-to-blue splotches and the skin is anesthetic and deep musculoskeletal sensation is lost.

The prevention and treatment of trench foot are similar to the treatment and prevention of frostbite. When possible, air-dry and elevate the feet, and exchange wet shoes and socks for dry ones to help prevent the development of trench foot.

Take the following steps:

- Thoroughly clean and dry your feet.
- Put on clean, dry socks daily.
- Treat the affected part by applying warm packs or soaking in warm water (102°F to 110°F) for approximately 5 min.
- When sleeping or resting, do not wear socks.
- Obtain medical assistance as soon as possible.

If you have a foot wound, your foot may be more prone to infection. Check your feet at least once a day for infections or worsening of symptoms.

Hypothermia

Hypothermia is a reduction of the body's core temperature below its normal (98°F), which results in a progressive deterioration in cerebral, musculoskeletal, and cardiac functions. There are three degrees of hypothermia, which are classified by their degree of severity. The three degrees of severity of hypothermia include mild, moderate, and severe hypothermia. Mild hypothermia ranges when the body temperature is 89.6°F–95°F and is initially characterized by violent shivering followed by virtual cessation of effective muscular activity, disorientation, and disinterest in surroundings. Moderate hypothermia results when the body temperature ranges from 78.8°F to 89.59°F, with cardiac irregularities occurring at approximately 86°F and corneal reflexes absent below 82.4°F. Severe hypothermia occurs at core temperatures of 78.79°F and lower, and with ventricular fibrillation a paramount risk below 80.6°F. At this temperature, the affected person may appear clinically dead. However, cold injury victims have been successfully resuscitated at core temperatures of 64°F, which created the medical saying, "No one is dead until he/she is warm and dead."[3]

Frostbite

Frostbite results from exposure to environmental temperatures below freezing. The speed of onset, depth, and severity of injury depend on temperature, wind chill, and the duration of exposure. Cellular injury and death occur from cellular trauma caused by the formation of ice crystals and from complex vascular reactions occurring in cold exposure. Superficial frostbite involves only the skin or the tissue immediately beneath it, while deep frostbite also affects the deep tissue beneath (including the bone). If the tissue has frozen, it appears "dead white" and is hard or even brittle. Differentiation of the types and severity of injury may be difficult.

First-degree frostbite is similar to mild chilblain with hyperemia (increase in blood flow to a tissue owing to the presence of metabolites and a change in general conditions), mild itching, and edema (swelling). There is no blistering or peeling of the skin.

Second-degree frostbite is characterized by blistering and desquamation (shedding of the outer layer of the skin).

Third-degree frostbite is associated with necrosis of skin and subcutaneous tissue with ulceration.

Fourth-degree frostbite includes destruction of connective tissues and bones, accompanied by gangrene. Secondary infections and the sequelae (pathological condition resulting from disease injury or trauma) noted for nonfreezing injuries are not infrequent, particularly if there is a history of freeze–thaw–refreeze.[3]

The susceptibility of persons to cold injuries include not being acclimatized, age, previous history of cold injuries, physical fitness, race or origin, nutrition, and use of medicine or alcohol. Full acclimation to cold climates occurs only after approximately 4 weeks.

Frostbite preventive measures include the following:

- Observance and adherence to the wind chill factor
- Proper protective clothing
- Proper nutrition and activity
- Fluid replacement or intake
- Training and discipline

Wind Chill Factor

Wind chill is defined as the number of calories lost during 1 h from a square meter of a surface kept at 91.4°F. The information provided in Table 11.5 shows the wind chill factor at various wind speeds and temperatures.

TABLE 11.5

Wind Chill Chart (NOAA)

Temp (°F)					Wind Speed (mph)							
Calm	5	10	15	20	25	30	35	40	45	50	55	60
40	36	34	32	30	29	28	28	27	26	26	25	25
35	31	27	25	24	23	22	21	20	19	19	18	17
30	25	21	19	17	16	15	14	13	12	12	11	10
25	19	15	13	11	9	8	7	6	5	4	4	3
20	13	9	6	4	3	1	0	−1	−2	−3	−3	−4
15	7	3	0	−2	−4	−5	−7	−8	−9	−10	−11	−11
10	1	−4	−7	−9	−11	−12	−14	−15	−16	−17	−18	−19
5	−5	−10	−13	−15	−17	−19	−21	−22	−23	−24	−25	−26
0	−11	−16	−19	−22	−24	−26	−27	−29	−30	−31	−32	−33
−5	−16	−22	−26	−29	−31	−33	−34	−36	−37	−38	−39	−40
−10	−22	−28	−32	−35	−37	−39	−41	−43	−44	−45	−46	−48
−15	−28	−35	−39	−42	−44	−46	−48	−50	−51	−52	−54	−55
−20	−34	−41	−45	−48	−51	−53	−55	−57	−58	−60	−61	−62
−25	−40	−47	−51	−55	−58	−60	−62	−64	−65	−67	−68	−69
−30	−46	−53	−58	−61	−64	−67	−69	−71	−72	−74	−75	−76
−35	−52	−59	−64	−68	−71	−73	−76	−78	−79	−81	−82	−84
−40	−57	−66	−71	−74	−78	−80	−82	−84	−86	−88	−89	−91
−45	−63	−72	−77	−81	−84	−87	−89	−91	−93	−95	−97	−98

To calculate the wind chill, the following equation can be used:

$$\text{Wind chill (°F)} = 35.74 + 0.6215\,T - 35.75(V^{0.16}) + 0.4275\,T(V^{0.16}),$$

where
   T = air temperature (°F)
   V = wind speed (mph)

For example, calculate the wind chill for the following: wind speed = 20 mph and a temperature of 10°F.

$$
\begin{aligned}
\text{Wind chill (°F)} &= 35.74 + 0.6215(10) - 35.75(20^{0.16}) + 0.4275(10)(20^{0.16}) \\
&= 35.74 + 6.215 - 35.75(1.62) + 4.275(1.62) \\
&= 35.74 + 6.215 - 57.92 + 6.93 \\
&= -9.
\end{aligned}
$$

Personal Protective Clothing

The appropriate amount and type of clothing is required to insulate the body from extreme temperatures and to prevent or slow down heat loss to the environment. Depending on the temperatures and the wind speed (wind chill factor), the clothing can also vary. In extreme temperatures, loose, layered clothing is recommended. The inner layer should be porous and allow for air flow. Other layers should protect against the wind and moisture. Insulated foot wear and dry socks are also recommended.

Nutrition and Activity

Nutrition is extremely important in preventing cold-related injuries. Normally, a calorie intake of 3500 cal/day will be adequate, but the work involved in heavy workloads and activities can use 4500–6000 cal/day. Carbohydrates and fats are the preferred sources for energy production. Ideally, the diet should consist of 20% protein, 45% carbohydrates, and 35% fats.

Fluid Replacement

Even in extremely cold climates, dehydration can occur. With the layering of clothing, the body continues to produce sweat. These fluids need to be replaced. On the average, a person exposed to extremely cold environments need to drink at least 2–3 L of water per day. Caffeine-containing drinks increase urine output and increase the risk of dehydration. Consuming warm liquids, such as broth or soups, is desirable to protect the core temperature.

Training and Discipline

Managers and supervisors should be aware that employees need to become acclimated to working in extremely cold environments and should enforce this acclimation process. Physical conditioning assists in preventing cold-related injuries and illnesses. In addition, personnel should be disciplined to maintain adequate personal hygiene, including the frequent changing of undergarments and socks. Training should include the signs, symptoms, and treatment of cold-related injuries.

Susceptible Groups

Greater protection and supervision should be provided for certain groups or individuals, including the fatigue group, the racial group, the geographic group, the previous cold injury group, and those with other injuries and illnesses.

Treatment of Cold-Related Injuries

Initial treatment of cold-related injuries should include the removal of the affected person to a warm environment, loosening of restricted clothing, and rewarming through blankets and warm environments. Drinking hot/warm fluids to reheat the core body temperature is highly encouraged. Frostbite victims should be seen by a licensed medical provider as soon as possible. Frostbite cases are treated by rapid rewarming in a water bath carefully controlled at 104°F (not to exceed 109°F). Rapid warming should not be continued beyond the time when thawing is complete and should not be instituted if thawing has already occurred. Nonfreezing injuries should not be warmed above 98.6°F. Smoking and the consumption of alcohol should be strictly prohibited.[3]

Key Information to Remember on Thermal Stressors

1. Sources of heat include radiation, convection, conduction, and metabolic.
2. Radiant heat is the amount of thermal radiation between two more objects.
3. Convective heat is thermal energy transferred by movement of a heated liquid or air.
4. Conductive heat is transferred by the contact of two objects.
5. Metabolic heat is generated from within the body through work.
6. Heat stroke is a medical emergency.

7. Heat-related injuries can be prevented through physical conditioning, fluid replacement, training, and adherence to a work/rest cycle.

8. The equation for an indoor WBGT is WBGT = 0.7 WB + 0.3 GT (no solar load).

9. The equation for outdoor WBGT is WBGT = 0.7 WB + 0.2 GT + 0.1 DB (solar load).

10. Body heat is lost through radiation, conduction, convection, and evaporation.

11. Hypothermia is a reduction of the body's core temperature below 98°F.

12. Frostbite occurs at temperatures below freezing.

References

1. Montana Department of Labor and Industry, Technical Program, *Thermal Stress: Heat & Cold Stress*, 2010. pp. 1–10.

2. U.S. Department of Labor, Occupational Safety and Health Administration, 2010. *Technical Manual (TED 01-00-015)*. Available at http://www.osha.gov /dts/osta/otm/otm_iii/otm_iii_4.html.

3. Navy Environmental Health Center, 1992. *Prevention and Treatment of Heat and Cold Injuries*, U.S. Navy, Norfolk, VA, Chapter 6.

12

Personal Protective Equipment

On November 15, 2007, the Occupational Safety and Health Administration (OSHA) announced a new rule clarifying employer responsibilities regarding payment for personal protective equipment (PPE). Many OSHA health, safety, maritime, and construction standards require employers to provide their employees with protective equipment, including PPE, when such equipment is necessary to protect employees from job-related injuries, illnesses, and fatalities. These requirements address PPE of many kinds: hard hats, gloves, goggles, safety shoes, safety glasses, welding helmets and goggles, face shields, chemical protective equipment, fall protection equipment, and so forth. The provisions in OSHA standards that require PPE generally state that the employer is to provide such PPE. However, some of these provisions do not specify that the employer is to provide such PPE at no cost to the employee. In this rulemaking, OSHA is requiring employers to pay for the PPE provided, with exceptions for specific items. The rule does not require employers to provide PPE where none has been required before. Instead, the rule merely stipulates that the employer must pay for required PPE, except in limited cases specified in the standard. This final rule became effective on February 13, 2008. The final rule must have been implemented by May 15, 2008.

It is not enough that a Safety Professional just know the regulatory requirements regarding PPE. Today's Safety Professional must have an in-depth knowledge of the uses and limitations of a varied array of PPE. In this chapter, we will look at hard hats, safety glasses and goggles, safety shoes and boots, gloves, fall protection and fall arrest systems, respirators, and other types of personal protective devices. The general industry standards for PPE are outlined in 29 CFR 1910.132-138.

Hazard Assessment

One of the first responsibilities in determining the types of PPE is to determine the hazards present in the work environment. The Safety Professional must conduct a thorough job hazard assessment to determine the specific hazards present. The job hazard assessment must be in writing and maintained by the employer. Once the job hazard assessment has been completed, then decisions on the types of PPE can be made.

Head Protection (29 CFR 1910.135)[1]

Most often, when we think of occupational head protection, we are referring to "hard hats." The OSHA standard addressing head protection can be located in 29 CFR 1910.135. Protective helmets purchased after July 5, 1994, shall comply with ANSI Z89.1-1986, "American National Standard for Personnel Protection—Protective Headwear for Industrial Workers—Requirements."

Prevention of head injuries is an important factor in every safety program. A recent Bureau of Labor and Statistics survey revealed that the majority of employees having head injuries were not using properly worn head protection.

Head injuries are caused by falling or flying objects, or by bumping the head against a fixed object. Head protection, in the form of hard hats, must do two things—resist penetration and absorb the shock of the blow. This is accomplished by making the shell of the hard hat strong enough to resist the impact and by utilizing a shock-absorbing lining composed of headband and crown straps to keep the shell away from the wearer's skull. In addition, protective hats may be designed to protect against electrical shock.

Classification

There are three basic categories or classifications of hard hats, as listed in Table 12.1.

The type of hard hat can be identified by looking inside the shell for the manufacturer, ANSI designation, and class. Helmets are date stamped by the manufacturer and should be replaced no later than the date recommended by the manufacturer, for example, 5 years.

TABLE 12.1

Hard Hat Classifications

Classification	Description of Use
E	Electrical helmets intended to reduce the danger of exposure to high-voltage electrical conductors, proof tested at 20,000 V. Class E is tested for force transmission first, and then tested at 20,000 V for 3 min, with 9 mA maximum current leakage, and then tested at 30,000 V, with no burn through permitted. (*Note: This classification was formerly classified as Class B hard hats.*)
G	General helmets intended to reduce the danger of exposure to low-voltage electrical conductors, proof tested to 2200 V. Class G is tested at 2200 V for 1 min, with 3 mA maximum leakage. (*Note: Class G classification was formerly classified as Class A.*)
C	Conductive helmets not intended to provide protection from electrical conductors. Class C is not tested for electrical resistance. There is no change in classification.

Proper Fit and Wear of Head Protection

Headbands are adjustable in 1/8" size increments. When the headband is adjusted to the right size, it provides sufficient clearance between the shell and the suspension. There should be a minimum of 1" to 1 1/4" clearance between the hard hat and the suspension. This space, known as the safety zone, is extremely important in the absorption of an impact to the hard hat.

Inspection and Maintenance

Manufacturers should be consulted with regard to paint or cleaning materials for their hard hats, because some paints and thinners may damage the shell and reduce protection by physically weakening it or negating electrical resistance. A common method of cleaning shells is dipping them in hot water (approximately 140°F) containing a detergent for at least 1 min. Shells should then be scrubbed and rinsed in clear hot water. After rinsing, the shell should be carefully inspected for any signs of damage. All components, shells, suspensions, headbands, sweatbands, and any accessories should be visually inspected daily for signs of dents, cracks, or penetrations and for any other damage that might reduce the degree of safety originally provided. Daily inspections should include the following as a minimum:

- Cracked, torn, frayed, or otherwise deteriorated suspension systems
- Deformed, cracked, or perforated brims or shells
- Flaking, chalking, or loss of surface gloss

Training

Each employee required to wear protective headgear should receive initial and periodic refresher training as needed. Employees should be trained in the following:

- Why head protection is necessary
- How the head protection will protect them
- The limitations of the head protection
- When they must wear the head protection
- How to properly wear the head protection
- How to adjust straps and other parts for a comfortable and effective fit
- How to check for signs of wear

Eye and Face Protection (29 CFR 1910.133)[2]

Suitable eye protection must be provided where there is a potential for injury to the eyes or face from flying particles, molten metal, liquid chemicals, acids or caustic liquids, chemical gases or vapors, potentially injurious light radiation, or a combination of these. Protection must meet the following minimum requirements:

- Provide adequate protection against the particular hazards for which they are designed
- Be reasonably comfortable when worn under the designated conditions
- Fit snugly without interfering with the movements or vision of the wearer
- Be durable
- Be capable of being disinfected
- Be easily cleanable
- Be kept clean and in good repair

Every protection shall be distinctly marked to facilitate identification of the manufacturer. Each affected employee shall use equipment with filter lenses that have a shade number appropriate for the work being performed for protection from injurious light radiation. Table 12.2a and b list the appropriate shade numbers for various work operations.

Each eye, face, or face-and-eye protector is designed for a particular hazard. In selecting the protector, consideration should be given to the kind and degree of hazard, and the protector should be selected on that basis. Where a choice of protectors is given, and the degree of protection required is not an important issue, worker comfort may be a deciding factor.

Fitting

Fitting of goggles and safety spectacles should be done by someone skilled in the procedure. Prescription safety spectacles should be fitted only by qualified optical personnel.

Inspection and Maintenance

It is essential that the lenses of eye protectors be kept clean. Continuous vision through dirty lenses can cause eye strain, often an excuse for not wearing the eye protectors. Daily inspection and cleaning of the eye protector with soap and hot water or with a cleaning solution and tissue is strongly recommended.

TABLE 12.2

Filter Lenses for Protection against Radiant Energy

(a)

Operation	Electrode Size (1/32 inch Diameter Standard)	Arc Current (A)	Minimum Protective Shade[a]
Shielded metal or welding	<3/32	<60	7
	3/32–5/32	60–160	9
	5/32–8/32	160–250	10
	>8/32	250–500	11
Gas metal arc welding and flux cored arc welding		<60	7
		60–160	10
		160–250	10
		250–500	10
Gas tungsten arc welding		>50	8
		50–150	8
		150–500	10
Air carbon arc cutting	(Light)	<500	10
	(Heavy)	500–1000	10
Plasma arc welding		<20	6
		20–100	8
		100–400	10
		400–800	11
Plasma arc cutting	(Light)[b]	<300	8
	(Medium)[b]	300–400	9
	(Heavy)[b]	400–800	10
Torch brazing			3
Torch soldering			2
Carbon arc welding			14

(b)

Operation	Plate Thickness		Minimum Protective Shade[a]
	in.	mm	
Gas welding:			
Light	<1/8	<3.2	4
Medium	1/8 to 1/2	3.2 to 12/5	5
Heavy	>1/2	>12.7	6
Oxygen cutting:			
Light	<1	<2.5	3
Medium	1 to 6	5 to 150	4
Heavy	>6	>150	5

Source: United States Department of Labor, Occupational Safety and Health Administration (OSHA), *OSHA Technical Bulletin-3077, Personal Protective Equipment*, p. 9.

[a] As a rule of thumb, start with a shade that is too dark to see the weld zone.

[b] These values apply where the actual arc is clearly seen.

Hearing Protection (29 CFR 1910.95)[4]

Exposure to high noise levels can cause hearing loss or impairment. It can create physical and psychological stress. There is no cure for noise-induced hearing loss; hence, the prevention of excessive noise exposure is the only way to avoid hearing damage. Specifically designed protection is required, depending on the type of noise encountered and the auditory condition of the employee.

Preformed or molded earplugs should be individually fitted by a professional. Waxed cotton, foam, or fiberglass wool earplugs are self-forming. When properly inserted, they work as well as most molded earplugs.

Some earplugs are disposable, to be used one time and then thrown away. The nondisposable type should be cleaned after each use for proper protection. Plain cotton is ineffective as protection against hazardous noise.

Earmuffs need to make a perfect seal around the ear to be effective. Glasses, long sideburns, long hair, and facial movements, such as chewing, can reduce protection. Special equipment is available for use with glasses or beards.

Hearing Protector Attenuation

The employer is responsible for evaluating hearing protector attenuation for the specific noise environments in which the protector will be used. Hearing protectors must attenuate employee exposure at least to an 8-h time-weighted average of 90 dB as required in 29 CFR 1910.95. For employees who have experienced a standard threshold shift, hearing protectors must attenuate employee exposure to an 8-h time-weighted average of 85 dB or below. The adequacy of hearing protector attenuation shall be reevaluated whenever employee noise exposures increase to the extent that the hearing protectors provided may no longer provide adequate attenuation. The employer shall provide more effective hearing protectors where necessary.

The Environmental Protection Agency requires that manufacturers of hearing protection devices put the noise reduction rating (NRR) on the package. When using the NRR to assess hearing protector adequacy, one of the following methods must be used:

1. When using a dosimeter that is capable of C-weighted measurements:
   a. Obtain the employee's C-weighted dose for the entire work shift, and convert to TWA.
   b. Subtract the NRR from the C-weighted TWA to obtain the estimated A-weighted TWA under the ear protector.

2. When using a dosimeter that is not capable of C-weighted measurements, the following method may be used:
   a. Convert the A-weighted dose to TWA.
   b. Subtract 7 dB from the NRR.
   c. Subtract the remainder from the A-weighted TWA to obtain the estimated A-weighted TWA under the ear protector.
3. When using a sound level meter set to the A-weighting network:
   a. Obtain the employee's A-weighted TWA.
   b. Subtract 7 dB from the NRR, and subtract the remainder from the A-weighted TWA to obtain the estimated A-weighted TWA under the protector.
4. When using a sound level meter set on the C-weighting network:
   a. Obtain a representative sample of the C-weighted sound levels in the employee's environment.
   b. Subtract the NRR from the C-weighted average sound level to obtain the estimated A-weighted TWA under the ear protector.
5. When using area monitoring procedures and a sound level meter set to the A-weighting network:
   a. Obtain a representative sound level for the area in question.
   b. Subtract 7 dB from the NRR and subtract the remainder from the A-weighted sound level for that area.
6. When using area monitoring procedures and a sound level meter set to the C-weighting network:
   a. Obtain a representative sound level for the area in question.
   b. Subtract the NRR from the C-weighted sound level for that area.

Training

The employer shall institute a training program for all employees who are exposed to noise at or above 85 dB and shall ensure employee participation in the program. The training program shall be repeated annually for each employee included in the hearing conservation program. Information provided in the training program shall be updated to be consistent with changes in protective equipment and work practices. The employer shall ensure that each employee is informed of the following:

- The effects of noise on hearing
- The purpose of hearing protectors, the advantages, disadvantages, and attenuation of various types, and instructions on selection, fitting, use and care, and the purpose of audiometric testing procedures

The employer shall make available to affected employees or their representatives copies of the OSHA standard and shall also post a copy in the workplace. The employer shall provide to affected employees any informational materials pertaining to the standard that are supplied to the employer by the Assistant Secretary of Labor. The employer shall provide, upon request, all materials related to the employer's training and education program pertaining to the OSHA standard to the Assistant Secretary of Labor or the OSHA Director.

Respiratory Protection (29 CFR 1910.134)[5]

The OSHA Standard for Respiratory Protection is located in 29 CFR 1910.134. Respirators shall be used in the following circumstances:

- Where exposure levels exceed the permissible exposure limit (PEL), during the period necessary to install or implement feasible engineering and work practice controls
- In those maintenance and repair activities and during those brief or intermittent operations where exposures exceed the PEL and engineering and work practice controls are not feasible or are not required
- In regulated areas
- Where the employer has implemented all feasible engineering and work practice controls and such controls are not sufficient to reduce exposures to or below the PEL
- In emergencies

Purpose

The purpose of the Respiratory Protection Standard is to control those occupational diseases caused by breathing air contaminated with harmful dusts, fogs, fumes, mists, gases, smokes, sprays, or vapors. This shall be accomplished as far as feasible by accepted engineering control measures. When effective engineering control measures are not feasible, or while they are being instituted, appropriate respirators shall be used.

Respirators shall be provided by the employer when such equipment is necessary to protect the health of the employee. The employer shall provide the respirators that are applicable and suitable for the purpose intended. The employer shall be responsible for the establishment and maintenance of a respiratory protection program that shall include the requirements of this standard.

Definitions

Air-purifying respirator means a respirator with an air-purifying filter, cartridge, or canister that removes specific air contaminants by passing ambient air through the air-purifying element.

Assigned protection factor means the protection factor assigned to the respirator type.

Atmosphere-supplying respirator means a respirator that supplies the respirator user with breathing air from a source independent of the ambient atmosphere and includes supplied-air respirators and self-contained breathing apparatus units.

Fit test means the use of a protocol to qualitatively or quantitatively evaluate the fit of a respirator on an individual.

Powered air-purifying respirator means an air-purifying respirator that uses a blower to force the ambient air through air-purifying elements to the inlet covering.

Qualitative fit test means a pass/fail test to assess the adequacy of respirator fit that relies on the individual's response to the test agent.

Quantitative fit test means an assessment of the adequacy of respirator fit by numerically measuring the amount of leakage into the respirator.

Self-contained breathing apparatus means an atmosphere-supplying respirator for which the breathing air source is designed to be carried by the user.

Supplied-air respirator or airline respirator means an atmosphere-supplying respirator for which the source of breathing air is not designed to be carried by the user.

Respiratory Protection Program

This standard requires the employer to develop and implement a written respiratory protection program with required worksite-specific procedures and elements for required respirator use. The program must be administered by a suitably trained program administrator. In addition, certain program elements may be required for voluntary use to prevent potential hazards associated with the use of the respirator. The employer shall include in the written program the following information:

- Procedures for selecting respirators for use in the workplace
- Medical evaluations of employees required to use respirators
- Fit testing procedures

- Procedures for proper use of respirators
- Procedures and schedules for cleaning, disinfecting, storing, inspecting, repairing, discarding, and otherwise maintaining respirators
- Procedures to ensure adequate air quality, quantity, and flow of breathing air for atmosphere-supplying respirators
- Training required for respirator usage
- Procedures for evaluating the effectiveness of the program

Training and Information

This standard requires the employer to provide effective training to employees who are required to use respirators. The training must be comprehensive, understandable, and recur annually, and more often, if necessary. Training must ensure that each employee can demonstrate knowledge and understanding of the topic and include

- Why respirator protection is necessary and how improper wearing or use can compromise the protection received
- Limitations and capabilities of the respirator or cartridge (filter)
- Inspection and maintenance procedures
- Cleaning, disinfecting, and storage procedures
- Proper wear of the respirator

Retraining shall be administered annually, or when a new process or procedure is implemented that the employee has not been previously trained.

Torso Protection

Many hazards can threaten the torso: heat, splashes from hot metals and liquids, impacts, cuts, acids, radiation, and high-pressure liquids (such as water blasting). A wide variety of protective clothing is available: vests, jackets, aprons, and full body suits. One of the better protective clothing used in protecting workers from high-pressure water blasting is the Water Armor or "Turtle Suit." This suit protects the chest, legs, and arms from glancing blows of up to 40,000 psi. Since many, even in the safety profession, are not familiar with hydro blasting, I have included a photograph of the "Turtle Skin."[6] See Figure 12.1.

FIGURE 12.1
Water armor protective suit. (Copyright Warwick Mills, Inc., 301 Turnpike Road, New Ipswich, New Hampshire 03071.)

Arm and Hand Protection (29 CFR 1910.138)[7]

When it comes to arm and hand protection, there is no single type that will protect the employee against all hazards. Therefore, it is extremely important to conduct a thorough job hazard analysis to determine the most appropriate arm and hand protection to provide employees. Examples of injuries to arms and hands are burns, cuts, electrical shock, amputation, and absorption of chemicals. There is a wide assortment of gloves, hand pads, sleeves, and wristlets for protection against various hazardous situations. Employers need to determine what hand protection their employees need through a thorough job hazard analysis. The work activities of the employees should be studied to determine the degree of dexterity required, the duration, frequency, and degree of exposure hazards and the physical stresses that will be applied. It is also important to know the performance characteristics of gloves relative to the specific hazard anticipated, for example, exposure to chemicals, heat, or flames. Gloves' performance characteristics should be assessed by using standard test procedures.

Before purchasing gloves, the employer should request documentation from the manufacturer that the gloves meet the appropriate test standard(s) for the hazard(s) anticipated. For example, for protection against chemical hazards, the toxic properties of the chemical(s) must be determined, particularly the ability of the chemical to pass through the skin and cause systemic effects. The protective devices should be selected to fit the job. For example, some

TABLE 12.3

ANSI Glove Requirements

Item	Standard
Rubber insulating gloves	ASTM D 120-87
Rubber matting for use around electrical apparatus	ASTDM D 178-88 or 178-93
Rubber insulating blankets	ASTDM D 1048-93 or 1048-88A
Rubber insulating hoods	ASTM D 1048-88 or 1049-93
Rubber insulating line hose	ASTM D 1050-90
Rubber insulating sleeves	ASTM D 1051-87

gloves are designed to protect against specific chemical hazards. Employees may need to use gloves (such as wire mesh, leather, and canvas) that have been tested and provide insulation from burns and cuts. The employee should become acquainted with the limitations of the clothing used.

Certain occupations require special protection. For example, electricians need special protection from shocks and burns. Rubber is considered the best material for insulating gloves and sleeves from these hazards. Rubber protective equipment for electrical workers must conform to the requirements established in ANSI as specified in Table 12.3.

Glove Selection Chart

The selection of the appropriate glove can be confusing. Table 12.4 is a compilation of several different charts. It will assist you in selecting the most appropriate personal protection for your employees. The legend for the chart is as follows:

E	Excellent (breakthrough times >8 h)
G	Good (breakthrough times >4 h)
F	Fair (breakthrough times >1 h)
P	Poor, Not Recommended (breakthrough times <1 h)

Foot and Leg Protection (29 CFR 1910.136)[8]

According to the BLS survey, most of the workers in selected occupations who suffered foot injuries were not wearing protective footwear. Furthermore, most of their employers did not require them to wear safety boots or shoes. The typical foot injury was caused by objects falling fewer than 4 ft and the median weight was about 65 lb. Most workers were injured while performing their normal job activities at their worksites.

For protection of feet and legs from falling or rolling objects, sharp objects, molten metal, hot surfaces, and wet slippery surfaces, workers should use

TABLE 12.4

Glove Selection Chart

Chemical	Natural Rubber	Neoprene	Butyl Rubber	PVC	Nitrile	Viton
Organic Acids						
Acetic acid	F	G	E	F	P	E
Formic acid	F	G	E	G	F	F
Lactic acid	E	E	E	G	E	E
Maleic acid	G	G	F	G	G	E
Oxalic acid	E	E	E	E	E	E
Inorganic Acids						
Chromic acid up to 70%	P	P	E	G	G	E
Hydrochloric acid up to 37%	G	G	E	G	G	G
Hydrofluoric acid up to 70%	F	F	G	P	P	?
Nitric acid 70+%	?	P	G	?	P	E
Perchloric acid up to 70%	E	E	G	E	E	E
Phosphoric acid up to 70+%	E	E	E	E	E	E
Sulfuric acid 70+%	P	F	E	F	P	F
Alkalis						
Ammonium hydroxide up to 70%	P	G	E	F	G	?
Potassium hydroxide up to 70%	E	E	E	E	E	E
Sodium hydroxide 70+%	E	E	E	E	G	G
Salt Solutions						
Ammonium nitrate	E	E	E	E	E	E
Calcium hypochlorite	P	G	E	E	G	E
Ferric chloride	E	E	E	E	E	E
Mercuric chloride	G	G	E	G	G	E
Potassium cyanide	E	E	E	E	E	E
Potassium dichromate	E	E	E	E	E	E
Potassium permanganate	E	E	?	E	E	?
Sodium cyanide	E	E	E	E	E	E
Sodium thiosulfate	E	E	E	E	E	E
Aromatic Hydrocarbons						
Benzene	P	P	P	P	P	G
Gasoline	P	P	P	P	E	E
Naphthalene	P	P	P	P	E	E
Toluene	P	P	P	P	P	E
Xylene	P	P	P	P	P	E
Aliphatic Hydrocarbons						
Diesel fuel	P	F	P	F	G	E
Hexanes	P	P	P	P	E	E

(*Continued*)

TABLE 12.4 (CONTINUED)

Glove Selection Chart

Chemical	Natural Rubber	Neoprene	Butyl Rubber	PVC	Nitrile	Viton
Kerosene	P	G	P	G	E	E
Naphtha	P	F	P	G	E	E
Pentane	P	P	P	P	G	E
Petroleum ether	P	P	P	F	G	E
Turpentine	P	P	P	P	F	E
Halogenated Hydrocarbons						
Carbon tetrachloride	P	P	P	P	P	E
Chloroform	P	P	P	P	P	E
Methylene chloride	P	P	P	P	F	G
Polychlorinated biphenyls (PCBs)	P	E	E	?	F	E
Perchloroethylene	P	P	P	P	F	E
Trichloroethylene	P	P	P	P	P	E
Esters						
Ethylene acetate	P	P	G	P	P	P
Butyl acetate	P	P	F	P	P	P
Methyl acetate	P	P	E	P	P	P
Isobutyl acrylate	P	P	E	P	P	P
Ethers/Glycols						
Diethyl ether	P	F	P	P	F	P
Ethylene glycol	P	F	E	P	F	E
Isopropyl ether	P	F	P	P	G	P
Propylene glycol	?	G	G	P	P	?
Tetrahydrofuran	P	P	F	P	P	P
Aldehydes						
Acetaldehyde	P	P	E	P	P	P
Acrolein	P	P	E	P	P	P
Benzaldehyde	P	P	E	P	P	G
Butyraldehyde	P	P	E	P	P	P
Formaldehyde	P	F	E	F	E	E
Glutaraldehyde	?	E	E	F	?	E
Ketones						
Acetone	P	P	E	P	P	P
Diisobutyl ketone	P	P	F	P	P	F
Methyl ethyl ketone	P	P	E	P	P	P
Alcohols						
Allyl alcohol	P	P	E	P	E	G
Butyl alcohol	P	G	E	F	G	E

TABLE 12.4 (CONTINUED)

Glove Selection Chart

Chemical	Natural Rubber	Neoprene	Butyl Rubber	PVC	Nitrile	Viton
Ethyl alcohol	P	F	G	P	G	E
Isopropyl alcohol	P	G	E	F	E	E
Methyl alcohol	P	P	E	P	P	E
Amines						
Aniline	P	P	E	P	P	F
Ethanolamine	F	E	E	G	E	E
Ethylamine	P	F	E	P	P	P
Methylamine	P	G	E	F	E	E
Triethanolamine	P	P	E	P	E	E
Elements						
Bromine	P	F	P	?	P	E
Chlorine aqueous	?	P	F	?	P	E
Iodine	?	P	G	?	G	E
Mercury	?	E	E	?	E	E
Miscellaneous						
Acetic anhydride	P	F	E	P	P	P
Acetonitrile	P	P	E	P	P	P
Acrylamide	P	P	G	P	F	G
Carbon disulfide	P	P	P	P	P	E
Cresols	P	G	E	?	F	E
Cutting fluid	?	F	?	F	G	?
Dimethyl sulfoxide	P	E	E	P	P	P
Hydraulic oil	?	?	P	F	G	?
Hydrazine	F	E	E	E	E	P
Hydrogen peroxide	E	F	E	G	E	E
Lubricating oil	G	G	?	?	E	G
Malathion	?	G	P	?	G	?
Nitrobenzene	P	P	E	P	P	E
Phenol	P	G	F	P	P	E
Photo solutions	G	E	?	G	E	?
Picric acid	P	F	G	P	F	E
Pyridine	P	P	E	P	P	P

Note: Viton is a registered trademark of DuPont Dow Elastomers.

appropriate footguards, safety shoes or boots, and leggings. Leggings protect the lower leg and feet from molten metal or welding sparks. Safety snaps permit their rapid removal.

Aluminum alloy, fiberglass, galvanized steel, or composite material footguards can be worn over usual work shoes, although they may present the possibility of catching on something and causing workers to trip. Heat-resistant soled shoes protect against hot surfaces like those found in the roofing, paving, and hot metal industries.

Safety shoes or boots should be sturdy and have an impact-resistant toe. In some shoes or boots, metal or composite insoles protect against puncture wounds. Additional protection, such as metatarsal guards, may be found in some types of footwear. Safety shoes or boots come in a variety of styles and materials, such as leather and rubber boots and oxfords.

Safety footwear is classified according to its ability to meet minimum requirements for both compression and impact tests. These requirements and testing procedures may be found in ANSI standards. Protective footwear purchased after July 5, 1994, must comply with ANSI Z41-1991.

Key Information to Remember on Personal Protective Equipment

1. On November 15, 2007, OSHA implemented a new rule clarifying employer responsibilities regarding payment for PPE. The final rule had a required implementation date of May 15, 2008.
2. The General Industry Standards for PPE are outlined in 29 CFR 1910.132-138.
3. A thorough job hazard assessment is required before issuing PPE.
4. Head Protection (29 CFR 1910.135).
5. There are three classifications of head protection, E (20,000 V), G (2200 V), and C (conductive helmets not intended for protection from electrical conductors).
6. Eye and Face Protection (29 CFR 1910.133).
7. Hearing Protection (29 CFR 1910.95).
8. Respiratory Protection (29 CFR 1910.134).
9. Arm and Hand Protection (29 CFR 1910.138).
10. Foot and Leg Protection (29 CFR 1910.136).

References

1. United States Department of Labor, Occupational Safety and Health Administration (OSHA), 2010. *29 CFR 1910.135, Hand Protection.*
2. United States Department of Labor, Occupational Safety and Health Administration (OSHA), 2010. *29 CFR 1910.133, Eye and Face Protection.*

3. United States Department of Labor, Occupational Safety and Health Administration (OSHA), 2010. *OSHA Technical Bulletin-3077, Personal Protective Equipment.*
4. United States Department of Labor, Occupational Safety and Health Administration (OSHA), 2010. *29 CFR 1910.95, Hearing Protectors.*
5. United States Department of Labor, Occupational Safety and Health Administration (OSHA), 2010. *29 CFR 1910.134, Respiratory Protection.*
6. Warwick Mills, Inc., 2010. *Image-Water Armor Protective Suit,* Warwick Mills, New Ipswich, NH.
7. United States Department of Labor, Occupational Safety and Health Administration (OSHA), 2010. *29 CFR 1910.138, Arm and Hand Protection.*
8. United States Department of Labor, Occupational Safety and Health Administration (OSHA), 2010. *29 CFR 1910.136, Foot and Leg Protection.*

13

Statistics for the Safety Professional

Statistics are extremely important in the career of the safety professional. We utilize them to report our annual injury and accident rates, calculate failure probability rates, and describe observed or experimental data. For our purposes, the definition of *statistics* is the study of obtaining meaningful information by analyzing data. These data can come from two different sources. They can come from experiments in a controlled environment or from observations made. Statistics can reduce large sums of data to a manageable form and allow the study and analysis of variance, thus allowing managers to maximize the use of information available in order to form an effective decision. The study and instruction of statistics can take up volumes of text. We will begin the review of statistics by describing *descriptive statistics*.

Descriptive Statistics

Descriptive statistics are just what the name implies. They are used to describe a set of data. For example, we've all heard polls on the local news that describes the data as 52% of all Americans are opposed to the latest congressional bill plus or minus 3%. What this is stating is that the range of the population that disapproves is 55% to 49%. This is a measure of central tendency.

Mean

The *mean* is also referred to as the "average" or arithmetic mean. For a data set, the mean is the sum of the observations divided by the number of observations. The *mean* is often quoted along with the *standard deviation*. The mean describes the central location of the data, and the standard deviation describes the spread. The mean is represented by the Greek letter (μ), which is mu, "*pronounced moo.*" It can also be represented by x. The mathematic representation for calculating the mean is as follows:

$$\mu = \bar{x} = \frac{x_1 + x_2 + \ldots x_n}{n},$$

where
  μ and $\bar{x}$ = average or arithmetic mean
  x_n = individual data values

Example

Given that there are four persons in a sample or data set, with the following weights: 125, 173, 108, and 211 lb. What is the mean of the weights?

$$\bar{x} = \frac{125 + 173 + 108 + 211}{4}$$

$$\bar{x} = 126.75$$

The average weight in this group is 126.75 or 127 lb.

Mode

The *mode* of a data sample is the variable that occurs most often in the collection. For example, the mode of the sample {1, 3, 6, 6, 6, 6, 7, 7, 12, 12, 17} is 6, since it occurs four times.

Median

The *median* is the middle value in the list of data. To find the median, your numbers have to be listed in numerical order, so you may have to rewrite your list first. If there is no "middle" number, because there is an even set of numbers, then the median is the mean (the usual average) of the middle two values. For example, given the data set {1, 2, 3, 4, 5, 6, 7, 8}, what is the median? The median would lie between 4 and 5; therefore, add 4 + 5 = 9, then divide by 2, giving a product of 4.5. Therefore, the median for the data set is 4.5, which is the central point of the data set.

NOTE: Mean, mode, and median are all measures of central tendency.

Variance

The *variance* can be described as the degree to which the variables in the data set are spread out. In other words, we would like to know how far away from the mean the variables are. In order to determine the variance from the mean, we would calculate each data point and the mean of the data set. For example, let's assume that you are the corporate safety manager for a corporation having six different and distinct locations. The following information in Table 13.1 shows the number of near misses for the previous year.

To utilize these data in a meaningful way, we cannot simply add the spread and divide by the number of points. By doing this, the positive and negative

TABLE 13.1

Example Information (Variance)

Company Location	No. of Near Misses (x_i)	($x_i - \mu$)
Illinois	43	11.33
Iowa	26	−5.67
Indiana	31	−0.67
Ohio	28	−3.67
Pennsylvania	38	6.33
Wisconsin	24	−7.67

numbers would simply cancel each other out and be equal to or near zero. Therefore, it is necessary that we square the distances from the mean (average). To do this, we utilize the following equation:

$$\text{Var}(x) = \frac{(x_1 - \bar{x})^2 + (x_2 - \bar{x})^2 + \dots (x_n - \bar{x})^2}{n}$$

Or written another way:

$$\text{Var}(x) = \sigma^2 = \frac{\sum_{i=1}^{n} n(x_i - \bar{x})^2}{n}.$$

The variance is represented by the Greek symbol σ^2 and is read as sigma squared.

The variance or sigma squared, in this scenario, is 45.56. The variance is considered to be the spread of the data (Table 13.2). When we take the square root of the variance, it is termed the *standard deviation*, which is represented mathematically by the Greek symbol σ. In this example, the standard deviation or σ is 6.75 or $\sqrt{45.56}$.

TABLE 13.2

Example Information—Variance Extension

Company Location	No. of Near Misses (x_i)	($x_i - \mu$)	($x_i - \mu$)^2
Illinois	43	11.33	128.37
Iowa	26	−5.67	32.15
Indiana	31	−0.67	0.45
Ohio	28	−3.67	13.47
Pennsylvania	38	6.33	40.07
Wisconsin	24	−7.67	58.83
Total	190	−0.02	273.34
Average (Mean)	31.67	0	45.56

The entire process listed above can be written in the following equation:

$$\sigma = \sqrt{\frac{\sum(x^2)}{N}},$$

where

$$x = \sum(x_i - \mu)^2$$

N = number of data points

The above equation is used whenever you are using data for the entire population. However, when you are taking data from a sample of the population, you should use the following equation:

$$s = \sqrt{\frac{\sum(x^2)}{N-1}},$$

where s = standard deviation or sigma of a sample population.

The only difference in this equation and the previous one is the deduction of 1 in the denominator.

Normal Distribution

Let's now discuss normal distribution. Most of us are familiar with the "bell curve." The bell curve is basically a graph of normal distribution, which has a single peak, and demonstrates that half of the data points are on the left side and half of the data points are on the right side of the curve. The mean lies in the center. The two tails extend indefinitely, never touching the horizontal axis. No matter what the value of the mean and the standard deviation, the area under the curve is 1.00. The variance and standard deviation are measures of the variability of a data set with respect to its mean. The graphic representation is shown in Figure 13.1.

As mentioned previously, the entire bell curve value is 1.00, regardless of the value of the data points. The numbers along the x axis of this graph represent the number of standard deviations in the distribution. Standard deviation values are listed below, which you should commit to memory, as there may be several questions on the examination related to standard deviations:

1 std dev = 68% of the data set

2 std dev = 95.45% of the data set

3 std dev = 99.73% of the data set

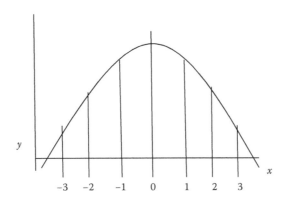

FIGURE 13.1
Graphic representation of standard deviation.

Calculating Correlation Coefficient

The *correlation coefficient* is also known as the Product Moment Coefficient or Pearson's Correlation and is described in the following equation:

$$r = \frac{N\sum (xy) - \left(\sum x\right)\left(\sum y\right)}{\sqrt{\left[N\sum (x^2) - \left(\sum x\right)^2\right]\left[N\sum (y^2) - \left(\sum y\right)^2\right]}},$$

where
 r = sample correlation coefficient
 N = number of data points
 $x = X - \bar{X}$
 $y = Y - \bar{Y}$

 This equation is used to determine if there is a correlation between variables. It does not necessarily mean that there is a causation. It simply means that there is a strong possibility that if x is this, then y can be predicted to be that, similar to regression analysis. It indicates the strength of a linear relationship between variables.

Example

Given the following set of data (Table 13.3), calculate the correlation between variable x and variable y.
 In order to insert the data into the equation, and for illustration purposes, let's arrange the needed data into the following format (Table 13.4):

TABLE 13.3

Example Information (Correlation)

x	y
2	1
4	3
6	5
8	7
10	9
12	11

TABLE 13.4

Example Information (Correlation) Extension

x	y	xy	x^2	y^2
2	1	2	4	1
4	3	12	16	9
6	5	30	36	25
8	7	56	64	49
10	9	90	100	81
12	11	132	144	121
Total 42	36	322	364	286
Average 7	6	53.67	60.67	47.67

There are a total of six samples (or data points in this sample set). Now that we have the data that we need, we can simply insert it into the equation, as follows:

$$r = \frac{N\sum(xy) - \left(\sum x\right)\left(\sum y\right)}{\sqrt{\left[N\sum(x^2) - \left(\sum x\right)^2\right]\left[N\sum(y^2) - \left(\sum y\right)^2\right]}}$$

$$r = \frac{6\sum(322) - (42)(36)}{\sqrt{[6(364) - (42)^2][6(286) - (36)^2]}}$$

$$r = \frac{420}{\sqrt{[420][420]}}$$

$$r = \frac{420}{\sqrt{176,400}}$$

$$r = \frac{420}{420}$$

$$r = 1.0.$$

The closer the correlation is to positive 1, the more the variables move together in same direction. The closer to negative 1, the more they move in opposite directions. The closer to zero the correlation between two variables are, the less related and more random their movement is.

Spearman Rank Coefficient of Correlation

One method of testing a hypothesis of a correlation between variables is to use the Spearman rank coefficient. The equation is shown mathematically as

$$r_s = 1 - \frac{6 \sum (D^2)}{N(N^2 - 1)},$$

where
r_s = Spearman ranking coefficient
D = Δ or difference between the rankings of the two variables
N = number of data points in the set

Example

As the corporate safety manager, you have five companies in your organization. It is your belief that there is a direct correlation between the safety audit scores and injury rates. Given the following data (Table 13.5), calculate the Spearman rank coefficient of correlation.

In order to obtain the data for the Spearman rank, we need to determine the difference in the rankings, as shown in Table 13.6.

TABLE 13.5

Example Information (Spearman Ranking)

	Plant A	Plant B	Plant C	Plant D	Plant E
Audit score	82.5	93	87.8	77.9	89.8
Injury rate	4.4	2.7	3.8	4.1	2.1

TABLE 13.6

Example Information (Spearman Ranking) Extension

Plant	Audit Score		Injury Rate	
	Rank	Rank	D or Δ	D² or Δ²
A	4	5	−1	1
B	1	2	−1	1
C	3	3	0	0
D	5	4	1	1
E	2	1	1	1

Now, we can insert our data into the equation as follows:

$$r_s = 1 - \frac{6(4)}{5(5^2 - 1)}$$

$$r_s = 1 - \frac{24}{5(25 - 1)}$$

$$r_s = 1 - \frac{24}{120}$$

$$r_s = 1 - 0.2$$

$$r_s = 0.8.$$

The Spearman rank is a reflection of correlation between two sets of numbers. The distance from 1 implies the correlation between the two variables, either a weak or strong correlation. If the number is positive, it may indicate a stronger correlation, whereas a negative number may imply a weaker correlation.

Calculating the *t* Test for Comparing Means

The *t* test is used to calculate the significance of observed differences between the means of two samples. It can be used to determine if there is a difference between two population parameters. Additionally, it can be used in situations where one of the conditions, the population standard deviation, is not known and the sample size is ≤30. It is generally used with scalar variables, such as length and width. The null hypothesis is that there are no significant differences between the means. To do this, we use the following equation:

$$t = \frac{\bar{X} - \mu}{s}\sqrt{N - 1} = \frac{\bar{x} - \mu}{\hat{s}}.$$

Example

A manufacturing process requires that a flange be manufactured to 25 cm. The quality control department collected five samples of the flange with the following results: 24.33 cm, 25.2 cm, 24.98 cm, 24.79 cm, and 24.88 cm. What is the *t*-statistic value?

The first step is to calculate the mean of the samples collected. The mean is calculated as follows:

$$\text{Mean} = \frac{24.33 + 25.2 + 24.98 + 24.79 + 24.88}{5}$$

$$\text{Mean} = \frac{124.18}{5}$$

$$\text{Mean} = 24.84 \text{ cm.}$$

From the mean, calculate the standard deviation, which is 0.3215.

$$t = \frac{24.84 - 25}{0.3215} \sqrt{5-1}$$

$$t = -0.9953.$$

Chi-Square (χ^2) Statistic

The *chi-square* (χ^2) *statistic* is useful in comparing observed distributions to theoretical ones. To calculate the *chi-square* (χ^2) *test*, we use the following equation:

$$\chi^2 = \sum_{j=1}^{k} \frac{(o_j - e_j)^2}{e_j},$$

where
   o = observed data
   e = expected data

Example

A review of 150 death certificates was conducted in a local community. Of the 150 deaths, 22 of the persons died of cancer, where 28.3 were expected under normal conditions. The rest died of other causes where 121.7 would normally be expected. What is the χ^2 value?

Solution

$$\chi^2 = \sum \frac{(22 - 28.3)^2}{28.3} + \frac{(128 - 121.7)^2}{121.7}$$

$$\chi^2 = \sum \frac{39.69}{28.3} + \frac{39.69}{121.7}$$

$$\chi^2 = 1.402 + 0.326$$

$$\chi^2 = 1.728.$$

Before explaining what the χ^2 value represents, it is first necessary to discuss *degrees of freedom* and *p values*.

Degrees of Freedom

Degrees of freedom can be described as the number of scores that are free to vary. For example, in the above scenario, you have two possible outcomes. One outcome is that employees were treated for cancer or they were treated for other illnesses. If an employee was treated for cancer, then they were included in the observed category. In summary, in this scenario, there is only one degree of freedom.

NOTE: In most studies, the degree of freedom is normally computed by the number of observed minus 1.

p Values

The p value is defined as the probability of obtaining a result as extreme as the one observed, assuming that the null hypothesis is true. The *lower* the p value, the *less* likely the result, assuming the null hypothesis, so the *more* "significant" the result. A p value of 0.05 represents that there is a 5% chance of an outcome that extreme, given the null hypothesis.

NOTE: Now that we have explained the degree of freedom and p value, we can take the χ^2 value from our example above (which is 1.128) and using the Chi-Square Distribution table, which can be found on the Internet or other published sources, we determine the p value for one degree of freedom to be between 0.2 and 0.1, but closer to 0.2. A p value of approximately 0.2 means that the data are statistically significant at the $p = 0.2$ level, or there is only a probability of 0.2 that the data are NOT significant.

Permutations and Combinations

There is no easy definition of permutation or combination, in regard to mathematics. Therefore, it is necessary that we explain the difference between the two. To illustrate what each are, an example of each is listed below:

Combination: A group of data or items where an order doesn't matter. For example: your meal consists of chicken, salad, bread, and vegetable. Bread, salad, vegetable, and chicken are still your meal. In other words, no matter what order you place the items in, it is still the original meal.

Permutation: A group of data or items where an order does matter. For example: the numbers to a combination lock are 5-8-6, which is not the same as 8-6-5. In this example, the order matters. A permutation is an ordered combination.

NOTE: If the order doesn't matter, it is a Combination. If the order matters, it is a Permutation.

There are two types of permutations. The first type is when repetition is allowed. For example, the lock mentioned above may have a data set of 4-4-4. The second type is when no repetition is allowed, as in describing the first three in an automobile race.

Permutations with Repetition

To calculate permutations with repetition, simply multiply the data points as described in the equation below:

$$n^r = n \times n \times \ldots n,$$

where
n^r = permutation when repetition is allowed
n = the number of possibilities to choose from

In this example, there are 10 possibilities to choose (0, 1, 2, 3, ... 9) from and you are selecting 3 of them; therefore, the permutation for this data set is solved mathematically as follows:

$$n^r = 10 \times 10 \times 10 = 10^3 = 1000 \text{ permutations.}$$

As you can see, calculating permutations with repetition is quite simple.

Permutations without Repetition

Calculating permutations without repetition is slightly more complicated. To calculate the permutation for the number of the combination lock mentioned previously where the numbers are 5-8-6, which are required to be in a specific order, we would use the following equation:

$$P_k^n = \frac{n!}{(n-k)!},$$

where
P_k^n = permutation
n = the number of data points to choose from
k = the number of data points you choose

Example

From the scenario above, you have 10 numbers to choose from (0, 1, 2 ... 9), and of these 10 numbers, you are actually choosing 3 of them (5-8-6). How many permutations can be present in this scenario?

$$P_3^{10} = \frac{10!}{(10-7)!}$$

$$P_3^{10} = \frac{3,628,800}{5040}$$

$$P_3^{10} = 720.$$

In this scenario, there are 720 permutations that can result.

Combinations

When calculating combinations, we use the following equation:

$$C_k^n = \frac{n!}{k!(n-k)!},$$

where
  C_k^n = combination
  n = the number of data points to choose from
  k = the number of data points you choose

Example

There are 10 plants in a corporation. Of these 10 plants, 4 are to be picked, at random (order doesn't matter), to be issued new safety equipment for evaluation. How many different combinations of plant representation are possible to receive the safety equipment?

$$C_4^{10} = \frac{10!}{4!(10-4)!}$$

$$C_4^{10} = \frac{3,628,800}{4!(6!)}$$

$$C_4^{10} = \frac{3,628,800}{24(720)}$$

$$C_4^{10} = \frac{3,628,800}{17,280}$$

$$C_4^{10} = 210.$$

In this scenario, there are a possible 210 combinations of plants that can be selected to receive the new safety equipment.

Z-Score

In statistics, a *standard score* indicates how many standard deviations an observation or datum is above or below the mean. It is a dimensionless quantity derived by subtracting the population mean from an individual raw score and then dividing the difference by the population standard deviation. This conversion process is called *standardizing* or *normalizing*; however, "normalizing" can refer to many types of ratios; see normalization (statistics) for more.

The standard deviation is the unit of measurement of the z-score. It allows comparison of observations from different normal distributions, which is done frequently in research.

Standard scores are also called *z-values, z-scores, normal scores,* and *standardized variables*; the use of "Z" is because the normal distribution is also known as the "Z distribution." They are most frequently used to compare a sample to a standard normal deviate (standard normal distribution, with $\mu = 0$ and $\sigma = 1$), although they can be defined without assumptions of normality.

The z-score is *only* defined if one knows the population parameters, as in standardized testing; if one only has a sample set, then the analogous computation with sample mean and sample standard deviation yields the Student's t statistic. Z-scores can be calculated using the following equation:

$$z = \frac{\chi^2 - \mu}{\sigma},$$

where
  z = score (unitless)
  χ^2 = chi-square value
  σ = standard deviation

Example

You just completed a statistical review of your safety training program, which showed a mean score on the final examination of 79.2% and a standard deviation of 8.3. One student scored 82.6%. Assuming a standard distribution, how many students scored higher that this one student on the exam?

Solution

$$z = \frac{82.6 - 79.2}{8.3}$$

$$z = \frac{3.4}{8.3}$$

$$z = 0.41.$$

Probability content from –00 to Z

Z	0.00	0.01	0.02	0.03	0.04	0.05	0.06	0.07	0.08	0.09
0.0	0.5000	0.5040	0.5080	0.5120	0.5160	0.5199	0.5239	0.5279	0.5319	0.5359
0.1	0.5398	0.5438	0.5478	0.5517	0.5557	0.5596	0.5636	0.5675	0.5714	0.5753
0.2	0.5793	0.5832	0.5871	0.5910	0.5948	0.5987	0.6026	0.6064	0.6103	0.6141
0.3	0.6179	0.6217	0.6255	0.6293	0.6331	0.6368	0.6406	0.6443	0.6480	0.6517
0.4	0.6554	0.6591	0.6628	0.6664	0.6700	0.6736	0.6772	0.6808	0.6844	0.6879
0.5	0.6915	0.6950	0.6985	0.7019	0.7054	0.7088	0.7123	0.7157	0.7190	0.7224
0.6	0.7257	0.7291	0.7324	0.7357	0.7389	0.7422	0.7454	0.7486	0.7517	0.7549
0.7	0.7580	0.7611	0.7642	0.7673	0.7704	0.7734	0.7764	0.7794	0.7823	0.7852
0.8	0.7881	0.7910	0.7939	0.7967	0.7995	0.8023	0.8051	0.8078	0.8106	0.8133
0.9	0.8159	0.8186	0.8212	0.8238	0.8264	0.8289	0.8315	0.8340	0.8365	0.8389
1.0	0.8413	0.8438	0.8461	0.8485	0.8508	0.8531	0.8554	0.8577	0.8599	0.8621
1.1	0.8643	0.8665	0.8686	0.8708	0.8729	0.8749	0.8770	0.8790	0.8810	0.8830
1.2	0.8849	0.8869	0.8888	0.8907	0.8925	0.8944	0.8962	0.8980	0.8997	0.9015
1.3	0.9032	0.9049	0.9066	0.9082	0.9099	0.9115	0.9131	0.9147	0.9162	0.9177
1.4	0.9192	0.9207	0.9222	0.9236	0.9251	0.9265	0.9279	0.9292	0.9306	0.9319
1.5	0.9332	0.9345	0.9357	0.9370	0.9382	0.9394	0.9406	0.9418	0.9429	0.9441
1.6	0.9452	0.9463	0.9474	0.9484	0.9495	0.9505	0.9515	0.9525	0.9535	0.9545
1.7	0.9554	0.9564	0.9573	0.9582	0.9591	0.9599	0.9608	0.9616	0.9625	0.9633
1.8	0.9641	0.9649	0.9656	0.9664	0.9671	0.9678	0.9686	0.9693	0.9699	0.9706
1.9	0.9713	0.9719	0.9726	0.9732	0.9738	0.9744	0.9750	0.9756	0.9761	0.9767
2.0	0.9772	0.9778	0.9783	0.9788	0.9793	0.9798	0.9803	0.9808	0.9812	0.9817
2.1	0.9821	0.9826	0.9830	0.9834	0.9838	0.9842	0.9846	0.9850	0.9854	0.9857
2.2	0.9861	0.9864	0.9868	0.9871	0.9875	0.9878	0.9881	0.9884	0.9887	0.9890
2.3	0.9893	0.9896	0.9898	0.9901	0.9904	0.9906	0.9909	0.9911	0.9913	0.9916
2.4	0.9918	0.9920	0.9922	0.9925	0.9927	0.9929	0.9931	0.9932	0.9934	0.9936
2.5	0.9938	0.9940	0.9941	0.9943	0.9945	0.9946	0.9948	0.9949	0.9951	0.9952
2.6	0.9953	0.9955	0.9956	0.9957	0.9959	0.9960	0.9961	0.9962	0.9963	0.9964
2.7	0.9965	0.9966	0.9967	0.9968	0.9969	0.9970	0.9971	0.9972	0.9973	0.9974
2.8	0.9974	0.9975	0.9976	0.9977	0.9977	0.9978	0.9979	0.9979	0.9980	0.9981
2.9	0.9981	0.9982	0.9982	0.9983	0.9984	0.9984	0.9985	0.9985	0.9986	0.9986
3.0	0.9987	0.9987	0.9987	0.9988	0.9988	0.9989	0.9989	0.9989	0.9990	0.9990

FIGURE 13.2
Table of normal distribution. This table is public domain and produced by William Knight, *APL Program*. Available at http://www.math.unb.ca/~knight/utility/NormTble.htm.

Now refer to Figure 13.2[1] and determine the area under the curve associated with a z-score of 0.41. Begin with 0.4 under the z column, then go over to Column 0.01 and the area beneath that represents a z-score of 0.41 is 0.6591.

To calculate the total number of persons taking the exam that scored higher than 82.6%, we subtract the total area from the whole as follows: $1 - 0.6591 = 0.3409$; $0.3409 \times 100 = 34.09\%$. Therefore, 34.09% of people taking the examination scored higher than 82.6%.

Coefficient of Determination and Coefficient of Correlation

In statistics, the *coefficient of determination*, r^2, is used in the context of statistical models whose main purpose is the prediction of future outcomes on the basis of other related information. It is the proportion of variability in a data set that is accounted for by the statistical model. The coefficient of determination can be calculated using the following equation:

$$r^2 = \frac{\text{explained variation}}{\text{total variation}}.$$

The coefficient of correlation can be calculated using the coefficient of determination, as follows:

Example

Your company president believes that unsafe behaviors lead to accidents and injuries. As a result, he requested that you, as the safety manager, conduct a study. Your study reveals the following data. In your study, you recorded total variations of 53.8 and explained variations of 39.2. Calculate the r^2 value and r value.

Solution

$$r^2 = \frac{39.2}{53.8}$$
$$r^2 = 0.7286$$

Now calculate the r value (correlation value):

$$r = \sqrt{0.7286}$$
$$r = 0.8536$$

An r value close to 1 indicates a strong linear relationship. A value closer to 0 indicates a weak linear relationship.

Reliability

Component Reliability

Reliability is the ability of a system or component to perform its required functions under stated conditions for a specified period. It is often reported as a probability. To calculate the reliability of a component, use the following equation:

$$R(t) = e^{-\lambda t},$$

where
  $R(t)$ = reliability
  t = time in which reliability is measured

$$\lambda = \frac{\text{\# of failures}}{\text{\# of time units exposed}}$$

Example

An inline valve has a failure rate of 2×10^{-6} failures per hour, what is its reliability in an operation period of 5000 h?
  Insert the data into the equation as follows:

$$R_{5000\,h} = e^{-2\times10^{-6}(5000\,h)}$$
$$R_{5000\,h} = e^{-0.01}$$
$$R_{5000\,h} = 0.99.$$

Based on the outcome of this equation, the component has a failure probability rate of 0.01 failures per 5000 h.

Probability of Failure (Component)

The probability that a component will fail in a projected time is equal to 1 minus the reliability of that period. Written mathematically, this statement is as follows:

$$P_f = 1 - R_t.$$

Example

Using the result from the reliability equation above, we insert the given data as follows:

$$P_f = 1 - 0.99$$
$$P_f = 0.01.$$

As previously stated, this component has a failure probability rate of 0.01 failures per 5000 h. This equation is used primarily for individual components within a system.

System Reliability

Mechanical and electrical systems consist of numerous individual components. The reliability of the system depends on how the components are arranged and upon their failure rates.

Series Reliability

If the components are arranged in such a manner that a failure of any individual component results in the entire system failing, then it is known as a *series system*. Use the following equation to calculate the overall system series reliability:

$$R_{system} = R_1 \times R_2 \times \ldots \times R_n.$$

NOTE: This equation is not listed on the BCSP exam reference sheet but is useful in calculating the overall system failure rates.

Example

You are asked to calculate the reliability of a system with six individual components. The reliability of the individual components is 0.98, 0.99, 0.98, 0.97, 0.97, and 0.98. What is the system reliability?

$$R_{system} = 0.98 \times 0.99 \times 0.98 \times 0.97 \times 0.97 \times 0.98$$
$$R_{system} = 0.88.$$

As you can see from the outcome of the calculation, the individual component reliability is relatively high, but when put in a parallel arrangement, the overall system reliability is reduced.

Parallel Reliability

A system where one individual component can fail and the system will still be functional is known as a parallel system. To calculate the reliability of a parallel system, use the following equation:

$$R_{system} = 1 - (1 - R_1)(1 - R_2)\ldots(1 - R_n).$$

Example

Given the same data provided in the series example, calculate the system reliability of the six components acting as a parallel system.

$$R_{system} = 1 - (1-0.98)(1-0.99)(1-0.98)(1-0.97)(1-0.97)(1-0.98)$$
$$R_{system} = 1 - (0.02)(0.01)(0.02)(0.03)(0.03)(0.02)$$
$$R_{system} = 1 - 7.2^{-11}$$
$$R_{system} = 0.9999999996 \text{ or } 1.0.$$

As you can see, the arrangement of the individual components is extremely important. Note that while the individual components have relatively high reliabilities, their arrangement increases the reliability of the system.

Probability of Failure (System)

When calculating the probability of failure of a system, use the following equation:

$$P_f = (1 - P_s),$$

where
P_f = probability of failure
P_s = probability of success or reliability of the system

Example

The reliability of a system is 0.88. What is the probability of failure rate for that system?

$$P_f = (1 - 0.88)$$
$$P_f = 0.12.$$

Reference

1. Knight, W., 2010. *APL Program*. Available at http://www.math.unb.ca/~knight/utility/NormTble.htm.

14

Electrical Safety

The average person is aware that electricity can be dangerous if not used or worked on properly, using safe work practices. Occupational hazards of electricity are increased, especially on construction sites, due to the fact that there are portable power tools, sometimes cluttered work sites, and the fast pace of production. Injuries resulting from electricity include electrocution (death from electrical shock), electrical shock, falls, and burns, which if left untreated can become infected. In this chapter we will discuss the basics of electricity and the basic practices to prevent injuries related to it.

Electricity Basics

Electricity flows in a similar fashion to water flowing through a garden hose. Water moves through the hose from an area of high pressure to an area of low pressure. Electricity operates in much the same way. Electrical currents move from high voltage to low voltage (area of least resistance). To understand electricity, three important concepts must be mastered, which include voltage, current, and resistance.

Voltage

Voltage between two points is basically a term used to describe the total amount of electrical force (energy) that drives the current between the two points. Voltage is measured in terms of volts (V).

Current

The electrical current is the flow of electric charge or the rate of flow of electric charge. Current, denoted by the letter I, is measured in amperes (A).

Resistance

Electrical resistance is a measure of the opposition to the flow of steady electrical current. Resistance (R) is measured in ohms (Ω).

Series and Parallel Circuits

Series Circuits

Components connected in series are connected along a single path; thus, the same current flows through all of the components. A simple series circuit is illustrated in Figure 14.1.

Parallel Circuits

In a parallel circuit, the electrical current to each element in the circuit is separate; hence, if one element was to burn out, the other resistors would still have power. Also, if you add an extra element, the other elements will still have the same amount of voltage as before. If you remove an element, the other elements will also still have the same amount of voltage as before. A simple parallel circuit is illustrated in Figure 14.2.

Direct and Alternating Currents

Direct Currents

Direct current (DC) is the unidirectional flow of an electric charge. In other words, the current flows through the circuit in the same direction at all times.

Alternating Currents

In alternating currents (AC), the flow of electric charge periodically reverses direction. The current flow passes through a regular succession of changing positive and negative values by periodically reverting its direction of flow. The total positive and negative values of current are equal.

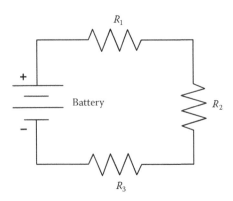

FIGURE 14.1
Illustration of a simple series circuit.

Parallel

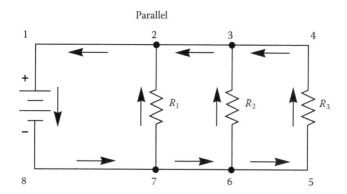

FIGURE 14.2
Illustration of a simple parallel circuit.

Calculating Values of Voltage, Current, and Resistance (Ohm's Laws)

Calculating Voltage in DC Circuits

The voltage of a circuit can be calculated using the following equation:

$$V = IR,$$

where
 V = volts
 I = current (A)
 R = resistance of the circuit (Ω)

For example, calculate the voltage for a DC circuit with the following data: (I = 0.5 A, 5 Ω)

$$V = (0.5\,\text{A})(5\,\Omega)$$
$$V = 2.5\,\text{V}.$$

Calculating Voltage in AC Circuits

To calculate the voltage in AC circuits, use the following equation:

$$V = \sqrt{\frac{PI}{\cos\phi}},$$

where
   V = voltage
   I = impedance (Ω)
   ϕ = the phase angle in degrees by which current lags voltage in an inductive circuit or leads in a captive circuit
   P = true power (W)

Calculating Power in DC Circuits

To calculate power in DC circuits, apply the following equation:

$$P = VI,$$

where
   P = power (W)
   V = volts
   I = current (A)

Calculating Power in AC Circuits

To calculate power in AC circuits, apply the following equation:

$$P = \frac{V^2 \times \cos\phi}{R \text{ (impedance)}},$$

where
   P = power (W)
   V = volts
   R = resistance (more specifically when discussing ACs, it is referred to as impedance) (Ω)
   ϕ = the phase angle in degrees by which current lags voltage in an inductive circuit or leads in a captive circuit

Calculating Resistance in DC Circuits

To calculate the resistance in DC circuits, use the following equation:

$$R = \rho\frac{L}{A},$$

where
   R = resistance of the circuit (Ω)
   ρ = Rho (resistivity of the metal) (intrinsic ability to resist the flow of electricity)

L = length (m)
A = cross-sectional area in square meters

Calculating Resistance in AC Circuits

The formula for calculating resistance in AC circuits is shown as

$$I = \frac{V}{A},$$

where
I = impedance (Ω)
V = volts
A = amperes

Resistors

Resistors are elements of electrical and electronic systems that are designed to intentionally resist the flow of electrical current through the system at a known measurement. The higher the value of resistance (measured in ohms), the lower the current will be. The primary characteristics of a resistor are the resistance, the tolerance, maximum working voltage, and the power rating.

Calculating the Resistance in a Series Circuit

To calculate the total resistance in a series circuit, use the following equation:

$$R_{series} = R_1 + R_2 + \dots R_n$$

where
R_{series} = total resistance (Ω)
R_x = resistance of individual resistor (Ω)

For example, you have three resistors in series, valued respectively at 2 Ω, 4 Ω, and 6 Ω. The total resistance of the circuit is calculated as follows:

$$R_{series} = 2\Omega + 4\Omega + 6\Omega$$
$$R_{series} = 12\Omega.$$

Calculating the Resistance in a Parallel Circuit

In a parallel circuit, the total resistance is calculated as follows:

$$\frac{1}{R_{parallel}} = \frac{1}{R_1} + \frac{1}{R_2} + \frac{1}{R_n}$$

where
$R_{parallel}$ = total resistance (Ω) in parallel
R_n = resistance of individual resistors (Ω)

For example, you have the same resistors as in the series circuit, but now they are in parallel. The values of each resistor are 2 Ω, 4 Ω, and 6 Ω. What is the total resistance of this parallel circuit?

$$\frac{1}{R_{parallel}} = \frac{1}{2\Omega} + \frac{1}{4\Omega} + \frac{1}{6\Omega}$$

$$\frac{1}{R_{parallel}} = 0.5\ \Omega + 0.25\Omega + 0.17\Omega$$

$$\frac{1}{R_{parallel}} = 0.92\Omega$$

$$R_{parallel} = 1.09\Omega.$$

Thus, the total resistance for this parallel circuit is 1.09 Ω.

Capacitors

A capacitor or condenser is a passive electronic component consisting of a pair of conductors separated by a dielectric (insulator). When a potential difference (voltage) exists across the conductors, an electric field is present in the dielectric. This field stores energy and produces a mechanical force between the conductors. Capacitor values are measured in farads (F).

Calculating Capacitance in a Series Circuit

To calculate the total capacitance in a series circuit, use the following equation:

$$\frac{1}{C_{series}} = \frac{1}{C_1} + \frac{1}{C_2} + \ldots \frac{1}{C_n}$$

where

C_{series} = total capacitance in a series circuit (F)

C_n = individual capacitor value (F)

For example, three capacitors with invidiual values of 0.1 µF, 0.2 µF, and 0.3 µF are aligned in series. What is the total capacitance?

$$\frac{1}{C_{series}} = \frac{1}{0.1\,\mu F} + \frac{1}{0.2\,\mu F} + \frac{1}{0.3\,\mu F}$$

$$\frac{1}{C_{series}} = 10\,\mu F + 5\,\mu F + 3.33\,\mu F$$

$$\frac{1}{C_{series}} = 18.33\,\mu F$$

$$C_{series} = 0.055\,\mu F.$$

The total capacitance for this series circuit is 0.055 µF. One important point to remember about capacitors connected in series is that the total capacitance will always be less than the value of the smallest capacitor value. In this particular case, the total value is 0.055 µF, which is less than 0.1 µF (smallest capacitor).

Calculating Capacitance in a Parallel Circuit

When calculating the capacitance of a parallel circuit, use the following equation:

$$C_{parallel} = C_1 + C_2 + \ldots C_n$$

where

$C_{parallel}$ = total capacitance in a parallel circuit (F)

C_n = individual capacitor value (F)

For example, using the values from the capacitors in the series circuit equation (0.1, 0.2, and 0.3 µF), calculate the capacitance in a parallel circuit.

$$C_{parallel} = 0.1\,\mu F + 0.2\,\mu F + 0.3\,\mu F$$

$$= 0.6\,\mu F.$$

Inductors

An *inductor* or a *reactor* is a passive electrical component that can store energy in a magnetic field created by the electric current passing through it. An inductor's ability to store magnetic energy is measured by its inductance, in units of henries (H). Typically, an inductor is a conducting wire shaped as a coil, the loops helping to create a strong magnetic field inside the coil owing to Faraday's law of induction. Inductors are one of the basic electronic components used in electronics where current and voltage change with time, because of the ability of inductors to delay and reshape ACs.

Calculating Inductance in a Series Circuit

To calculate the inductance in a series circuit, use the following equation:

$$L_{series} = L_1 + L_2 + \dots L_n$$

where
L_{series} = the total series circuit inductance (H)
L_n = individual inductor value (H)

For example, there are three inductors in series, with the individual values of 0.35 µH, 0.4 µH, and 0.5 µH. What is the total inductance of the series circuit?

$$L_{series} = 0.35\,\mu H + 0.4\,\mu H + 0.5\,\mu H$$
$$L_{series} = 1.25\,\mu H.$$

Calculating Inductance in a Parallel Circuit

To calculate the inductance in a parallel circuit, use the following equation:

$$\frac{1}{L_{parallel}} = \frac{1}{L_1} + \frac{1}{L_2} + \dots \frac{1}{L_n},$$

where
$L_{parallel}$ = the total parallel circuit inductance (H)
L_n = individual inductor value (H)

Using the values described in the series circuit, calculate the total inductance if the inductors were placed in a parallel circuit.

$$\frac{1}{L_{series}} = \frac{1}{0.35\,\mu H} + \frac{1}{0.4\,\mu H} + \frac{1}{0.5\,\mu H}$$

$$\frac{1}{L_{series}} = 2.86\,\mu H + 2.5\,\mu H + 2\,\mu H$$

$$\frac{1}{L_{series}} = 7.36\,\mu H$$

$$L_{series} = 0.14\,\mu H.$$

Electrical Shock Hazard

An electrical shock is received when electrical current passes through the body. Current will pass through the body in a variety of situations. Whenever two wires are at different voltages, current will pass between them if they are connected. The body can connect the wires if you touch both of them at the same time, and the current will then pass through your body. For example, in the normal household wiring, the black and red wires are at 120 V, while the white wire is at 0 V (since it is connected to ground). If you come into contact with an energized black wire and are also in contact with the neutral white wire, current will pass through your body, and thus, you will receive an electrical shock.[1] Moisture (sweat, pools of standing water) increases the risk of receiving an electrical shock.

You can also receive an electrical shock even if you're not in contact with an electrical ground. For example, contact with both live wires of a 240-V cable will deliver a shock. The primary reason for this type of electrical shock is that one wire may be at 120 V and the other at –120 V (a difference of 240 V). Electrical shocks can be caused by electrical components that are not properly grounded.[2]

The severity of injury from an electrical shock depends on the amount of electrical current and the length of time the current passes through the body. For example 1/10 of an ampere of electricity going through the body for just 2 s is enough to cause death. The amount of internal current a person can withstand and still be able to control the muscles of the arm and hand can be less than 10 mA. Currents greater than 75 mA cause ventricular fibrillation. This condition will cause death within a few minutes unless a special device called a defibrillator is used to save the victim. Heart paralysis occurs at 4 A, which means the heart does not pump at all. Tissue is burned with currents greater than 5 A.[3] Table 14.1 shows the effects of electrical current on the human body.[4]

TABLE 14.1

Effects of Electrical Currents on the Human Body

Current	Reaction
<1 mA	Generally not perceptible.
1 mA	Faint tingle.
5 mA	Slight shock felt; not painful but disturbing. Average individual can let go. Strong voluntary reactions can lead to other injuries.
6–25 mA (women) 9–20 mA (men)	Painful shock, loss of muscular control. The freezing current (let go) range. Individual cannot let go, but can be thrown away from the circuit if extensor muscles are stimulated.
50–150 mA	Extreme pain, respiratory arrest, severe muscular contractions. Death is possible.
1000–4300 mA	Rhythmic pumping action of the heart ceases. Muscular contraction and nerve damage occur; death likely.
10,000 mA	Cardiac arrest and severe burns occur. Death is probable.
15,000 mA	Lowest overcurrent at which a typical fuse or circuit breaker opens a current.

Source: National Institute of Occupational Safety and Health, 2009. *Electrical Safety, Safety and Health for Electrical Trades, Student Manual,* Department of Health and Human Services, Centers for Disease Control, NIOSH, Cincinnati, OH, http://www.cdc.gov/niosh/docs/2002-123/pdfs/02-123.pdf, p. 7.

Burns Caused by Electricity

The most common shock-related, nonfatal injury from electricity is burns. Burns caused by electricty may be of three types: electrical burns, arc burns, and thermal contact burns. Electrical burns can result when a person touches electrical wiring or equipment that is used or maintained improperly. Typically, such burns occur on the hands. Electrical burns are one of the most serious injuries you can receive and need to be given immediate medical attention. Additionally, clothing may catch fire and a thermal burn may result from the heat of the fire.[5]

Arc Blasts

Arc blasts occur when powerful, high-amperage currents arc through the air. Arcing is the luminous electrical discharge that occurs when high voltages exist across a gap between conductors and current travels through the air. This situation is often caused by equipment failure owing to abuse or fatigue. Temperatures as high as 35,000°F have been reached in arc blasts. There are three primary hazards associated with an air blast.

1. Arcing gives off thermal radiation (heat) and intense light, which can cause burns.
2. A high-voltage arc can produce a considerable pressure wave blast.

3. A high-voltage arc can also cause many of the copper and aluminum components in electrical equipment to melt. These droplets of molten metal can be blasted great distances by the pressure wave.[6]

Electrical Fires

Electricity is one of the most common causes of fires and thermal burns in homes and workplaces. Defective or misused electrical equipment is a major cause of electrical fires. If there is a small electrical fire, use only a Class C or multipurpose (ABC) fire extinguisher.[7]

Controlling Electrical Hazards

Electrical accidents are caused by a combination of three factors:

- Unsafe equipment or installation
- Workplaces made unsafe by the environment
- Unsafe work practices

Exposed Electrical Parts

Electrical hazards exist when wires or other electrical parts are exposed. Wires and parts can be exposed if a cover is removed from a wiring or breaker box. The overhead wires coming into a home may be exposed. Electrical terminals in motors, appliances, and electronic equipment may be exposed. Older equipment may have exposed electrical parts. If you contact exposed live electrical parts, you will be shocked.[8]

Overhead Power Lines

Overhead power lines are not usually insulated. Some examples of equipment that come into contact with power lines include cranes, ladders, scaffolding, backhoes, scissor lifts, raised dump truck beds, and aluminum paint rollers. To control the hazards presented by overhead power lines,

- A worker can stay at least 10 ft away.
- A worker can post warning signs.
- A worker can assume that lines are energized.

- A worker can use wood or fiberglass ladders, not metal.
- Power line workers need special training and personal protective equipment.

Inadequate Wiring

An electrical hazard exists when the wire is too small a gauge for the current it will carry. Normally, the circuit breaker in a circuit is matched to the wire size. However, in older wiring, branch lines to permanent ceiling light fixtures could be wired with a small gauge than the supply cable. When you use an extension cord, the size of the wire you are placing into the circuit may be too small for the equipment. The circuit breaker could be the right size for the circuit but not right for the smaller-gauge extension cord. A tool plugged into the extension cord may use more current than the cord can handle without tripping the circuit breaker. The wire will overheat and could cause a fire.[9]

Defective or Damaged Cords and Wires

All too often on construction sites, cords become damaged through aging, door or window edges, staples or fastenings, abrasion from adjacent materials, or activity in the area. Improper use of these cords can cause shocks, burns, or fires. To control the hazards associated with damaged or defective cords or wires, use one or more of the following control methods:

- Insulate live wires.
- Inspect all cords and wires before each use.
- Use only cords that are 3-wire type.
- Use only cords marked for hard or extra-hard usage.
- Use only cords, connection devices, and fitting equipped with strain relief.
- Remove cords by pulling on the plugs, not the cord.
- Cords not marked for hard or extra-hard use, or which have been modified or damaged, must be taken out of service immediately.

Use of Flexible Cords

Do not use flexible wiring where frequent inspection would be difficult or where damage would be likely to occur. Flexible cords must not be

- Run through holes in walls, ceilings, or floors
- Run through doorways, windows, or similar openings unless physically protected
- Hidden in walls, ceilings, floors, conduits, or other raceways

Improper Grounding

When an electrical system is not grounded properly, a hazard exists. The most common Occupational Safety and Health Administration electrical violation is improper grounding of equipment and circuitry. The metal parts of an electrical wiring system that we touch (switch plates, ceiling light fixtures, conduit, etc.) should be grounded and at 0 V. If the system is not grounded properly, these parts may become energized. Metal parts of motors, appliances, or electronics that are plugged into improperly grounded circuits may be energized. When a circuit is not grounded properly, a hazard exists because unwanted voltage cannot be safely eliminated. If there is no safe path to ground for fault currents, exposed metal parts in damaged appliances can become energized.[10]

Ground Fault Circuit Interrupters

A ground fault circuit interrupter (GFCI) is an inexpensive lifesaving device. GFCIs detect any difference in current between the two circuit wires (the black and white wires). This difference in current could happen when electrical equipment is not working correctly, causing leakage current. If leakage current (a ground fault) is detected in a GFCI-protected circuit, the GFCI switches off the current in the circuit, protecting you from a dangerous shock. GFCIs are set at approximately 5 mA and are designed to protect workers from electrocution. GFCIs are able to detect the loss of current resulting from leakage through a person who is beginning to be shocked. If this situation occurs, the GFCI switches off the current in the circuit. GFCIs are different from circuit breakers because they detect leakage currents rather than overloads.[11]

Assured Equipment Grounding Conductor Program

An Assured Equipment Grounding Conductor Program must cover

- All cord sets
- Receptacles not part of a building or structure
- Equipment connected by plug and cord

The program requirements include

- Specific procedures adopted by the employer
- Competent person to implement the program
- Visual inspection for damage of equipment connected by cord and plug

Overloaded Circuits

Overloads in an electrical system are hazardous because they can produce heat or arcing. Wires and other components in an electrical system or circuit have a maximum amount of current they can carry safely. If too many devices are plugged into a circuit, the electrical current will heat the wires to a very high temperature. If any one tool uses too much current, the wires will heat up. The temperature of the wires can be high enough to cause a fire. If their insulation melts, arcing may occur. Arcing can cause a fire in the area where the overload exists, even inside a wall. In order to prevent too much current in a circuit, a circuit breaker or fuse is placed in the circuit. If there is too much current in the circuit, the breaker "trips" and opens like a switch. If an overloaded circuit is equipped with a fuse, an internal part of the fuse melts, opening the circuit. Both breakers and fuses do the same thing: open the circuit to shut off the electrical current.[12]

Safety-Related Work Practices

To protect workers from electrical shock, the following work practices should be enforced at all times:

- Use barriers and guards to prevent passage through areas of exposed energized equipment.
- Preplan work and post hazard warnings.
- Keep working spaces and walkways clear of cords.
- Use special insulated tools when working on fuses with energized terminals.
- Don't use worn or frayed cords and cables.
- Don't fasten extension cords with staples, hang from nails, or suspend by wire.

Planning

One of the first steps in preventing electrical injuries and accidents is to plan your work and to others that you are working with or around be aware of your proposed activities. Plan to avoid falls that may result from your work. When working around electrical equipment, strictly follow the lockout/tagout program. Ensure that you or others working around electricity remove all jewelry and avoid wet conditions and overhead power lines.

Training

All employees working with electric equipment should be trained in safe work practices, which include the following:

- Deenergize electric equipment prior to inspecting or repairing.
- Using cords, cables, and electrical tools that are in good repair.

- Lockout/tagout recognition and procedures.
- Use appropriate personal protective equipment.

Key Information to Remember on Electrical Safety

1. Electrical currents move from high voltage to low voltage.
2. Voltage is measured in volts (V).
3. Current is the flow of electric charge and is measured in amperes (A).
4. Resistance is a measure of the opposition to the flow of steady electrical current and is measured in ohms (Ω).
5. Components connected in series are connected along a single path; thus, the same current flows through all of the components.
6. In a parallel circuit, the electrical current to each element in the circuit is separate.
7. Resistors are elements of electrical and electronic systems that are designed to intentionally resist the flow of electrical current through the system at a known measurement.
8. Capacitors or condensors are passive electronic components consisting of a pair of conductors separated by dielectric insulators, creating an electrical field that stores energy and produces a mechanical force between the conductors.
9. Inductors are passive electrical components that can store energy in a magnetic field created by the electric current passing through it.
10. The most common shock-related, nonfatal injury from electricity is burns.
11. Arc blasts occur when powerful, high-amperage currents arc through the air.
12. Electricity is one of the most common causes of fires and thermal burns in homes and workplaces.

References

1. National Institute of Occupational Safety and Health, 2009. *Electrical Safety, Safety and Health for Electrical Trades, Student Manual,* Department of Health and Human Services, Centers for Disease Control, NIOSH, Cincinnati, OH, p. 2. Available at http://www.cdc.gov/niosh/docs/2002-123/pdfs/02-123.pdf.

2. National Institute of Occupational Safety and Health, 2009. *Electrical Safety, Safety and Health for Electrical Trades, Student Manual*, Department of Health and Human Services, Centers for Disease Control, NIOSH, Cincinnati, OH, p. 5. Available at http://www.cdc.gov/niosh/docs/2002-123/pdfs/02-123.pdf.

3. National Institute of Occupational Safety and Health, 2009. *Electrical Safety, Safety and Health for Electrical Trades, Student Manual*, Department of Health and Human Services, Centers for Disease Control, NIOSH, Cincinnati, OH, p. 6. Available at http://www.cdc.gov/niosh/docs/2002-123/pdfs/02-123.pdf.

4. National Institute of Occupational Safety and Health, 2009. *Electrical Safety, Safety and Health for Electrical Trades, Student Manual*, Department of Health and Human Services, Centers for Disease Control, NIOSH, Cincinnati, OH, p. 7. Available at http://www.cdc.gov/niosh/docs/2002-123/pdfs/02-123.pdf.

5. National Institute of Occupational Safety and Health, 2009. *Electrical Safety, Safety and Health for Electrical Trades, Student Manual*, Department of Health and Human Services, Centers for Disease Control, NIOSH, Cincinnati, OH, p. 12. Available at http://www.cdc.gov/niosh/docs/2002-123/pdfs/02-123.pdf.

6. National Institute of Occupational Safety and Health, 2009. *Electrical Safety, Safety and Health for Electrical Trades, Student Manual*, Department of Health and Human Services, Centers for Disease Control, NIOSH, Cincinnati, OH, pp. 12–13. Available at http://www.cdc.gov/niosh/docs/2002-123/pdfs/02-123.pdf.

7. National Institute of Occupational Safety and Health, 2009. *Electrical Safety, Safety and Health for Electrical Trades, Student Manual*, Department of Health and Human Services, Centers for Disease Control, NIOSH, Cincinnati, OH, p. 14. Available at http://www.cdc.gov/niosh/docs/2002-123/pdfs/02-123.pdf.

8. National Institute of Occupational Safety and Health, 2009. *Electrical Safety, Safety and Health for Electrical Trades, Student Manual*, Department of Health and Human Services, Centers for Disease Control, NIOSH, Cincinnati, OH, p. 24. Available at http://www.cdc.gov/niosh/docs/2002-123/pdfs/02-123.pdf.

9. National Institute of Occupational Safety and Health, 2009. *Electrical Safety, Safety and Health for Electrical Trades, Student Manual*, Department of Health and Human Services, Centers for Disease Control, NIOSH, Cincinnati, OH, pp. 24–25. Available at http://www.cdc.gov/niosh/docs/2002-123/pdfs/02-123.pdf.

10. National Institute of Occupational Safety and Health, 2009. *Electrical Safety, Safety and Health for Electrical Trades, Student Manual*, Department of Health and Human Services, Centers for Disease Control, NIOSH, Cincinnati, OH, p. 29. Available at http://www.cdc.gov/niosh/docs/2002-123/pdfs/02-123.pdf.

11. National Institute of Occupational Safety and Health, 2009. *Electrical Safety, Safety and Health for Electrical Trades, Student Manual*, Department of Health and Human Services, Centers for Disease Control, NIOSH, Cincinnati, OH, p. 30. Available at http://www.cdc.gov/niosh/docs/2002-123/pdfs/02-123.pdf.

12. National Institute of Occupational Safety and Health, 2009. *Electrical Safety, Safety and Health for Electrical Trades, Student Manual*, Department of Health and Human Services, Centers for Disease Control, NIOSH, Cincinnati, OH, pp. 30–31. Available at http://www.cdc.gov/niosh/docs/2002-123/pdfs/02-123.pdf.

15

Mechanics

In this chapter, we will discuss and review the equations listed on the Board of Certified Safety Professionals (BCSP) examination reference sheet[1] related to Mechanics. Our primary focus will be on the equations related to energy, work, force, and velocity. *Mechanics* can be defined as the study of the relationships between motion, forces, and energy. As a safety professional, you must have a thorough working knowledge of this science in order to adequately perform the basic responsibilities of your position.

Energy

We will discuss two types of energy in this chapter, *potential energy* and *kinetic energy*. The laws of energy are as follows:

- The first law of thermodynamics says that energy under normal conditions cannot be created or destroyed, simply transformed from one type of energy to another (also known as the *law of conservation*).

- The second law of thermodynamics is a bit more complex than the first law, but basically states that any time you do work, including any time you make an energy transformation, some of the starting energy is going to be lost as heat.

Kinetic Energy

Kinetic energy is the energy of motion. An object that has motion, whether it be vertical or horizontal motion, has kinetic energy. There are many forms of kinetic energy, which include vibrational, rotational, and translational (energy due to motion from one location to another). The basic equation to determine kinetic energy is written as

$$\text{K.E.} = \frac{mv^2}{2},$$

or written another way:

$$K.E. = \frac{1}{2}mv^2,$$

where
  K.E. = kinetic energy (N)
    m = mass of the object
    v = speed of the object (velocity)

Example

Determine the kinetic energy of a 625-kg roller coaster car that is moving with a speed of 18.3 m/s.

$$K.E. = \frac{625 \text{ kg}(18.3 \text{ m/s})^2}{2}$$

$$K.E. = \frac{625 \text{ kg}(334.89 \text{ m/s}^2)}{2}$$

$$K.E. = \frac{209,306.25}{2}$$

$$K.E. = 104,653.13 \text{ m/s}^2$$

or

$$1.05 \times 10^5 \text{ N}.$$

NOTE: Kinetic energy is measured in joules. The conversion factor for joules is

$$1 \text{ joule} = 1 \text{ kg} \cdot \text{m/s}^2.$$

Therefore, in this example, there is 1.05×10^5 J (or joules) of kinetic energy.

Potential Energy

Potential energy is the same as stored energy. Potential energy exists whenever an object that has mass has a position within a force field. The most everyday example of this is the position of objects in the earth's gravitational field. Another good example of potential energy is a round object balanced atop a pyramid. The round object is not in motion but has the potential to be in motion. The equation to determine potential energy is listed as

$$P.E. = mgh,$$

where
P.E. = potential energy (J)
m = mass of the object (kg)
g = gravitational acceleration of the earth (9.8 m/s²)
h = height above earth's surface (m)

Example

In the example of the round object balanced on the pyramid, imagine that the object weighs 2.5 kg and is balanced at a height of 6 m above the earth's surface. What is the potential energy of the object?

$$P.E. = (2.5 \text{ kg})(9.8 \text{ m/s}^2)(6 \text{ m})$$

$$P.E. = 147 \text{ J}.$$

NOTE: It must be mentioned that in both of the previous equations, there is the assumption that no surface friction exists.

Elastic Potential Energy

Another type of potential energy important to a safety professional is *elastic potential energy*. Elastic potential energy is the energy stored in elastic materials as the result of their stretching or compressing. Elastic potential energy can be stored in rubber bands, bungee cords, trampolines, springs, an arrow drawn into a bow, and so on. The amount of elastic potential energy stored in such a device is related to the amount of stretch of the device. The more stretch of the material, the more energy that is stored. It is important to introduce the reader to *Hooke's law*, which states that if a spring is not stretched or compressed, then there is no elastic potential energy stored in it. The spring is said to be at its *equilibrium position*. The equilibrium position is the position that the spring naturally assumes when there is no force applied to it. In terms of potential energy, the equilibrium position could be called the zero-potential energy position. There is a special equation for springs, which relates the amount of elastic potential energy to the amount of stretch (or compression) and the spring constant. According to Hooke's law, the force required to stretch the spring will be directly proportional to the amount of stretch. This equation is written as follows:

$$P.E._{\cdot elastic} = \frac{kx^2}{2},$$

where

P.E. = potential energy (elastic) (J)
k = spring constant (N/m²)
x = amount of compression (distance in meters)

Example

A force of 100 N/m² is required to compress an automobile suspension spring 0.45 m. Determine the potential energy of the spring.

$$P.E._{\cdot elastic} = \frac{(100 \text{ N/m}^2)(0.45^2 \text{ m})}{2}$$

$$P.E._{\cdot elastic} = \frac{(100 \text{ N/m}^2)(0.2025 \text{ m})}{2}$$

$$P.E._{\cdot elastic} = 10.13 \text{ J}.$$

Force

Force is a push or pull upon an object resulting from the object's interaction with another object. Whenever there is an interaction between two objects, there is a force upon each of the objects. Force is a quantity that is measured using the SI unit known as the newton. A *newton* is abbreviated as "N." To say "10 N" means 10.0 newtons of force. One newton is the amount of force required to give a 1-kg mass an acceleration of 1 m/s².

There are two types of forces: contact forces and action-at-a-distance forces.

Contact Forces

Contact forces include the following:

- *Frictional force* is a force that resists the relative motion of objects that are in contact with each other.
- *Tension force* is the force required to pull an object (opposite of compression).

- *Normal force* is the force on an object caused by the normal interaction between two objects.
- *Air resistance force* is the force between an object traveling through air and the contact with the air. As with all forces, air resistance force opposes the motion of the object.
- *Applied force* is the force applied to an object by a person or another object.
- *Spring force* is the force exerted by a compressed or stretched spring.

Action-at-a-Distance Forces

Action-at-a-distance forces include the following:

- *Gravitation force* is the force caused by the earth's gravitational pull.
- *Electrical force* is the force caused by an electrical field.
- *Magnetic force* is the force caused by a magnetic field.

Defining Mass and Weight

Before discussing force equations, it is necessary to clearly define the difference between mass and weight. The force of gravity acting upon an object is sometimes referred to as the weight of the object. The weight of an object (*measured in newtons*) will vary according to where in the universe the object is. The mass of an object refers to the amount of matter that is contained by the object (*measured in kilograms*) and will be the same no matter where in the universe the object is located. Weight, being equivalent to the force of gravity, is dependent upon the value of *g*. On the earth's surface, *g* is 9.8 m/s² (often approximated as 10 m/s²). The weight of an object can be determined using the following equation:

$$W = mg,$$

where
W = amount of work done on or to an object due to gravity
m = mass (kg)
g = gravity (9.8 m/s²) (constant)

Example

What is the amount of work done on or to an object having a mass of 2 kg?

$$W = 2 \text{ kg} \times 9.8 \text{ m/s}^2$$

$$W = 19.6 \text{ N}.$$

Amount of Force

The amount of force is calculated using the following equation:

$$F = ma,$$

where
  F = amount of force
  m = mass (kg)
  a = acceleration (m/s^2)

Example

Determine the amount of force of a 2500-lb automobile with an acceleration rate of 20 m/s^2.

NOTE: Remember that our unit of mass is measured in kilograms for this equation. Convert pounds to kilograms.

$$kg = 2500 \ lb \times \frac{kg}{2.2} \ lb$$

$$kg = 1136.4.$$

Now, insert the data into the equation, as follows:

$$F = 1136.4 \ kg \times 20 \ m/s^2$$

$$F = 22{,}727.3 \ N.$$

Frictional Force

All objects on earth have some type of frictional force. The equation for determining the frictional force is as follows:

$$F = \mu N,$$

where
  F = frictional force (parallel to the surface) (can also be written as F_{fric}) (N)
  μ = coefficient of friction
  N = force acting on the surface in a direction that is normal (perpendicular) to the surface (N)

Example

Determine the frictional force that results from an object having a coefficient of friction of 0.3 and 200 N.

$$F_{fric} = 0.3(200 \text{ N})$$

$$F_{fric} = 60 \text{ N}$$

Force and Distance

When a force acts upon an object, there is usually some movement (displacement) of the object. The relationship between force and distance is that they have a direct proportion relationship. To calculate the effects of force on distance, we can use the following equation:

$$F_1 D_1 = F_2 D_2,$$

where
  F = force (N)
  D = distance

Example

A simple beam is balanced on a fulcrum at the center of the beam. If a 100-kg man walks a distance of 2.3 m from the center and causes the beam to move upward 0.5 m at the opposite end, how far will a 25-kg child have to walk in order to cause the beam to move upward at the opposite end a distance of 0.5 m?

First, convert the weight of the man and child into newtons, by using the following equation:

$$N = 100 \text{ kg} \times 9.8 \text{ m/s}^2$$

$$N = 980.$$

The force of the 100-kg man and assuming a gravitational force of 9.8 m/s² would be 980 N. For the child, it would be 245 N. Therefore, we can now insert the data into our equation, as follows:

$$980 \text{ N}(2.3 \text{ m}) = 245 \text{ N}(D_2)$$

$$\frac{980 \text{ N}(2.3 \text{ m})}{245 \text{ N}} = \frac{245 \text{ N}(D_2)}{245 \text{ N}}$$

$$9.2 \text{ m} = D_2.$$

The 25-kg child would have to walk 9.2 m away from the center point of the beam to move the beam upward at the opposite end 0.5 m.

Momentum

Momentum is a measure of the motion of a body equal to the product of its mass and velocity. We calculate momentum by using the following equation:

$$\rho = mv,$$

where
  ρ = momentum
  m = mass (kg)
  v = velocity (m/s)

Example

An object of mass 76 kg rolls down a hill at 8.2 m/s. Calculate the momentum (which is downhill).

$$\rho = 76\ \text{kg} \left(\frac{8.2\ \text{m}}{\text{s}} \right)$$

$$\rho = 623\ \text{kg} \times \text{m/s}.$$

Work

When a force acts upon an object to cause a displacement of the object, it is said that *work* is done upon the object. There are three key ingredients to work—force, displacement, and cause. In order for a force to qualify as having done work on an object, there must be a displacement and the force must cause the displacement. Force and displacement are vectors. When a force (F) acting on an object causes displacement (s) in a direction different from the one along where the force acts, the work done is calculated:

$$W = Fs,$$

where
  W = work done on or to a system (usually in joules or newtons) (1 J = 1 N × 1 m)
  F = magnitude of the force (N)
  s = displacement (m)

NOTE: This equation is for a horizontal object with no angle.

Example

A box is sitting on the floor. A force of 250 N is applied horizontally to the side of the box and the box is moved horizontally 9 m. What is the amount of work done?

$$W = (250 \text{ N})(9 \text{ m}).$$

$$W = 2250 \text{ J}.$$

Modified Work

As mentioned previously, the equation $W = Fs$ was written for a horizontal object with no angle, but what if you had to determine the amount of work done on an object at an angle? In order to determine the amount of work done on an object by a force applied at an angle, we use the following equation, which is NOT listed on the BCSP examination reference sheet:

$$W = Fs \times \cos\theta,$$

where

$W =$ amount of work done on or to an object (in joules or newtons) (1 J = 1 N × 1 m)
$F =$ magnitude of force (N)
$s =$ displacement (m)
$\theta =$ angle between the directions of force and displacement

Example

A force is acting at 30° with the horizontal on an object that is displaced 3 m along the horizontal direction. The normal force on the surface is 400 N and the coefficient of friction is 0.35. Calculate the work done by the force.

Step 1: Determine the amount of force based on the friction coefficient of 0.35.

$$F_{fric} = 400 \text{ N}(0.35)$$

$$F_{fric} = 140 \text{ N}$$

Step 2: Insert the data into the modified work equation, as follows:

$$W_{\text{modified}} = (140 \text{ N})(3 \text{ m})(\cos 30°)$$

$$W_{\text{modified}} = 420 \text{ N/m}(0.8660)$$

$$W_{\text{modified}} = 363.7 \text{ N.}$$

Newton's Laws of Motion

Newton's first law of motion states: "A body continues to maintain its state of rest or of uniform motion unless acted upon by an external unbalanced force." The second law of motion ($f = ma$) states that the net force on an object is equal to the mass of the object multiplied by its acceleration. The third law of motion states, "To every action there is an equal and opposite reaction."

Speed

Speed is a quantity of scale that determines how fast an object is moving. A fast-moving object has a high speed, while a slow-moving object has a low speed. An object with zero speed is not moving.

Velocity

Velocity is defined as the rate at which an object changes its position and is a vector quantity. In other words, velocity must have a direction. It can be written mathematically as follows:

$$v = v_0 + at,$$

where
  v = velocity
  v_0 = original velocity at the start of the acceleration

a = acceleration

t = time (s)

Example

A dragster is traveling at 15 m/s and then accelerates to 20 m/s for 4 s. How fast is the dragster traveling?

$$v = 15\frac{m}{s} + \left(20\frac{m}{s}\right)(4\ s)$$

$$v = 15\frac{m}{s}(80\ m)$$

$$v = 95\ m/s.$$

The velocity of the dragster is 95 m/s. Now, convert this velocity to miles per hour. We do this as follows:

Step 1: Convert meters per second to feet per second.

$$\frac{95\ m}{s} \times \frac{3.281\ ft}{m} = \frac{311.7\ ft}{s}$$

Step 2: Now, convert feet per second to feet per mile.

$$\frac{311.7\ ft}{s} \times \frac{3600\ s}{h} \times \frac{mi}{5280\ ft} = \frac{212.5\ mi}{h}$$

The dragster traveling at 95 m/s is equivalent to traveling at 212.5 mi/h.

Calculating Final Velocity

The equation for calculating terminal velocity is written as

$$v^2 = v_0^2 + 2as,$$

where

v = final velocity

v_0 = initial velocity

a = acceleration of the object (m/s)

s = displacement of the object (change in position—normally described in distance from original position)

Example

An over-the-road truck driver working for your company is traveling at 42 m/s when he notices that the traffic light is red. The truck driver reduces his speed at a rate of −9 m/s². He continues this deceleration for 82 m before the light changes to green and begins to accelerate. At the point of acceleration, what is the velocity? We can calculate this velocity by inserting the known variables, as follows:

$$v^2 = (42 \text{ m/s})^2 + 2\left(\frac{-9 \text{ m}}{s^2}\right)(82 \text{ m})$$

$$v^2 = \left(\frac{1764 \text{ m}}{s^2}\right) + 2(-728 \text{ m})$$

$$v^2 = \frac{1764 \text{ m}}{s^2} + (-1476 \text{ m})$$

$$v^2 = 288 \text{ m/s}^2$$

$$\sqrt{v^2} = \sqrt{288 \text{ m/s}}$$

$$v = 16.97 \text{ m/s}.$$

Calculating Displacement

Displacement has been identified as change in position of an object. We calculate displacement by utilizing the following equation:

$$s = v_0 t + \frac{at^2}{2},$$

where

s = displacement of the object (change in position—normally described in distance from its original position)

v_0 = initial velocity

t = time

a = acceleration

Example

Using the same scenario from the "Calculating Velocity" section, where your truck driver is approaching the traffic light at a velocity of 42 m/s, when he notices the traffic light turn yellow. He immediately applies the brakes at a velocity of −9.0 m/s and comes to a complete stop after 3 s. What is the distance traveled (displacement) from the time he applied the brakes until he comes to a complete stop? Insert the data into the equation as follows:

$$s = \frac{42 \text{ m}}{\text{s}}(3 \text{ s}) + \frac{\left(-9.0\dfrac{\text{m}}{\text{s}}\right)(3 \text{ s})^2}{2}$$

$$s = 126 \text{ m} + (-40.5 \text{ m})$$

$$s = 85.5 \text{ m.}$$

The distance traveled or displacement of the truck was 85.5 m.

Key Information to Remember on Mechanics

1. The first law of thermodynamics states that energy under normal conditions cannot be created or destroyed (also known as the *law of conservation*).

2. The second law of thermodynamics states that any time work is done, some of the starting energy will be lost as heat.

3. Kinetic energy is the energy of motion.

4. Kinetic energy is measured in newtons.

5. $1 \text{ N} = 1 \text{ kg} \cdot \text{m/s}^2$.

6. Potential energy is the same as stored energy.

7. Force is a push or pull upon an object resulting from the object's interaction with another object.

8. Mass is always constant, regardless of its location in the universe. Weight is variable depending on the gravitational force.

9. Earth's gravity = 9.8 m/s^2.

10. Three key ingredients to work are force, displacement, and cause.

11. Newton's first law of motion states: "A body continues to maintain its state of rest unless acted upon by an external, unbalanced force."

12. Newton's second law of motion states: "The net force on an object is equal to the mass of the object multiplied by its acceleration."

13. Newton's third law of motion states: "To every action there is an equal and opposite reaction."

14. Speed is a scalar quantity that determines how fast an object is moving.

15. Velocity is a vector quantity that describes the rate at which an object changes its position.

16. Displacement has been identified as change in position of an object.

Reference

1. Board of Certified Safety Professionals, *Comprehensive Examination Equation Reference Sheet*. Available at http://www.bcsp.org/pdf/ASPCSP/ExamRef5.pdf.

16

Hydrostatics and Hydraulics

In preparing for the Associate Safety Professional/Certified Safety Professional examinations, the candidate must have a fundamental understanding of hydraulics and hydrostatics. In the everyday "safety world," you may not automatically understand the importance of having this understanding. However, if you just take a look around, you will see that having this knowledge can make a difference in the safety and health of employees. A question you may be asking yourself is "What is hydrostatics and hydraulics?" *Hydrostatics*, also known as fluid statics, is the science of fluids at rest and is a subfield within fluid mechanics. It embraces the study of the conditions under which fluids are at rest in stable equilibrium. Hydrostatics is about the pressures exerted by a fluid at rest. Any fluid is meant, not just water.[1] The use of fluid to do work is called *hydraulics*, and the science of fluids in motion is fluid dynamics. *Hydraulics* is a topic in applied science and engineering dealing with the mechanical properties of liquids. Hydraulics is used for the generation, control, and transmission of power by the use of pressurized liquids.

The study of hydrostatics and hydraulics goes back many centuries to early civilization. Early engineers and scientists succeeded in making water flow from one location to another. The problems they encountered usually involved hydraulics in the pipe flow. Whenever velocity, flow direction, or elevation changes in liquids, forces and pressures are produced.

Water Properties

Water, or wastewater, weighs 8.34 lb/gal. There are 7.48 gal/ft^3; therefore, 1 ft^3 of water will weigh 62.4 lb (8.34 lb/gal × 7.48 gal/ft^3 = 62.4 lb/ft^3). When using metric units, the weight of water is 9.8 kN/m^3. Another important property of water related to the study of hydraulics is the pressure exerted on a column of water. One foot of water height is approximately equal to 0.434 pounds per square inch (psi). This measurement will hold true, no matter what the diameter of the container (column) is.

Hydrostatic Pressure

As mentioned previously, liquids will exert pressure and force against the walls of the container, regardless of whether they are stored in a tank or flowing through pipes. Pressure is defined as force per unit area. Pressure exerted by water is known as hydrostatic pressure. The unit of measurement in the US system is usually pounds per square inch. In the metric system, pressure is usually denoted by newtons per square meter. To convert from the US unit to the metric unit, use the conversion of $0.000145 \text{ psi} = 1 \text{ N/m}^2$. Newtons per square meter can also be termed pascals ($1 \text{ N/m}^2 = 1 \text{ Pa}$). To calculate pressure, use the following equation:

$$P = \frac{F}{a},$$

where
  P = pressure (psi)
  F = force or weight (water: 62.4 lb/ft^3)
  a = area (in^2)

Example

A piping system repair must be made. The pipe is 40 ft long, with a diameter of 3 ft and is installed vertically. It is filled with liquid (water), which is leaking at a point located 25 ft from the top. For operational reasons, the liquid cannot be drained before making the repairs. Before the repairs can be made, a suitable method and material decision should be made. What is the pressure exerted at the location of the pipe repair?

Solution

To solve this problem, you must determine the force at the repair location. We know that the repair location is 25 ft from the top and the diameter of the pipe is 3 ft. Therefore, we must obtain a volume (ft^3). To do this, we use the following equation:

$$V_{cylinder} = \pi r^2 \times h,$$

where
  r = radius of cylinder
  h = height of cylinder

In this case, the height is 25 ft and the radius of the pipe is 1.5 ft or 18 in.

$$V_{\text{cylinder}} = (31.4)(1.5)^2 \times 25 \text{ ft}$$
$$V_{\text{cylinder}} = (3.14)(2.25 \text{ ft}^2)(25 \text{ ft})$$
$$V_{\text{cylinder}} = 176 \text{ ft.}$$

There is 176 ft³ of water at this location of the pipe and water weighs 62.4 lb/ft³; therefore, the weight of the water at the point of repairs is 11,027 lb. Now, we must calculate the area for the pipe, which is determined by

$$\text{Area}_{\text{pipe}} = \pi r^2$$
$$\text{Area}_{\text{pipe}} = (3.14)(18 \text{ in.})^2$$
$$\text{Area}_{\text{pipe}} = (3.14)(324 \text{ in.}^2)$$
$$\text{Area}_{\text{pipe}} = 1017.36 \text{ in.}^2.$$

We can now solve for the pressure, using the following equation:

$$P = \frac{F}{a}$$
$$P = \frac{11,027 \text{ lb}}{1017.36 \text{ in.}^2}$$
$$P = 10.84 \text{ lb/in.}^2.$$

The pressure at the point of repairs is 43.36 psi.

Torricelli's Law

Torricelli's law, also known as *Torricelli's theorem*,[2] is a theorem in fluid dynamics relating the speed of fluid flowing out of an opening to the height of fluid above the opening. Torricelli's law states that the velocity, (*v*), of a fluid through a sharp-edged hole at the bottom of a tank filled to a depth *h* is the same as the speed that a body (in this case, a drop of water) would acquire in falling freely from a height (head, in feet) *h*, that is, where *g* is the acceleration due to gravity.

Torricelli's theorem, also called *Torricelli's law*, *Torricelli's principle*, or *Torricelli's equation*, states that the speed, *v*, of a liquid flowing under the force of gravity out of an opening in a tank is proportional jointly to the square root of the vertical distance, h, between the liquid surface and the center of the opening and to the square root of twice the acceleration caused by gravity, $2g$, or simply $v = (2gh)^{1/2}$. (The value of the acceleration caused by gravity at the earth's surface is approximately 32.2 ft/s², or 9.8 m/s².) The theorem is named after Evangelista Torricelli, who discovered it in 1643.

The speed of a portion of water flowing through an opening in a tank a given distance, h, below the water surface is the same as the speed that would be attained by a drop of water falling freely under the force of gravity alone (i.e., neglecting effects of air) through the same distance, h. The speed of efflux is independent of the direction of flow; at the point of the opening, the speed is given by this equation, whether the opening is directed upward, downward, or horizontally.

The equation is written as follows[2]:

$$v = \sqrt{2gh},$$

where

v = velocity (ft/s)
g = acceleration due to gravity (9.8 m/s or 32 ft/s)
h = head (ft)

$$P = 0.433 \times h,$$

where

P = pressure (psi)
h = head (ft)

Example

What is the velocity at discharge, if the nozzle of a fire suppression system is 50 psi?

Solution

This problem requires us to solve for head (h).

$$P = 0.433 \times h$$

$$50 \text{ psi} = 0.433 \times h$$

$$\frac{50 \text{ psi}}{0.433} = \frac{0.433 \times h}{0.433}$$

$$115.47 \text{ ft} = h.$$

Head Pressure

In fluid dynamics, *head* is a concept that relates the energy in an incompressible fluid to the height of an equivalent static column of that fluid. From Bernoulli's principle, the total energy at a given point in a fluid is the energy associated with the movement of the fluid, plus energy from pressure in the fluid, plus energy from the height of the fluid relative to an arbitrary datum (a reference from which measurements are made). Head is expressed in units of height such as meters or feet. The hydraulic head can be used to determine a *hydraulic gradient* between two or more points. We can calculate the head pressure using the following equation:

$$h_p = \frac{p}{w},$$

where

h_p = head pressure (ft)
p = gauge pressure
w = weight of water

Example

Given a pressure of 20 psi and the weight of water at 62.4 lb/ft³, calculate the head pressure.

Solution

To solve this equation, you must convert psi to lb/ft², as follows:

$$20\,\frac{\text{lb}}{\text{in.}^2} \times \left(12\,\frac{\text{in.}}{\text{ft}}\right) \times \left(12\,\frac{\text{in.}}{\text{ft}}\right) = 2880 \text{ lb/ft}^2.$$

We now insert the data into our original equation.

$$h_p = \frac{2880 \text{ lb/ft}^2}{62.4 \text{ lb/ft}^3}$$

$$h_p = 46.15 \text{ ft.}$$

It means that the water, under this pressure, will rise vertically 46.15 ft.

Velocity Head

Velocity head can be described as the velocity of a fluid expressed in terms of the head or static pressure required to produce that velocity. *Velocity head* is attributed to the motion of a fluid (kinetic energy).

Velocity head can be calculated as follows:

$$h_v = \frac{v^2}{2g},$$

where
   h_v = velocity expressed in head (ft)
   v = velocity (ft/s, m/s, etc.)
   g = acceleration of gravity (9.8 m/s² or 32.2 ft/s²)

Example

Calculate the head velocity if water is flowing through a pipe at 5 ft/s.

$$h_v = \frac{(5 \text{ ft/s})^2}{(2)32.2 \text{ ft/s}^2}$$

$$h_v = \frac{25 \text{ ft/s}}{64.4 \text{ ft/s}^2}$$

$$h_v = 0.388 \text{ ft.}$$

Velocity Pressure at Constant Laminar Velocity

Laminar flow, sometimes known as streamline flow, occurs when a fluid flows in parallel layers, with no disruption between the layers. At low velocities, the fluid tends to flow without lateral mixing, and adjacent layers slide past one another like playing cards. There are no cross currents perpendicular to the direction of flow; neither are there eddies or swirls of fluids. In laminar flow, the motion of the particles of fluid is very orderly with all particles moving in straight lines parallel to the pipe walls. In fluid dynamics, laminar flow is a flow regime characterized by high momentum diffusion and low momentum convection.

When a fluid is flowing through a closed channel such as a pipe or between two flat plates, either two types of flow may occur depending on the velocity of the fluid: laminar flow or turbulent flow. Laminar flow is the opposite of turbulent flow, which occurs at higher velocities where eddies or small packets of fluid particles form leading to lateral mixing. In nonscientific terms, laminar flow is "smooth," while turbulent flow is "rough."

The type of flow occurring in a fluid in a channel is important in fluid dynamics problems. The dimensionless Reynolds number is an important parameter in the equations that describe whether flow conditions lead to laminar or turbulent flow. In the case of flow through a straight pipe with a circular cross section, Reynolds numbers of less than 2100 are generally considered to be of a laminar type; however, the Reynolds number upon which laminar flows become turbulent is dependent upon the flow geometry. When the Reynolds number is much less than 1, creeping motion or Stokes flow occurs. This is an extreme case of laminar flow where viscous (friction) effects are much greater than inertial forces. The common application of laminar flow would be in the smooth flow of a viscous liquid through a tube or pipe. In that case, the velocity of flow varies from zero at the walls to a maximum along the centerline of the vessel. The flow profile of laminar flow in a tube can be calculated by dividing the flow into thin cylindrical elements and applying the viscous force to them.[3]

The Reynolds number classification is listed as follows:

- Laminar flow: Re < 2000
- Transitional flow: 2000 < Re < 4000
- Turbulent flow: Re > 4000

The pressure at constant laminar velocity can be calculated using the following equation:

$$p_v = \frac{Q^2}{891d^4},$$

where
p_v = pressure at constant laminar flow (psi)
Q = flow rate (gpm)
d = pipe diameter (in., ft, m)

Example

Calculate the pressure of a pipe measuring 18 in. in diameter with a flow rate of 2500 gpm.

$$p_v = \frac{(2500 \text{ gpm})^2}{(891)(18 \text{ in.})^4}$$

$$p_v = \frac{6,250,000 \text{ gpm}}{93,533,616 \text{ in.}^4}$$

$$p_v = 0.067 \text{ psi.}$$

Flow Rates and Pressure Drops

A piping system, such as a fire water supply system, can operate at different flow rates (Q). If you were to operate at one pressure, increase or decrease the pressure to achieve the maximum operating flow rate, and then the flow rate would change. For example, if you were operating at $Q = 350$ gpm and 20 psi, the measured pressure would be under the conditions for 350 gpm. However, if you increase the flow pressure to 25 psi, then the flow rate would also change.

You can calculate this difference in pressure by using the following equation:

$$Q_2 = Q_1 \left[\frac{(S-R_2)^{0.54}}{(S-R_1)^{0.54}} \right],$$

where

Q_2 = final flow rate after increasing the flow rate of system 1 (gpm)
Q_1 = flow rate under original conditions (gpm)
S = highest pressure in the system (psi)
R_1 = pressure that the system dropped to as a result of running at specific conditions (1) (psi)
R_2 = pressure that the system dropped to as a result of running at specific conditions (2) (psi)

Example

A fire system is operating at 350 gpm and at 20 psi (0.03 psi drop). If the operator were to increase the pressure to 25 psi (0.086 psi drop), what is the final flow rate?

$$Q_2 = 350 \text{ gpm} \left[\frac{(25 \text{ psi} - 0.86 \text{ psi})^{0.54}}{(25 \text{ psi} - 0.03 \text{ psi})^{0.54}} \right]$$

$$Q_2 = 350 \text{ gpm} \left[\frac{5.58 \text{ psi}}{5.68 \text{ psi}} \right]$$

$$Q_2 = 350 \text{ gpm} \left[0.98 \right]$$

$$Q_2 = 343.84 \text{ gpm}.$$

The amount of pressure used in a system is calculated using the following equation:

$$P = \left(\frac{Q}{K} \right)^2,$$

where

P = pressure (psi)
Q = flow rate (gpm)
K = discharge coefficient (0–1)

In a nozzle or other constriction, the discharge coefficient is the ratio of the mass flow rate at the discharge end of the nozzle to that of an ideal nozzle that expands an identical working fluid from the same initial conditions to the same exit pressure. It is also known as coefficient of discharge (Sci-Tech[4]).

Example

A fire system has a wide-angle full cone nozzle with a flow rate of 5 gpm and a discharge coefficient of 0.79. What is the pressure from the nozzle?

$$P = \left(\frac{5 \text{ gpm}}{0.79} \right)^2$$

$$P = 6.33^2$$

$$P = 40.05 \text{ psi.}$$

Flow Rates and Pressures

There is a direct relationship between flow rates and pressures in the system. This relationship is calculated using the following equation:

$$\frac{Q_1}{Q_2} = \frac{\sqrt{P_1}}{\sqrt{P_2}},$$

where

Q = flow rate (gpm)
P = pressure (psi)

Example

A system is operating at 350 gpm at 25 psi. If the system pressure were increased to 30 psi, calculate the flow rate.

Solution

$$\frac{350 \text{ gpm}}{Q_2} = \frac{\sqrt{25 \text{ psi}}}{\sqrt{30 \text{ psi}}}$$

$$\frac{350 \text{ gpm}}{Q_2} = \frac{5 \text{ psi}}{5.48 \text{ psi}}$$

$$\frac{350 \text{ gpm}}{Q_2} = 0.91$$

$$(Q_2)\frac{350 \text{ gpm}}{Q_2} = (0.91)(Q_2)$$

$$\frac{350 \text{ gpm}}{0.91} = \frac{0.91 \, Q_2}{0.91}$$

$$384.62 \text{ gpm} = Q_2.$$

Calculating Pressure Loss Due to Friction

To calculate the pressure loss due to friction, use the following equation:

$$P_d = \frac{4.52Q^{1.85}}{C^{1.85}d^{4.87}},$$

where
P_d = pressure lost to friction (psi/ft)
Q = flow rate (gpm)
C = coefficient of roughness of the pipe
d = inside pipe diameter (in.)

Example

A fire system has a piping system with 6 in. inside diameter, with a coefficient of friction of 120 and a flow rate of 800 gpm. What is the total friction loss in 500 ft of pipe?

Solution

$$P_d = \frac{4.52(800 \text{ gpm})^{1.85}}{(120)^{1.85} d(6 \text{ in.})^{4.87}}$$

$$P_d = \frac{4.52(2,344,809.9 \text{ gpm})}{(7022.4)(6160.2 \text{ in.})}$$

$$P_d = \frac{10,598,540.75 \text{ gpm}}{43,256,924.4 \text{ in.}}$$

$$P_d = 0.006 \text{ psi/ft}.$$

The total friction loss in 500 ft of piping is 0.006 psi/ft × 500 ft = 3.02 psi.

Bernoulli's Principle

In fluid dynamics, *Bernoulli's principle* states that for an inviscid flow, an increase in the speed of the fluid occurs simultaneously with a decrease in pressure or a decrease in the fluid's potential energy.[1,2] Bernoulli's principle is named after the Dutch–Swiss mathematician Daniel Bernoulli who published his principle in his book *Hydrodynamica* in 1738.

Bernoulli's principle can be applied to various types of fluid flow, resulting in what is loosely denoted as *Bernoulli's equation*. In fact, there are different forms of the Bernoulli equation for different types of flow. The simple form of Bernoulli's principle is valid for incompressible flows (e.g., most liquid flows) and also for compressible flows (e.g., gases) moving at low Mach numbers. More advanced forms may in some cases be applied to compressible flows at higher Mach numbers (see the derivations of the Bernoulli equation).

Bernoulli's principle can be derived from the principle of conservation of energy. This states that, in a steady flow, the sum of all forms of mechanical energy in a fluid along a streamline is the same at all points on that stream-line. This requires that the sum of kinetic energy and potential energy remain constant. If the fluid is flowing out of a reservoir, the sum of all forms of energy is the same on all streamlines because in a reservoir, the energy per unit mass (the sum of pressure and gravitational potential ρgh) is the same everywhere.

Fluid particles are subject only to pressure and their own weight. If a fluid is flowing horizontally and along a section of a streamline, where the speed increases, it can only be because the fluid on that section has moved from a region of higher pressure to a region of lower pressure, and if its speed decreases, it can only be because it has moved from a region of lower pres-sure to a region of higher pressure. Consequently, within a fluid flowing horizontally, the highest speed occurs where the pressure is lowest, and the lowest speed occurs where the pressure is highest.[5]

The form of the equation supplied on the examination is as follows:

$$\frac{v_A^2}{2g} + \frac{p_A}{w} + z_A = \frac{v_B^2}{2g} + \frac{p_B}{w} + z_B + h_{AB},$$

where

v_A = velocity at point A (ft/s, m/s)
v_B = velocity at point B (ft/s, ms/s)
g = gravity (9.8 m/s or 32.2 ft/s)
p_A = pressure at point A (psi)
p_B = pressure at point B (psi)
z_A = elevation at point A (ft)
z_B = elevation at point B (ft)
w = weight (not mass) ($\rho \times g$)
h_{AB} = the piezometric head (the sum of the elevation z and the pressure head between points A and B)

Example

In a horizontally installed fire system, water (density of 62 lb/ft³) flows through a 6-in. pipe. The head loss in a 1000-ft section of the system has a head loss of 50 ft. The residual pressure at point A is 45 psi and the velocity at point A is 5 ft/s. The velocity at point B is 7 ft/s. What is the residual pressure at point B?

Solution

$$\frac{5^2 \text{ ft/s}}{2\left(32.2\dfrac{\text{ft}}{\text{s}^2}\right)} + \frac{45 \text{ psi}}{62} + 0 \text{ ft} = \frac{7^2 \text{ ft/s}}{2\left(32.2\dfrac{\text{ft}}{\text{s}^2}\right)} + \frac{p_B}{62} + 0 \text{ ft} + 50 \text{ ft}$$

$$\frac{25 \text{ ft/s}}{64.4 \frac{\text{ft}}{\text{s}^2}} + \frac{45 \text{ psi}}{62}\left(144 \frac{\text{in.}^2}{\text{ft}^2}\right) + 0 \text{ ft} = \frac{49 \text{ ft/s}}{62 \frac{\text{ft}}{\text{s}^2}} + \frac{p_B}{62} + 50 \text{ ft}$$

$$0.388 + 104.52 \text{ psi} + 0 \text{ ft} = 0.766 + \frac{p_B}{62} + 50 \text{ ft}$$

$$104.91 \frac{\text{psi}}{\text{ft}} = 50.766 + \frac{p_B}{62}$$

$$104.91 - 50.766 = 50.766 - 50.766 + \frac{p_B}{62}$$

$$54.14 \left(62\right) = \left(62\right)\frac{p_B}{62}$$

$$3356.93 \text{ psf} = p_B.$$

To get the actual psi value, you must divide by 144 in.², which results in 23.33 psi.

Key Information to Remember on Hydrostatics and Hydraulics

1. Hydrostatics, also known as fluid statics, is the science of fluids at rest and is a subfield within fluid mechanics.

2. Hydraulics is the science of fluids in motion, also called fluid dynamics.

3. Water weight, 8.34 lb/gal or 62.4 lb/ft³.

4. Pressure is defined as force per unit area.

5. Torricelli's law is a theorem in fluid dynamics relating the speed of fluid flowing out of an opening to the height of fluid above the opening.

6. In fluid dynamics, head is a concept that relates energy in an incompressible fluid to the height of an equivalent static column of that fluid.

7. Velocity head can be described as the velocity of a fluid expressed in terms of the head or static pressure required to produce that velocity.

8. Reynolds number: <2000 is laminar, 2000 < Re < 4000 is transitional, and >4000 is turbulent.

9. Bernoulli's principle states that for an inviscid flow, an increase in the speed of the fluid occurs simultaneously with a decrease in pressure or a decrease in the fluid's potential energy.

References

1. http://mysite.du.edu/~jcalvert/tech/fluids/hydstat.
2. http://www.britannica.com/EBchecked/topic/600154/Torricellis-theorem.
3. https://www.princeton.edu/~achaney/tmve/wiki100k/docs/Laminar_flow.html.
4. Sci-Tech, 2010. Available at http://www.answers.com/topic/discharge-coefficient (accessed July 27, 2010).
5. https://www.princeton.edu/~achaney/tmve/wiki100k/docs/Bernoulli_s_principle.html.

17

Training

One of the most fundamental and important roles that any safety professional will have is that of *training*. The majority of safety professionals conduct training every day, whether in a formal classroom or conducting safety inspections or audits. Nearly every aspect of the Occupational Safety and Health Administration's regulatory program requires that employees be adequately trained. It is for this reason that a chapter on training has been included.

Principles of Adult Learning

The overwhelming majority of persons that you will be teaching are adults, and adults learn differently compared to children. Therefore, it is necessary to understand the principles of adult learning. Adult learners have special needs. There are six characteristics of adult learners. Adult learners

- Are autonomous and self-directed
- Have a foundation of life experiences and knowledge
- Are goal oriented
- Are relevancy oriented
- Are practical in nature
- Need to be shown respect

These characteristics have implications that the trainer must understand. For each of the characteristics, the implications are as follows:

Adult learners are autonomous and self-directed.
  Implications:

- Involve participants in the training.
- Instructors should serve as facilitators.
- Instructors must determine the interests of the students.

Adult learners have a foundation of life experiences and knowledge.
  Implications:

- Trainers must recognize the expertise of the students.
- Trainers must encourage participants to share their experiences and knowledge with other students in the training.

Adult learners are goal oriented.
  Implications:

- The instructor must be organized in his or her training session.
- The training must have clear and understandable objectives.

Adult learners are relevancy oriented.
  Implications:

- The instructor/trainer must explain how the training objectives relate to the training activities.

Adult learners are practical.
  Implications:

- The instructor should demonstrate the relevance of the training to the actual job.

Adult learners need to be respected.
  Implications:

- The instructor should acknowledge the wealth of knowledge and experiences the participants bring to the training.
- Treat all participants in the training as equals rather than subordinates.

Safety Training Program

Most companies have a safety training program. Contrary to the way many programs are put into place, the safety training program should be well thought out and a needs analysis should be performed prior to developing a training program. The primary purpose of training is to solve an actual problem in the workplace or to modify specific behaviors. To structure a formal training program, the following steps must be implemented:

- *Performance Analysis*: This first step in the process is to determine, first and foremost, if training is the right solution to the actual workplace problem. If it is determined that training is at least part of the solution, then extensive research into what specific job skills or knowledge is needed.

- *Instructional Design*: On the basis of the needs analysis, the safety professional should develop what methods will be used to train employees, what materials are to be used in the training, and what order the training events will happen.

- *Materials Acquisition or Development*: The instructor of the training course should acquire through purchase or develop the training materials to be used in the training. It is important in this phase to target the specific audience. "One size does not fit all." The materials should reinforce the training objectives being sought. Keep in mind the adult learning characteristics when you are acquiring or developing the training materials.

- *Delivery of Training*: The training is delivered via the chosen method. Training delivery methods will be discussed in more depth within this chapter.

- *Course Evaluation*: Once the training has been delivered, course evaluation should be conducted. The purpose of the course evaluation is to obtain feedback from the participants to ensure that future training sessions are improved. In addition to the evaluation of the course, the actual work site should be evaluated to determine if the training is effective.

Delivery Methods

There are three major types of delivery methods used to deliver the training. An important aspect of delivery methods to consider when determining specific delivery methods is the retention of information. Figure 17.1 illustrates the average retention of information based on delivery methods.

The three major types of delivery methods include instructor-led training, self-paced training, and structured on-the-job training. Each are discussed in the following.

Instructor-Led Training

This type of delivery method has traditionally been conducted by an instructor who presents the material in a classroom. However, in today's society, the Internet has become an integral tool for almost everyone, including safety

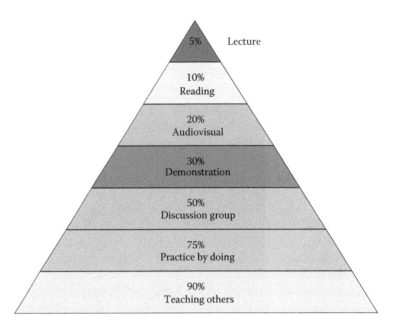

FIGURE 17.1
Retention rates based on delivery methods.

professionals. As such, one instructor can conduct training from a central location and broadcast via the Internet or through an online meeting room. The Internet variation to the instructor-led delivery method has unlimited possibilities and a variety of ways that it can be set up. In either case, an instructor serves as a facilitator and keeps the instruction flowing toward an established objective.

Self-Paced Learning

In self-paced learning, the student typically learns at their own pace by using training materials such as workbooks or textbooks. A newer version of the self-paced learning methodology is the inclusion of computer-based training (CBT) modules. In this method the participant goes through a series of training modules on a computer or through an online service. At the end of the course, testing is completed and a certificate can be printed out for documentation purposes. In the early stages, CBT was not a staple in the safety training program. However, with numerous improvements, CBT is now a viable option for providing safety training. When selecting a CBT, it is advisable to purchase or develop one that requires the student to complete all courses of study prior to taking the test. Some CBT modules will allow the student to go straight to the test. This type of CBT should be avoided.

Structured On-the-Job Training

This type of training delivery is usually conducted by a work supervisor who acts as a coach or guide in the training process. When conducting structured on-the-job training, a clear set of goals and objectives should be established prior to the training. The supervisor, acting as the instructor, should document that the training has been completed.

Training Needs Analysis

As discussed previously, a training needs analysis should be performed prior to developing and implementing a training program. Part of the needs analysis is to determine what the level of worker performance is and compare it to a desired performance level. Training needs can be analyzed by observing performance, interviewing employees, reviewing quality scores, and completing employee questionnaires.

Type of Needs Analysis

Needs analysis can be conducted using an assortment of methods. Some of the major types of needs analysis are listed as follows:

Context Analysis. A context analysis is performed to determine the desired training needed by the organization. The major objective in this analysis is to answer why training is the recommended solution, who decided that the training would be conducted, and what is the history of the organization in regard to employee training.

User Analysis. User analysis focuses on the prospective students and instructors. It determines who is going to receive or instruct the training and what their existing knowledge level of the subject matter is. In addition, user analysis assists in determining what the most appropriate learning style of potential participants is.

Work Analysis. Work analysis focuses on the desired skill and performance requirements of the job being performed. It can also be referred to as a task or job analysis. A good place to start in this type of analysis is a review of the various job descriptions and requirements.

Content Analysis. A thorough analysis of the documents used in the job is conducted. Such analysis may include a review of laws or written procedures that are directly related to the task or job. An example of the type of document that may be reviewed is confined space entry requirements.

Training Suitability Analysis. A thorough analysis of the task, job, or project is performed to determine if training is the desired (or only) solution to performance problems. Training may not necessarily be the answer to correcting or improving employee behaviors. Training may be part of the solution. In either case, it should be recognized for what it is. This is important, especially when the evaluation phase is conducted.

Cost–Benefit Analysis. Cost–benefit analyses are conducted to determine the return on investment to the company from the original investment of training costs.

Techniques

As mentioned above, there are several techniques that can be used to conduct a training needs analysis. A few of them are listed below:

- Direct observation
- Questionnaires
- Consultation with employees having knowledge of the task
- Review of documents and relevant literature
- Interviews
- Focus groups
- Tests
- Records and reports (i.e., quality assurance/quality control reports)
- Representative samples of behaviors

Training Program Development

Once a needs analysis has been conducted and the decision has been made that training is the answer, or part of the answer, a training program must be developed. Training program development involves a systematic approach, which includes the following steps or phases:

- Written performance objectives
- Content outline
- Selecting the training delivery method
- Selecting the materials to be used in the training
- Testing and evaluation

Each of the steps or phases is discussed in more detail within this chapter.

Written Performance Objectives

Once the training needs have been determined, you must identify specific goals and objectives of the training program. These goals and objectives should be clearly identified and must be written. The goals and objectives should identify exactly what employees are expected to do to perform their tasks, to improve their performance, and to improve their behavior.

Objectives need to be well thought out and planned in order for the training to be successful. In addition, objectives should be clear, concise, measureable, and shared with the employees. The four elements to consider when developing the objectives are as follows:

- Target audience—who will receive the training?
- Behavior—what observable action will the employee exhibit during evaluation?
- Condition—under what circumstances will the desired action be performed? What materials and equipment will be used?
- Degree/standard—how well must a task be accomplished? What score must be achieved?[1]

Developing Course Outline

A course outline that clearly identifies how the course is to be conducted should be developed. The course outline will serve as the basis for the curriculum development. The following are some of the items included in the course outline:

- General course description
  - What is and isn't covered in the course
  - Teaching approach (i.e., lecture, discussion, hands-on, etc.)
  - Prerequisites (other courses, advanced job skills, etc.)
- Goals and objectives
  - List the goals and objectives of successfully completing the course
- Outline of course structure
  - How the course will be broken down (i.e., modules, hands-on, etc.)
  - Time required to complete the training
- Course requirements
  - Evaluation criteria

Selection of Training Delivery Method

The primary question for a training developer to ask is, "Will the chosen training method deliver the desired results?" Therefore, it is necessary to select a method of delivery that meets the desired objectives of the training program and is economical to the organization. In addition, the method of delivery must be adequate for the target audience's skill and literacy level.

Development of Course Materials

Materials used in the course of instruction should be well thought out. Keeping in mind the learning retention of most adults, as identified in Figure 17.1, materials should be presented in a manner that is instrumental in achieving the goals and objectives of the training. Materials should include an introduction, presentation, practice and feedback, and a summary of the materials presented. Materials may include text, graphics, charts, examples of "real world" experiences, and a host of other items. Media presentations will be discussed later in this chapter.

Testing and Evaluation

When evaluating the success or failure of a training program, instructors often perform appraisals of their students. This section focuses on the testing and evaluation of the students in a course, not on the evaluation of the overall safety program. Testing and evaluation can be performed by using pretests, review tests, or posttests.

Pretests

Pretests are administered to determine the skill level of students prior to being presented the course materials. It is a valuable tool that can be used to assist in quantitative measurements of the course.

Review Tests

Review tests are given during the presentation phase of the training, but before the actual posttest. This type of test is beneficial whenever courses are more than a couple of hours in length. For example, an instructor of a 40-h training course may give review tests at the end of each day or training period. Review tests are beneficial tools for the instructor and the students, by letting the instructor know if students are receiving the materials and letting students know what type of evaluation is expected.

Posttests

Posttests should be designed to determine if the participant can perform the learning objectives. Often, the posttests contain the same information presented in the pretests and review tests. The final results can be quantitatively measured to determine if the course materials or delivery methods are adequate to achieve the desired goals and objectives of the training course.

Media Presentations

Some safety professionals are old enough to remember attending classes where professors made use of overhead slide transparencies. This method of presenting materials was considered "state-of-the-art" at the time. If we were to use transparencies today, most students would wonder how up-to-date the training would be. Technologies have advanced in the education field to the point that training materials and media are virtually endless. Computers serve as great tools for communicating ideas and information. As we mentioned previously, training can be conducted via web-based training programs or online meeting rooms.

PowerPoint Presentations

One of the most common and useful tools for presenting classroom-based training is Microsoft PowerPoint software and a projector. This method of presentation can provide both visual and audible formats. In the graphic presentation using this method, font styles and sizes are important. There is a wide variety of font styles that can be used including serifs (fonts with curves or lines at the end of the letters), sans serifs (fonts without curves or lines at the end of letters), script fonts (cursive writing style), and Gothic or Helvetica (fonts without curves or lines). Font size is extremely important in visual presentations. After all, if a participant cannot see the information, then the information cannot be retained. As a general rule, fonts should be at least 16 points, with at least a 2-point size difference in the text and the headline. Block letters tend to work best in this visual presentation. Table 17.1 shows the different sizes of font for various distances.

TABLE 17.1

Height, Thickness, and Distance of PowerPoint Fonts

Distance (ft)	Letter Height (in.)	Letter Thickness (in.)
25	1	1/8
50	1 3/4	1/4
75	3	1/3–1/2
100	3 1/2	1/2

Charts and Graphs

Charts and graphs can be inserted into PowerPoint presentations or presented separately. A chart is a visual representation of data, in which "the data are represented by symbols, such as bars in a bar chart, lines in a line chart, or slices in a pie chart." A chart can represent tabular numeric data, functions, or some kinds of qualitative structures. A graph may refer to a graphic (such as a chart or diagram) depicting the relationship between two or more variables used, for instance, in visualizing scientific data.

Key Information to Remember on Training

1. Adult learners are autonomous and self-directed, have a foundation of life experiences and knowledge, are goal oriented, are relevancy oriented, are practical in nature, and need to be shown respect.
2. Training program development includes the following steps: performance analysis, instructional design, materials acquisition or development, training delivery, and course evaluation.
3. The three basic types of delivery methods include instructor-led training, self-paced learning, and structured on-the-job training.
4. A training needs analysis is the first step in developing a training program.
5. Training program development includes written performance objectives, preparing a course outline, selecting the training delivery method, selecting the materials to be used in the training, and testing and evaluation.
6. Pretests are administered prior to the presentation of course information and are designed to provide the starting point of knowledge.
7. Review tests are useful in longer courses.
8. Posttests should be designed to determine if the participant can perform the learning objectives.
9. Font size (height and width) varies with distances from the screen.

Reference

1. Texas Department of Insurance, Division of Workers' Compensation Workplace Safety, 2010. Available at http://www.tdi.state.tx.us/pubs/videoresource/stptrainingprod.pdf.

18

Engineering Economics

Engineering economics is a subcategory of economics that is specifically designed for the engineering professions. The essential idea behind engineering economics is that money generates money. You cannot compare $10.00 today to $10.00 a year from now without adjusting for the investment potential.[1] Some may be curious at this point as to what economics has to do with the safety profession. Regardless of what some safety professionals may believe, safety is a business unit. Some examples of this are (a) as the safety manager, you are requesting that your company authorize funds to purchase safety equipment to reduce the number of specific types of injuries; (b) you are requesting to purchase a computer-based training program; or (c) the decision has been made to purchase an additional piece of equipment and your company can either manufacture it in-house or purchase it from an outside vendor and you are asked to conduct a cost–benefit analysis. As a safety professional, you will be asked on many occasions to justify your expenditures and not just in terms of "it's the right thing to do." When you are equipped with viable information on the associated costs, including the "cost of money," your requests are more likely to be approved. At this point, it is worth mentioning that not all safety expenditures are based entirely on financial constraints. However, financial considerations are an integral part of any decision-making process. In this chapter, we will focus primarily on the present and future value of money, interest, and various payment structures as they relate to safety.

Simple Interest

When borrowing money, the lender will, under normal circumstances, charge interest on that money (*cost of money*). When the loan is paid back, the borrowed amount (*principal amount*) and the interest rate are paid back. To calculate simple interest, we use the following equation:

$$I = Pni,$$

where
$I =$ the amount of interest paid
$P =$ principal (amount borrowed)

n = number of years (or period)
i = interest rate

NOTE: This equation is not listed on the Board of Certified Safety Professionals (BCSP) examination reference sheet.

Example

Your company would like to refit one of its processing lines and has determined that the cost to do so would require $50,000. You are asked to calculate the total cost of the project, if the interest rate is 6% annually and the loan repaid in 3 years. What is the total cost of the project?

$$I = (\$50,000)(3 \text{ years})(0.06 \text{ per year})$$
$$I = \$9000.00.$$

The cost of borrowing $50,000 is $9000. Add the interest to the principal amount and the total project cost is $59,000.

Compound Interest

Most lenders charge interest by using compound interest methods, which means that in addition to the principal, the amount of interest charged also is charged interest. This addition is called compounding. For example, a loan of $10,000 is charged an interest rate of 5%, which is compounded monthly. Therefore, at the end of the first month, the loan balance would be $10,500. At the end of the second month, the loan balance would be $11,025, assuming there have been no payments to reduce the principal amount. An easy equation to calculate compound interest is as follows:

$$F = P\left(1 + \frac{i}{n}\right)^{nt},$$

where
F = future value of the loan
P = initial loan amount
i = interest rate (expressed in decimal)
n = number of times per year interest is compounded
t = number of years invested

NOTE: This equation is not listed on the BCSP examination reference sheet but is included to provide you an understanding of how financial calculations are derived.

Example

Your company has borrowed $10,000 from the lender, who charges an interest rate of 3%, compounded semiannually (twice a year). The loan will run for 3 years. Calculate the entire loan cost.

$$F = \$10,000\left(1 + \frac{0.03}{2}\right)^{2(3)}$$
$$F = \$10,000(1.015)^6$$
$$F = \$10,000(1.093)$$
$$F = \$10,934.34.$$

Therefore, the 3-year loan cost would be $10,934.34.

Future Value of Money

Similar to the compound interest equation, but asking the question in another way, let's look at the future value of the same $10,000. In other words, what is the value of the investment if we were to put the $10,000 into an interest bearing account at 3%? We can calculate this by using the following equation:

$$F = P(1 + i)^n,$$

where
F = future value of money
P = present value of the money (principal)
i = interest rate (annual percentage rate [APR])
n = number of years (in this case, number of years invested)

$$F = \$10,000(1 + 0.03)^3$$
$$F = \$10,000(1.03)^3$$
$$F = \$10,927.27.$$

Example

Company XYZ has purchased a tract of land for $182,000. What would be the value of this land in 20 years, assuming an APR of 9%?

$$F = \$182,000(1 + 0.09)^{10}$$
$$F = \$182,000(1.09)^{10}$$
$$F = \$430,860.19.$$

Loan Balance

Most loans, however, are not based on an end of loan payment. Most are made on the terms that you would begin making equal payments through-out the term of the loan. For example, on the basis of the previous scenario, your company borrowed $10,000, you can determine your loan balance at the end of any given year (or period) using the following equation:

$$B = A\left(1 + \frac{i}{n}\right)^{nt} - P\frac{\left(1 + \frac{i}{n}\right)^{nt} - 1}{\left(1 + \frac{i}{n}\right) - 1},$$

where

B = balance after t years
A = original loan amount (principal)
n = number of payments per year
P = amount paid per payment
t = number of years
i = APR

NOTE: This equation is not listed on the BCSP examination reference sheet but is helpful in determining the loan balances.

Example

Your company has been made a loan of $10,000 from the bank and agrees to make equal installment payments of $100 monthly until the loan is repaid. The APR is 4%. What is the balance at the end of 3 years?

$$B = \$10,000\left(1 + \frac{0.04}{12}\right)^{12(3)} - \$100\frac{\left(1 + \frac{0.04}{12}\right)^{12(3)} - 1}{\left(1 + \frac{0.04}{12}\right) - 1}$$

$$B = \$10,000(1.003)^{36} - \$100\frac{1.114 - 1}{1.003 - 1}$$

$$B = \$10,000(1.114) - \$100(38)$$

$$B = \$11,140 - 3800$$

$$B = \$7340.00.$$

If your company repays the loan at $100 per month for 3 years, the com-pany would still owe $7340.00 toward the loan.

Time Value of Money

Many times you will be asked to evaluate when an investment will pay for itself, based on savings (i.e., injury/accident prevention, etc.). We can predict the time value of money using the following equation:

$$P = F(1 + i)^{-n},$$

where

P = present worth of money (principal)
F = future worth (or savings)
i = annual interest rate (APR) in decimal

Example

If XYZ Corporation makes an investment of $30,000 to purchase safety equipment and by purchasing this equipment the company predicts it would prevent $10,000 in injury-related costs, when will the investment pay for itself? The annual interest rate is 12%. To calculate this, you must calculate each year individually, starting with the first year.

$$P = \$10,000(1+0.12)^{-1(\text{year } 1)}$$
$$P = \$10,000(1.12)^{-1}$$
$$P = \$8928.57.$$

At the end of 1 year, the value of the investment is $8928.57, leaving a balance of $21,071.43. Therefore, you would continue to calculate for the second year, as follows:

$$P = \$10,000(1+0.12)^{-2(\text{year } 2)}$$
$$P = \$10,000(1.12)^{-2}$$
$$P = \$7971.94.$$

At the end of 2 years, the value of the investment is $7971.94 plus the value of year 1 of $8928.57 for a total value of $16,900.51. As you can see, the value of the loan has not paid for itself, so you will need to continue calculating.

$$P = \$10,000(1+0.12)^{-3(\text{year } 3)}$$
$$P = \$10,000(1.12)^{-3}$$
$$P = \$7117.80.$$

The value of the investment at the end of 3 years is equal to the value of all 3 years combined or $24,018.31. Continue calculating until the original investment of $30,000 has paid for itself.

$$P = \$10,000(1 + 0.12)^{-4}$$
$$P = \$10,000(1.12)^{-4}$$
$$P = \$6255.18.$$

The value at the end of 4 years is $30,373.49. Therefore, it would require almost 4 years for the original investment of $30,000 to pay for itself.

Series Compound Amount Factor

To calculate the future value of money, given a series of regular payments, also known as the *series compound amount factor*, we can utilize the following equation:

$$F = A(1 + i)^n,$$

where
F = future value of money
A = each payment (amount in \$) (end of period)
i = interest rate (APR)
n^{-1} = number of period (in this case, at the end of the first year; if it is 2 years, then we represent it as n^{-2})

or, represented differently, by the equation listed below and on the BCSP examination reference sheet:

$$F = A\left(\frac{(1+i)^n - 1}{i} \right).$$

Example

What would be the future value of a loan if you made equal payments of $100 at the end of the loan period and the interest rate was 8% per year for 3 years?

$$F = 100\left(\frac{(1+0.08)^3 - 1}{0.08} \right)$$
$$F = 100\left(\frac{1.26 - 1}{0.08} \right)$$
$$F = 100(3.25)$$
$$F = \$325.$$

Sinking Fund Factor

The *sinking fund factor* is considered to be the inverse of the series compound factor. This equation is used whenever you would like to determine the amount of each payment. This equation is listed below:

$$A = F\left(\frac{i}{[(1+i)^n - 1]}\right).$$

Example

You have determined that the future value of a loan is $10,285.00 at an interest rate of 7.5% over 3 years. You wish to repay the loan at the end of each year. What are your yearly payments?

$$A = \$10,285\left(\frac{0.075}{[(1+0.075)^3 - 1]}\right)$$

$$A = \$10,285\left(\frac{0.075}{(1.24) - 1}\right)$$

$$A = \$10,285(0.31)$$

$$A = \$3188.35 \text{ per year.}$$

Capital Recovery Factor

The *capital recover factor* is similar to the two previous equations. You may want to know the payment amount based on the present value (or present worth) of the loan given a certain percentage rate. To do this, we utilize the following equation:

$$A = P\left(\frac{i(1+i)^n}{(1+i)^n - 1}\right),$$

where
  A = period payment amount ($)
  P = present worth of money
  i = APR in decimal
  n = number of periods

Example

You are asked to assist with a capital improvement project that includes numerous safety-related items, such as life safety and NFPA code upgrades. The total cost of the project is estimated by your engineers to $5,350,800. The owner currently has $640,000 to invest in the project and will need to borrow the remaining funds. There is a 2.5% (of the total loan amount) loan origination fee associated with securing the loan. Determine the amount of money that the company must pay at the end of each year to repay the loan if it is locked in at an annual rate of 6.375%, compounded annually, and loses all of its value in 10 years.

The solution to this problem is multifaceted. The first step would be to determine the total loan needed.

Total project funds needed:	$5,350,800
Company funds ("Down payment")	($640,000)
Total funding required:	$4,710,800

Now, you must determine the loan amount, based on needed funds of $4,710,800. Remember that you have the loan origination fee, which will also be financed. The loan origination fee is 2.5% of $4,710,800. When added together, the total loan amount will be $4,828,570.

Now, we can insert our data into the equation as follows:

$$A = \$4,828,570 \left(\frac{0.06375(1+0.06375)^{10}}{(1+0.06375)^{10} - 1} \right)$$

$$A = \$4,828,570 \left(\frac{0.118}{0.855} \right)$$

$$A = \$4,828,570(0.138)$$

$$A = \$666,342.66 \text{ per year.}$$

Series Present Worth Factor

The *series present worth factor* is the corresponding inverse of the capital recovery factor and is demonstrated in the equation below:

$$P = A \frac{\left[(1+i)^n - 1\right]}{i(1+i)^n}.$$

Example

Using the information listed in the capital recovery section, you have secured a loan at 0.575% annual interest, compounded annually and have agreed to make annual payments of $500,000. How much money will be repaid on this loan at the end of 10 years?

$$P = \$500,000 \frac{\left[(1+0.0575)^{10} - 1\right]}{0.0575(1+0.0575)^{10}}$$

$$P = \$500,000 \frac{0.749}{0.1}$$

$$P = \$3,745,000.$$

Therefore, at the end of 10 years, you have paid $3,745,000 of the $4,828,570 loan, leaving a balance of $1,083,570.

Summary

Engineering economics continues to become an increasing factor in the day-to-day operations of a safety professional. As you can see from the few equations listed in this chapter, the structuring of a loan is an important financial aspect of any business owner or financial manager. In most cases, financial planners will provide the necessary financial information needed to finance a project. However, it is important to know the basic financial information in order for you, as Safety Managers, to justify some of your recommendations. A well thought out recommendation will be received in more favorable light by the final decision makers.

Reference

1. http://www2.latech.edu/~sajones/Senior%20Design%20Web%20Pages/Home work%207%20on%20Engineering%20Economics.htm.

19

Management Theories

Management in all business areas and human organization activity is the act of getting people together to accomplish desired goals and objectives. Management consists of *planning, organizing, staffing, leading* or *directing*, and *controlling* an organization (a group of one or more people or entities) or effort for the purpose of accomplishing a goal. Safety professionals, at all levels, should have a thorough understanding of management principles, theories, and styles, as they will undoubtedly either become managers or become managed.

There are many management theories in the business world today. In this chapter, we will discuss the traditional management styles and theories. There may be a few questions on the Associate Safety Professional or Certified Safety Professional examinations regarding management theories.

Management Theories

Maslow's Hierarchy of Needs

In 1943, Abraham Maslow introduced his *Hierarchy of Needs Theory*. In this theory, Maslow proposed that all human beings are motivated by unsatisfied needs and that certain lower factors need to be satisfied before higher needs can be satisfied. Figure 19.1[1] illustrates Maslow's Hierarchy of Needs (represented in pyramid form to illustrate the lower to higher factors).

Maslow theorized that the lower needs have to be satisfied before the next need level serves as a motivator. Furthermore, once the lower level need has been satisfied, it no longer serves as a motivator. As a safety professional, it is important to understand this theory and the potential motivation of employees.

Physiological Needs

These needs are fairly obvious, as they are the basic requirements for human survival. Physiological needs include breathing, food, water, sexual activity, homeostasis, and excretion.

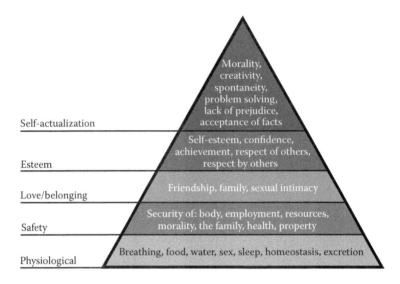

FIGURE 19.1
Maslow's Hierarchy of Needs. (From Wikipedia, 2010. Available at http://en.wikipedia.org/wiki/Maslow's_hierarchy_of_needs.)

Safety Needs

Once the physiological needs are satisfied, safety and security take precedence and tend to dominate a person's behavior. Safety and security are not limited to physical safety. They also involve the need for a predictable, orderly world where there is very little inconsistency or uncontrollable events, including job security. Safety and security can include

- Personal security
- Financial security
- Health and well-being
- Safety net against accidents/illnesses and their adverse effects

Love and Belonging

This third layer of human needs is social and involves feelings of belonging. This aspect involves emotionally based relationships in general such as

- Friendship
- Intimacy
- Family

Esteem

All humans have a need to be respected and to have self-esteem and self-respect. It represents the basic human desire to be accepted and valued by others. Included in this step is the personal need to contribute and feel appreciated. Esteem is considered an ongoing need.

Self-Actualization

This level represents a person's full potential and the prospect of reaching that potential. In other words, what a person is meant to achieve takes an increasing desire to achieve that potential. In order to reach the level of self-actualization, a person must have all of the lower-level needs met. Once a person has achieved this level, they can accept their own human nature, with all its shortcomings and discrepancies from the ideal image without feeling any real concern. Self-actualized persons tend to focus on solving problems for the good of something greater than themselves. Problems encountered by the self-actualized person are considered only as "a task that they must do." Self-actualization needs are to be considered continuing.

McGregor's Theory X and Theory Y

In 1960, Douglas McGregor announced his two theories of organization management and employee motivation, calling them *Theory X* and *Theory Y*. In both of these theories, McGregor assumed that the manager's role was to organize resources for the benefit of the company. However, McGregor's characterization of employees and their motivation is quite different between the two theories.[2]

Theory X

In Theory X, McGregor states that leadership assumes the following:

- Work is inherently distasteful to most people, and they will attempt to avoid work whenever possible.
- Most people are not ambitious, have little desire for responsibility, and prefer to be directed.
- Most people have little aptitude for creativity in solving organizational problems.
- Motivation occurs only at the physiological and security levels of Maslow's Hierarchy of Needs.
- Most people are self-centered. As a result, they must be closely controlled and often coerced to achieve organizational objectives.

- Most people resist change.
- Most people are gullible and unintelligent.

To summarize Theory X, McGregor believed that the main source of most employee motivation is monetary, with security as a strong second. The manager of Theory X employees can manage in one of two ways. He or she can manage by coercion, threats, or micromanagement. The other approach is to be permissive with his or her employees and seek harmony. As you can plainly see, neither of these approaches is optimal. The best management method, under this theory, lies somewhere between these two approaches.

Theory Y

In Theory Y, McGregor theorizes that employees are motivated primarily at the esteem and self-actualization levels. Almost in contrast to Theory X, Theory Y leadership makes the following general assumptions:

- Work can be as natural as play if the conditions are favorable.
- People will be self-directed and creative to meet their work and organizational objectives if they are committed to them.
- People will be committed to their quality and productivity objectives if rewards that address higher needs such as self-fulfillment are in place.
- The capacity for creativity spreads throughout organizations.
- Most people can handle responsibility because creativity and ingenuity are common in the population.
- Under these conditions, people will seek out responsibility.

Under Theory Y, an organization can apply the following scientific management principles to improve employee motivation:

- *Decentralization and delegation*: Decentralize control and reduce the number of management levels required to operate.
- *Job enlargement*: Broadening the scope of an employee's job, which adds variety and opportunities.
- *Participative management*: Organizations consult with employees in the decision-making process.
- *Performance appraisals*: Having employees set objectives and participate in the process of evaluating how well they were met.

Herzberg Motivational Theory

In 1959, Frederick Herzberg introduced his Motivational Theory. In this motivational theory, Herzberg attempted to explain factors that motivate individuals through identifying and satisfying their individual needs, desires, and the aims pursued to satisfy these desires. In his theory, Herzberg stated that motivation can be split into two major categories: hygiene factors and motivation factors. Hygiene factors affect the level of dissatisfaction but are rarely noted as creators of job satisfaction. However, if these factors are not present or satisfied, they can demotivate a person.[3] Hygiene factors include the following:

- Supervision
- Interpersonal relationships
- Physical working conditions
- Salary

Job dissatisfaction, under normal circumstances, is not normally attributed to motivation factors. However, when they are present, they serve as motivational factors. Motivation factors include

- Achievement
- Advancement
- Recognition
- Responsibility

Whenever there is a shortage of motivation factors present in the work environment, the employee will focus on other factors, such as the hygiene factors.

The Deming Cycle

The Deming Cycle or PDCA cycle is a continuous improvement quality model that consists of four repetitive steps conducted in a logical sequence. The four steps include Plan, Do, Check, and Act. Deming proposed that if these four steps are continuously followed, quality will improve. Figure 19.2[4] illustrates Deming's PDCA Cycle.

Management by Objectives

Peter Drucker first used the term *management by objectives* in 1954. Management by objectives (MBO) is the process of agreeing upon objectives within an organization so that management and employees agree to the objectives and

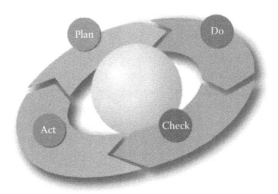

FIGURE 19.2
The Deming Cycle (PDCA). (From http://en.wikipedia.org/wiki/PDCA.)

understand what they are in the organization.[5] The main focus of MBO is to set established goals for the organization by allowing both management and employees to participate in the process. By doing this, the employee achieves personal goals and has a direct impact on the success of the organization. Some of the important features and advantages of MBO are as follows:

- *Motivation*: It involves employees in the whole process of goal setting and increases empowerment and employee job satisfaction and comittment.

- *Improved communication and coordination*: In the MBO process, there are frequent evaluations of the objectives and goals. This evaluation process, which is conducted by both management and employees, assists in maintaining harmonious relationships. When goals are lacking, both employees and management participate in the problem-solving process.

- *Clarity of goals.*

Contingency Theory

Contingency theory is a class of behavioral theory that claims that there is no one best way to organize a corporation, lead a company, or make decisions. Instead, the optimal course of action is contingent upon the internal and external situation.

Systems Theory

Systems theory is an interdisciplinary theory about the nature of complex systems in nature, society, and science and is a framework by which one can investigate or describe any group of objects that work together to produce some result. In this theory, an organization is considered and treated as a system.

Chaos Theory

Chaos theory is a field of study in mathematics, physics, and philosophy studying the behavior of dynamical systems that are highly sensitive to initial conditions. This sensitivity is popularly referred to as the butterfly effect. Small differences in initial conditions yield widely diverging outcomes for chaotic systems, rendering long-term prediction impossible in general. This happens even though these systems are deterministic, meaning that their future behavior is fully determined by their initial conditions, with no random elements involved. In other words, the deterministic nature of these systems does not make them predictable.

Management Styles

Managers perform many roles within an organization. How they function in these roles depends entirely on their management style. A management style is the overriding method of leadership used by a manager. The two main categories of management styles include

- Autocratic
- Permissive

These two styles are in direct contrast to each other. Each can be divided into subgroups, as discussed in the following.

An autocratic leader or manager makes all decisions unilaterally, while the permissive leader permits subordinates to take part in the decision making and gives them a considerable degree of autonomy in completing routine tasks.

Directive Democrat

The directive democrat leader allows subordinates to participate in the decision-making process but closely supervises employees.

Directive Autocrat

The directive autocrat leader makes decisions unilaterally and closely supervises employees.

Permissive Democrat

The permissive democrat leader allows employees to participate in the decision-making process and gives subordinates some latitude in carrying out their work.

Permissive Autocrat

The permissive autocrat leader makes decisions unilaterally but gives employees latitude in carrying out the work.

As you might well imagine, the situation dictates the style of leadership that a manager may wish to employ. Obviously, no one style of management will work in every situation. For example, if you are a leader of military troops that have been assigned to attack a village, you may wish to employ a directive autocrat style of leadership, whereas if you are the manager of a department that is about to undergo significant process changes, it would be advisable to use more of a permissive democrat style of management. The style of management must change with the task or job at hand.

Key Information to Remember on Management Theories

1. Maslow's Hierarchy of Needs includes physiological, safety, belonging, esteem, and self-actualization.
2. McGregor's Theory X states that employees do not want to work and are only motivated by money.
3. McGregor's Theory Y states that employees like to work and, when their needs are met, actually seek out responsibility.
4. Herzberg's Motivation Theory classifies factors into two categories: hygiene factors and motivation factors.
5. The Deming Cycle has four steps, which are continuous, and include Plan, Do, Check, and Act.
6. Autocratic leaders make decisions unilaterally.
7. Permissive leaders permit participation in the decision-making process.

References

1. Wikipedia, 2010. Available at http://en.wikipedia.org/wiki/Maslow's_hierarchy _of_needs.
2. http://www.mindtools.com/pages/article/newLDR_74.htm.
3. http://www.businessballs.com/herzberg.htm.
4. http://en.wikipedia.org/wiki/PDCA.
5. http://www.1000ventures.com/business_guide/mgmt_mbo_main.html.

20

Accident Causation and Investigation Techniques

It is understood throughout the safety profession that the most important role a safety professional plays is the prevention of accidents and injuries. There have been varied and differing discussions as to whether accidents are totally preventable. It is my personal belief that all accidents are preventable. However, when an accident or injury occurs, it is the responsibility of the safety professional to ensure that a thorough investigation is conducted. This gives rise to the question, "Why do we conduct accident investigations?"

The major reasons for conducting accident investigations are listed as follows:

- To prevent future occurrences of the same incident, by determining the preventable actions that can be taken.
- Accidents and occupational illnesses severely limit efficiency and productivity.
- Accidents and injuries cause severe detriment to employee morale.
- Federal and state occupational safety and health regulations require an employer to provide a safe and healthy work environment free of recognized hazards.
- Preventing accidents and injuries greatly lowers workers' compensation insurance premiums, thereby increasing profits.

Accident investigations are a huge responsibility for any safety professional and one that should not be taken lightly. Whether you have a minor or a major injury, the safety professional should have the same passion for finding the "root cause" of the accident. During the course of my professional safety career, I have had the misfortune of conducting accident investigations involving personal injuries. I say misfortune because once you look an injured employee in the face, lying in a hospital bed, an accident investigation becomes more than just a function of your profession. It is now a personal endeavor to insure that it never happens again.

With this said, let's discuss how to conduct an accident investigation. In order to do this, the safety professional should have a thorough understanding of the causes of accidents. There are many traditional accident causation theories that have been proven outdated and ineffectual in preventing accidents. Some of them include the following bases:

- Some workers are simply "accident prone."
- Workers are careless. Worker carelessness is the cause of most accidents.
- Some workers are more susceptible to injury or disease than others. This is the primary purpose of preemployment screening to ensure that only the healthiest, fittest applicants are hired. Despite all efforts, research has shown that preemployment medicals are rarely predictive of future illness.
- Lifestyle choices cause most diseases.
- All activities, including work, contain an element of risk.
- Accidents are just a cost of doing business.

As you can see, each of these theories has its own pitfalls and can be considered to be somewhat insulting.

There are several more modern and scientific theories concerning accident causation, each of which has some explanatory and predictive value. The following theories of accident causation will be discussed inside this chapter:

- Domino Theory
- Human Factors Theory
- Accident/Incident Theory
- Epidemiological Theory
- Systems Theory
- Energy Release Theory
- Behavior Theory

By understanding the accident theories, the safety professional is guided through the course of the accident/incident investigation. The accident theory describes the scope of the investigation.

Domino Theory

The Domino Theory, also known as Heinrich's Domino Theory, was developed by H.W. Heinrich in 1932 (Figure 20.1)[1]. It is considered the first scientific approach to accident prevention. According to Heinrich, an "accident" is one factor in a sequence that may lead to an injury. The factors can be

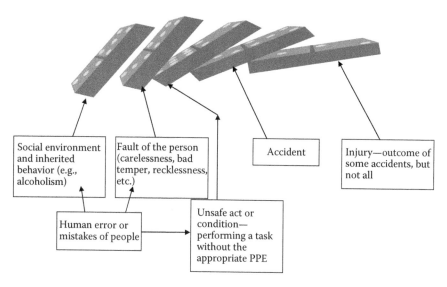

FIGURE 20.1

Graphical representation of the Domino Theory. (From United States Army, 2009. www
.armystudyguide.com/content/bm~doc/accident-causation.ppt.)

visualized as a series of dominoes standing on edge; when one falls, the link-
age required for a chain reaction is completed. Each of the factors is dependent on the preceding factor.

According to the Domino Theory, a personal injury (the final domino)
occurs only as a result of an accident. An accident occurs only as a result of a
personal or mechanical hazard. Personal and mechanical hazards exist only
through the fault of careless persons or poorly designed or improperly maintained equipment. Faults of persons are inherited or acquired as a result of
their social environment or acquired by ancestry. The environment is where
and how a person was raised and educated. The factor preceding the accident (the unsafe act or the mechanical or physical hazard) should receive
the most attention. Heinrich felt that the person responsible at a company
for loss control should be interested in all five factors, but be concerned primarily with accidents and the proximate causes of those accidents. He also
emphasized that accidents, not injuries or property damage, should be the
point of attack. He defined an accident as any unplanned, uncontrolled event
that *could* result in personal injury or property damage. For example, if a
person slips and falls, an injury may or may not result, but an accident has
taken place.

Heinrich's Axioms of Industrial Safety

- Injuries result from a completed series of factors, one of which is the
  accident itself.

- Accidents occur only as a result of an unsafe act by a person or a physical or mechanical hazard.
- Most accidents result from unsafe behaviors of people.
- Unsafe acts do not always result in an accident/injury.
- Reasons why people do unsafe acts can aid in the corrective action.
- Severity of the accident is largely fortuitous, and the cause is largely preventable.
- Accident prevention techniques are analogous with best quality and production techniques.
- Management should assume responsibility for safety as it is in the best position to get results.
- The supervisor is the key person in accident prevention.
- Any accident results in direct and indirect costs.

In Heinrich's theory, he proposed three corrective sequence actions ("The Three E's"), as follows:

- Engineering
  - Control hazards through product design or process change
- Education
  - Train workers regarding all facets of safety
  - Impose on management that attention to safety pays off
- Enforcement
  - Insure that internal and external rules, regulations, and standard operating procedures are followed by workers, as well as management

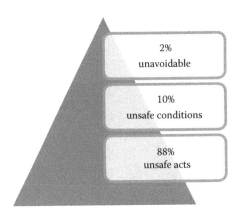

FIGURE 20.2
Heinrich's conclusions.

Heinrich concluded that accidents were primarily the result of unsafe acts or unsafe conditions or were unavoidable. In Figure 20.2, we see them presented graphically.

Human Factors Theory

On the basis of the Human Factors Theory, accidents are entirely the result of human error. These errors are categorized broadly as

- Overload
  - The work task is beyond the capability of the worker
    - Environmental factors (noise, distractions, etc.)
    - Internal factors (personal problems, emotional stress)
    - Situational factors (unclear instructions, risk level)

Worker capacity is measured in terms of productivity or load; be it in the form of manufacturing or clerical load, load can be measured to evaluate one's productivity worth. Many factors affect worker load capacity: natural ability, training, and state of mind are but a few. If these factors affect worker load capacity, overload is reached. Overload is a syndrome where the workers' natural capacity for productivity is overwhelmed. Most often, accidents are a result of worker inexperience or lack of progressive training. Other factors that influence a workers' capacity are fatigue level, stress, and physical conditioning. Noise, climatic lighting, and distractions only add to the overload foundation. Internal factors or limits on the mental state of worker productivity are personal problems, emotional stress, and worry. Situational factors are the physical working environment surrounding workers. These can be stressors such as level of risk and unclear instructions, or just the novelty of the job can adversely affect worker load.

- Inappropriate Worker Response
  - To hazards and safety measures (workers' fault)
    - Detecting a hazard, but not correcting it
    - Removing safeguards from machines and equipment
    - Ignoring safety rules and regulations
  - To incompatible work station (management, environment faults)
    - Ignoring safety rules and regulations for production

How a person responds in a given situation can be the determining factor in whether an accident could be prevented or could lead to the cause of one. If he or she ignores a suspected hazard and does nothing to correct the situation, he or she has responded inappropriately. If a person willingly disregards a safety procedure or circumvents safety locks on equipment, that person has acted with an inappropriate response.

- Inappropriate Activities
  - Performing tasks without the requisite training
  - Misjudging the degree of risk involved with a given task

Human error can be the result of inappropriate activities. Oftentimes, undertaking a task without the requisite training can lead to injuries. Being unfamiliar with equipment and procedures and misjudging the degree of risk associated with the task are examples of inappropriate activities.[2]

Accident/Incident Theory

This theory is also known as *Petersen's Model* or *Petersen's Theory* (Figure 20.3). It is basically an extension of the Human Factors accident causation model. Dan Petersen introduced additional elements such as ergonomic traps, the decision to err, and system failures. He stated that a decision to err by an employee may be an unconscious and based on logic, or it could be a conscious decision. Factors such as deadlines, peer pressure, and budget factors could make a person decide to behave in an unsafe manner. One important factor in the Petersen model that causes a person to make a logical decision to disregard procedures is the "Superman Syndrome." The Superman Syndrome leads the person to believe that he is invincible or bulletproof, simply because "it won't happen to me or accidents happen to others who don't pay attention."

His addition of *system failure* is an important step in identifying the potential for causal relationship between management decisions or management behaviors regarding safety. System failure helps establish and solidify management's role in the accident prevention process. It also helps identify the avenues in which the system can fail, such as clearly defining areas of responsibility, inspections, measurements, training, and orientation of employees.

As part of Petersen's subclass of System Failure, management's role is multilayered. Management is responsible for setting policies, placing responsibility, training employees, following up on training and enforcement of policies with inspections to ensure compliance, and finally enforcing standards with corrective actions. If, at any time, management attention is diverted from the system, a failure will occur.[3]

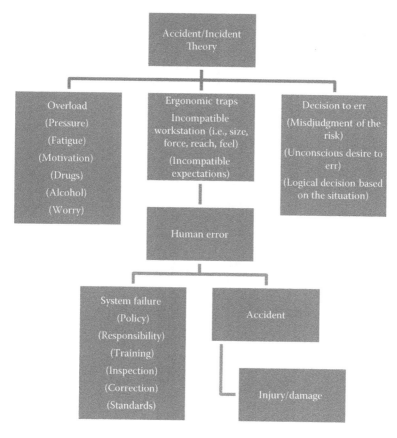

FIGURE 20.3
Petersen's model of the Accident/Incident Theory.

Epidemiological Theory

The Epidemiological Theory focuses primarily on industrial hygiene and the causal relationship between environmental factors and disease. The Epidemiological Theory is appropriate for studying the causal relationships between environmental factors and accidents. The Epidemiological Theory is represented graphically in Figure 20.4.

As you can see from the model above, the two key components are predispositional characteristics and situational characteristics. Predispositional characteristics are those in which a person is very susceptible to peer pressure. In other words, the person is controlled by the moods and emotional state of others around them. Situational characteristics can best be described as the current state of affairs people are immersed in, whether the situation

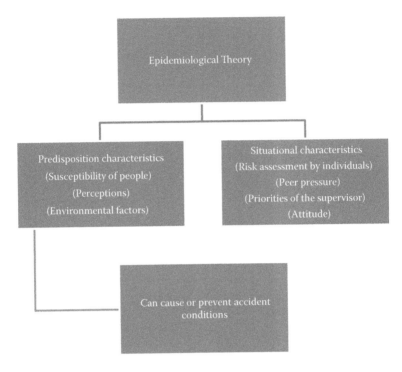

FIGURE 20.4
Epidemiological Theory of accident causation.

has them driving to work or doing a goal-oriented task. When these two characteristics are combined and a bad decision is made, the outcome is potentially the result of an accident occurrence.

Systems Theory

Looking at Figure 20.5, one can see that the Systems Theory of accident causation is a relationship between man, machine systems, and the surroundings, which function as a unit or a whole (system). Man, while the most valuable component of the system, is also the most flexible. The graphical representation shows smooth lines between the relationships. However, it would be better to envision them as jagged lines to account for the wavering attributes/characteristics of man. In the Systems Theory, the *likelihood* of a system failure increases with increasing complexity in the hardware. The *severity* of the potential outcome increases if time for error correction is short, information to the operator is ambiguous or indirect, and subsystems are interconnected so that one failure can lead to many others in an unpredictable fashion.

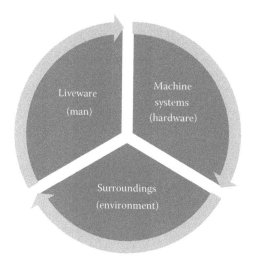

FIGURE 20.5
Systems Theory graphical representation.

Energy Release Theory[4]

William Haddon, a physician and one-time director of the National Highway Transportation Safety Administration, was heavily involved in vehicular safety. As such, he helped impose the following regulations for new cars:

- Seat belts for all occupants
- Energy-absorbing steering column
- Penetration-resistant windshield
- Dual braking systems
- Padded instrument panel
- All measures correspond with the energy and barrier concept

In the Energy Release Theory, Dr. Haddon portrays accidents in terms of energy and transference. This transfer of energy, in large amounts and at rapid rates, can adversely affect living and nonliving objects, causing injury and damage. Thus, an accident is caused by out-of-control energy. Various techniques, according to this theory, can be employed to reduce accidents, including

- Preventing the buildup of energy
- Reducing the initial amount of energy
- Preventing the release of energy

- Carefully controlling the release of energy
- Separating the energy being released from the living or nonliving object

Behavior Theory

The Behavior Theory is often referred to as *behavior-based safety (BBS)* and has captured an increasing amount of attention recently. There are seven basic principles involved in this theory:

- Intervention
- Identification of internal factors
- Motivation to behave in the desired manner
- Focus on the positive consequences of appropriate behavior
- Application of the scientific method
- Integration of information
- Planned interventions

It is primarily based on a larger scientific field called "organizational behavior." BBS is not a new concept. It has been "tossed around" by many professionals since the 1930s. A BBS program should consist of the following:

- Common goals of the employee and managerial involvement in the process
- Definition of what is expected (clear definitions of target behaviors derived from safety assessments)
- Observational data collection (safety sampling)
- Decisions about how best to proceed based on those data
- Feedback to associates being observed
- Review

Combination Theory

The cause of many accidents cannot be neatly classified into one of the theories mentioned above. The Combination Theory allows a safety professional to use any and all of the theories mentioned in this chapter to determine the root cause or causes of an accident. Using a combination of theories and models may be the optimal approach toward problem solution, by drawing the best conclusion from each of the theories mentioned above.

Modern Causation Model

The modern causation model parallels Heinrich's theory to a certain point (Figure 20.6). Injury is called RESULT, indicating it could involve damage as well as personal injury and the result can range from no damage to the very severe. The word MISHAP is used rather than Accident to avoid the popular misunderstanding that an accident necessarily involves injury or damage. The term OPERATING ERROR is used instead of Unsafe Act and Unsafe Condition. There are several aspects to the modern causation model, including operating errors, systems defects, command errors, safety program defects, and safety management errors, which will be discussed.

Operating Errors

Examples of operating errors include the following:

- Being in an unsafe position
- Stacking supplies in unstable stacks
- Poor housekeeping
- Removing a guard

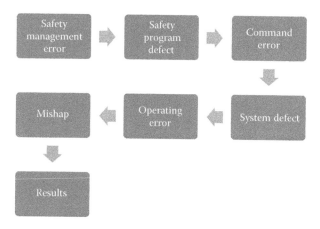

FIGURE 20.6
Modern causation model.

Systems Defects

Systems defects can be defined as a weakness in the design or operation of a system or program. Examples of a systems defect include the following:

- Improper assignment of responsibility
- Improper climate of motivation
- Inadequate training and education
- Inadequate equipment and supplies
- Improper procedures for the selection and assignment of personnel
- Improper allocation of funds

Command Error

System defects occur as a result of management or command error.

Safety Program Defect

A safety program defect is a defect in some aspect of the safety program that allows an avoidable error to exist, such as the following:

- Ineffective information collection
- Weak causation analysis
- Poor countermeasures
- Inadequate implementation procedures
- Inadequate control

Safety Management Errors

Safety management errors are weaknesses in the knowledge or motivation of the safety manager that permit a preventable defect in the safety program to exist.

Seven Avenues

There are seven avenues through which we can initiate countermeasures to a result:

- Safety management error
- Safety program defect

- Management/command error
- System defect
- Operating error
- Mishap
- Result

Countermeasures for each avenue include the following.

Safety Management Error Countermeasures

- Training
- Education
- Motivation
- Task design

Safety Program Defect Countermeasures

- Revise information collection
- Collection
- Analysis
- Implementation

Command Error Defect Countermeasures

- Training
- Education
- Motivation
- Task design

System Defects Countermeasures

Design revision through

- Standard operating procedures
- Regulations
- Policy letters
- Statements

Operating Errors Countermeasures

- Engineering

- Training
- Motivation

Mishap Countermeasures

- Protective equipment
- Barriers
- Separation

Result Countermeasures

- Containment
- Firefighting
- Rescue
- Evacuation
- First aid

Near-Miss Relationship

In the *near-miss relationship*, initial studies have shown that for each disabling injury, there were 29 minor injuries and 300 close calls/no injury (Figure 20.7). More recent studies indicate that for each serious injury, there are 59 minor and 600 near misses.

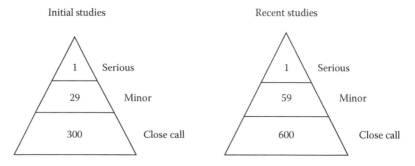

FIGURE 20.7
Near-miss relationship.

Accident Investigation Procedures

The type and complexity of any accident determine the particular procedures involved in the investigation. With this being said, most, if not all, accident investigations involve common procedures. Those procedures will be discussed in this section. Specific analysis methods will be discussed later in this chapter.

Purpose of the Investigation

The primary purpose of an accident investigation should be to prevent a recurrence of the same accident. It should not be to exonerate individuals or management, satisfy insurance requirements, defend a position for legal argument, or assign blame. All accidents and near misses (close calls) should be investigated.

Investigation Procedures

1. Secure the scene of the accident, if possible.
   a. Eliminate the hazards.
      i. Control chemicals.
      ii. Deenergize energy sources.
      iii. Shoring, if required.
      iv. Depressurize, if required.
2. Provide care to the injured.
   a. Ensure that medical care is provided to the injured people before proceeding with the investigation.
3. Isolate the scene of the accident.
   a. Barricade the area of the accident, and keep everyone out.
   b. The only persons allowed inside the barricade should be Rescue/ EMS personnel, law enforcement, and investigators.
   c. Protect the evidence until investigation is completed.
4. Lead investigator selects team, assigning specific tasks to each member.
5. Present a preliminary briefing to the investigating team, including the following:
   a. Description of the accident, with damage estimates
   b. Normal operating procedures
   c. Maps (local and general)
   d. Location of the accident site

     e. List of witnesses

     f. Events that preceded the accident

6. Visit the accident site to get updated information.

7. Inspect the accident site.

     a. Do not disturb the scene, unless a hazard exists.

     b. Prepare the necessary sketches and photographs. Label each carefully and keep accurate records.

8. Interview each victim and witness. Also interview those who were present before the accident and those who arrived at the site shortly after the accident. Keep accurate records of each interview. It is recommended that witnesses be interviewed as soon as possible and provide written, signed statements.

9. Determine

     a. What was not normal before the accident

     b. Where the abnormality occurred

     c. When it was first noted

     d. How it occurred

10. Check each sequence against the data obtained from #9 above.

11. Determine the most likely sequence of events and the most probable cause(s).

12. Conduct a postinvestigation briefing.

13. Prepare a summary report, including the recommended actions to prevent reoccurrence.

14. Distribute the report according to your company instructions or guidance procedures.

Fact Finding

Information should be gathered from a variety of sources during the investigation. Information can be obtained from witnesses and reports, as well as observation. Inspect the accident site before any changes occur and to verify witness statements. Take photographs and make sketches of the accident scene. Record all information and data accurately. Standard operating procedures are a good source to determine if correct procedures were followed.

Interviews

As mentioned previously, interviews should be conducted as soon as possible following an accident. Their memories of events are fresher in their minds, and therefore, information obtained from them can be more valuable. From experience, it is best to get written, signed statements from witnesses as soon as possible, prior to interviewing them. Once the statements have been obtained, the interviewer or interview team should determine a convenient time and location to conduct the interviews. Interviewers should be experienced in interview techniques. The witness should be made to feel comfortable. The lead interviewer should explain the purpose of the investigation and put the witness at ease. Listen carefully to each witness. Let them talk freely. Some member of the interview team should be assigned the task of taking notes, without distracting the witness. Tape recorders should only be used with the knowledge and consent of the witness. Use sketches and diagrams to assist the witness. Emphasize areas of direct observation. Word each question carefully and be sure the witness understands. Identify the qualification of each witness (name, address, occupation, years of experience, etc.). Supply each witness with a copy of his or her statements.

Once all interviews have been completed, the team should analyze them. It is recommended that statements be verified with the actual facts. Sometimes, it may be necessary to "reinterview" some of the witnesses to confirm or clarify key points.

Problem-Solving Techniques (Accident Investigation Techniques)

Accidents represent problems that must be solved through investigations. Several formal procedures solve problems of any degree of complexity. There is no one right or wrong technique. Each accident or accident type will present the best method to use.

The Scientific Method

The scientific method forms the basis of nearly all problem-solving techniques. It is used for conducting research. In its simplest form, it involves the following sequence: making observations, developing hypotheses, and testing the hypotheses.

Even a simple research project may involve many observations. A researcher records all observations immediately. A good investigator must do the same thing. Where possible, the observations should involve quantitative measurements. Quantitative data are often important in later development and testing of the hypotheses. Such measurements may require the use of many instruments in the field as well as in the laboratory.

When making observations, the investigator develops one or more hypotheses that explain the observations. The hypothesis may explain only a few of the observations or it may try to explain all of them. At this stage, the hypothesis is merely a preliminary idea. Even if rejected later, the investigator has a goal toward which to proceed.

Test the hypothesis against the original observations. A series of controlled experiments is often useful in performing this evaluation. If the hypothesis explains all of the observations, testing may be a simple process. If not, either make additional observations, change the hypothesis, or develop additional hypotheses.

As with scientific research, the most difficult part of any investigation is the formulation of worthwhile hypotheses. Use the following three principles to simplify this step:

1. *The principle agreement.* An investigator uses this principle to find one factor that associates with each observation.

2. *The principle differences.* This principle is based on the idea that variations in observations are due only to differences in one or more factors.

3. *The principle of concomitant variation.* This principle is the most important because it combines the ideas of both of the preceding principles. In using this principle, the investigator is interested in the factors that are common as well as those that are different in the observations.

In using the scientific method, the investigator must be careful to eliminate personal bias. The investigator must be willing to consider a range of alternatives. Finally, he or she must recognize that accidents often result from the chance occurrence of factors that are too numerous to evaluate fully.

Gross Hazard Analysis

Perform a gross hazard analysis (GHA) to get a rough assessment of the risks involved in performing a task. It is "gross" because it requires further study. It is particularly useful in the early stages of an accident investigation in developing hypotheses. A GHA will usually take the form of a logic diagram or table. In either case, it will contain a brief description of the problem or accident and a list of the situations that can lead to a problem. In some cases, analysis goes a step further to determine how the problem could occur. A GHA diagram or table thus shows at a glance the potential causes of an accident. One of the following analysis techniques can then expand upon a GHA.

Job Safety Analysis

Job safety analysis (JSA) is part of many existing accident prevention programs. In general, JSA breaks a job into basic steps and identifies the hazards associated with each step. The JSA also prescribes controls for each hazard. A JSA is a chart listing these steps, hazards, and controls. Review the JSA during the investigation, if a JSA has been conducted for the job involved in an accident. Perform a JSA if one is not available. Perform a JSA as part of the investigation to determine the events and conditions that led to the accident.

Failure Modes and Effects Analysis

Failure modes and effects analysis (FMEA) determines where failures occurred. Consider all items used in the task involved in the accident. These items include people, equipment, machine parts, materials, and so on. In the usual procedure, FMEA lists each item on a chart. The chart lists the manner or mode in which each item can fail and determines the effects of each failure. Included in the analysis are the effects on other items and on overall task performance. In addition, make evaluations about the risks associated with each failure. That is, project the chance of each failure and the severity of its effects. Determine the most likely failures that led to the accident. This is done by comparing these projected effects and risks with actual accident results.

An example of an FMEA is shown in Table 20.1.

TABLE 20.1

Example of FMEA

#	Component Description	Failure Mode	Effects	Means of Detection and Safeguards	Actions and Recommendations
1	Master cylinder for brakes in auto	No brake fluid Piston sticks in closed position	No braking when main brake pedal is pressed Brakes lock	Emergency brake lever/ handle	Add fluid level indicator Scheduled inspection and maintenance every 12 months
2	Impact sensor in auto airbag system	No signal to airbag on impact Spurious signal to airbag	Airbag doesn't inflate No protection for driver Airbag inflates at the wrong time	System malfunction warning light 2 out of 3 voting logic for airbag deployment	Include redundant sensors Add sensor testing to 20,000 km service schedule

Fault Tree Analysis

Fault tree analysis is a logic diagram. It shows all the potential causes of an accident or other undesired event. The undesired event is at the top of a "tree." Reasoning backward from this event, determine the circumstances that can lead to the problem. These circumstances are then broken down into the events that can lead to them, and so on. Continue the process until the identification of all events can produce the undesired event. Use a logic tree to describe each of these events and the manner in which they combine. This information determines the most probable sequence of events that led to the accident. Fault tree diagrams use a graphic model of the pathways within a system that can lead to a foreseeable, undesirable loss event (or a failure). The pathways interconnect contributory events and conditions, using standard logic symbols (AND, OR, etc.). The basic constructs in a fault tree diagram are gates and events. The more common symbols of a fault tree diagram are shown in Table 20.2.

While fault tree diagrams can become very detailed in more complicated systems, the example shown in Figure 20.8 demonstrates a simple car collision at an intersection.

Multilinear Events Sequencing Method

The multilinear events sequencing (MES) method, conceived by Ludwig Benner, is an integrated body of concepts and procedures for investigating and analyzing a wide range of desired and undesired processes before or after they happen. It can best be viewed as a process investigation and analysis method.

TABLE 20.2

Common Fault Tree Design Symbols

Name of Gate	FTD Symbol	Description
AND		The output event occurs if all input events occur.
OR		The output event occurs if at least one of the input events occurs.
Voting OR (k-out-of-n)	k	The output event occurs if k or more of the input events occur.
Inhibit		The input event occurs if all input events occur and an additional conditional event occurs.
Priority AND		The output event occurs if all input events occur in a specific sequence.
XOR		The output event occurs if exactly one input event occurs.

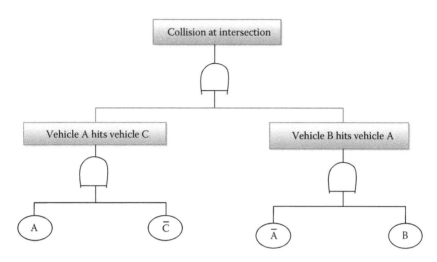

FIGURE 20.8
Example of a fault tree diagram.

The MES method uses a timeline chart to display the sequence of events that contributed to or was the direct cause of the accident (Figure 20.9). Every event is a single action by a single actor. The actor is something that brings about events, while actions are acts performed by the actor. A timeline is displayed at the bottom of the chart to show the timing sequence of the events while conditions that influence the events are inserted in the time flow in logical order to show the flow relationship. With this chart, countermeasures can be formulated by examination of each individual event to see where changes can be introduced to alter the process.

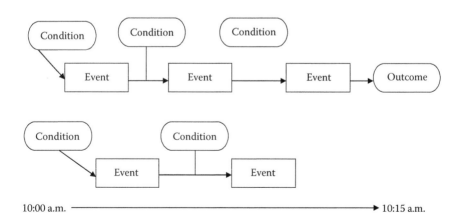

FIGURE 20.9
Example of a MES chart.

Report of Investigation

An accident investigation is not complete until a report is prepared and submitted to the proper authorities, such as the company president. A final report should be prepared to include the following sections:

- Cover page
- Title page
- Executive summary
- Table of contents
- Narrative (body)
- Conclusions and recommendations

The following outline for the Narrative or Text Body is useful in developing the report:

- Background information
  - Where and when the accident occurred
  - Who and what were involved
  - Operating personnel and other witnesses
- Account of the accident (what happened?)
  - Sequence of events
  - Extent of damage
  - Accident type
  - Source of energy or hazardous materials involved
- Discussion (analysis of the accident—how/why)
  - Direct causes (energy sources, hazardous materials)
  - Indirect causes (unsafe acts and conditions)
  - Basic causes (management policies, personal or environmental factors)
- Recommendations to prevent a recurrence for immediate and long-range action to remedy
  - Direct causes
  - Indirect causes
  - Basic causes
- Summary

Key Information to Remember on Accident Causation and Investigation Techniques

1. The primary purpose of an accident investigation is to prevent the recurrence of the same event.

2. The Domino Theory, also known as Heinrich's Domino Theory, is considered the first scientific approach to accident prevention.

3. According to Heinrich, an injury is caused by the social environment and inherited behavior, fault of the person, unsafe acts or conditions, and a resulting accident.

4. Injuries result from a completed series of factors, one of which is the accident itself (Heinrich's Domino Theory).

5. Heinrich's "Three E's" are Engineering, Education, and Enforcement.

6. Heinrich's conclusions are that 2% of accidents are unavoidable, 10% are attributed to unsafe conditions, and 88% are attributed to unsafe acts.

7. The Human Factors Theory states that all accidents are the result of human error, categorized as overload, inappropriate worker response, or inappropriate activities.

8. Petersen's Accident/Incident Theory is a basic extension of the Human Factors Theory, except that he introduced ergonomic traps and that a decision to err may be based on logic.

9. The Epidemiological Theory focuses primarily on industrial hygiene aspects.

10. The Systems Theory states that there is a relationship between man, machine systems, and the surroundings that make up a whole system.

11. The Energy Release Theory, developed by William Haddon, portrays accidents in terms of energy and transference.

12. Behavior Theory is also known as behavior-based safety.

13. The Combination Theory allows an investigator to use parts or all of any of the theories to solve a problem.

14. The modern causation model uses a series of seven avenues to demonstrate the cause of accidents: safety management error, safety program defect, command error, system defect, operating error, mishap, and results.

References

1. United States Army, 2009. Available at www.armystudyguide.com/content /bm~doc/accident-causation.ppt.
2. http://www.segurancaetrabalho.com.br/download/accident-causation.ppt.
3. http:///www.ceet.niu.edu/depts/tech/asse/tech434/AccCausation.doc.
4. Cleveland State University, 2010. Available at http://academic.csuohio.edu /duffy_s/Section_03.pdf.

21

Workers' Compensation

Worker compensation laws have been implemented in each state and are specific to that state. The laws of each state vary in so many ways that it would be impossible to discuss each and every state individually. Therefore, the intent of this chapter is to provide the reader with a basic understanding of workers' compensation law. As a practicing safety professional, you will, no doubt, be heavily involved in major aspects of the workers' compensation program.

Most states require that employers with five or more employees maintain workers' compensation insurance or prove financial stability for self-insurance. Workers' compensation laws provide employees who are injured or disabled on the job fixed monetary awards, without having to endure litigation. Benefits are also provided for dependents of those workers who are killed because of work-related injuries or illnesses. In addition, it is designed to protect employers and coworkers by limiting recovery amounts and liabilities of both the employers and coworkers. Both state and federal workers' compensation statutes provide this structure.

History of Workers' Compensation in the United States

Workers' Compensation (formerly referred to as Workmen's Compensation) as a legal issue emerged in the United States in the early 1900s. Prior to the establishment of codified workers' compensation laws, the only compensation or remedy for an injured employee was to bring a negligence suit against his or her employer. If the employee won the suit, then he or she was compensated for damages, including pain and suffering. It must also be mentioned that during this time, there was a significant increase in the legal profession. Oftentimes, these damages were substantial in nature. The major drawback to this type of remedy was that the employee had to prove negligence on the part of the employer, which was sometimes difficult to do. In addition, the employer used several legal defenses, which included

- *Assumption of Risk*: The employees accepted the risk they were facing when they accepted the job. By doing so, they gave up any right to collect compensation for injuries.

- *Contributory Negligence*: Since the employees contributed to their injuries, regardless of how little, the employees are not permitted to recover compensation for their injuries.
- *Fellow-Servant Rule*: The employer is not at fault because the accident was the fault of another employee or other employees.

After the implementation of workmen's compensation laws, these three common defenses were eliminated by each state.[1]

The very first workmen's compensation law passed in the United States was the Federal Employer's Liability Act. It was adopted in 1908 and signed into law by President Theodore Roosevelt. He was instrumental in getting this law passed, because he demonstrated to congress that "the entire burden of an accident fell upon the helpless man, his wife and children" and that was "an outrage."

In 1911, the State of Wisconsin passed the first official state law covering workers' compensation in the United States. In this legislation, the employer agreed to provide medical and indemnity (wage replacement) benefits and the injured employee agreed to give up his or her right to sue the employer. Pressure was now on the business community, as a result of increasing awards to injured employees. Also, in 1911, 10 more states enacted "workmen's compensation" laws. By 1948, all 48 states and the territories of Hawaii and Alaska had implemented some form of workmen's compensation law. Today, workers' compensation is the exclusive remedy for the injured worker. These laws also protect employers from damage suits filed by the injured worker, as well as providing employers with a basis for calculating production costs.

Current Workers' Compensation Laws

In these workers' compensation laws, most states have prohibited employees from suing businesses that comply with them, except for deliberate assaults and conditions so flagrantly unsafe as to make injury virtually certain. This has been a win/win situation for a number of years. Lately, however, employees have retained attorneys that use this loophole to successfully sue an employer for increased compensation. In these cases, the employee repays the workers' compensation fund for the money already received. These cases are successful primarily because of sympathetic juries who see employers as an endless source of money and the fact that attorneys work on a "contingency fee," meaning that if no award is made, then the attorney collects no fee. While this practice is in direct conflict with the spirit of the workers' compensation laws, it is one of the major flaws in today's system.

If an employee is injured in a work-related accident, he or she must undergo a waiting period from 3 to 14 days, depending on the state law. Benefits are typically limited to two-thirds of the employee's wages with a cap, which is also dependent on the state. Injured employees living on two-thirds of their wages will suffer a significant decrease in their standard of living, and therefore, litigation becomes more attractive. On the other hand, if full wages were paid, the employee would be reluctant to return to work. It has been determined that even at two-thirds of their wages, some employees have been found to be malingering so that they can receive benefits without working.

Injuries are placed into one of four categories listed below:

- *Partial*: when the employee can still work but is unable to perform all duties of the job because of the injury, as would often be the case with a broken finger or a severed toe.
- *Total*: when the employee is unable to work or perform substantial duties on the job, as would often be the case with a severe back injury or blindness.
- *Temporary*: when the employee is expected to fully recover, as would be the case with a broken limb or a sprain.
- *Permanent*: when the employee will suffer the effects of the injury from now on, as would be the case with a severed limb, blindness, or permanent hearing loss.

Coverage Exemptions

The following employees are exempt from the coverage of workers' compensation:

- Any person employed as a domestic servant in a private home
- Any person employed for not exceeding 20 consecutive work days in or about the private home of the employer
- Any person performing services in return for aid or sustenance only, received from any religious or charitable organization
- Any person employed in agriculture
- Any person participating as a driver or passenger in a voluntary vanpool or carpool program while that person is on the way to or from his place of employment
- Any person who would otherwise be covered but elects not to be covered

Workers' Compensation Premiums

Workers' compensation insurance premiums represent a significant line item on any business' profit and loss statement. Therefore, it is necessary to keep these costs to a bare minimum. Today's safety professional is or should be heavily involved in the insurance carrier selection process. In order to do this, you should have some knowledge of how the premiums are calculated. On the surface, the calculation seems simple. However, the truth is that it is anything but simple. Historically, workers' compensation premiums were calculated using the Standard Industrial Classification codes for a particular business or industry. Workers' compensation premiums are usually expressed in dollars per $100 of payroll. Insurance companies who provide workers' compensation coverage now use an additional multiplier known as the *Experience Modification Rate* (*EMR*) or *Experience Modification Factor* (*EMF*) to assist in the determination of the cost of workers' compensation costs.

Calculating Experience Modification Rates

The EMR is not simply a comparison of actual losses with expected losses. If premiums were based solely on prior losses, the average insured would pay no premiums after a loss-free period. A period is usually a rolling, 3-year average. Contrary to this, they would pay huge premiums after a major accident involving its employees. In addition, only part of the premium goes to paying losses. The insurance company has expenses that involve underwriting and marketing, which must be paid for.

The experience modification formula is summarized as follows:

$$\text{Experience Modification Rate} = \frac{\text{Adjusted Actual Losses} + \text{Ballast}}{\text{Expected Losses} + \text{Ballast}}.$$

These terms are defined below:

Adjusted Actual Losses: Adjusted actual losses are determined by an additional formula that has a stabilizing effect on the resulting value. The stabilizing effect is accomplished by (1) separating actual losses into primary losses and excess losses and then using only a percentage of excess losses in the computation and (2) including a portion of the insured's expected losses in the computation. Limiting the amount of excess losses has a moderating effect if the insured suffers an unusually high level of losses during the experience period. Including a portion of the insured's expected losses guarantees that no insured can have a zero modification. Even if the insured has no actual losses during the experience period, the formula will always produce an adjusted actual loss value greater than zero because a portion of the adjusted losses is calculated as a percentage of expected losses, not just the insured's actual losses.

Expected Losses: Expected losses are calculated by multiplying the payrolls in each applicable classification of the insured's employees by an expected loss ratio factor published for each classification in each state.

Ballast: Ballast is an amount that is added to both the numerator and denominator of the experience modification formula to dampen the swings between large credits and large debits. The result is that all modifications come closer to a modification factor of 1.0, also known as "unity modification." A unity modification means that no credit or debit is applied to the insured's rates.

It is particularly important that each employee not be classified into a single category. Office personnel do not have the same risks (expected losses) as say an over-the-road driver. Therefore, when requesting coverage by an insurance carrier, businesses must do a thorough analysis of the job classifications for each employee.

In recent years, I have experienced a tightening of client requirements that include such provisions in the contract that a vendor must maintain an EMR of 1.0 or below. I have even experienced some clients requiring an EMR of less than or equal to 0.90. Therefore, the EMR becomes more important than just determining the cost of insurance premiums.

Waiver of Subrogation

In recent years, contracts have required a waiver of subrogation statement be put on certificates of insurance before a company is allowed to work on a host employer site. To accommodate this requirement, the *Waiver of Our Right to Recover from Others Endorsement* provides that the insurer will not enforce its right to recover against organizations listed in the endorsement's schedule when the insured is performing work under contract that requires a waiver. Insurers usually charge additional premium for providing this waiver. In a few states, contractual waiver of subrogation is prohibited by statute, rendering the endorsement invalid.

Safety Professional's Role and Responsibilities in Workers' Compensation

The primary role or responsibility that a safety professional has in relation to workers' compensation is to ensure that there exists a viable safety program that ensures the safety and health of the employee. Depending on the size of the company, the safety professional may also be called upon to report injuries and file insurance claims, manage injury cases, and coordinate a return-to-work program.

Reporting of Injuries and Claims

Each individual insurance carrier will have its own specific reporting requirements. However, in general, each claim will use a Form 45-Employer's First Report of Injury. This First Report of Injury should be completed and filed at the earliest opportunity. Each carrier will also provide a claims reporting number, once a policy has been issued. It is advisable to report all injuries, regardless of the intention to file a claim. For example, an employee has a minor sprain to the ankle. The physician advises the employer that the employee is to be on light or limited duty for 3 days, but that a full recovery is expected. The employer elects to pay the medical expenses out-of-pocket, without filing an insurance claim. This injury should still be reported to the insurance carrier as "FOR RECORD ONLY." This type of reporting does not increase your future premiums, but does put the carrier on notice in case the injury goes beyond initial first aid.

When reporting a claim, the following information should be available:

- Employer's FEIN
- Employer's name and address
- Injured employee's full name and address
- Injured employee's date of birth
- Injured employee's date of hire
- Injured employee's specific job title
- Date and time of accident
- Time injured employee began work on the date of the accident
- General description of the accident (where, how, when, what, etc.)
- Name, address, and telephone number of the treating physician or health care provider

It is important for the safety professional to develop a personal relationship and trust with the insurance carrier or agent. The carrier or agent can be a valuable resource in how to manage many complex cases. If there are some questionable reports of injuries by employees or additional information available, the insurance carrier should be notified at the time of claim filing or as soon as possible after discovery.

Case Management

Prompt Reporting

Managing workers' compensation claims begins with prompt reporting of the injury to the insurance carrier. Statistics show that claims reported late cost, on the average, 60% more than claims reported early. The majority of carriers have a 24-h claim reporting number.

It is well documented that lost time accidents and injuries cost substantially more than accidents and injuries that result in no lost time. The workers' compensation carrier will incur the cost of the medical expenses and wage loss. The employer, however, incurs hidden costs, consisting of loss of production, reduced employee morale, and possibly loss of customers. Each of these items is not covered under the insurance. One of the best ways to manage workers' compensation costs is to develop a Return-To-Work/Light Duty Policy/Program for your specific company.

Return-to-Work/Light Duty Program

A return-to-work/light duty program is a program to enhance the injured employee's rehabilitation and facilitate his or her return to normal job duties by providing temporary modified work. Some of the employee benefits of a return-to-work/light duty program are as follows:

- Generally gets employees back to work 50% faster
- Speeds recovery up to three times
- Reduces vocational rehabilitation
- Reduces degree of permanent partial disability
- Reduces possibility of reinjury upon return
- Employee's wage loss drastically minimized (full wages vs. 66.7%)

The employer benefits of a return-to-work/light duty program are as follows:

- Helps control medical costs by as much as 70%
- Reduces indemnity costs
- Less abuse of workers' compensation system
- Improves morale/employee relations
- Improved work ethic
- Enhanced company image
- Less litigation

The specifics of the return-to-work/light duty policy/program depend entirely on the restrictions placed on the employee by the medical provider. As one might well imagine, not every injury is the same, which makes it difficult to write the policy specifics. However, an effective return-to-work/light duty program involves the following:

- Identification of potential jobs for modified duty
- Early reporting of injuries and illnesses
- Accident investigation
- Communications between the employer, the injured employee, the medical provider, and the insurance company
- Education of employees and supervisors

The importance of communications between the employer, employee, medical provider, and insurance company cannot be overstated. It is essential that a trusting relationship be established with the medical provider. The medical provider's main focus should be to insure that the injured employee is treated promptly and properly. His or her secondary focus is to assist them in returning to work as soon as possible. If a medical provider does not have trust that an employer will strictly adhere to the work restrictions placed on an employee, the provider may have a tendency to keep the injured employee on full restrictions longer than necessary. A written policy goes a long way in reassuring the medical provider that you have the best interest of the employee in mind.

The return-to-work/light duty program should be put in writing and communicated to all employees. When communicating the program/policy to the employees, it is helpful to emphasize the company's commitment to get injured employees back to productive work as quickly as possible.

Key Information to Remember on Workers' Compensation

1. Most states require employers with five or more employees to maintain workers' compensation insurance or be self-insured.
2. The first state to pass a workers' compensation law was Wisconsin, which enacted a law in 1911. Ten additional states passed laws in the same year.
3. By 1948, all 48 states and two territories (Alaska and Hawaii) had enacted workers' compensation laws.
4. Work-related injuries are classified into four main categories: partial, total, temporary, and permanent.

5. Workers' compensation insurance premiums are based on dollars per $100 of payroll and an experience modification rate or factor.

6. A return-to-work/light duty program can be one of the most effective tools in reducing workers' compensation premiums and experience modification rates.

Reference

1. Friend, M. A. and Kohn, J. P., 2001. *Fundamentals of Occupational Safety and Health*, 2nd Edition, Government Institutes, Lanham, MD, p. 44.

22

Ergonomics

As a safety professional, you have or will, undoubtedly, encounter low back pain and work-related strains. Despite your best efforts at controlling these injuries, they will still account for a significant proportion of employee injuries and economic cost to the employer. A significant amount of your energy is or will be spent in attempting to develop and implement programs to prevent these types of injuries. Ergonomics plays an important role in the prevention of strains, injuries, and repetitive motion disorders. The International Ergonomics Association defines *ergonomics* as follows: *Ergonomics (or human factors) is the scientific discipline concerned with the understanding of interactions among humans and other elements of a system, and the profession that applies theory, principles, data and methods to design in order to optimize human well-being and overall system performance.*

One of the first professional societies in the field of ergonomics was the Ergonomics Research Society in the United Kingdom, which was started in 1949. This society later became the Ergonomics Society. In 1959, the European Productivity Agency's steering committee on ergonomics became known as "International Ergonomics Association."[1]

There are numerous and varied subdisciplines within the field of ergonomics. This chapter will focus primarily on the *National Institute for Occupational Safety and Health (NIOSH) Lifting Equation.*

Definitions

Before proceeding, it is first necessary to understand some of the basic ergonomic terminologies.

Recommended Weight Limit (RWL)[2]: The RWL is the primary product in the NIOSH lifting equation. It is defined for a specific set of task conditions as the weight of the load that nearly all healthy workers could perform over a substantial period (e.g., up to 8 h) without an increased risk of developing lifting-related low back pain or injury.

Lifting Index[2]: The lifting index is a term that provides a relative estimate of the level of physical stress associated with a particular manual lifting task. The estimate of the level of physical stress is defined by the relationship of the weight of the load lifted and the RWL.

Lifting Task[2]: The act of manually grasping an object of definable size and mass with two hands and vertically moving the object without mechanical assistance.

Load Weight (L)[2]: Weight of the object to be lifted, in pounds or kilograms, including the container.

Horizontal Location (H)[2]: Distance of the hands away from the midpoint between the ankles, in inches or centimeters (measure at the origin and destination of the lift).

Vertical Location (V)[2]: Distance of the hands above the floor, in inches or centimeters (measure at the origin and destination of lift).

Vertical Travel Distance (D)[2]: Absolute value of the difference between the vertical heights at the destination and origin of the lift, in inches or centimeters.

Asymmetry Angle (A)[2]: Angular measure of how far the object is displaced from the front (midsagittal plane) of the workers' body at the beginning or ending of the lift, in degrees (measure at the origin and destination of the lift). The asymmetry angle is defined by the location of the load relative to the workers' midsagittal plane, as defined by the neutral body posture, rather than the position of the feet or the extent of body twist (see Figure 22.1).

Neutral Body Position[2]: Describes the position of the body when the hands are directly in front of the body and there is minimal twisting at the legs, torso, or shoulders.

Lifting Frequency (F)[2]: Average number of lifts per minute over a 15-min period.

Lifting Duration[2]: Three-tiered classification of lifting duration specified by the distribution of work time and recovery time (work pattern). Duration is classified as either short (1 h), moderate (1–2 h), or long (2–8 h), depending on the work pattern.

Coupling Classification[2]: Classification of the quality of the hand-to-object coupling (e.g., handle, cutout, or grip). Coupling quality is classified as good, fair, or poor.

Significant Control[2]: Significant control is defined as a condition requiring precision placement of the load at the destination of the lift. This is usually the case when (1) the worker has to regrasp the load near the destination of the lift, (2) the worker has to momentarily hold the object at the destination, or (3) the worker has to carefully position or guide the load at the destination.

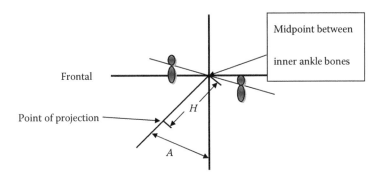

FIGURE 22.1
Top view of sagittal line. (From DHHS [NIOSH] Publication No. 94-110.)

NIOSH Lifting Equation[2]

The lifting equation is a tool for assessing the phyical stress of two-handed manual lifting tasks. As with any tool, its application is limited to those conditions for which it was designed. Specifically, the lifting equation was designed to meet specific lifting-related criteria that encompass biomechanical, work physiology, and psychophysical assumptions and data. The revised NIOSH lifting equation is based on the assumption that manual handling activities other than lifting are minimal and do not require significant energy expenditure, especially when repetitive lifting tasks are performed. Examples of nonlifting tasks include holding, pushing, pulling, carrying, walking, and climbing. If such nonlifting activities account for more than approximately 10% of the total worker activity, then measures of workers' energy expenditures and heart rate may be required to assess the metabolic demands of the different tasks.

The NIOSH lifting equation is listed as follows:

$$RWL = LC \times HM \times VM \times DM \times AM \times FM \times CM.$$

The NIOSH lifting equation for metric measurements is as follows:

$$RWL(kg) = (23)\left(\frac{25}{H}\right)\left[1 - 0.003|V - 75|\right]\left[0.82 + \left(\frac{4.5}{D}\right)\right](1 - 0.0032A)(FM)(CM)$$

TABLE 22.1

NIOSH Lifting Equation Standards

Symbol	Meaning	Metric System	US System
RWL	Recommended weight limit	kg	lb
LC	Load constant	23 kg	51 lb
HM	Horizontal multiplier	25/h	10/h
		See Table 22.2	See Table 22.2
VM	Vertical multiplier	$1-(0.003\|V-75\|)$	$1-(0.0075\|V-30\|)$
		See Table 22.3	See Table 22.3
DM	Distance multiplier	$0.82+\left(\dfrac{4.5}{D}\right)$	$0.82+\left(\dfrac{1.8}{D}\right)$
		D = distance in centimeters	D = distance in inches
		See Table 22.4	See Table 22.4
AM	Asymmetric multiplier	$1-(0.0032A)$	$1-(0.0032A)$
		See Table 22.5	See Table 22.5
FM	Frequency multiplier	See Table 22.6	See Table 22.6
CM	Coupling multiplier	See Table 22.7	See Table 22.7

The NIOSH lifting equation for the US system is as follows:

$$RWL\ (lb) = (51)\left(\frac{10}{H}\right)\left[1-\left(0.0075|V-30|\right)\right]\left[0.82+\left(\frac{1.8}{D}\right)\right]$$
$$\times(1-0.0032A)(FM)(CM),$$

where the following standards are shown in Table 22.1.

The term *multiplier* refers to the reduction coefficients in the equation. Obtaining the multipliers from the appropriate table or computation, it may be necessary to interpolate. For example, when the measured frequency is not a whole number, the appropriate multiplier must be interpolated between the frequency values in the table for the two values that are closest to the actual frequency.

Horizontal Component[2]

As mentioned previously in the Definitions section, the horizontal location is measured from the midpoint of the line joining the inner ankle bones to a point projected on the floor directly below the midpoint of the hand grasps (load center). In some cases, the H value cannot be measured. In these cases, the H value can be approximated using the following equation(s):

Metric (H value) estimation equation (all distances in centimeters):

$$H = 20 + \frac{W}{2} \text{ for } V \geq 25 \text{ cm}$$

or

$$H = 25 + \frac{W}{2} \text{ for } V < 25 \text{ cm,}$$

where
W = the width of the container in the sagittal plane
V = the vertical location of the hands from the floor

US system (H value) estimation equation (all distances in inches):

$$H = 8 + \frac{W}{2} \text{ for } V \geq 10 \text{ in.}$$

or

$$H = 10 + \frac{W}{2} \text{ for } V < 10 \text{ in.}$$

Horizontal Multiplier

See Table 22.2.

Vertical Component[2]

In the Definitions section, we defined the vertical location as the vertical height of the hands above the floor. The vertical location (V) is measured vertically from the floor to the midpoint between the hand grasps, as defined by the large middle knuckle.

Vertical Multiplier

See Table 22.3.

TABLE 22.2

Horizontal Multiplier

Metric System		US System	
H (cm)	HM	H (in.)	HM
≤25	1.00	≤10	1.00
28	0.89	11	0.91
30	0.83	12	0.83
32	0.78	13	0.77
34	0.74	14	0.71
36	0.69	15	0.67
38	0.66	16	0.63
40	0.63	17	0.59
42	0.60	18	0.56
44	0.57	19	0.53
46	0.54	20	0.50
48	0.52	21	0.48
50	0.50	22	0.46
52	0.48	23	0.44
54	0.46	24	0.42
56	0.45	25	0.40
58	0.43	>25	0.00
60	0.42		
63	0.40		
>63	0.00		

Source: Developed from DHHS (NIOSH) Publication No. 94-110.

Distance Component[2]

The vertical travel distance variable (D) is defined as the vertical travel distance of the hands between the origin and destination of the lift. The distance multiplier can be determined by the following data found in Table 22.4.

Asymmetry Component[2]

Asymmetry refers to a lift that begins or ends outside the midsagittal plane as shown in Figure 22.1. Asymmetric lifting should be avoided whenever possible. If asymmetric lifting cannot be avoided, however, the RWLs are significantly less than those limits used for symmetrical lifting.

TABLE 22.3

Vertical Multiplier

Metric System		US System	
V (cm)	VM	V (in.)	VM
0	0.78	0	0.78
10	0.81	5	0.81
20	0.84	10	0.85
30	0.87	15	0.89
40	0.90	20	0.93
50	0.93	25	0.96
60	0.96	30	1.00
70	0.99	35	0.96
80	0.99	40	0.93
90	0.96	45	0.89
100	0.93	50	0.85
110	0.90	55	0.81
110	0.87	60	0.78
120	0.84	65	0.74
130	0.81	70	0.70
140	0.78	>70	0.00
150	0.75		
160	0.75		
170	0.72		
175	0.70		
>175	0.00		

Source: Developed from DHHS (NIOSH) Publication No. 94-110.

Table 22.5 shows the asymmetric multipliers for various angles. The asymmetric multipliers apply to both the metric and US systems.

Frequency Component[2]

The frequency multiplier is defined by (a) the number of lifts per minute (frequency), (b) the amount of time engaged in the lifting activity (duration), and (c) the vertical height of the lift from the floor. Lifting frequency (*F*) refers to the average number of lifts made per minute, as measured over a 15-min period.

TABLE 22.4

Distance Multiplier

Metric System		US System	
D (cm)	DM	*D* (in.)	DM
≤25	1.00	≤10	1.00
40	0.93	15	0.94
55	0.90	20	0.91
70	0.88	25	0.89
85	0.87	30	0.88
100	0.87	35	0.87
115	0.86	40	0.87
130	0.86	45	0.86
145	0.85	50	0.86
160	0.85	55	0.85
175	0.85	60	0.85
>175	0.00	70	0.85
		>70	0.00

Source: Developed from DHHS (NIOSH) Publication No. 94-110.

TABLE 22.5

Asymmetric Multiplier

A (Angle-Degree)	AM
0	1.00
15	0.95
30	0.90
45	0.86
60	0.81
75	0.76
90	0.71
105	0.66
120	0.62
135	0.57
>135	0.00

Source: Developed from DHHS (NIOSH) Publication No. 94-110.

Lifting Duration

Lifting duration is classified into three categories: short, moderate, and long duration. *Short duration* pertains to lifting tasks that have a work duration of 1 h or less, followed by a recovery time equal to 1.2 times the work time. *Moderate duration* concerns lifting tasks that have a duration of more than 1 h, but less than 2 h, followed by a recovery period of at least 0.3 times the work time. *Long duration* pertains to lifting tasks that have a duration of between 2 and 8 h, with standard industrial rest allowances.

Frequency Multiplier

See Table 22.6.

TABLE 22.6

Frequency Multiplier

	Work Duration					
	≤1 h		>1 h but ≤2 h		>2 h but <8 h	
Frequency, Lifts/min (F)[a]	V < 30[b]	V ≥ 30	V < 30	V ≥ 30	V < 30	V ≥ 30
≤0.2	1.00	1.00	0.95	0.95	0.85	0.85
0.5	0.97	0.97	0.92	0.92	0.81	0.81
1	0.94	0.94	0.88	0.88	0.75	0.75
2	0.91	0.91	0.84	0.84	0.65	0.65
3	0.88	0.88	0.79	0.79	0.55	0.55
4	0.84	0.84	0.72	0.72	0.45	0.45
5	0.80	0.80	0.60	0.60	0.35	0.35
6	0.75	0.75	0.50	0.50	0.27	0.27
7	0.70	0.70	0.42	0.42	0.22	0.22
8	0.60	0.60	0.35	0.35	0.18	0.18
9	0.52	0.52	0.30	0.30	0.00	0.15
10	0.45	0.45	0.26	0.26	0.00	0.13
11	0.41	0.41	0.00	0.23	0.00	0.00
12	0.37	0.37	0.00	0.21	0.00	0.00
13	0.00	0.34	0.00	0.00	0.00	0.00
14	0.00	0.31	0.00	0.00	0.00	0.00
15	0.00	0.28	0.00	0.00	0.00	0.00
>15	0.00	0.00	0.00	0.00	0.00	0.00

Source: Developed from DHHS (NIOSH) Publication No. 94-110.
[a] For lifting less frequently than once per 5 min, set $F = 0.2$.
[b] Values of V are in inches.

TABLE 22.7

Coupling Multiplier

Coupling Type	Coupling Multiplier	
	V < 30 in. or 75 cm	V ≥ 30 in. or 75 cm
Good	1.00	1.00
Fair	0.95	1.00
Poor	0.90	0.90

Source: Developed from DHHS (NIOSH) Publication No. 94-110.

Coupling Component[2]

The nature of the hand-to-object coupling or gripping method can affect not only the maximum force a worker can or must exert on the object but also the vertical location of the hands during the lift. A *good* coupling will reduce the maximum grasp forces required and increase the acceptable weight for lifting, while a *poor* coupling will generally require higher maximum grasp forces and decrease the acceptable weight for lifting.

There are three hand-to-object classifications: good, fair and poor. *Good* is defined as handles or hand-hold cutouts of optimal design or a comfortable grip in which the hand can be easily wrapped around the object. *Fair* can be defined as handles or hand-hold cutouts of less than optimal design or a grip in which the hand can be flexed approximately 90°. *Poor* can be defined as containers with less than optimal design or loose parts, which are hard to handle or nonrigid bags that may sag in the middle.

Coupling Multiplier

See Table 22.7.

Lifting Index[2]

As mentioned previously in the Definitions section, the lifting index is a term that provides a relative estimate of the level of physical stress associated with a particular manual lifting task. The estimate of the level of physical stress is defined by the relationship of the weight of the load lifted and the RWL. To calculate the lifting index, we use the following equation:

$$\text{Lifting Index (LI)} = \frac{\text{Object Weight } (L)}{\text{RWL}}.$$

JOB ANALYSIS WORKSHEET

Department _____ Job Description _____

Job Title _____ _____

Analyst Name _____ _____

Date _____ _____

Measure and Record Task Variables								
Object Weight		Hanc Location			Asymmetric Angle (Degrees)		Frequency Rate	Duration	
				Vertical					Object
L (avg)	L (Max)	Origin	Dest.	Distance	Origin	Dest.	Lifts/min	(hrs)	Coupling

Determine the Multipliers and Compute the RWL's
RWL = LC x HM x VM x DM x AM x FM x CM

Origin: RWL = ☐ x ☐ x ☐ x ☐ x ☐ x ☐ x ☐ = ☐

Destination: RWL = ☐ x ☐ x ☐ x ☐ x ☐ x ☐ x ☐ = ☐

Compute the Lifting Index

Origin: $Lifting\ Index = \dfrac{Object\ Weight\ (L)}{RWL} = -\, -\, = ☐$

Destination: $Lifting\ Index = \dfrac{Object\ Weight\ (L)}{RWL} = -\, -\, = ☐$

FIGURE 22.2
Job analysis worksheet.

Job Analysis Worksheet

The above worksheet (Figure 22.2) can be easily used to evaluate most processes.

Example Problems

Problem 1

A typical operation involves a punch press operator routinely handling small parts, feeding them into a press, and removing them. Occasionally (once per shift), the operator must load a heavy reel of supply stock from the floor onto the machine. The diameter of the reel is 30 in., the width of the reel between the workers' hands is 12 in., and the reel weighs 44 lb. Significant control of the load is required at the destination of the lift owing to the design of the machine (see Figure 22.3).

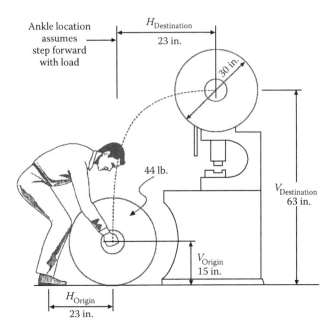

FIGURE 22.3
Loading punch press stock. (From DHHS (NIOSH) Publication No. 94-110.)

Determine the RWL and the lifting index using the revised NIOSH lifting equation. The multipliers can be determined from the tables shown previously (Figure 22.4).

By evaluating this process, we have determined the following:

- At its origin, the object is 2.7 times the RWL.
- At its destination, the object is 3 times the RWL.

The question now is "What can be done to prevent injuries arising from this process?"

These lifting indices indicated that the lift would be hazardous for a majority of healthy industrial employees. There are numerous possibilities for reducing or eliminating potential injuries in this process. A few of them are listed as follows:

- Engineer the lifting out of the operation, if possible.
- Bring the object closer to the worker at the destination to increase the HM value.
- Lower the destination of the lift to increase the VM value.

Example 1 JOB ANALYSIS WORKSHEET								
Department	Operations			Job Description Loading supply stock				
Job Title	Punch Press Operator			onto punch press machine				
Analyst Name	Bob Jones							
Date	XX/XX/XXXX							
Measure and Record Task Variables								
Object Weight		Hand Location			Asymmetric Angle (Degrees)		Frequency Rate	Duration	
		Origin	Dest.	Vertical Distance	Origin	Dest.	Lifts/min	(hrs)	Object Coupling
L (avg)	L (Max)								
44	44	23 | 15	23 | 63	48	0	0	<0.2	<1	Fair

Determine the Multipliers and Compute the RWL's

$$RWL = LC \times HM \times VM \times DM \times AM \times FM \times CM$$

Origin: RWL = | 51 | x | .44 | x | .89 | x | .86 | x | 1.0 | x | 1.0 | x | .95 | = | 16.3 lbs |

Destination: RWL = | 51 | x | .44 | x | .75 | x | .86 | x | 1.0 | x | 1.0 | x | 1.0 | = | 14.5 lbs |

Compute the Lifting Index

Origin: $\quad$ Lifting Index $= \dfrac{Object\ Weight\ (L)}{RWL} = \dfrac{44}{16.3} = \boxed{2.7}$

Destination: $\quad$ Lifting Index $= \dfrac{Object\ Weight\ (L)}{RWL} = \dfrac{44}{14.5} = \boxed{3.0}$

FIGURE 22.4
Example 1. Job analysis worksheet.

- Reduce the vertical distance between the origin and the destination of the lift to increase the DM value.
- Modify the job so that significant control of the object at the destination is not required. This will eliminate the need to use the lower RWL value at the destination.

Problem 2

An employee is assigned to load bags into a hopper. The employee positions himself midway between the hand truck and the mixing hopper (see Figure 22.5). Without moving his feet, he twists to the right and picks up a bag off the hand truck. In one continuous motion, he then twists to his left to place the bag on the rim of the hopper. A sharp-edged blade within the hopper cuts open the bag to allow the contents to fall into the hopper. This task is done infrequently (1–12 times per shift) with large recovery periods

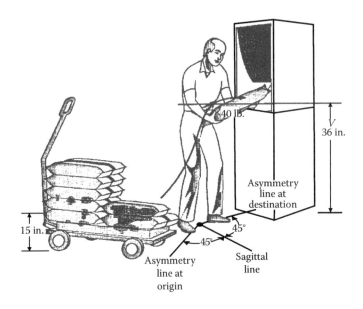

FIGURE 22.5
Loading bags into a hopper. (From DHHS (NIOSH) Publication No. 94-110.)

between lifts (>1.2 h recovery time). In observing the worker perform the job, it was determined that the nonlifting activities could be disregarded because they require minimal force and energy. Significant control is not required at the destination, but the worker twists at the origin and destination of the lift. Although several bags are stacked on the hand truck, the highest risk of overexertion injury is associated with the bottom bag (Figure 22.6).

By evaluating this process, we have determined the following:

* At its origin, the object is 2.1 times the RWL.
* This lift would be physically stressful for most industrial workers.

Some redesign suggestions to improve this process and potentially prevent employee injuries are as follows:

* Bringing the load closer to the worker to increase the HM.
* Reducing the angle of asymmetry to increase the AM. This could be accomplished by moving the origin and destination points closer together or further apart.
* Raising the height at the origin to increase VM.

Hopper Example JOB ANALYSIS WORKSHEET								
Department		Operations			Job Description Loading supply stock			
Job Title		Punch Press Operator			onto punch press machine			
Analyst Name		Bob Jones						
Date		XX/XX/XXXX						

Measure and Record Task Variables								
Object Weight		Hanc Location		Vertical Distance	Asymmetric Angle (Degrees)		Frequency Rate	Duration
								Object Coupling
L (avg)	L (Max)	Origin	Dest.		Origin	Dest.	Lifts/min	(hrs)
40	40	18 15 10 36		21	45	45	<0.2	<1	Fair

Determine the Multipliers and Compute the RWL's

$$RWL = LC \times HM \times VM \times DM \times AM \times FM \times CM$$

Origin: RWL = | 51 | × | .56 | × | .89 | × | .91 | × | .86 | × | 1.0 | × | .95 | = | 18.9 lbs |

Destination: RWL = | | × | | × | | × | | × | | × | | × | | = | |

Compute the Lifting Index

Origin: $Lifting\ Index = \dfrac{Object\ Weight\ (L)}{RWL} = \dfrac{40}{18.9} = \boxed{2.1}$

Destination: $Lifting\ Index = \dfrac{Object\ Weight\ (L)}{RWL} = -- = \square$

FIGURE 22.6
Hopper example. Job analysis worksheet.

Key Information to Remember on Ergonomics

1. The recommended weight limit (RWL) is defined for a specific set of task conditions as the weight of the load that nearly all healthy workers could perform over a substantial period (up to 8 h) without an increased risk of developing lifting-related low back pain or injury.

2. The lifting index is a term that provides a relative estimate of the level of physical stress associated with a particular manual lifting task. It is a ratio between the weight of the object and the RWL.

3. Horizontal location is the distance of the hands away from the midpoint between the ankles (measured in inches or centimeters).

4. The vertical location is the distance of the hands above the floor.

5. The vertical travel distance is the absolute value of the difference between the vertical heights at the destination and origin of the lift.

6. Asymmetry angle is the angular measure of how far the object is displaced from the front (midsagittal plane) of the workers' body at the beginning or ending of a lift (measured in degrees).

7. Lifting frequency is the average number of lifts in a 15-min period.

8. Lifting duration is classified as short, moderate, or long duration.

9. Coupling classifications are good, fair, or poor.

10. Significant control is defined as a condition requiring precision placement of the load at the destination of the lift.

References

1. Karwowski, W. and Marras, W. S., 1999. *Occupational Ergonomics Handbook*, CRC Press, Boca Raton, FL, p. 4.
2. U.S. Department of Health and Human Services, 1994. NIOSH Revised Lifting Equation. National Institute for Occupational Safety and Health (NIOSH), Cincinnati, OH.

23

Construction Safety

Although construction safety is a specialized area of safety, every safety professional should have more than a general knowledge of the requirements. As a general safety practitioner, you will eventually become involved in some facet of construction safety. Construction safety regulations are outlined in 29 CFR 1926. Construction sites often consist of multiple employers, who are primarily concerned with the safety and health of their employees. However, many of their activities involve other contractors working in the area. Therefore, it is necessary to have a person responsible for the site safety and health program.

Over the past 10 years, the safety and health of the construction industry has significantly improved. This improvement can be attributed to a number of issues, including increased training focus of employees/employers, greater knowledge of safety professionals, increased oversight by insurance companies, and focused inspections by the Occupational Safety and Health Administration (OSHA). Table 23.1 shows the 2008 statistics for fatal occupational industries compared with all industries:

The construction fatalities by event are shown in Table 23.2.

As you can see from Tables 23.1 and 23.2, construction fatalities account for nearly 20% of all work-related fatalities.[1]

Excavation

An excavation is defined as any man-made cut, cavity, trench, or depression in the earth's surface. A trench is considered an excavation. Employees working on or near construction sites must be protected from cave-in when the excavation is 4 ft or more in depth. Cave-in protection is not required when (a) excavations are made entirely of stable rock or (b) they are less than 4 ft in depth and examination of the ground by a competent person provides no indication of a potential cave-in.

The hazards associated with an excavation are numerous and include weight (nearby vehicles/equipment), vibration, underground utilities, water, and soil erosion. The "competent person" must take all of these factors into consideration when evaluating an excavation. In addition, the hazards and

TABLE 23.1

Bureau of Labor and Statistics Fatalities for 2008

	All Industries	Construction Industry	Percentage of Construction Industry (%)
Total fatalities	5071	1005	19.8
Contact with objects and equipment	923	209	22.6
Falls	680	333	48.9
Exposure to harmful substances	432	133	30.8
Transportation-related fatalities	2053	261	12.7
Fires and explosions	173	26	15.0
Assaults and violent acts	794	41	5.2

TABLE 23.2

Bureau of Labor and Statistics Construction Fatalities for 2008

Event	No. of Fatalities	Percentage of Construction Fatalities (%)
Total fatalities	1005	100
Contact with objects and equipment	209	20.8
Falls	333	33.1
Exposure to harmful substances	133	13.2
Transportation-related fatalities	261	25.9
Fires and explosions	26	2.6
Assaults and violent acts	41	4.1

conditions at an excavation site change frequently and should be continuously reevaluated.

The spoils pile (materials removed from the excavation) and the equipment being too close to the vertical walls of the trench are called "surcharge loads," which increase the likelihood of collapse. Additionally, equipment vibration, adverse weather conditions, and groundwater can change the condition and classification of soils.

A competent person is someone who:

- Can identify existing or predictable hazards in excavations
- Has authority to take corrective actions as necessary
- Is familiar with the OSHA regulations and standards for excavations
- Is knowledgeable in soil analysis and classification, as well as the erection, use, and precautions for the protective system on site

A Registered Professional Engineer (RPE) is a person registered as an engineer in the state where the excavation is being performed.

The procedures for safe trench work include the following:

- Identify the knowledgeable competent person.
- Check and verify above- and below-ground utility locations, any adjacent structures or surface encumbrances, and the water table.
- Determine the soil classification through testing.
- Verify protective system installation and setup.
- Provide safe access.
- Comply with requirements specified in the "Excavation, Trenching, and Shoring" regulations.
- Conduct daily inspections prior to the start of work, after any weather event and as needed.

NOTE: Each city, state, or municipality has a toll-free number to call regarding the location of underground utilities.

Soil Classifications

It is crucial to correctly classify soil types before selecting and using a protective system. Soil types are classified as Type A, B, or C. In order to classify soil, at least one visual test and one manual test are required. If soil is classified as Type C, then no testing is required. Table 23.3 shows the characteristics of each soil type.

TABLE 23.3

Soil Classification Characteristics

Soil Type	Soil Characteristics
A	Good cohesive soil with a high compressive strength such as clay, silty clay, sandy clay, clay loam, and cemented soils, such as caliche, duricrust, and hardpan. • Fine grained • Does not crumble • Hard to break up when dry
B	Cohesive soil with a moderate compressive strength such as silt, silty clay, sandy clay, sandy loam, angular gravel (similar to crushed rock), any previously disturbed fissure or soil subject to vibration. • Granular–coarse grains • Little or no clay content • Crumbles easily when dry
C	Cohesive soil with a low compressive strength, such as granular soils including gravel, sand, and loamy sand or submerged soil or rock that is not stable or soil from which water is freely seeping. • Granular soil–very coarse • Minimal cohesion

Protective Systems

The choices of protective systems include sloping and benching, shoring, and shield systems. Protective systems are required under the following conditions:

- Under 4 ft deep when there is a potential for cave-ins.
- Four to 20 ft deep.
  - Sloping or benching (benching is not an option in Type C soils).
  - Shield or shoring systems.
- Over 20 ft deep—protective system must be designed by an RPE or approved in the manufacturer's tabulated data.

NOTE: Protective system is not required for stable rock.

Sloping

Sloping is the process of removing soil to eliminate the chance of cave-in. The required maximum allowable slope is determined by the soil classification. The maximum allowable slope for each soil type is listed in Table 23.4.

An example of sloping in Type C soil is shown in Figure 23.1.

Shoring

Shoring is one of the most commonly used methods of worker protection in excavations. It is lightweight and easy to install. The manufacturer provides tabulated data with the shoring that provides the limitations, precautions, required spacing, and proper use.

Shields (Trench Boxes)

Shields are manufactured by a number of companies and are designed to protect workers working within the confines of the shields. Check the tabulated

TABLE 23.4

Maximum Allowable Slope

Soil Type	H:V Ratio	Angle (°)
Stable rock	Vertical	90
A	3/4:1	53
B	1:1	45
C	1 1/2:1	34

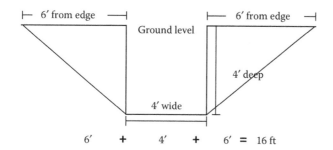

FIGURE 23.1
Example of slope required for Type C soil.

data for the maximum allowable depth it can be used. The tabulated data must accompany the shield when it is being used. Additionally, the shield must be designed by an RPE, be in good condition, and be used properly.

When shields that don't go all the way to the bottom of the trench are used, the excavation should be no deeper than 2 ft from the bottom of the shield. Furthermore, the system must be designed to resist forces calculated for the full depth of the trench and there are no indications while the trench is open of a possible loss of soil from behind or below the bottom of the support system.

RPE-Designed Protective Systems

An RPE can design a protective system for use on a specific project. The RPE should consider the soil type and conditions, as well as other concerns that might exist at the excavation site. The system must be designed by the RPE. The competent person must still perform daily inspections to verify site conditions and any changes that have occurred.

Safe Entry and Exit

A trench that is 4 ft or more in depth must have a safe means for workers to get in and out of the trench. A means of egress is required to be within 25 ft of lateral travel. The most common method for access is a straight ladder or an extension ladder. If a ladder is used, it must extend a minimum of 3 ft above the landing. The use of step ladders is not permitted. Other means of access could be a stairway, ramp, or other means as designed by an RPE. Locating the method of access/egress outside the protective system is strictly prohibited. Buckets of excavators, backhoes, and so on, are not to be used as a means of egress.

Most Commonly Cited Trenching Violations

Some of the most commonly cited trenching violations include the following:

- No cave-in protective system being used when required by soil classification, depth, or code
- Excessive surcharge load (spoils pile too close to the excavation wall or equipment traffic too close to excavation operations)
- No access/egress for excavations 4 ft or more in depth
- No competent person on-site

Electrical Safety in Construction

Approximately five workers in construction are electrocuted each week. Electricity causes 12% of young worker work-related deaths each year. It takes very little electricity to cause harm. Electricity as low as 50 mA can cause death. In addition, where electricity is present, there is a significant risk of causing fires.[2]

Electricity is the flow of energy from one location to another. It requires a source of energy, usually a generating station. The flow of electrons (current) travels through a conductor and travels in a closed circuit.

Electrical Definitions

The following definitions are provided for your use throughout this chapter:

Current: electrical movement (measured in amperes)

Circuit: complete path of the current; includes electricity source, a conductor, and the output device or load (such as a lamp, tool, or heater)

Resistance: restriction to electrical flow

Conductor: substances, such as metals, with little resistance to electricity that allow electricity to flow

Grounding: a conductive connection to the earth that acts as a protective measure

Insulator: substances with high resistance to electricity (like glass, porcelain, plastic, and dry wood) that prevent electricity from getting to unwanted places

Electrical Injuries

There are four main types of electrical injuries, categorized as either direct or indirect injury. Each of these is listed below:

- Direct
    - Electrocution or death because of electrical shock
    - Electrical shock
    - Burns
- Indirect
    - Falls

Electrical Shock

An electrical shock is received when electrical current passes through the body. A person will get an electrical shock if a part of his or her body completes an electrical circuit by either touching a live wire and an electrical ground or touching a live wire and another wire at a different voltage. The severity of the shock depends on (a) the path of the current through the body, (b) amount of current flowing through the body (amperes), and (c) the duration of the shocking current through the body.

NOTE: Low voltage does not mean low hazard.

Currents above 10 mA can paralyze or freeze the muscles. Currents more than 75 mA can cause a rapid, ineffective heartbeat and death will occur in a few minutes unless a defibrillator is used. Electrical current as low as 75 mA can cause death. A milliampere (mA) is equal to 1/1000 A.

Burns

Burns are the most common shock-related injury. Burns occur when you touch electrical wiring or equipment that is improperly used or maintained. They typically occur on the hands. Burns are very serious injuries that require immediate medical attention.

Falls

Electric shock can also cause indirect injuries. Workers working in elevated locations who experience a shock may fall, resulting in serious injury or death.

Controlling Electrical Hazards

Electrical accidents are caused by a combination of three factors:

- Unsafe equipment or installation
- Workplaces made unsafe by the environment
- Unsafe work practices

Exposed Electrical Parts

Exposed electrical parts present the most common type of hazard on a construction site. To prevent injuries from exposed electrical parts, you must isolate the parts by the use of guards or barriers and replace missing covers. Conductors going into electrical cabinets, boxes, or fittings must be protected and unused openings must be closed. Junction boxes, pull boxes, and fittings must have approved covers. Unused opening in cabinets, boxes, and fitting must be closed. There must be no missing knockouts.

Overhead Power Lines

Overhead power lines are not usually insulated. Some examples of equipment that come into contact with power lines include cranes, ladders, scaffolding, backhoes, scissor lifts, raised dump truck beds, and aluminum paint rollers. To control the hazards presented by overhead power lines:

- A worker should stay at least 10 ft away.
- A worker can post warning signs.
- A worker should assume that lines are energized.
- A worker should use wood or fiberglass ladders, not metal.
- Power line workers need special training and personal protective equipment.

Inadequate Wiring

Inadequate wiring exists whenever the wiring is too small for the amount of current. For example, a portable tool used with an extension cord that has a wire too small for the tool presents a significant hazard. The tool will draw more current than the cord can handle, causing overheating and a possible fire without tripping the circuit breaker. The circuit breaker could be the right size for the circuit but not for the smaller extension cord. To control the hazard of inadequate wiring, use the correct sized wire. The wire used depends on operation, building materials, electrical load, and environmental factors. Used fixed cords rather than flexible cords and use the correct extension cord, which must be 3-wire type and designed for hard or extra-hard use.

Defective or Damaged Cords and Wires

All too often on construction sites, cords become damaged through aging, door or window edges, staples or fastenings, abrasion from adjacent materials, or activity in the area. Improper use of these cords can cause shocks, burns, or fires. To control the hazards associated with damaged or defective cords or wires, use one or more of the following control methods:

- Insulate live wires.
- Inspect all cords and wires before each use.
- Use only cords that are 3-wire type.
- Use only cords marked for hard or extra-hard usage.
- Use only cords, connection devices, and fitting equipped with strain relief.
- Remove cords by pulling on the plugs, not the cord.
- Cords not marked for hard or extra-hard use, or which have been modified or damaged, must be taken out of service immediately.

Use of Flexible Cords

Do not use flexible wiring where frequent inspection would be difficult or where damage would like occur. Flexible cords must not be

- Run through holes in walls, ceilings, or floors
- Run through doorways, windows, or similar openings unless physically protected
- Hidden in walls, ceilings, floors, conduit, or other raceways.

Grounding

Grounding creates a low-resistance path from a tool to the earth to disperse unwanted current. When a short or lightning occurs, energy flows to the ground, protecting you from electrical shock or death. Tools plugged into improperly grounded circuits may become energized. Broken wires or plugs on an extension cord can create an improper grounding situation. Improper grounding is one of the most frequently cited violations of the OSHA standard (Table 23.5).

To properly control improper grounding, use one or more of the following control methods:

TABLE 23.5

Minimum Specification for Bonding and Grounding

Bonding wires	4.1 mm (#6 AWG) copper wire
Bonding connectors	12.7 mm (1/2″) copper bonding straps, solder lugs, pressure connectors, or bolts and star washer
Bonding conductors	19 mm (3/4″) copper braid with pressure-type clamps
Grounding wires	5.2 mm (#4 AWG) copper wire
Grounding connectors	At the equipment: approved connectors securely bolted to the machine. Solder lugs are acceptable, but pressure connectors are preferable At the earth: an approved ground clamp attached to a ground electrode

- Ground power supply systems, electrical circuits, and electrical equipment.
- Frequently inspect electrical systems to ensure path to ground is continuous.
- Inspect electrical equipment before each use.
- Do not remove ground prongs from tools or extension cords.
- Ground exposed metal parts of equipment.

Ground Fault Circuit Interrupters

Ground fault circuit interrupters (GFCIs) protect from electrical shock by detecting the difference in current between the black and white wires. If ground fault is detected, the GFCI shuts off electricity in 1/40th of a second. Use GFCIs on all 120-V, single-phase, 15- and 20-A receptacles, or have an assured equipment grounding conductor program.

Assured Equipment Grounding Conductor Program

An Assured Equipment Grounding Conductor Program must cover

- All cord sets
- Receptacles not part of a building or structure
- Equipment connected by plug and cord

The program requirements include the following:

- Specific procedures adopted by the employer
- Competent person to implement the program
- Visual inspection for damage of equipment connected by cord and plug

Overloaded Circuits

Another frequent occurrence on construction sites is overloaded circuits. The hazards from overloaded circuits can result from too many devices plugged into a circuit, causing heated wires and possibly fires. Hazards can also result from damaged tools overheating, lack of overcurrent protection, and wire insulation melting, which may cause arcing and a fire in the area where the overload exists, even inside a wall. Overloaded circuit hazards can be controlled through the use of electrical protective devices that automatically opens the circuit if excess current from overload or ground fault is detected, shutting off the electricity. Electrical protective devices include GFCIs, fuses, and circuit breakers. Fuses and circuit breakers are overcurrent devices. When too much current is present, the fuses melt and the circuit breakers trip open (opening the circuit, which prevents the flow of current).

Safety-Related Work Practices

To protect workers from electrical shock, the following work practices should be enforced at all times:

- Use barriers and guards to prevent passage through areas of exposed energized equipment.
- Preplan work and post hazard warnings.
- Keep working spaces and walkways clear of cords.
- Use special insulated tools when working on fuses with energized terminals.
- Do not use worn or frayed cords and cables.
- Do not fasten extension cords with staples, hang from nails, or suspend by wire.

Planning

One of the first steps in preventing electrical injuries and accidents is to plan your work. Let others that you are working with or around be aware of your proposed activities. Plan to avoid falls that may result from your work. When working around electrical equipment, strictly follow the lockout/tagout program. Insure that you or others working around electricity remove all jewelry and avoid wet conditions and overhead power lines.

Training

All employees working with electric equipment should be trained in safe work practices, which include the following:

TABLE 23.6

Electrical Hazards and Control Measures

Hazard	Control Measures
• Inadequate wiring	• Proper grounding
• Exposed electrical parts	• Use GFCIs
• Ungrounded electrical systems and tools	• Use fuses and circuit breakers
• Overloaded circuits	• Guard live parts
• Damaged power tools and equipment	• Lockout/tagout
• Using the wrong PPE and tools	• Proper use of flexible cords
• Overhead power lines	• Close electric panels
• All hazards are made worse in wet conditions	• Training
	• Planning

- Deenergizing electric equipment prior to inspecting or repairing
- Using cords, cables, and electrical tools that are in good repair
- Lockout/tagout recognition and procedures
- Using appropriate personal protective equipment

Summary

The hazards and control measures associated with electricity are summarized in Table 23.6.

Static Electricity

One form of electricity can be a major problem on construction sites, if not addressed. Static electricity should be addressed around fuel storage areas, other explosive environments (such as areas with high dust concentrations), and high-pressure vacuuming.

What is static electricity? At its simplest, static electricity is an electrical charge that cannot move. It is created when two objects or materials that have been in contact with each other are separated. When in contact, the charges try to balance each other (free flow of electrons – negatively charged particles). When separated, they are left with either an excess or a shortage of electrons. This causes both objects to become electrically charged. If these charges don't have a path to ground, they are unable to move and become "static." If static electricity is not rapidly eliminated, the charge will build up. It will eventually develop enough energy to jump, in the form of a spark to some nearby grounded or less highly charged object in an attempt to balance the charge.

Sources of static electricity include the following:

- Liquid flows through a pipe or hose
- Spraying or coating
- Blending and mixing
- Filling drums, cans, or pails
- Dry powdered material passing through chutes or hoses
- Nonconductive conveyor belts or drive belts

The main hazard from static electricity is the creation of sparks in an explosive or flammable atmosphere. The danger is greatest when flammable liquids are being poured or transferred. Static electricity can also build up to where injury or even death can result from electrical shock.

There are four main conditions that must be present for static electricity to be a hazard. These conditions are listed as follows:

1. There must be a means for a static charge to develop.
2. Enough energy must build up to cause ignition.
3. There must be a discharge of this energy (a spark).
4. The spark must occur in an ignitable vapor or dust mixture.

Static electricity is controlled through (a) bonding and ground, (b) humidification, (c) static controllers, or (d) additives. Each of these will be discussed individually.

Bonding and Grounding

Bonding and grounding are the most common methods of controlling static electricity. *Bonding* is connecting two or more conductive objects with a conductor, such as a copper wire, to equalize the potential charge between them. It should be noted that bonding does not eliminate the static charge. *Grounding* is connecting one or more conductive objects directly to the earth using ground rods, cold water pipes, or building steel. Unlike bonding, grounding drains the static charges away as quickly as they are produced.

Humidification

A high relative humidity is no guarantee against the accumulation of static electricity. Therefore, it should not be relied upon solely as a control measure in areas where there are flammable liquids, gases, or dusts.

Static Controllers (Collectors)

Devices that collect static electricity can be used on moving belts, plastic film, and similar nonconductive materials. Examples of static collectors include needle-pointed copper combs, spring copper brushes, and metallic tinsel bars. To be effective, collectors must be properly grounded.

Additives

Antistatic additives (as in fuels) can be used to control static electricity. The additive increases the conductivity or lowers the resistance of the liquid. Additives also reduce the time it takes for the static charge to leak through the wall of the container and to the ground.

Scaffolds

No discussion on construction safety would be complete without discussing scaffolding. The standards and regulations on scaffolding discussed in this chapter can be found in 29 CFR 1926.450–454.[3] These requirements, however, do not apply to crane or derrick suspended platforms, which are covered by 29 CFR 1926.550.[4] Furthermore, the requirements for aerial lifts are located in 29 CFR 1926.453.

A scaffold is an elevated, temporary work platform. There are three basic types of scaffolds, described in the following:

- *Supported Scaffolds*: platforms supported by rigid, load-bearing members, such as poles, legs, frames, and outriggers
- *Suspended Scaffolds*: platforms suspended by ropes or other nonrigid, overhead support
- *Aerial Lifts*: working platforms such as "cherry pickers" or "boom trucks"

Hazards associated with scaffolds include

- Falls from elevations that are caused by slipping, unsafe access, and the lack of fall protection
- Being struck by falling tools or debris
- Scaffold collapse caused by instability or overloading
- Bad planking giving way

Fall Hazards

Falls may occur

- While climbing on or off the scaffold
- Working on unguarded scaffold platforms
- When scaffold platforms or planks fail

In general, workers on a scaffold who can potentially fall more than 10 ft must be protected by guardrails or personal fall arrest systems (PFASs). The specifics of fall prevention are discussed in the Fall Protection and Prevention in Construction section.

Falling Object (Struck by) Protection

Workers below a scaffold have too often been struck by objects such as tools or materials that fall from the scaffold. To mitigate the potential of falling objects striking workers below the scaffold, one or all of the methods listed below are recommended.

- All employees or visitors to the work site should be required to wear hard hats.
- The area beneath the scaffold should be barricaded to prohibit entry into the area.
- Use panels or screens, if material is stacked higher than the toe board.
- Build a canopy or erect a net below the scaffold that will contain or deflect falling objects.

Elements of Safe Scaffold Construction

When scaffolding is required, it is to be designed and supervised by a *competent person*. The competent person shall ensure that appropriate scaffold construction methods are used and the scaffold has proper access. A competent person is considered as one that is capable of identifying and promptly correcting hazards. The competent person determines if it is safe to work on a scaffold during storms or high winds. In addition, the competent person selects qualified workers to conduct work on scaffolds and trains these workers to recognize hazards. The competent person is responsible for inspecting the scaffold for visible defects before each shift or after alterations. Defective parts must be immediately repaired prior to using the scaffold.

The platform of the scaffold must be fully planked or decked with no more than 1 in. gap. In addition, the platform must be at least 18 in. wide and be capable of supporting its weight and four times the maximum load. The height of the scaffold should not be more than four times its minimum base dimension, unless guys, ties, or braces are used. Additional requirements of scaffold construction include the following:

- No large gaps in front edge of the platform.
- Each abutted end of plank must rest on a separate support surface.
- Overlap platforms at least 12 in. over supports, unless restrained to prevent movement.
- No paint on wood platforms.
- Fully planked between front upright and guardrail support.
- Component pieces used must match and be of the same type.
- Erect on stable and level ground.
- Lock wheels and braces.
- Each end of a platform, unless cleated or otherwise restrained by hooks, must extend 6 in. over its support.

Supported scaffold platforms should be restrained from tipping by guys, ties, or braces. Scaffold poles, legs, posts, frames, and uprights must be on base plates and mud sill or other firm foundation. Scaffolds can only be erected, moved, dismantled, or altered under the supervision of a competent person. The competent person selects and directs trained workers and determines the feasibility of a fall.

Suspension scaffolds are platforms suspended by ropes or wires. The rope or wire must be capable of supporting six times the load. Employees must be trained to recognize hazards. The platform must be secured in order to prevent swaying. The support devices must rest on surfaces that can support four times the load. The competent person (a) must evaluate connections to ensure that the supporting surfaces can support the loads and (b) should inspect ropes and wires for defects before each shift. PFASs must have anchors that are independent of the scaffold support system.

Employees can't be on a moving scaffold unless

- Surface is level.
- Height to base ratio is 2:1.
- Outriggers are installed on both sides of the scaffold.
- Competent person is on site to supervise the movement.
- Employees are on the scaffold portion inside the wheels.

Training Requirements

All persons including the competent person, scaffold erectors, and scaffold users are to be trained on scaffold hazards and procedures to control the hazards. The training must include

- Nature of electrical, fall, and falling objects hazards
- How to deal with electrical hazards and fall protections systems
- Proper use of the scaffold
- Scaffold load capacities

Scaffold erectors and competent persons should receive additional training in erecting, disassembling, moving, operating, repairing, maintaining, and inspecting a scaffold.

Fall Protection and Prevention in Construction

As stated in the introduction to this chapter, falls account for approximately 33% of all construction-related fatalities. Therefore, it is incumbent upon the safety professional to have an extensive knowledge of fall protection and prevention. The OSHA regulations covering fall protection is provided in 29 CFR 1926 (Subpart M)[5] (discussed in detail in Chapter 2). To further illustrate the significance of preventing falls, the following statistics are provided:

- Forty percent of the fatalities were from heights over 40 ft.
- Most falls occur on scaffolds or roofs.
- Twenty-five percent of fall fatalities were from heights of 11–20 ft.
- Twenty-five percent of fall fatalities were from heights of 20–30 ft.

Physics of a Fall

On average, it takes most people approximately 1/3 of a second to become aware. It takes another 1/3 of a second for the body to react to what is happening. Consider that the body can fall 7 ft in 2/3 (or 0.67) s or 64 ft in 2 s. By the time a person becomes aware and begins to react, they have already fallen 7 ft. Given these numbers and the likelihood of death or serious injury, one can determine that the potential for falls in the construction industry requires the attention of safety professionals.

Fall Prevention and Protection

When discussing falls, it is incumbent upon the safety professional to understand the difference in fall prevention and protection philosophies. The two philosophies are *fall prevention* and *fall protection*. Fall prevention is the process, equipment, and methods used to actually prevent a fall from occurring, whereas fall protection is the process, equipment, and methods used to protect a worker once a fall has occurred. These philosophies are illustrated in Table 23.7.

Each of these will be discussed individually.

Guardrails (29 CFR 1926.502)

Guardrails are the most commonly used method to prevent a fall from occurring. The top edge height of top rails, or equivalent guardrail system members, shall be 42 ± 3 in. above the walking/working surface. Under certain conditions, when warranted, the top edge height may exceed 45 in. However, the guardrail system must still comply with all other requirements of 29 CFR 1926.502.

NOTE: When employees are using stilts, the top edge of the rail must be increased to an amount equal to the height of the stilts.

Midrails, screens, mesh, intermediate vertical members, or equivalent intermediate structural members shall be installed between the top edge of the guardrail system and the walking/working surface when there is no wall or parapet wall at least 21 in. high. Midrails, when used, shall extend from the top rail to the walking/working level and along the entire opening between top rail supports. Intermediate members, such as balusters, when used between posts, shall be not more than 19 in. apart. Other structural members (such as additional midrails and architectural panels) shall be installed such that there are no openings in the guardrail system that are greater than 19 in. wide.

Guardrail systems shall be capable of withstanding, without failure, a force of at least 200 lb applied within 2 in. of the top edge, in any

TABLE 23.7

Fall Prevention and Fall Protection Systems

Philosophy	Process, Equipment, or Methods
Fall prevention	• Guardrails • Restraint/positioning • Warning lines • Controlled access zones • Safety monitors
Fall protection	• Fall arrest systems • Safety nets

outward or downward direction, at any point along the top edge. When the 200-lb test load is applied in a downward direction, the top edge of the guardrail shall not deflect to a height less than 39 in. above the walking/working level. Top rails and midrails shall be at least 1/4 in. nominal diameter or thickness to prevent cuts and lacerations. If wire rope is used for top rails, it shall be flagged at not more than 6-ft intervals with high-visibility material. When guardrail systems are used at hoisting areas, a chain, gate, or removable guardrail section shall be placed across the access opening between guardrail sections when hoisting operations are not taking place.

Positioning Device System (29 CFR 1926.502(e))

Restraint Systems

A restraint system prevents a worker from being exposed to any fall. If the employee is protected by a restraint system, either a body belt or a harness may be used. When a restraint system is used for fall protection from an aerial lift or a boom-type elevating work platform, the employer must ensure that the lanyard and anchor are arranged so that the employee is not potentially exposed to falling *any* distance.

Positioning Devices: Construction Work

The only time a body belt may be used where there may be a fall is when an employee is using a "positioning device." In 29 CFR 1926.500 of the construction standards for fall protection, a "positioning device system" is defined as a body belt or body harness system rigged to allow an employee to be supported on an elevated vertical surface, such as a wall (or a pole), and work with both hands free while leaning. Therefore, in construction work, a positioning device may be used only to protect a worker on a *vertical* work surface. These devices may permit a fall of up to 2 ft. They may be used in concrete form work, installation of reinforcing steel, and certain telecommunications work. Since construction workers in bucket trucks, scissor lifts, and boom-type elevating work platforms are on a *horizontal* surface, a positioning device may not be used for those workers (OSHA Standard Interpretation—Dated August 14, 2000).

Positioning device systems shall be rigged such that an employee cannot free fall more than 2 ft and shall be secured to an anchorage capable of support at least twice the potential impact load of an employee's fall or 3000 lb, whichever is greater. Connectors shall be drop forged, pressed, or formed steel or made of equivalent materials. Connectors shall have a corrosion-resistant finish, and all surfaces and edges shall be smooth to prevent damage to interfacing parts of this system. Connecting assemblies shall have a minimum tensile strength of 5000 lb. Dee-rings and snap hooks shall be proof-tested to a minimum tensile strength of 3600 lb without cracking, breaking, or taking permanent deformation. Unless the snap hook is a locking type and designed for the following connections, snap hooks shall not be engaged

- Directly to webbing, rope, or wire rope
- To each other
- To a Dee-ring to which another snap hook or other connector is attached
- To a horizontal lifeline
- To any object that is incompatibly shaped or dimensioned in relation to the snap hook such that unintentional disengagement could occur by the connected object being able to depress the snap hook keeper and release itself

Positioning device systems shall be inspected prior to each use for wear, damage, and other deterioration, and defective components shall be removed from service. Body belts, harnesses, and components shall be used only for employee protection and not to hoist materials.

Warning Line Systems (29 CFR 1926.502(f))

A warning line system means a barrier erected on a roof to warn employees that they are approaching an unprotected roof side or edge and which designates an area in which roofing work may take place without the use of guardrail, body belt, or safety net systems to protect employees in the area. When warning line systems are used, they shall be erected around all sides of the roof work area. When mechanical equipment is not being used, the warning line shall be erected not less than 6 ft from the roof edge. When mechanical equipment is being used, the warning line shall be erected not less than 6 ft from the roof edge that is parallel to the direction of mechanical equipment operation and not less than 10 ft from the roof edge that is perpendicular to the direction of mechanical equipment operation.

Points of access, materials handling areas, storage areas, and hoisting areas shall be connected to the work area by an access path formed by two warning lines. When the path to a point of access is not in use, a rope, wire, chain, or other barricade, equivalent in strength and height to the warning line, shall be placed across the path at the point where the path intersects the warning line erected around the work area, or the path shall be offset such that a person cannot walk directly into the work area. Warning lines shall consist of ropes, wires, or chains and supporting stanchions erected as follows:

- The rope, wire, or chain shall be flagged at not more than 6-ft intervals with high-visibility material.

- The rope, wire, or chain shall be rigged and supported in such a way that its lowest point (including sag) is no less than 34 in. from the walking/working surface and its highest point is no more than 39 in. from the walking/working surface.
- After being erected, with the rope, wire, or chain attached, stanchions shall be capable of resisting, without tipping over, a force of at least 16 lb applied horizontally against the stanchion, 30 in. above the walking/working surface, perpendicular to the warning line, and in the direction of the floor.
- The rope, wire, or chain shall have a minimum tensile strength of 500 lb and, after being attached to the stanchions, shall be capable of supporting, without breaking, the loads applied to the stanchions.
- The line shall be attached at each stanchion in such a way that pulling on one section of the line between stanchions will not result in slack being taken up in adjacent sections before the stanchion tips over.

No employee shall be allowed in the area between a roof edge and a warning line unless the employee is performing roofing work in that area. Mechanical equipment on roofs shall be used or stored only in areas where employees are protected by a warning line system, guardrail system, or PFAS.

Controlled Access Zone System (29 CFR 1926.502(g))

Controlled access zone means an area in which certain work (e.g., overhand bricklaying) may take place without the use of guardrail systems, PFASs, or safety net systems and access to the zone is controlled. When used to control access to areas where leading edge and other operations are taking place, the controlled access zone shall be defined by a control line or by any other means that restricts access and shall be erected not less than 6 ft or not more than 25 ft from the unprotected or leading edge, except when erecting precast concrete members. When erecting precast concrete members, the control line shall be erected not less than 6 ft nor more than 60 ft or half the length of the member being erected, whichever is less, from the leading edge. The control line shall extend along the entire length of the unprotected or leading edge and shall be approximately parallel to the unprotected or leading edge. The control line shall be connected on each side to a guardrail system or wall. When used to control access to areas where overhand bricklaying and related work are taking place, the controlled access zone shall be defined by a control line erected not less than 10 ft nor more than 15 ft from the working edge. On floors and roofs where guardrail systems are in place, but need to be removed to allow overhand bricklaying work or leading edge work to take place, only that portion of the guardrail necessary to accomplish that day's work shall be removed.

Safety Monitoring System (29 CFR 1926.502(h))

A safety monitoring system consists of a competent person capable of recognizing fall hazards and who monitors the safety of other employees from the same walking/working surface. The safety monitor must be within visual sighting distance of the employee being monitored and close enough to communicate orally. Safety monitors should have no other responsibilities that could take their attention away from monitoring. The safety monitor is responsible for warning the employee when it appears that the employee is unaware of a fall hazard or is acting unsafely.

Personal Fall Arrest System (29 CFR 1926.502(d))

A PFAS is a system used to arrest an employee in a fall from a working level. It consists of an anchorage, connectors, and body harness and may include a lanyard, deceleration device, lifeline, or suitable combinations of these. PFAS connectors shall meet the same requirements described in the Positioning Device System section. There are various types and components of a PFAS, including harnesses, lanyards, beam wraps, carabiners, rope grabs, and so on. The main thing to remember concerning PFAS is that there is no "one size fits all" solution. For example, if a worker is working 12 ft off the ground, a harness and a 6-ft shock absorbing lanyard will not protect an employee from injury. Consider a 6-ft man and a 6-ft lanyard with a 3-ft shock absorber. The total length of the system is 15 ft. Therefore, the employee wearing this system at this height would actually hit the ground. There are numerous and varied fall protection products on the market. Check with the manufacturer to determine the specific uses and limitations before purchasing.

Suspension Trauma Related to Fall PFASs

Suspension trauma, also known as "harness hang" or "orthostatic incompetence," is the medical effect of immobilization in a vertical position. Suspension trauma presents an immediate threat of death to anyone immobilized in a vertical position. All persons working at heights and wearing PFASs should be trained in how to recognize, manage, and prevent suspension trauma. The danger of suspension trauma occurs only when an employee who experiences a fall is unable or unwilling to move.

Signs and symptoms of suspension trauma are as follows:

- General feeling of unease
  - Dizzy, sweaty, and other signs of shock
  - Increased pulse and respiratory rates
- Sudden drop in pulse and blood pressure
- Instant loss of consciousness

- Death, if not rescued, from suffocation owing to a closed airway or from lack of blood flow and oxygen to the brain

The veins in your legs are intertwined within the skeletal muscles, and when you move your legs, these muscles squeeze the veins, pushing the blood out of the way. These veins have one-way valves, so each squeeze can pump the blood a short distance toward the heart. Providing you're moving around, these veins continue to pump blood. However, if you're immobilized, the muscles are not pumping blood upward, and the blood pools in the leg. The brain tries to "shock" the system for a while, but the blood still remains pooled in the legs. Eventually, the brain is deprived of oxygen and the person faints. Under normal circumstances, the person that faints goes from an upright position to a horizontal position and the blood begins to flow back toward the heart and brain. However, someone who has fallen in a body harness stays upright and therefore the brain is not receiving any oxygen supply, the airway becomes restricted, and death can occur after 10 min.

The answer to preventing death resulting from suspension trauma may seem simple—*rescue within 10 min.* Rapid responsible rescue is needed. However, there is an issue known as "reflow syndrome" that must be dealt with. The pooled blood that is trapped in the legs may not be in very good condition and may even kill the person if we let it all pour back into the brain. Reflow syndrome is medically complicated. Once the blood begins to flow, it cannot be controlled and the patient may die. Pooled blood in the legs is "stale" after 10–20 min, since it is drained of oxygen and saturated with carbon dioxide. It is also loaded with toxic wastes. Once suspension trauma has occurred, the blood flow must be trickled back into the system. Anyone released from immobile suspension should be kept in a sitting position for at least 30 min.

The question now arises, "How do we prevent suspension trauma?" If you're in a harness by choice, keep your legs moving as much as you possibly can and take regular rest breaks. If you fall accidentally and are suspended:

- Avoid using your legs. You don't want blood sent there.
- Lift your knees into a sitting position or at least higher than your hips.
- Relax as much as possible. Panic makes things worse.
- If you can, every few minutes swing yourself upside down.

If you're trapped and cannot move:

- Strain your leg muscles as hard as you can every 5 s.
- Breathe slowly and deeply.
- Await rescue.

Safety Net Systems (29 CFR 1926.502(c))

When using a safety net system, they should be installed as close as practicable under the walking/working surface on which employees are working, but in no case more than 30 ft below such level. Safety nets shall extend outward from the outermost projection of the work surface as described in Table 23.8.

Safety nets shall be installed with sufficient clearance under them to prevent contact with the surface or structures below when subjected to an impact force equal to the drop test. Safety nets and their installations shall be capable of absorbing an impact force equal to that produced by the drop test. Safety nets and safety net installations shall be drop tested at the job site after initial installation and before being used as a fall protection system, whenever relocated, after major repair, and at 6-month intervals if left in one place. The drop test shall consist of a 400-lb bag of sand 30 ± 2 in. in diameter dropped into the net from the highest walking/working surface at which employees are exposed to fall hazards, but not from less than 42 in. above that level. Defective nets shall not be used. Safety nets shall be inspected at least once a week for wear, damage, and other deterioration. Defective components shall be removed from service. Safety nets shall also be inspected after any occurrence that could affect the integrity of the safety net system.

Materials, scrap pieces, equipment, and tools that have fallen into the safety net shall be removed as soon as possible from the net and at least before the next work shift. The maximum size of each safety net mesh opening shall not exceed 36 in.[2] nor be longer than 6 in. on any side, and the opening, measured center to center of mesh ropes or webbing, shall not be longer than 6 in. All mesh crossings shall be secured to prevent enlargement of the mesh opening. Each safety net shall have a border rope for webbing with a minimum breaking strength of 5000 lb. Connections between safety net panels shall be as strong as integral net components and shall be spaced not more than 6 in. apart.

TABLE 23.8

Minimum Required Horizontal Distance of Nets from Working Surfaces

Vertical Distance from Working Level to Horizontal Plane of Net	Minimum Required Horizontal Distance of Outer Edge of Net from the Edge of the Working Surface
Up to 5 ft	8 ft
>5 ft up to 10 ft	10 ft
>10 ft	13 ft

Cranes and Derrick Safety (29 CFR 1926.550[4])

Cranes and derricks represent a significant hazard on any construction site. The specifics of cranes and derricks are voluminous. Therefore, this section is intended to provide the general requirements and practices for the majority of cranes and derricks that may be used on any construction site.

First, the operator of any crane shall comply with the manufacturer's specifications and limitations. Rated load capacities, recommended operating speeds, and special hazard warning or instructions shall be conspicuously posted on all equipment. Instructions or warnings shall be visible to the operator while he is at his or her control station. Hand signals to crane and derrick operators shall be those prescribed by the applicable ANSI standard for the type of crane in use. An illustration of the signals shall be posted at the job site. The employer shall designate a competent person who shall inspect all machinery and equipment prior to each use, and during use, to make sure it is in safe operating condition. Any deficiencies shall be repaired, or defective parts replaced, before continued use. A thorough annual inspection of the hoisting machinery shall be made by a competent person or by a government or private agency recognized by the US Department of Labor. The employer shall maintain a record of the dates and results of inspections for each hoisting machine and piece of equipment.

Wire rope shall be taken out of service when any of the following conditions exist:

- In running ropes, six randomly distributed broken wires in one lay or three broken wires in one strand in one lay
- Wear of one-third the original diameter of outside individual wires. Kinking, crushing, bird caging, or any other damage resulting in distortion of the rope structure
- Evidence of any heat damage from any cause
- Reductions from nominal diameter of more than 1/64 in. for diameters up to and including 5/16 in., 1/32 in. for diameters 3/8 in. to and including 1/2 in., 3/64 in. for diameters 9/16 in. to and including 3/4 in., 1/16 in. for diameters 7/8 in. to 1 1/8 in. inclusive, and 3/32 in. for diameters 1 1/4 to 1 1/2 in. inclusive
- In standing ropes, more than two broken wires in one lay in sections beyond end connections or more than one broken wire at an end connection

TABLE 23.9

Minimum Clearance for Power Lines (Rating)

Line Rating	Minimum Clearance
50 kV or below	10 ft
>50 kV	10 ft plus 0.4 in. for each kilovolt over 50 kV

Belts, gears, shafts, pulleys, sprockets, spindles, drums, fly wheels, chains, or other reciprocating, rotating, or other moving parts or equipment shall be guarded if such parts are exposed to contact by employees, or otherwise create a hazard. Accessible areas within the swing radius of the rear of the rotating superstructure of the crane, either permanently or temporarily mounted, shall be barricaded in such a manner as to prevent an employee from being struck or crushed by the crane. All exhaust pipes shall be guarded or insulated in areas where contact by employees is possible in the performance of normal duties.

Fuel tank filler pipes shall be located in such a position or protected in such a manner as to not allow spill or overflow to run onto the engine, exhaust, or electrical equipment of any machine being fueled. An accessible fire extinguisher of 5BC rating, or higher, shall be available at all operator stations or cabs of equipment.

Except where electrical distribution and transmission lines have been deenergized and visibly grounded at point of work or where insulating barriers, not a part of or an attachment to the equipment or machinery, have been erected to prevent physical contact with the lines, equipment or machines shall be operated proximate to power lines only as listed in Table 23.9.

In transit with no load and boom lowered, the equipment clearance shall be a minimum of 4 ft for voltages less than 50 kV, and 10 ft for voltages over 50 kV, up to and including 345 kV, and 16 ft for voltages up to and including 750 kV.

Welding, Cutting, and Brazing (29 CFR 1926.350–353[6])

As a safety professional, it is incumbent that you have basic knowledge of welding, cutting, and brazing operations. Hazards associated with welding, cutting, and brazing operations are endless, including, but not limited to

- Health hazards (air contaminants from welding operations [iron oxide, zinc, beryllium, copper, etc.], burns, burns to the eyes from flashes)
- Physical hazards (tripping from hoses)
- Fires and explosions (igniting gases, improper storage)

Therefore, the basics of welding operations will be discussed, with the primary focus being on the OSHA regulations. There are three basic types of welding, which includes (1) oxygen-fuel gas welding and cutting, (2) arc welding and cutting, and (3) resistance welding.

Oxygen-Fuel Gas Welding and Cutting

Oxygen-fuel gas welding is a process that uses fuel gases to weld, by heating metals to a temperature that produces a shared pool of molten metal, which cools into a common metal. Oxygen-fuel gas cutting is a process that uses fuel gases to cut metals, by heating the metal to a temperature that produces molten metal, which is then "blown" away.

The general requirements, for oxygen-fuel welding and cutting, focus on the safe use of acetylene. Therefore, the following standards are provided when transporting, moving, and storing compressed gas cylinders:

- When cylinders are hoisted, they shall be secured in a cradle, slingboard, or pallet. They shall not be hoisted or transported by means of magnets or choker slings.
- Cylinders shall be moved by tilting and rolling them on their bottom edges and not be intentionally dropped, struck, or permitted to strike each other violently.
- When cylinders are transported by powered vehicles, they shall be secured in a vertical position.
- Valve protection caps shall not be used for lifting cylinders from one vertical position to another. Bars shall not be used under valves or valve protection caps to pry cylinders loose when frozen. Warm, not boiling, water shall be used to thaw cylinders loose.
- Unless cylinders are firmly secured on a special carrier intended for this purpose, regulators shall be removed and valve protection caps put in place before cylinders are removed.
- A suitable cylinder truck, chain, or other steadying device shall be used to keep cylinders from being knocked over while in use.
- When work is finished, when cylinders are empty, or when cylinders are moved at any time, the cylinder valve shall be closed.
- Compressed gas cylinders shall be secured in an upright position at all times except, if necessary, for short periods while cylinders are actually being hoisted or carried.
- Oxygen cylinders in storage shall be separated from fuel-gas cylinders or combustible materials (especially oil or grease), a minimum distance of 20 ft or by a noncombustible barrier of at least 5 ft high having a fire-resistance rating of at least 1/2 h.

- Inside of buildings, cylinders shall be stored in a well-protected, well-ventilated, dry location at least 20 ft from highly combustible materials such as oil or excelsior. Cylinders should be stored in definitely assigned places away from elevators, stairs, or gangways. Assigned storage places shall be located where cylinders will not be knocked over or damaged by passing or falling objects, or subject to tampering by unauthorized persons. Cylinders shall not be kept in unventilated enclosures such as lockers or cupboards.

The safe use of fuel gas requires that the employer thoroughly instruct employees in the safe use of fuel gas as follows:

- Before a regulator to a cylinder valve is connected, the valve shall be opened slightly and closed immediately. This action is generally termed "cracking" and is intended to clear the valve of dust or dirt that might otherwise enter the regulator. The person cracking the valve shall stand to one side of the outlet, not in front of it. The valve of a fuel gas cylinder shall not be cracked where the gas would reach welding work, sparks, flame, or other possible sources of ignition.
- The cylinder valve shall always be opened slowly to prevent damage to the regulator. For quick closing, valves on fuel gas cylinders shall not be opened more than 1 1/2 turns.
- Fuel gas shall not be used from cylinder through torches or other devices that are equipped with shutoff valves without reducing the pressure through a suitable regulator attached to the cylinder valve or manifold.
- Before a regulator is removed from a cylinder valve, the cylinder valve shall always be closed and the gas released from the regulator.
- If, when the valve on a fuel gas cylinder is opened, there is a leak found around the valve stem, the valve shall be closed and the gland nut tightened. If this action does not stop the leak, the use of the cylinder should be discontinued, and it shall be properly tagged and removed from the work area.
- If a leak should develop at a fuse plug or other safety device, the cylinder shall be removed from the work area.

Fuel Gas and Oxygen Manifolds (29 CFR 1926.350(e))

Fuel gas and oxygen manifolds shall bear the name of the substance they contain in letters at least 1 in. high, which shall be painted either on the manifold or on a sign permanently attached to it. Fuel gas and oxygen manifolds shall be placed in safe, well-ventilated, and accessible locations. They shall not be located within enclosed spaces. Manifold hose connections, including both ends of the supply hose that lead to the manifold, shall be such that the hose cannot be interchanged between fuel gas and oxygen manifolds and supply header connections. Adapters shall not be used to permit the interchange of hose. Hose connections shall be kept free of grease and oil. When not in use, manifold and header hose connections shall be capped.

Hoses (29 CFR 1926.350(f))

Fuel gas hoses and oxygen hoses shall be easily distinguishable from each other. The contrast may be made by different colors or by surface characteristics readily distinguishable by the sense of touch. Oxygen and fuel gas hoses shall not be interchangeable. A single hose having more than one gas passage shall not be used.

Torches (29 CFR 1926.350(g))

Clogged torch tip openings shall be cleaned with suitable cleaning wires, drills, or other devices designed for such purpose. Torches in use shall be inspected at the beginning of each work shift for leaking shutoff valves, hose couplings, and tip connections. Defective torches shall not be used. Torches shall be lighted by friction lighters or other approved devices, and not by matches of from other hot work.

Regulators and Gauges (29 CFR 1926.350(h))

Oxygen and fuel gas pressure regulators, including their related gauges, shall be in proper working order while in use.

Oil and Grease Hazards (29 CFR 1926.350(i))

Oxygen cylinders and fittings shall be kept away from oil or grease. Cylinders, cylinder caps and valves, couplings, regulators, hose, and apparatus shall be kept free from oil or greasy substances and shall not be handled with oily hands or gloves. Oxygen shall not be directed at oil surfaces and greasy clothes or within a fuel oil or other storage tank or vessel.

Arc Welding and Cutting

Arc welding or cutting is a fusion process wherein the coalescence of the metals is achieved from the heat of an electrode and formed between an element. Three key terms to be familiar with in this type of welding include *application, installation,* and *operation and maintenance.* Application applies to a large and varied group of processes that use an electric arc as the source of heat to melt and join metals. Proper installation of equipment is required for arc welding. A key safety component of the installation process is ensuring that proper grounding is completed. For the operation and maintenance of the arc welding equipment, all connections to the machine shall be checked to make certain that they are properly made.

Manual Electrode Holders (29 CFR 1926.351(a))

Only manual electrode holders that are specifically designed for arc welding and cutting and are of a capacity capable of safely handling the maximum rated current required by the electrodes shall be used. Any current-carrying parts passing through the portion of the holder, which the arc welder or cutter grips in his hand, and the outer surfaces of the jaws of the holder, shall be insulated against the maximum voltage encountered to ground.

Welding Cables and Connectors (29 CFR 1926.351(b))

All arc welding and cutting cables shall be of the completely insulated, flexible type, capable of handling the maximum current requirements of the work in progress, taking into account the duty cycle under which the arc welder or cutter is working. Only cable free from repair or splices for a minimum distance of 10 ft from the cable end to which the electrode holder is connected shall be used, except that cables with standard insulated connectors or with splices whose insulating quality is equal to that of the cable are permitted. Cables in need of repair shall not be used.

Ground Returns and Machine Grounding (29 CFR 1926.351(c))

A ground return cable shall have a safe current-carrying capacity equal to or exceeding the specified maximum output capacity of the arc welding or cutting unit that it services. When a single ground return cable services more than one unit, its safe current-carrying capacity shall equal or exceed the total specified maximum output capacities of all the units that it serves.

Resistance Welding

Resistance welding utilizes pressure and heat that is generated in the pieces to be welded by resistance to an electrical current. When conducting resistance welding, all equipment shall be installed by a qualified electrician in conformance with 29 CFR 1926 Subpart S. If spot and seam welding machines are used, the following precautions must be taken:

- All doors and access panels of all resistance welding machines and control panels shall be kept locked and interlocked to prevent access to live portions of the equipment by unauthorized persons.

- In all press welding operations, if there is a possibility of the operator's fingers being under the point of operation, effective guards must be used.

Fire Prevention (29 CFR 1926.352)

When practical, objects to be welded, cut, or heated shall be moved to a designated safe location or, if the objects to be welded, cut, or heated cannot be readily moved, all movable fire hazards in the vicinity shall be taken to a safe place, or otherwise protected. If the object to be welded, cut, or heated cannot be moved and if all the fire hazards cannot be removed, positive means shall be taken to confine the heat, sparks, and slag and to protect the immovable fire hazards from them. No welding, cutting, or heating shall be done where the application of flammable paints or the presence of other flammable compounds or heavy dust concentrations creates a hazard. Suitable fire extinguishing equipment shall be immediately available in the work area and shall be maintained in a state of readiness for instant use. For most companies, the welding, cutting, and brazing policy or procedure requires the use of a fire watch, which is to be present at all times of the operation, including 30 min after all "hot work" operations have been concluded.

Ventilation and Protection in Welding, Cutting, and Heating (29 CFR 1926.353[6])

Ensuring adequate ventilation in welding, cutting, and heating operations is extremely important since many of the metals being welded require special or specific control measures in order to prevent illnesses or other adverse health effects to the welder. These may include the following:

- Fluorine compounds
- Zinc
- Lead
- Beryllium
- Cadmium
- Mercury
- Cleaning compounds
- Stainless steel

General ventilation shall be of sufficient capacity and so arranged as to produce the number of air changes necessary to maintain welding fumes and smoke within safe limits. Local exhaust ventilation shall consist of freely movable hoods intended to be placed by the welder or burner as close as practicable to the work. This system shall be of sufficient capacity and so arranged as to remove fumes and smoke at the source and keep the concentration of them in the breathing zone within safe limits. Contaminated air exhausted from a working space shall be discharged into the open air or otherwise clear of the source of intake air.

Hand and Power Tool Safety (29 CFR 1910 Subpart P[7])

OSHA requires that each employer be responsible for the safe condition of tools and equipment used by employees, including tools and equipment that may be furnished by employees. Employers shall not issue or permit the use of unsafe hand or power tools. Employees who use hand and power tools and who are exposed to the hazards of falling, flying, abrasive, and splashing objects or exposed to harmful dusts, fumes, mists, vapors, or gases must be provided with the particular personal protective equipment necessary to protect them from the hazard. Employees and employers have a responsibility to work together to establish safe working procedures. If a hazardous situation is encountered, it should be brought to the attention of the proper individual immediately.

Hand tools are nonpowered and include anything from axes to wrenches. The greatest hazards posed by hand tools result from misuse and improper maintenance. For example, dull chisels cause many accidents that could have been avoided. Other hazards associated with the misuse or improper maintenance may include

- Using a screwdriver as a chisel, causing the tip of the screwdriver to break and hit an employee in the eye or cause a puncture wound.
- Wooden handles on a hammers or axes may become loose and fly off and strike a bystander.
- A wrench whose jaws are sprung might slip, creating "busted knuckle."

Hand tools used around flammable substances may create sparks and serve as an ignition source. Where flammable substances are present or present the possibility of being ignited, nonsparking tools should be used. Nonsparking tools are manufactured from brass, bronze, Monel metal, copper aluminum alloys, copper-beryllium, and titanium. However, it must be noted that nonsparking tools have less tensile strength and may be more likely to break and wear more quickly than other tools. Therefore, frequent inspections to ensure proper maintenance are required.

All hazards involved in the use of power tools can be prevented by following the five basic safety rules:

- Keep all tools in good condition with regular maintenance.
- Use the right tool for the job.
- Examine each tool for damage before use.
- Operate all tools according to the manufacturer's instructions.
- Provide and use the proper personal protective equipment.

It would be extremely lengthy to discuss every hand or power tool available. A few of the more common power tools will be discussed in more detail and include abrasive grinders, portable circular saws, and compressed air systems.

Abrasive Grinders (29 CFR 1910.243)

Both portable and fixed grinding wheels are designed to operate at very high speeds. Shattered wheels can travel at several hundreds of miles per hour. Therefore, the potential for serious injury or property damage from the fragments is high. Employees operating grinding wheels must know the potential hazards, and more importantly, they should know the necessary precautions to prevent the occurrence of accidents.

The specific hazards associated with the use of grinding wheels include the following:

- Breathing in dusts from grinded metals.
- Contact with lubricating oils and metallic dusts can irritate the skin.
- Vibration can cause various ergonomic-related injuries, such as "white finger."
- Shattered wheels can cause physical injuries.
- Contact with a wheel in motion can cut, mangle, or even amputate a body part.
- Sparks from grinding metals can ignite nearby flammable materials.

Wheel Testing

Of particular importance to grinding wheels is the presence of cracks that may or may not be visible to the naked eye. There are two basic methods used to check for cracks in a grinding wheel. They are the "ring test" and the "vibration test."

The Ring Test

The ring test evaluates the sound coming from the wheel when lightly tapped with a nonmetallic material. An undamaged wheel sends out a clear ringing tone. The ring test should not be performed on wheels that have a diameter of 10 cm or less, plugs and cones, mounted wheels, segment wheels, or inserted nut and projecting stud disc wheels. To perform the ring test, follow the steps listed below:

- Suspend the wheel from the hole on a small pin or finger. If the wheel is too heavy, rest it on its outer edge on a clean, hard floor.
- Tap the wheel gently with a nonmetallic tool. For light wheels, use the wooden handle of a screwdriver. For heavy wheels, use a wood mallet. The best spot to tap a wheel is approximately 45° from either side of the vertical centerline, approximately 2.5–5 cm (1–2 in.) from the outside edge. Tapping on the centerline, even an undamaged wheel may give off a muffled sound.
- Listen for the sound that comes from the wheel when it is tapped.
- Turn the wheel 45° to the right or left and repeat the test.
- Compare the sounds from the wheel being tested with those from other wheels of the same lot and type.
- Set aside any wheel that is suspect.

The Vibration Test

The vibration test evaluates how dry sand moves on the side of a vibrating wheel. If the wheel is in good condition, the sand will remain evenly spread out over the entire surface of the wheel. Use the vibration test on all bonded wheels. Unlike the ring test, the vibration test can be performed in noisy areas.

To perform the vibration test, follow these steps:

- Set the abrasive wheel on its side on a test fixture.
- Coat the wheel with a thin layer of fine, dry sand.
- Turn on the test fixture to get the wheel to gently vibrate.
- Watch the grains of sand as the wheel vibrates. If the sand moves away from an area of the wheel, this indicates a crack. If the sand remains evenly distributed, the wheel is acceptable.
- Repeat these steps for the other side of the wheel.

Understand that both the ring test and the vibration test are not 100% accurate. Therefore, it is necessary when grinding that all persons stand clear of any grinding operations.

Portable Circular Saws (29 CFR 1910.243)

In the construction industry, probably the most commonly used power tool is the portable circular saw. Unfortunately, it is also one of the most commonly abused tools. Familiarity with its use can cause carelessness on the part of the user. The following are specific safety practices that must be followed when using portable circular saws:

- Always use the appropriate personal protective equipment including safety glasses or goggles with side shields that comply with the current standard. In addition to safety glasses or goggles, a face shield is highly recommended. The use of hearing protection is also required.
- Do not wear loose clothing, jewelry, or dangling objects, including long hair, which may catch on rotating parts or accessories.
- Do not use a circular saw that is too heavy for you to easily control.
- Be sure that the switch actuates properly and never tape the switch or "trigger" down on any power tool.
- Use sharp blades. Dull blades can cause binding, stalling, or kickback.
- Use the correct blade for the application.
- Always inspect the portable circular saw prior to use to ensure that all safety features provided by the manufacturer are installed, including the blade guard. Be sure to inspect the power cord and ensure that it is connected in proper order to a GFCI.
- For maximum control of the circular saw, be sure to operate it with both hands.
- Avoid cutting small pieces that cannot be properly secured and material on which the saw cannot rest.
- ALWAYS ensure that the surface you are cutting on is stable. NEVER hold or balance a piece of material in one hand, while trying to operate the circular saw with the other.

Compressed Air Systems (29 CFR 1910.242)

Pneumatic tools used on a construction site are powered by compressed air and can include chippers, drills, hammers, and sanders. The dangers associated with the use of pneumatic tools and compressed air systems include getting struck by one of the tool's attachments or fastener that the employee is using with the tool, such as nails in a nail gun. Other hazards exist as a result of carelessness or misuse on the part of the operator. For example, one of the more frequent violations is the taping down of the triggers (especially on pneumatic hammers or nailers). NEVER bypass a safety feature and ALWAYS follow the manufacturer's operating specifications.

Using pneumatic tools requires an efficient compressor system to adequately power the tools. Use the following guidelines to properly maintain an efficient compressor system:

- Before making or breaking any air connection, always turn off the air supply. Use the valve to turn off the air and never "kink" the hose as a shortcut.
- Protect the air hose from damage.
- Be sure to use the proper size air hose and fittings to keep air pressure at a maximum throughout the entire line.
- Clear any dirt off the nipple before connecting the air hose to the tool.
- When the tool is connected, check the hose and all connections for leaks or damage before using the tool.
- Maintain a clean, dry, regulated source of air to operate air tools at peak performance.

Whenever there is compressed air used on a work site, employees invariably attempt to utilize the air from the system to "clean" or blow off the area or even on their persons. OSHA states in 29 CFR 1910.242(b), "...Compressed air shall not be used for cleaning except where reduced to less than 30 psi (pounds per square inch)."

Housekeeping (29 CFR 1926.25[8])

One of the most common conditions, and one of the most commonly cited, on a work site is poor general housekeeping. The benefits of good housekeeping include the following:

- Eliminating accident and fire hazards
- Maintaining safe, healthy work conditions
- Saves time, money, materials, space, and effort
- Improved productivity and quality
- Boosts morale of employees
- Reflects a well-run organization

The costs associated with poor housekeeping include the following:

- Slips, trips, and falls
- Fires

- Chemical and machine accidents
- Injuries from electrical problems
- Collisions and falling objects
- Health problems

OSHA's standard on housekeeping requires the following:

- During the course of construction, alteration, or repairs, form and scrap lumber with protruding nails, and all other debris, shall be cleared from work areas, passageways, and stairs, in and around buildings or other structures.
- Combustible scrap and debris shall be removed at regular intervals during the course of construction. Safe means shall be provided to facilitate such removal.
- Containers shall be provided for the collection and separation of waste, trash, oily and used rags, and other refuse. Containers used for garbage and other oily, flammable, or hazardous wastes, such as caustics, acids, harmful dusts, and so on, shall be equipped with covers. Garbage and other waste shall be disposed of at frequent and regular intervals.

Key Information to Remember on Construction Safety

1. Excavations must be protected from cave-in if a trench is more than 4 ft deep.
2. Soil classifications are A, B, and C. Soil Type A is good cohesive soil with a high compressive strength. Soil Type B is cohesive soil with a moderate compressive strength, and Soil Type C is cohesive soil with low compressive strength.
3. Electrical current as low as 75 mA can cause death.
4. Bonding is connecting two or more conductive objects with a conductor.
5. Grounding is connecting one or more conductive objects directly to the earth using ground rods, cold pipes, or building steel.
6. Scaffolding requirements are addressed in 29 CFR 1926.450–454.[3]
7. Anchor points for personal fall arrests should be capable of holding 5000 lb per person.
8. Cranes and derricks must be clear of power lines by at least 10 ft for voltages up to 50 kV or 10 ft plus 0.4 in. for each kilovolt over 50 kV.

9. It is the employer's responsibility to ensure that only well-maintained and operable hand and power tools are utilized on the job site.

10. When using abrasive grinders, either the ring test or the vibration test must be performed on the wheel prior to using.

11. Only compressed air less than 30 psi may be used for cleaning operations.

12. Good housekeeping practices outlined in 29 CFR 1926.25[8] should be adhered to on all projects.

References

1. United States Department of Labor, Bureau of Labor Statistics, 2009. *Injuries, Illnesses and Fatalities for 2008.* Available at http://www.bls.gov/iif.
2. Occupational Safety and Health Administration, 2008. *29 CFR 1926.416.*
3. Occupational Safety and Health Administration, 2008. *29 CFR 1926.450–454.*
4. Occupational Safety and Health Administration, 2008. *29 CFR 1926.550.*
5. Occupational Safety and Health Administration, 2008. *29 CFR 1926 Subpart M.*
6. Occupational Safety and Health Administration, 2008. *29 CFR 1926.350–353.*
7. Occupational Safety and Health Administration, 2008. *29 CFR 1926 Subpart P.*
8. Occupational Safety and Health Administration, 2008. *29 CFR 1926.25.*

24

Risk Management

Every safety professional should be familiar with the risk management process. After all, managing risks is one of our primary functions as safety professionals. Risk management can be defined as the identification, assessment, and prioritization of risks followed by coordinated and economical application of resources to minimize, monitor, and control the probability and impact of unfortunate events or to maximize the realization of opportunities. Risks can come from liabilities, credit risk, accidents, natural causes, and disasters as well as deliberate attacks from an adversary.[1]

There are a wide variety of specific methods used in risk management depending on the type of industry or service where the risk is being assessed. However, in this chapter, we will discuss the five basic steps of risk management.

Definitions

Risk: The chance or probability of occurrence of an injury, loss, or a hazard or potential hazard.

Risk Assessment: The process of assessing the risks associated with each identified hazard, in order to make decisions and implement appropriate control measures to prevent the hazard from occurring.

Hazard: A condition with the potential to cause injury, illness, or death of personnel; damage to or loss of equipment or property; or mission degradation.

Hazard Identification: The process of examining each work area to identify the hazards associated with each job or task.

Probability: The likelihood that a given event will occur.

Severity: The degree of undesired consequences.

Risk Management Process

In this chapter, we will discuss the risk management process in terms of five basic steps, which are illustrated in Figure 24.1.

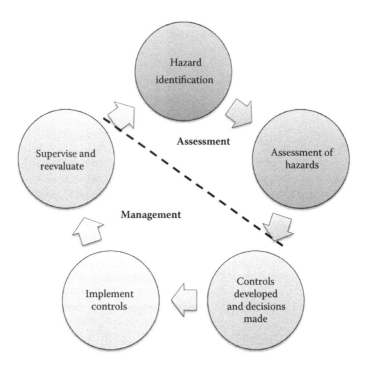

FIGURE 24.1
Risk management process.

Each of these steps is discussed in more detail within the scope of this chapter.

Hazard Identification

As mentioned previously, a hazard is an actual or potential condition, situation, or event that can result in injury, illness, or death of personnel. In addition, a hazard can result in damage, loss, or destruction of equipment or a serious degradation of capabilities or total failure of the project or mission. Hazards are found in all environments and in all situations. Hazards are identified through experience, historical data, intuitive analysis, judgement, standards, brainstorming, and a large variety of other means and methods. Hazard identification is, without a doubt, the most important step, in the risk management process. It must also be noted that hazard identification is a continuous process and should be repeated throughout the work period. The specific method of hazard identification will change from site to site and from task to task. However, the basic process to systematically identify hazards in the work place include the following:

- Identify specific work areas.
- Review previous documents or data involved in the operation to determine if prior injuries or accidents have occurred as a result of this task.

- Conduct an on-site, visual inspection of the work area.
- Determine the individual job tasks of individuals.
- Break the individual job tasks into steps.
- Analyze each job task and identify the hazards or potential hazards involved in performing these tasks.

An important step, often overlooked, is to get the employee performing the actual work involved in the hazard identification process. No one knows more about the job task being performed than the employee performing the work. Useful information that may save lives or prevent property and equipment damage can be obtained.

Hazard Assessment

Once all known hazards have been identified, they must be assessed in terms of their probability and severity to determine the risk level of one or more of the hazardous incidents that can result from exposure to the hazard. In this step of the process, the hazards are assessed individually. This process is systematic in nature and uses charts, codes, and numbers to present a methodology to assess probability and severity to obtain a standardized risk level. *The charts and graphs presented in this chapter are only included to illustrate the risk management process and uses. They can be modified to fit your individual needs.* The incident must be credible in that it must have a reasonable expectation of happening. The end result is an estimate of risk from each hazard and an estimate of the overall risk to the job task or project caused by hazards that cannot be eliminated.[2]

Two major components in the hazard assessment step are the *probability* of the hazard occurring and the *severity* of the hazard. As we have mentioned in this chapter, *probability* is defined as the likelihood that a given event or incident will occur. The incident must be credible in that it must have a reasonable expectation of happening. The end result is an estimate of risk from each hazard and an estimate of the overall risk to the job task or project caused by hazards that cannot be eliminated.[2] *Severity*, for our purposes, can be defined as the degree of undesired consequences.

Probability

Probability is the likelihood of an event actually occurring. This is a subjective estimate of the probability, given the information that you know. The probability levels estimated for each hazard are based on the job or project being performed at the time. There are five levels of probability that we will use in our assessment of the hazards. They include the following ratings: frequent, likely, occasional, seldom, and unlikely. A description of each of these is listed in Table 24.1.

TABLE 24.1

Hazard Probability Rating

Probability Rating	Description
Frequent (A)	Occurs very often, known to happen regularly. For example, out of 500 exposures, it will happen at least once (numerically, 1/500).
Likely (B)	Occurs several times, a common occurrence. For example, out of 1000 exposures, it could happen once (numerically, 1/1000).
Occasional (C)	Occurs sporadically, but is not uncommon. You may or may not complete a job or project without this incident occurring to someone.
Seldom (D)	Remotely possible, could occur at some time. Usually several things must go wrong in order for it to occur.
Unlikely (E)	Can assume that the incident will not occur.

Source: United States Army, July, 2006. *FM No. 5-19 (Formerly FM 100-14), Composite Risk Management.* http://www.cid.army.mil/documents/Safety/Safety%20ReferencesFM%205-19%20Composite%20Risk%20Management.pdf.

Severity

Severity is expressed in terms of the degree to which an incident will affect the safety and health of employees or the project being performed. The degree of severity is somewhat subjective but is estimated for each hazard based on the knowledge of results from similar past events. Four levels of severity are illustrated in Table 24.2.

TABLE 24.2

Severity Rating

Severity Rating	Description
Catastrophic	Death or permanent total disability; complete project failure or the loss of ability to complete the project; loss of major critical systems or equipment; severe environmental damage, unacceptable collateral damage.
Critical	Permanent partial disability or temporary total disability of employees; severely degraded project capability; extensive major damage to equipment or systems; significant property or environmental damage and significant collateral damage.
Marginal	Lost work day injuries or illnesses; degraded project capabilities; minor damage to equipment or systems or minor damage to the environment.
Negligible	First aid or minor medical treatment; little or no adverse impact on project capability; slight equipment or system damage, but fully functional or serviceable; little or no property or environmental damage.

Source: United States Army, July, 2006. *FM No. 5-19 (Formerly FM 100-14), Composite Risk Management.* http://www.cid.army.mil/documents/Safety/Safety%20ReferencesFM%205-19%20Composite%20Risk%20Management.pdf.

TABLE 24.3

Risk Assessment Matrix

		Probability				
Severity	Category	Frequent (A)	Likely (B)	Occasional (C)	Seldom (D)	Unlikely (E)
Catastrophic	I	E	E	H	H	M
Critical	II	E	H	H	M	L
Marginal	III	H	M	M	L	L
Negligible	IV	M	L	L	L	L

Source: United States Army, July, 2006. *FM No. 5-19 (Formerly FM 100-14), Composite Risk Management.* http://www.cid.army.mil/documents/Safety/Safety%20ReferencesFM%205-19%20Composite%20Risk%20Management.pdf.

Note: E, extremely high risk; H, high risk; L, low risk; M, moderate risk.

As mentioned previously in this section, the end result is an estimate of risk from each hazard and an estimate of the overall risk to the job task or project caused by hazards that cannot be eliminated. Table 24.3 illustrates a risk assessment of both the probability and severity of the hazards using a ranking system of extremely high risk, high risk, moderate risk, and low risk.

Once all hazards have been identified and a risk assessment has been performed on each hazard, the highest level of risk will determine the overall risk assessment. Executive management should determine the level of risk the company should assume. For example, the decision to proceed with the project or task for a low-risk project/task can be made by a line supervisor, whereas an extremely high-risk decision is to be made by the chief executive officer within the organization. When all risks are considered, the overall risk should be the highest risk level in the hazard assessment.

Risk Assessment Scenario

Using the following scenario, we will perform a risk assessment. Company XYZ has been contracted to provide industrial cleaning services, which involves the high-pressure water jetting of tube bundles within a condenser. The process involves project mobilization, project setup, performance of high-pressure water jetting, project cleanup and demobilization. We will look at some of the major hazards associated with this project for illustration purposes, but the list in Table 24.4 is not to be considered as all-encompassing. The first step in the risk management process is to identify all hazards.

On the basis of the initial assessment, the highest rating in each category resulted in this project being an extremely high rating. One of the things to

TABLE 24.4

Hazard Identification Worksheet

Locaton: Plant ABC
Date: March 29, 2010
Prepared by: Bill Smith

Hazard Identification	Probability	Consequence	Risk Rating
Back strains from manual lifting during loading and unloading of vehicles	Frequent	Marginal	High
Abrasions and lacerations resulting from pinch points and metals	Frequent	Negligible	Moderate
Vehicular accident during travel to and from the project site	Seldom	Catastrophic	High
Water blasting and jetting injuries resulting from "whipping" nozzles	Occasional	Critical	High
Eye injuries from flying debris	Frequent	Critical	Extremely high
Injuries from rolling vehicles	Seldom	Critical	Moderate
Total rating (highest in each category)	Frequent	Catastrophic	Extremely high

remember in regard to a simplified risk assessment rating system is that the rater determines the rating. No two raters will rate each category the same. For example, one rater may determine that the probability of occurrence is "frequent," while another may rate it as "likely" or "occasional." Neither one would be wrong, if they based their decisions on the criteria listed previously. More in-depth methods may attempt to quantify each method to reduce the variations in assessments. As you can see from the above assessment, most projects will have one or more of the hazards listed as an "extremely high risk." If this were the case, most managers would not attempt to assume that level of risk, which means that little or no work would ever be accomplished. Therefore, step 3 in the process is intended to reduce the level of risk, by implementing controls and making decisions.

Controls Development and Decision Making

As mentioned in the previous step, most projects or tasks have at least one step that has an extremely high risk, which is unacceptable to most everyone. Therefore, in this phase of the risk management process, it is necessary to develop controls to reduce the level of risk posed by the hazards. After assessing each hazard, supervisors and managers develop one or more control that either eliminate the hazard or reduce the risk of a hazardous incident. When developing controls, the supervisors or managers consider the reason for the hazard, not just the hazard itself.

Types of Controls

The types of controls can take many forms, but fall into three main categories: *educational controls, physical controls,* and *avoidance.*

- *Educational controls.* These controls are based on the knowledge and skills of the employees or individuals performing the task. Effective control is implemented through individual and collective training that ensures performance to a standard.
- *Physical controls.* These controls may take the form of barriers and guards or signs to warn employees and others that a hazard exists. Additionally, special controller or supervisory personnel responsible for locating specific hazards fall into this category.
- *Avoidance.* These controls are applied when supervisors and managers take positive action to prevent contact or exposure with the identified hazard.

Criteria for Controls

To be effective, each control developed must meet the following criteria:

- *Support.* Availability of adequate personnel, equipment, supplies, and facilities necessary to implement suitable controls.
- *Standards.* Guidance and procedures for implementing a control are clear, practical, and specific.
- *Training.* Knowledge and skills are adequate to implement a control.
- *Leadership.* Supervisors and managers are competent to implement a control.
- *Individual.* Individual employees are sufficiently self-disciplined to implement a control measure.[2]

Some examples of control measures include the following:

1. Engineering or designing to eliminate or control hazards
2. Avoidance of identified hazards
3. Limiting the number of personnel and the amount of time they are exposed to hazards
4. Providing protective clothing, equipment, and safety devices
5. Providing warning signs and signals

A key element in developing and implementing control measures is to specify who, what, when, where, and how each control is to be used. Taking the scenario above under the heading Controls Development and Decision

Making section, we will continue to expand upon the worksheet to illustrate the development of controls. Table 24.5 illustrates the reevaluation of hazards after control measures have been developed.

As you can see in Table 24.5, after controls have been developed, the hazards and, subsequently, the risks associated with this project or task have been reduced to a moderate risk. This level of risk is acceptable to most managers and supervisors. Once the control measures have been accepted, the responsible level of authority determines the overall residual risk, as listed

TABLE 24.5

Controls Development Worksheet

Locaton: Plant ABC
Date: March 29, 2010
Prepared by: Bill Smith

Hazard Identification	Controls	Probability	Consequence	Risk Rating
Back strains from manual lifting during loading and unloading of vehicles	1. Toolbox talks 2. Buddy system when loading and unloading 3. Use of materials handling equipment (mechanical)	Seldom	Marginal	Low
Abrasions and lacerations resulting from pinch points and metals	1. Use of personal protective equipment (i.e., gloves) 2. Toolbox talks	Seldom	Negligible	Low
Vehicular accident during travel to and from the project site	1. Driver training 2. Supervisors insuring that drivers are not fatigued 3. Toolbox talks	Seldom	Critical	Moderate
Water blasting and jetting injuries resulting from "whipping" nozzles	1. Protective clothing including body armor suits 2. Use of antiwithdrawal devices 3. Toolbox talks	Unlikely	Critical	Low
Eye injuries from flying debris	1. Wearing of face shields and goggles 2. Toolbox talks	Seldom	Marginal	Low
Injuries from rolling vehicles	1. Use of wheel chocks 2. Toolbox talks 3. Enforce emergency brake usage	Unlikely	Critical	Low
Total rating (highest in each category)		Seldom	Critical	Moderate

in Table 24.5. The residual risk is defined as the risk remaining after controls have been selected for the hazards.[2]

Decision Making

A key element of the risk decision is determining if the risk is justified. The supervisor or manager responsible for the project must compare and balance the risk against the benefits. If the residual levels are too high, then the senior supervisor or manager on the project may elect to do one of the following:

- Add additional control measures to further reduce the risks
- Limit the scope of work, which eliminates the high risk tasks
- Make the decision to discontinue the project

Implement Controls

Managers and supervisors must ensure that controls are integrated into the standard operating procedures, written and verbal instructions, and toolbox talks prior to the beginning of the project or task. The critical check for this step, with oversight, is to ensure that controls are converted into clear, simple instructions understood at all levels. Implementing controls includes coordination and communication with appropriate superiors and employees.

Supervise and Evaluate

During project or task preparation, managers and supervisors must ensure that employees understand how to execute the risk controls. In addition, the manager or supervisor of the project or task must continually assess the risks during the operational phase of the project. Methods to supervise and evaluate the effectiveness of control measures include spot checks, inspections, daily reports, and close, direct supervision. After the project or task has been completed, it is recommended that a postproject evaluation, which includes all employees on the project, be conducted. The results of this postproject evaluation, and any changes, should be maintained in the project file, for future use.

Key Information to Remember on Risk Management

1. Risk is defined as the chance or probability of occurrence of an injury, loss, or a hazard or potential hazard.
2. Risk assessment is the process of assessing the risks associated with each identified hazard, in order to make decisions and implement appropriate control measures to prevent the hazard from occurring.

3. Hazard is a condition with the potential to cause injury, illness, or death of personnel; damage to or loss of equipment or property; or mission degradation.

4. Hazard identification is the process of examining each work area to identify the hazards associated with each job or task.

5. Probability is defined as the likelihood that a given event will occur.

6. Severity is defined as the degree of undesired consequences.

7. The five basic steps in the risk management process are *hazard identification, hazard assessment, development of controls and decision making, implementation,* and *supervision and evaluation.*

8. The types of controls can take many forms, but fall into three main categories: *educational controls, physical controls,* and *avoidance.*

9. A key element in developing and implementing control measures is to specify who, what, when, where, and how each control is to be used.

10. A key element of the risk decision is determining if the risk is justified.

11. The critical check for controls implementation, with oversight, is to ensure that controls are converted into clear, simple instructions understood at all levels.

References

1. Wikipedia, 2010. Available at http://en.wikipedia.org/wiki/Risk_management.
2. United States Army, 2006. *FM No. 5-19 (Formerly FM 100-14), Composite Risk Management.* Available at http://www.cid.army.mil/documents/Safety/Safety%20ReferencesFM% 205-19%20Composite%20Risk%20Management.pdf.

25

Hazardous Materials Management

One of the most difficult areas of the safety profession to manage is one concerning hazardous materials. The difficulty arises from the process of almost continous hazardous materials inventory and control of incoming hazardous materials. However, this task is not without hope. In this chapter, we will discuss the hazards associated with chemicals, impacts of hazardous materials, fundamentals of hazardous materials management, hazardous materials disposal, hazardous materials response procedures, and substances with special standards. The information available on hazardous materials management is voluminous. Therefore, information provided in this chapter is basic and superficial. It is intended to provide the Associate Safety Professional/Certified Safety Professional candidate with information to understand the basic requirements of a hazardous materials management program.

Hazardous Materials

What is a hazardous material? A hazardous material (Hazmat or HAZMAT) is any solid, liquid, or gas that can harm people, other living organisms, property, or the environment. The term *hazardous material* is used in this context almost exclusively in the United States. The equivalent term in the rest of the English-speaking world is *dangerous goods*. A hazardous material may be radioactive, flammable, explosive, toxic, corrosive, a biohazard, an oxidizer, an asphyxiant, or an allergen, or may have other characteristics that make it hazardous in specific circumstances.[1]

Hazardous Waste

Hazardous waste is defined as a "solid waste" that, because of its quantity, concentration, or physical, chemical, or infectious characteristics, may

- Pose a substantial present or potential hazard to human health or the environment when improperly treated, stored or disposed of, or otherwise mismanaged
- Cause or contribute to an increase in mortality or an increase in irreversible or incapacitating illness

A solid waste is defined as any discarded material that is abandoned by being disposed of, burned or incinerated, recycled, or considered "waste-like." A solid waste can physically be a solid, liquid, semisolid, or container of gaseous material.[2]

Basic Chemistry Review

In order to understand hazardous materials and hazardous wastes, and subsequently their management, the safety professional must have a basic understanding of chemicals and chemistry. As we have all learned, all matter is classified according to its physical state, which is solid, liquid, or gas.

Solid

A solid is characterized by structural rigidity and resistance to changes of shape or volume. Unlike a liquid, a solid object does not flow to take on the shape of its container, nor does it expand to fill the entire volume available to it like a gas does. The atoms in a solid are tightly bound to each other, either in a regular geometric lattice (crystalline solids, which include metals and ordinary water ice) or irregularly (an amorphous solid such as common window glass).

Liquid

Liquid is one of the three classic states of matter. Like a gas, a liquid is able to flow and take the shape of a container, but, like a solid, it resists compression. Unlike a gas, a liquid does not disperse to fill every space of a container and maintains a fairly constant density.

Gas

Gas is one of three classic states of matter. Near absolute zero, a substance exists as a solid. As heat is added to this substance, it melts into a liquid at its melting point, boils into a gas at its boiling point, and if heated high enough would enter a plasma state in which the electrons are so energized that they leave their parent atoms from within the gas. A pure gas may be made up of individual atoms (e.g., a noble gas or atomic gas like neon), elemental molecules made from one type of atom (e.g., oxygen), or compound molecules made from a variety of atoms (e.g., carbon dioxide). A gas mixture would contain a variety of pure gases much like the air. What distinguishes a gas from liquids and solids is the vast separation of the individual gas particles.

Plasma

A fourth state of matter is plasma, which is only mentioned as it is considered distinct from a gas. Plasma is a gas in which a certain portion of the particles are ionized. The presence of a nonnegligible number of charge carriers makes the plasma electrically conductive so that it responds strongly to electromagnetic fields. Plasma, therefore, has properties quite unlike those of solids, liquids, or gases and is considered to be a distinct state of matter. Like gas, plasma does not have a definite shape or a definite volume unless enclosed in a container; unlike gas, in the influence of a magnetic field, it may form structures such as filaments, beams, and double layers.

Atomic Mass

As we discussed in Chapter 3, the atomic mass or weight of an atom includes protons, neutrons, and electrons. This weight is measured in grams. The atomic weight of any element is the average weight compared to carbon. The mass in grams of any element that is equal to that element's atomic weight is called 1 gram-molecular weight, or 1 mole (6.02×10^{23} atoms = Avogadro's number).

Atoms

Also discussed in Chapter 3, an atom is a basic unit of all matter, which contains a nucleus. The nucleus contains positively charged protons and neutrally charged neutrons. Surrounding the nucleus is a cloud of negatively charged electrons. Refer to Chapter 3 for a review of the atom.

Definitions Related to Matter

Density: The density of matter is equal to its mass per unit volume ($d = m/v$).

Specific Gravity: The ratio of the density of a material to the density of water ($SG_x = d_x/d_{H_2O}$). Those substances with a specific gravity of <1 are considered lighter than water and will float, while those with a specific gravity of >1 will sink.

Vapor Density: Weight of a unit volume of gas or vapor compared to the weight of an equal volume of air (or, sometimes, hydrogen). Substances with vapor densities <1 are lighter than air and those with vapor densities >1 are heavier than air. All gases and vapors mix with air, the lighter substances tend to rise and dissipate, and the heavier substances tend to concentrate in low places along floors, sewers, and trenches and may create fire and health hazards.

Solubility: Solubility is the property of a solid, liquid, or gaseous chemical substance called solute to dissolve in a liquid solvent to form a homogeneous solution. The solubility of a substance strongly depends on the used solvent as well as on temperature and pressure. Water solubility is usually measured in parts per million.

Periodic Table of the Elements

Refer to Chapter 3 for a detailed discussion on the Periodic Table of Elements. Currently, there are 118 individual elements listed on the table. The table lists the element's symbol, atomic number, atomic mass (gram), and the isotopes.

Measurements of Concentrations

Hazardous chemical concentrations are usually measured in parts per million, parts per billion, or milligrams per cubic meter. In Chapter 3, we discussed the calculations for each of these. For more information, refer back to Chapter 3.

Hazardous Materials/Hazardous Waste Properties

In this section, we will discuss the properties that make any material hazardous. Anyone who has taken a basic college course in toxicology has heard the term, "the dose makes the poison." This is true for all substances, including water. By drinking too much water, you can overhydrate and cause water intoxication. Hazardous materials and hazardous wastes also have physical hazards that we will also discuss in this section.

Physical Hazards

Physical hazards associated with materials/waste include engulfment, over-pressurization, slips, trips and falls, crushing hazards, fire and explosion hazards, corrosion hazards, thermal decomposition, and water reactivity hazards. Each of these is discussed in more detail.

Engulfment

Engulfment results when a worker is surrounded and overcome by a granular substance such as soil, sand, gravel, sawdust, seed, grain, or flour or if submerged in a liquid such as water or a chemical. Engulfment causes physical harm when the material has enough force on the body to cause injury or death by constriction, crushing, or strangulation. Respiratory hazards associated with engulfment include suffocation from breathing in a fine substance that fills the lungs or from drowning in a liquid. When working in confined spaces, it is important to strictly adhere to the Occupational Safety and Health Administration's confined space requirements (29 CFR 1910.146) and control of hazardous energy requirements (29 CFR 1910.147). By fully complying with these regulations, the risk of engulfment can be minimized.

Overpressurization

Overpressurization can occur as a result of an increase in the ambient temperature surrounding a container or other systematic failures. As noted in Chapter 3, combined gas law, an increase in temperature will result in an increase in pressure. For example, a container has 208.197 L of chemical stored in an unventilated warehouse at 14.7 psi and at 299.67 K. The temperature in the storage location increases to 312.44 K. Inserting this information into the combined gas law equation shown in Chapter 3, the pressure increased to 15.33 psi. If the container was not designed to handle this pressure, then overpressurization can occur. The results of overpressurization can be catastrophic, such as fires and explosions.

Other Physical Safety Hazards

Chemicals in liquid or solid state can cause employees to slip, trip, or fall. Crushing or mechanical injuries may also occur as a result of materials handling and storage. Pinch points are created by handling and storage of drums, which can cause crushing, soft tissue injuries.

Fires and Explosions

There are many potential causes of explosions and fires associated with hazardous materials, which include the following:

- Chemical reactions that produce explosions, fires, or heat
- Ignition of explosive or flammable chemicals
- Ignition of materials caused by oxygen enrichment
- Agitation of shock- or friction-sensitive compounds
- Sudden release of materials under pressure

Explosions and fires may arise spontaneously. However, more commonly, they result from activities, such as moving drums, accidentally mixing incompatible chemicals, or introducing an ignition source into an explosive or flammable environment.[3] Fires and explosions pose a threat to personnel, equipment, and facilities. When working with potentially flammable or combustible materials, always use nonsparking tools, such as brass.

Corrosion

Corrosion is the disintegration of an engineered material into its constituent atoms owing to chemical reactions with its surroundings. In the most common use of the word, this means electrochemical oxidation of metals in reaction with an oxidant such as oxygen. Formation of an oxide of iron caused by oxidation of the iron atoms in solid solution is a well-known example of electrochemical corrosion, commonly known as rusting. This type of damage typically produces oxide(s) or salt(s) of the original metal.

In other words, corrosion is the wearing away of metals because of a chemical reaction. Many structural alloys corrode merely from exposure to moisture in the air, but the process can be strongly affected by exposure to certain substances such as to acids or bases. Corrosion can be concentrated locally to form a pit or crack, or it can extend across a wide area more or less uniformly corroding the surface. Because corrosion is a diffusion-controlled process, it occurs on exposed surfaces.

Thermal Decomposition

Thermal decomposition refers to the by-products of incomplete combustion such as carbon monoxide gases. Other examples include reversions of chemicals exposed to heat to materials that were used to synthesize them.

Water-Reactive Material Hazards

Water-sensitive chemicals are chemicals that react vigorously with moisture. This reaction can result in extreme heats and can be potentially flammable, corrosive, toxic, or reactive. The most common water-sensitive chemicals include sodium, potassium, lithium metals, and aluminum alkyls. Examples of water-reactive chemicals include the following:

- Alkali metals, such as Na, Li, and K
- Alkali metal hydrides, such as LiH, CaH_2, $LiAlH_4$, and $NaBH_4$
- Alkali metal amides, such as $NaNH_2$
- Metal alkyls, such as lithium and aluminum alkyls
- Halides of nonmetals
- Inorganic acid halides
- Anhydrous metal halides
- Phosphorus pentoxide
- Calcium carbide
- Organic acid halides and anhydrides of low molecular weight

Health Hazards Associated with Hazardous Materials

As we discussed in Chapter 4, chemicals can enter the human body through several routes, including inhalation, ingestion, absorption, and percutaneous and intravenous injections (refer to Chapter 4).

Preventing exposure to toxic chemicals is a primary concern when working with hazardous materials, in order to prevent entry into the body. A contaminant can cause damage at the point of contact or can act systemically, causing a toxic effect at a part of the body distant from the point of initial contact.

Chemical exposures are generally divided into two major categories: acute and chronic. Symptoms resulting from acute exposures usually occur during or shortly after exposure to a sufficiently high concentration of a contaminant. The term *chronic exposure* generally refers to exposures to "low concentrations" of a contaminant over a long period. The "low concentrations" required to produce symptoms of chronic exposure depend upon the chemical, the duration of exposure, and the number of exposures.[4]

Key Regulations Governing Hazardous Materials and Hazardous Wastes

Resource Conservation and Recovery Act (1976)

The Resource Conservation and Recovery Act, enacted in 1976, is the principal Federal law in the United States governing the disposal of solid waste and hazardous waste. In 1984, Congress expanded the law with the Hazardous and Solid Waste Amendments of 1984. The amendments strengthened the law by covering small quantity generators of hazardous waste and establishing requirements for hazardous waste incinerators, and the closing of substandard landfills. In 1986, the law was expanded further to regulate underground storage tanks and other leaking waste storage facilities. This law regulates hazardous waste "from cradle to grave," including generation, treatment, storage, and disposal of hazardous wastes.

Hazardous Waste Generators

The US Environmental Protection Agency (EPA) classifies hazardous generators according to the amount of generated hazardous waste. Table 25.1 shows the classifications and quantities allowed.

TABLE 25.1

EPA Categories of Hazardous Waste Generators

Type of Generator	Generation Rate (per Calendar Month), Ordinary Waste	Generation Rate (per Calendar Month), Acute Hazardous Waste	Maximum Storage Amount	Maximum Storage Time
Large quantity generator (LQG)	>1000 kg (2200 lb)	>1 kg (2.2 lb)	No limit	90 days for most wastes
Small quantity generator (SQG)	>100 kg (220 lb)	≤1 kg (2.2 lb)	≥6000 kg (13,200 lb)	180 days 270 days if the TSDF is more than 200 miles away
Conditionally exempt small quantity generator (CESQG)	≤100 kg (220 lb)	≤1 kg (2.2 lb)	≤1000 kg (2200 lb)	No limit

Comprehensive Environmental Response, Compensation, and Liability Act (1980)

The Comprehensive Environmental Response, Compensation, and Liability Act (CERCLA), commonly known as Superfund, was enacted by Congress on December 11, 1980. This law created a tax on the chemical and petroleum industries and provided broad Federal authority to respond directly to releases or threatened releases of hazardous substances that may endanger public health or the environment.
   CERCLA

- Established prohibitions and requirements concerning closed and abandoned hazardous waste sites
- Provided for liability of persons responsible for releases of hazardous waste at these sites
- Established a trust fund to provide for cleanup when no responsible party could be identified

The law authorizes two kinds of response actions:

- Short-term removals, where actions may be taken to address releases or threatened releases requiring prompt response.
- Long-term remedial response actions, which permanently and significantly reduce the dangers associated with releases or threats of releases of hazardous substances that are serious, but not immediately life threatening. These actions can be conducted only at sites listed on EPA's National Priorities List (NPL).

CERCLA also enabled the revision of the National Contingency Plan (NCP). The NCP provided the guidelines and procedures needed to respond to releases and threatened releases of hazardous substances, pollutants, or contaminants. The NCP also established the NPL.[5]

Superfund Amendment and Reauthorization Act of 1986

The Superfund Amendment and Reauthorization Act (SARA) amended the CERCLA on October 17, 1986. SARA reflected EPA's experience in administering the complex Superfund program during its first 6 years and made several important changes and additions to the program. SARA

- Stressed the importance of permanent remedies and innovative treatment technologies in cleaning up hazardous waste sites
- Required Superfund actions to consider the standards and requirements found in other State and Federal environmental laws and regulations

- Provided new enforcement authorities and settlement tools
- Increased State involvement in every phase of the Superfund program
- Increased the focus on human health problems posed by hazardous waste sites
- Encouraged greater citizen participation in making decisions on how sites should be cleaned up
- Increased the size of the trust fund to $8.5 billion

SARA also required EPA to revise the Hazard Ranking System to ensure that it accurately assessed the relative degree of risk to human health and the environment posed by uncontrolled hazardous waste sites that may be placed on the NPL.[6]

Toxic Substances Control Act (1976)

The *Toxic Substances Control Act* (TSCA) became law on October 11, 1976. The Act authorized EPA to secure information on all new and existing chemical substances, as well as to control any of the substances that were determined to cause unreasonable risk to public health or the environment. Congress later added additional titles to the Act, with this original part designated at Title I—Control of Hazardous Substances.[7] Other titles included Title II—Asbestos Hazard Emergency Response Act, Title III—Indoor Air Radon Abatement, and Title IV—Lead Based Paint Exposure.

Emergency Planning and Community Right-to-Know Act (1986)

The *Emergency Planning and Community Right-to-Know Act* (EPCRA) was enacted by Congress on October 17, 1986, as an outgrowth of concern over the protection of the public from chemical emergencies and dangers. Previously, this had been covered by state and local regulatory authorities. After the catastrophic accidental release of methyl isocyanate at Union Carbide's Bhopal, India facility in December 1984, and a later toxic release from a West Virginia chemical plant, it was evident that national public disclosure of emergency information was needed. EPCRA was enacted as a stand-alone provision, Title III, in the SARA of 1986.[7]

Federal Insecticide, Fungicide, and Rodenticide Act (1972)

Congress enacted the Federal Environmental Pesticide Control Act of 1972, which amended the Federal Insecticide, Fungicide, and Rodenticide Act (FIFRA) by specifying methods and standards of control in greater detail. Subsequent amendments have clarified the duties and responsibilities of the EPA. In general, there has been a shift toward greater emphasis on minimizing risks associated with toxicity and environmental degradation, and away from pesticide efficacy issues.

Important FIFRA requirements are as follows:

- No one may sell, distribute, or use a pesticide unless it is registered by the EPA or meets a specific exemption *as described in the regulations*. Registration includes approval by the EPA of the pesticide's label, which must give detailed instructions for its safe use.
- EPA must classify each pesticide as either "general use," "restricted use," or both. General-use pesticides may be applied by anyone, but restricted-use pesticides may only be applied by certified applicators or persons working under the direct supervision of a certified applicator. Because there are only limited data for new chemicals, most pesticides are initially classified as restricted use. Applicators are certified by a state if the state operates a certification program approved by the EPA.[8]

Asbestos Hazard Emergency Response Act (1986)

In 1986, the Asbestos Hazard Emergency Response Act was signed into law as Title II of the TSCA. Additionally, the Asbestos School Hazard Abatement Reauthorization Act (ASHARA), passed in 1990, required accreditation of personnel working on asbestos activities in schools and public and commercial buildings.

Specifically, Asbestos-Containing Materials in Schools outlines a detailed process that ensures the safe management of all asbestos-containing building materials (ACBM) by a designated person for a local education agency (LEA).

Additionally, the ASHARA, passed in 1990, required accreditation of personnel working on asbestos activities in schools and public and commercial buildings. Specifically, the Asbestos Model Accreditation Plan (40 CFR Part 763, Appendix C) required the use of accredited inspectors, workers, supervisors, project designers, and management planners (schools only) when conducting asbestos activities at schools and public and commercial buildings.

Although asbestos is hazardous when inhaled, the risk of exposure to airborne fibers is very low. Therefore, removal of asbestos from schools is often not the best course of action. It may even create a dangerous situation when none previously existed. The EPA only requires removal of asbestos to prevent significant public exposure during demolition or renovation. EPA does, however, require an in-place, proactive asbestos management program for all LEAs in order to ensure that ACBM remains in good condition and is undisturbed by students, faculty, and staff.[9]

Hazard Communication Standard (29 CFR 1910.1200)

The purpose of this standard is to ensure that the hazards of all chemicals produced or imported are evaluated and that information concerning their hazards is transmitted to employers and employees. This transmittal of information is to be accomplished by means of comprehensive hazard communication programs, which are to include container labeling and other forms of warning, material safety data sheets, and employee training.[10] The specifics of this standard are discussed in Chapter 2.

Classification of Hazardous Materials

The United States Department of Transportation identifies hazardous materials, specifically for transportation purposes, into nine different classifications. Table 25.2 identifies each classification.

Hazardous Waste Operations

Hazardous waste operations, and specifically cleanup operations, pose a multitude of health and safety concerns to workers and to the general public. The types of hazards associated with hazardous materials and hazardous wastes have been discussed previously in this chapter. However, cleanup operations pose special threats to the safety and health of employees. Several factors distinguish the hazardous waste site environment from other occupational situations involving hazardous substances. A key factor is the uncontrolled condition of the site. Oftentimes, cleanup operations involve abandoned or seriously degraded sites. Additional factors include the large variety and number of substances that may be present on the site. Frequently, an accurate assessment of all chemical hazards is impossible owing to the large number of substances and the potential interactions among the substances. Many times, the identity of the substances on site is unknown, especially in the early stages of the project.

Planning and Organization

Adequate planning is the first and most critical element of hazardous waste operations. By anticipating and taking steps to prevent potential hazards to health and safety, work at a waste site can proceed with minimum risk to workers and the general public. Planning involves three aspects: developing an overall organizational structure for site operations, establishing a comprehensive work plan that considers each specific phase of the operation, and developing and implementing a site safety and health plan.[11]

Training

Employees should not enter a hazardous waste site until they have been trained to a level commensurate with their job function and responsibilities

TABLE 25.2

Department of Transportation Hazardous Materials Classification (OSU[19])

Classification Division	Hazard Name	Description	Placard
1.1	Explosive	Articles and substances having a mass explosion hazard	
1.2	Explosive	Articles and substances having a projection hazard, but not a mass explosion hazard	
1.3	Explosive	Articles and substances having a fire hazard, a minor blast hazard, or a minor projection hazard, but not a mass explosion hazard	
1.4	Explosive	Articles and substances presenting no significant hazard (explosion limited to package)	
1.5	Explosive	Very insensitive substances having a mass explosion hazard	
1.6	Explosives	Extremely insensitive articles that do not have a mass explosion hazard	
2.1	Flammable gas	Flammable gas	
2.2	Nonflammable compressed gas	Nonflammable, nontoxic, inert gas under pressure	

(*Continued*)

TABLE 25.2 (CONTINUED)

Department of Transportation Hazardous Materials Classification (OSU[19])

Classification Division	Hazard Name	Description	Placard
2.3	Poison gas	Toxic gas	POISON GAS 2
3	Flammable liquid	Liquid having a flash point of 141°F or less	FLAMMABLE 3
3	Combustible liquid	Liquid having a flash point above 141°F but less than 200°F	COMBUSTIBLE 3
4.1	Flammable solid	Flammable solid	FLAMMABLE SOLID 4
4.2	Spontaneously combustible	Substances liable to spontaneous combustion	SPONTANEOUSLY COMBUSTIBLE
4.3	Dangerous when wet	Substances that, in contact with water, emit flammable gases	DANGEROUS WHEN WET
5.1	Oxidizers	Oxidizers	OXIDIZER 5.1
5.2	Organic peroxides	Organic peroxides	ORGANIC PEROXIDE 5.2

TABLE 25.2 (CONTINUED)

Department of Transportation Hazardous Materials Classification (OSU[19])

Classification Division	Hazard Name	Description	Placard
6.1	Toxic substances	Toxic substances	
6.2	Infectious substances	Infectious substances	
7	Radioactive	Substances having a specific activity >70 Bq/g	
8	Corrosives	Corrosives	
9	Misc. hazardous materials	Magnetized materials, elevated temperature goods, dry ice, asbestos, environmentally hazardous substances, life-saving appliances, internal combustion engines, battery-powered equipment or vehicle, zinc dithionite, and so on	

and with the degree of anticipated hazards.[12] General site workers should be trained in the following subject areas, at a minimum:

- Site safety plan
- Safe work practices
- Nature of anticipated hazards
- Handling emergencies and self-rescue
- Rules and regulations for vehicle
- Safe use of field equipment
- Handling, storage, and transportation of hazardous materials

- Employee rights and responsibilities
- Use, care, and limitations of personal protective clothing and equipment
- Safe sampling techniques

Medical Program

A medical program should be developed for each site on the basis of the specific needs, location, and potential exposures of employees at the site. The program should be designed by an experienced occupational health physician or another qualified occupation health consultant in conjunction with the Site Safety Officer. A site medical program should include the following components:[13]

- Surveillance
  - Preemployment screening
  - Periodic medical examinations
  - Termination examination
- Treatment
  - Emergency
  - Nonemergency
- Record Keeping
- Program Review

Site Characterization

Site characterization provides the information needed to identify site hazards and to select worker protection methods. The more accurate, detailed, and comprehensive the information available about the site, the more the protective measures can be tailored to the actual hazards that workers may encounter. Site characterization proceeds in three phases:[14]

- Prior to site entry, conduct offsite characterization: gather information away from the site and conduct reconnaissance from the site perimeter.
- Next, conduct onsite surveys. During this phase, restrict site entry to reconnaissance personnel.
- Once the site has been determined safe for commencement of other activities, perform ongoing monitoring to provide a continuous source of information about site conditions.

Air Monitoring

Airborne contaminants can present a significant threat to worker health and safety. Therefore, identification and quantification of these contaminants through air monitoring is an essential component of a health and safety program at a hazardous waste site. Reliable measurements of airborne contaminants are useful for[15]

- Selecting personal protective equipment
- Delineating areas where protection is needed
- Assessing the potential health effects of exposure
- Determining the need for specific medical monitoring

Personal Protective Equipment

Anyone entering a hazardous waste site must be protected against potential hazards. The purpose of personal protective clothing and equipment (PPE) is to shield or isolate individuals from the chemical, physical, and biologic hazards that may be encountered at a hazardous waste site. Careful selection and use of adequate PPE should protect the respiratory system, skin, eyes, face, hands, feet, head, body, and hearing.[16]

Site Control

The purpose of site control is to minimize potential contamination of workers, protect the public from the site's hazards, and prevent vandalism. Site control is especially important in emergency situations, such as overturned tractor/trailers transporting hazardous materials or wastes.

Decontamination

Decontamination is the process of removing or neutralizing contaminants that have accumulated on personnel or equipment and is critical to health and safety at hazardous waste sites. Decontamination protects workers from hazardous substances that may contaminate and eventually permeate the protective clothing, respiratory equipment, tools, vehicles, and other equipment used on site.[17]

Key Information to Remember on Hazardous Materials Management

1. A hazardous material is any solid, liquid, or gas that can harm people, other living organisms, property, or the environment.
2. A hazardous waste is defined as a "solid waste" that, because of its quantity, concentration, or physical, chemical, or infectious characteristics, may (a) pose a substantial present or potential hazard to human health or the environment or (b) cause or contribute to an

increase in mortality or an increase in irreversible or incapacitating illness.

3. All matter is classified according to its physical state, which is solid, liquid, or gas.

4. The physical hazards associated with materials include engulfment, overpressurization, fires and explosions, corrosion, thermal decomposition, and other physical safety hazards.

5. Routes of entry into the body are inhalation, ingestion, absorption, and percutaneous or intravenous injections.

6. The Resource Conservation and Recovery Act regulates hazardous waste from cradle to grave, including generation, treatment, storage, and disposal of hazardous wastes.

7. Large quantity generators (LQGs) generate >1000 kg of ordinary waste, or 1 kg acute hazardous waste per month, and can store it for up to 90 days. There is no limit on the storage amount.

8. Small quantity generators (SQGs) generate >100 kg of ordinary waste and ≤1 kg of acute hazardous waste and can store a maximum amount of 6000 kg for up to 180 days or 270 days if the treatment, storage, and disposal facility is located greater than 200 miles away.

9. Conditionally exempt small quantity generators (CESQGs) generate less than or equal to 100 kg of ordinary waste and less than or equal to 1 kg of acute hazardous waste and may store less than or equal to 1000 kg indefinitely.

10. The Comprehensive Environmental Response, Compensation, and Liability Act created a tax on the chemical and petroleum industries. CERCLA established prohibitions and requirements concerning closed and abandoned hazardous waste sites and provided for liability of persons responsible for releases of hazardous waste at these sites. In addition, it established a trust fund to provide for cleanup when no responsible party could be identified.

11. The Toxic Substances Control Act authorized EPA to secure information on all new and existing chemical substances, as well as to control any of the substances that were determined to cause unreasonable risk to public health or the environment.

12. The US Department of Transportation classifies hazardous materials into nine different classifications including explosives, flammable gases, flammable liquids, flammable solids, oxidizers, toxic substances, radioactive, corrosives, and miscellaneous hazardous materials.

13. Hazardous waste operations pose a multitude of health and safety concerns to workers and to the general public.

14. Adequate planning is the first and most critical element of hazardous waste operations.

15. Employees should not enter a hazardous waste site until they have been trained to a level commensurate with their job function and responsibilities.

16. A medical program should be developed for each site on the basis of the specific needs, location, and potential exposures of employees at the site.

17. Site characterization provides the information needed to identify site hazards and to select worker protection methods.

18. Identification and quantification of air contaminants are made through air monitoring, which is essential in selecting PPE, delineating areas where protection is needed, assessing the potential health effects of exposure, and determining the need for specific medical monitoring.

19. The purpose of site control is to minimize potential contamination of workers, protect the public from the site's hazards, and prevent vandalism.

20. Decontamination protects workers from hazardous substances that may contaminate and eventually permeate the protective clothing, respiratory equipment, tools, vehicles, and other equipment used on hazardous waste sites.

References

1. Search.com, 2010. Available at http://www.search.com/reference/Hazardous_material.
2. ThinkQuest, 2010. Available at http://library.thinkquest.org/10625/HAZDEFIN.HTM.
3. National Institute of Occupational Safety and Health (NIOSH), 1985. *Occupational Safety and Health Guidance Manual for Hazardous Waste Site Activities*, OSHA, Washington, DC, p. 2–3. Available at https://www.osha.gov/Publications/complinks/OSHG-HazWaste/all-in-one.pdf.
4. National Institute of Occupational Safety and Health (NIOSH), 1985. *Occupational Safety and Health Guidance Manual for Hazardous Waste Site Activities*, OSHA, Washington, DC, p. 2–2. Available at https://www.osha.gov/Publications/complinks/OSHG-HazWaste/all-in-one.pdf.
5. U.S. Environmental Protection Agency (EPA), 2010. Available at http://epa.gov/superfund/policy/cercla.htm.
6. U.S. Environmental Protection Agency (EPA), 2010. Available at http://www.epa.gov/superfund/policy/sara.htm.

7. U.S. Environmental Protection Agency (EPA), 2010. Available at http://www.epa.gov/oecaerth/civil/epcra/epcraenfstatreq.html.

8. U.S. Environmental Protection Agency (EPA), 2010. Available at http://www.epa.gov/Enforcement/civil/fifra/fifraenfstatreq.html.

9. U.S. Environmental Protection Agency (EPA), 2010. Available at http://www.epa.gov/region2/ahera/ahera.htm.

10. Occupational Safety and Health Administration (OSHA), 2010. Available at http://www.osha.gov/pls/oshaweb/owadisp.show_document?p_table=standards&p_id=10099.

11. National Institute of Occupational Safety and Health (NIOSH), 1985. *Occupational Safety and Health Guidance Manual for Hazardous Waste Site Activities*, U.S. Environmental Protection Agency, p. 3–1.

12. National Institute of Occupational Safety and Health (NIOSH), 1985. *Occupational Safety and Health Guidance Manual for Hazardous Waste Site Activities*, U.S. Environmental Protection Agency, p. 4–2.

13. National Institute of Occupational Safety and Health (NIOSH), 1985. *Occupational Safety and Health Guidance Manual for Hazardous Waste Site Activities*, U.S. Environmental Protection Agency, p. 5–2.

14. National Institute of Occupational Safety and Health (NIOSH), 1985. *Occupational Safety and Health Guidance Manual for Hazardous Waste Site Activities*, U.S. Environmental Protection Agency, p. 6–1.

15. National Institute of Occupational Safety and Health (NIOSH), 1985. *Occupational Safety and Health Guidance Manual for Hazardous Waste Site Activities*, U.S. Environmental Protection Agency, p. 7–1.

16. National Institute of Occupational Safety and Health (NIOSH), 1985. *Occupational Safety and Health Guidance Manual for Hazardous Waste Site Activities*, U.S. Environmental Protection Agency, p. 8–1.

17. National Institute of Occupational Safety and Health (NIOSH), 1985. *Occupational Safety and Health Guidance Manual for Hazardous Waste Site Activities*, U.S. Environmental Protection Agency, p. 10–1.

26

Radiation Safety

In order to discuss the safety aspects of any issue, it is necessary to understand the basics of the hazard or issue. In this chapter, we will discuss the fundamentals of both ionizing and nonionizing radiation.

Ionizing Radiation

Ionizing radiation occurs as the result of particles or electromagnetic waves having enough energy to detach electrons from atoms or molecules, therby causing *ionization* of the atom. Ionization is defined as the process of converting a stable atom or molecule into a charged one through the gain or loss of electrons. Ionizing radiation is produced by the natural decay of radioactive material. This occurrence depends entirely on the energy of the particles or waves and not on the number. Ionizing radiation comes from radioactive materials, x-ray tubes, and particle accelerators and is present in the natural environment. There are two ways to cause ionization: direct and indirect. Direct ionization occurs as a result of charged particle interaction with matter. Indirect ionization occurs as a result of uncharged particle interaction with matter through collision. As mentioned, ionizing radiation comes in the form of particles or electromagnetic waves.

Particle Radiation

The three types of particle radiation include

- Alpha (α)
- Beta (β)
- Neutron

Alpha (α) Radiation

Alpha radiation is a helium nucleus that has two neutrons and two protons. It causes radiation through direct ionization. The approximate range in air is only 2 cm, since it is a highly charged particle. It will not penetrate a layer of skin and does not pose any known external radiation hazards. The major hazard associated with alpha radiation is the potential for internal damage. As the alpha particles travel through matter, they create many ions. Therefore, alpha radiation is a serious hazard to the lungs and tissues internally. An example of alpha radiation is a naturally occurring source of radon. Whenever a nuclide emits an alpha particle, a new nuclide is created with a change in atomic mass of 4 and an atomic number of 2.[1] Refer to the following example:

$$^{226}_{88}\text{Ra} \quad ^{4}_{2}\text{He}^{+2} \quad ^{222}_{86}\text{Rn}.$$

Beta (β) Radiation

Beta particles are excess electrons. They are formed when an atom with one excess neutron transforms the neutron to a proton and ejects the extra electron. Particles can be low or high energy emitters. Low energy emitters can be shielded by cardboard, while high energy emitters need a more dense shielding material. Beta radiation results from a high-speed electron with a −1 charge overall. Beta particles cannot penetrate the body to irradiate internal organs. They can penetrate dead outer-layer skin and result in damage to live skin cells. In addition, beta particles can cause damage to the lenses of the eyes. Ingestion, inhalation, or absorption through the skin might result in internal exposure.

Neutron (n) Radiation

Since neutrons do not have a charge, they do not ionize atoms in the same way that charged particles such as protons and electrons do. Neutron radiation is an indirect ionizing radiation, which consists of free neutrons. Neutrons are emitted during either spontaneous or induced nuclear fission or nuclear fusion processes and very high energy reactions.[2]

Electromagnetic Radiation

The two types of electromagnetic radiation are

- Gamma radiation
- X-rays

Gamma (γ) Radiation

Gamma (γ) rays are released when an atomic nucleus releases excess energy after a decay reaction. Many beta emitters also emit gamma rays. There are no pure gamma emitters. Gamma rays have no charge and are highly penetrating to tissue. The approximate range in air is 500 m. There are three ways in which gamma rays interact with matter (photoelectric effect, Compton effect, and pair production). In the photoelectric effect, a gamma photon will eject an electron and transfer all of its energy to the electron. In the

Compton effect, a gamma photon ejects an electron and a gamma photon and the gamma photon may cause further ionizations. In pair product, a gamma photon comes into the vicinity of and without coming into contact with a nucleus, causing an electron and positron to be created.

X-rays

X-rays are produced when an atomic nucleus stabilizes itself by taking an electron from an electron cloud. The properties of x-rays are the same as for gamma rays. However, x-rays have less energy than gamma rays, typically in the 500 eV to 500 keV range.

Radiation Basics

As discussed in Chapter 4, all matter is composed of atoms, and each atom is made up of three fundamental particles (*protons, neutrons,* and *electrons*). The *proton* is a subatomic particle with an electric charge of +1 elementary charge. It is found in the nucleus of each atom, along with neutrons, but is also stable by itself and has a second identity as the hydrogen ion, H^+. The mass of a proton is considered to be 1 atomic mass unit (amu). The *neutron* is a subatomic particle with no net electric charge and a mass slightly larger than that of a proton or mathematically 1 amu. They are usually found in atomic nuclei. The *nuclei* of most atoms consist of protons and neutrons, which are therefore collectively referred to as nucleons.[2] The number of protons in a nucleus is the atomic number and defines the type of element the atom forms. The *electron* is a subatomic particle that carries an electric charge of –1. Its mass is approximately 0.0005 amu.

An atomic mass unit is approximately equal to the mass of a proton or neutron and numerically equal to 1.66E–24 g. The mass of an electron is negligible. As discussed in Chapter 4, atoms are symbolized as follows:

$$^A X_{z'}$$

where
   X = the chemical symbol of the element
   A = the mass number (number of protons plus number of neutrons)
   Z = the atomic number (number of protons in the nucleus)

Biological Effects of Ionizing Radiation

General

Biological effects are attributed to the ionization process that destroys the capacity for cell reproduction or division or causes cell mutation. The effects of one type of radiation can be reproduced by any other type. A given total

TABLE 26.1

Acute Dose (Rad) Effects

Rad	Effect
0–25	No observable effect.
25–50	Minor temporary blood changes.
50–150	Possible nausea and vomiting and reduced white blood cells.
150–300	Increased severity of above and diarrhea, malaise, loss of appetite. Some death.
300–500	Increase severity of above and hemorrhaging, depilation. LD_{50} at 450–500 rad.
>500	Symptoms appear sooner. LD_{100} at approximately 600 rad.

Source: U.S. Department of Labor (OSHA), *Introduction to Ionizing Radiation—Lecture Outline*, http://www.osha.gov/SLTC/radiationionizing/introtoionizing/ionizing handout.html, p. 4.

dose will cause more damage if received in a shorter period. A fatal dose (600 R) causes a temperature rise of only 0.001°C and ionization of 1 atom in 100 million.

Acute Somatic Effects

All cells in the human body are considered somatic, with the exception of the sperm and egg cells. The effects of ionizing radiation are relatively immediate to a person acutely exposed and the severity depends upon the dose. Death usually results from damage to bone marrow or intestinal wall. Acute radiodermatitis is common in radiotherapy, while chronic cases occur mostly in industry. Table 26.1 shows the acute dose (rad) effects.[3]

Delayed Somatic Effects

Genetic effects to offspring of exposed persons are irreversible and nearly always harmful. Doubling dose for mutation rate is approximately 50–80 rem. Spontaneous mutation rate is approximately 10–100 mutations per million population per generation.

Critical Organs

Organs that are generally most susceptible to radiation damage include lymphocytes, bone marrow, gastrointestinal cells, gonads, and other fast-growing cells. The central nervous system is resistant. Many nuclides concentrate in certain organs rather than being uniformly distributed over the body, and the organs may be particularly sensitive to radiation damage.

Description of Ionizing Radiation Units

Table 26.2 provides the units used in ionizing radiation and a description of each.

TABLE 26.2

Radiation Unit Description

Usage	Unit	Description
Activity	Curie (Ci)	The number of atoms disintegrating per unit of time. 1 Ci = 3.7×10^{10} disintegrations per second
Activity	Becquerel (Bq)	1 Bq is equivalent to one decay per second
Exposure	Roentgen (R)	A Roentgen is the amount of gamma or x-ray radiation necessary to produce 1 esu (electrostatic unit) of charge in 1 cm^3 of dry air at 760 mm Hg (equivalent to 0.258×10^3 Coulombs/kg air or 5.4×10^7 MeV/g air)
Exposure	Roentgen (R)	Amount of x-ray or gamma radiation that produces ionization resulting in one electrostatic unit (esu) of charge in 1 cm^3 of dry air at STP. Measured in mR/h
Dose	rad (rad)	Radiation absorbed dose 1 rad = 100 erg/g 1 rad = approximately 1 R
Biological dose	Radiation equivalent to man (rem)	• Dose equivalent • Biological effect on a person varies with different types of radiation • rem = #rads × QF • QF = quality factor • QF (x-ray, gamma, and beta radiation = 1) • QF (thermal neutrons = 3) • QF (high-speed neutrons = 10) • QF (alpha particles, energy dependent = 1–20)
Biological dose	Sievert (Sv)	1 Sv = 100 rem

Types of Radioactive Decay

As part of the natural process, radioactive materials spontaneously emit various combinations of ionizing particles (alpha and beta) and gamma or x-rays of ionizing radiation to become more stable. The first type of decay is known as *alpha decay*. In alpha decay, the atomic mass is reduced by 4 and the protons are reduced by 2. For example, radium has an atomic mass of 226 and has 88 protons. During alpha decay, a separate helium nuclide is produced that has an atomic mass of 4 and has 2 protons, leaving a +2 charge. In addition, the radium now becomes radon, with an atomic mass of 222 and with 86 protons. This is represented in the formula below.

$$\ce{^{226}_{88}Ra} \quad \rightarrow \quad \ce{^{4}_{2}He^{+2}} \quad \rightarrow \quad \ce{^{222}_{86}Rn}$$

The second type of decay is known as *beta decay*. During beta decay, there is no change in the atomic mass. However, the protons increase by 1. For example, the element strontium has an atomic mass of 90 and has 38 protons.

During beta decay, one beta electron is added, creating yttrium, which has an atomic weight of 90 and has 39 protons. Beta decay is represented as follows from strontium:

$$^{90}_{38}Sr \rightarrow \beta^- \text{ electron} \rightarrow {}^{90}_{39}Y.$$

Calculating Radioactive Decay

Radioactive decay can be calculated using the following equation:

$$N_t = N_o e^{-\gamma(t)},$$

where
  N_t = number of atoms remaining at time t
  N_o = number of original atoms
  γ = disintegration constant $(0.693/T_{1/2})$
  t = time of decay
  $T_{1/2}$ = half-life (time in which half of the original atoms will disintegrate)

For example, Strontium-90 has an initial activity of 5050 GBq/g. What would be the activity after 14 years of decay, given that the half-life value is 29.12 years?

$$N_t = N_o e^{-\gamma(t)}$$

$$N_{14 \text{ years}} = (5050 \text{ GBq/g}) e^{-\left(\frac{0.693}{29.12 \text{ years}}\right)14 \text{ years}}$$

$$N_{14 \text{ years}} = \left(5050 \frac{\text{GBq}}{\text{g}}\right)(0.7167)$$

$$N_{14} = 3619.06 \text{ GBq/g}.$$

Therefore, after 14 years, Strontium-90 would have a 71.66% of the fraction of the original activity.

Radioactive Half-Life

The half-life of a radioactive element is the time that it takes for one-half of the atoms of that substance to disintegrate into another nuclear form. These can range from fractions of a second, to many billions of years. In addition, the half-life of a particular radionuclide is unique to that radionuclide, meaning that knowledge of the half-life leads to the identity of the radionuclide. Table 26.3 provides a list of half-life values for commonly encountered radionuclides.

TABLE 26.3

List of Half-Life Values for Commonly Encountered Radionuclides

Element	Symbol	Half-Life Value
Actinium	Ac-225	10.0 days
	Ac-227	21.773 years
	Ac-228	6.13 h
Americium	Am-241	432.2 years
	Am-242	16.02 h
	Am-242m	152 years
	Am-243	7380 years
Antimony	Sb-124	60.2 days
	Sb-125	2.77 years
	Sb-126	12.4 days
	Sb-126m	19.0 min
	Sb-127	3.85 days
Argon	Ar-41	1.827 h
Beryllium	Be-10	1.6E6 years
	Be-7	53.44 days
Cadmium	Cd-113m	13.6 years
	Cd-115m	44.6 days
Cerium	Ce-141	32.5 days
	Ce-143	33.0 h
	Ce-144	284.3 days
Californium	Cf-252	2.638 years
Carbon	C-11	20.38 min
	C-14	5730 years
Cesium	Cs-134	2.062 years
	Cs-134m	2.90 h
	Cs-135	2.3E6 years
	Cs-136	13.1 days
	Cs-137	30.0 years
	Cs-138	32.2 min
Chromium	Cr-51	27.704 days
Cobalt	Co-56	78.76 days
	Co-57	270.9 days
	Co-58	70.8 days
	Co-60	5.27 years
Copper	Cu-61	3.408 h
	Cu-64	12.701 h
Curium	Cm-242	162.8 days
	Cm-243	28.5 years
	Cm-244	18.11 years
	Cm-245	8500 years
	Cm-246	4730 years

(Continued)

TABLE 26.3 (CONTINUED)

List of Half-Life Values for Commonly Encountered Radionuclides

Element	Symbol	Half-Life Value
	Cm-247	1.56E7 years
	Cm-248	3.39E5 years
Iridium	Ir-192	74.02 days
Krypton	Kr-83m	1.83 h
	Kr-85	10.72 years
	Kr-85m	4.48 h
	Kr-87	76.3 min
	Kr-88	2.84 min
Plutonium	Pu-238	87.74 years
	Pu-239	24,065 years
	Pu-240	6537 years
	Pu-241	14.4 years
	Pu-242	3.76E5 years
	Pu-243	4.956 h
	Pu-244	8.26E6 years
Radium	Ra-223	11.434 days
	Ra-224	3.66 days
	Ra-225	14.8 days
	Ra-226	1600 years
	Ra-228	5.74 years
Radon	Rn-219	3.96 s
	Rn-220	55.6 s
	Rn-222	3.824 days
Selenium	Se-75	119.78 days
	Se-79	65,000 years
Strontium	Sr-85	64.84 days
	Sr-87m	2.81 h
	Sr-89	50.5 days
	Sr-90	29.12 years
	Sr-91	9.5 h
	Sr-92	2.71 h
Thallium	Tl-201	73.06 h
	Tl-207	4.77 min
	Tl-208	3.07 min
	Tl-209	2.20 min
Uranium	U-232	72 years
	U-233	1.59E5 years
	U-234	2.445E5 years
	U-235	7.03E8 years
	U-236	2.34E7 years
	U-237	6.75 days

TABLE 26.3 (CONTINUED)

List of Half-Life Values for Commonly Encountered Radionuclides

Element	Symbol	Half-Life Value
	U-238	4.47E9 years
	U-240	14.1 h
Yttrium	Y-90	64.0 h
	Y-91	58.51 days
	Y-91m	49.71 min
	Y-92	3.54 h
	Y-93	10.1 h

Radiation Control Methods

There are three primary controls for external radiation. These three controls are *time, distance,* and *shielding.*

Time

Exposure to radiation is a function of time. Therefore, reducing the time means less exposure. The time equation, used in radiation work, is as follows:

$$\text{Dose} = D \times T,$$

where
  D = dose rate
  T = time of exposure

For example, a radiation worker is working in an area, in which the radiation measures 50 mrem/h for a period of 6 h. What is the employee's total exposure for this period?

$$\text{Dose} = 50 \frac{\text{mrem}}{\text{h}} \times 6 \, \text{h}$$

$$\text{Dose} = 300 \, \text{mrem}.$$

Therefore, the time of exposure to radioactive materials should be minimized as much as possible.

Distance

The dose received by employees is inversely proportional to distance; therefore, the greater the distance from the source means the less the dose. The maximum distance from the source to properly control or manipulate the materials should be determined. To calculate the dose at a distance, use the following equation, known as the Inverse Square Law:

$$I_2 = I_1 \frac{(d_1)^2}{(d_2)^2},$$

where
  I_1 = intensity at distance 1
  I_2 = intensity at distance 2
  $d_{1\text{ or }2}$ = distances at location 1 or 2

For example, an employee is working approximately 1 ft from the source, which measures 148 mrem/h. If the employee were to work 3 ft from the source, what would the dose rate be?

$$I_2 = 148\,\text{mrem/h} \frac{(1\,\text{ft})^2}{(3\,\text{ft})^2}$$

$$I_2 = 148\,\text{mrem/h} \frac{1}{9}$$

$$I_2 = 16.44\,\text{mrem/h}.$$

By moving the employee an additional 2 ft from the source, the exposure would be 16.44 mrem/h versus 148 mrem/h.

Source Strength for Gamma (γ) Radiation

To calculate the source strength from gamma (γ) radiation, use the following equation:

$$S = 6CE,$$

where
  S = R/h/1 ft
  C = Curie strength
  E = energy (MeV)

Shielding

By placing an appropriate shield between the radioactive source and the employee, radiation is attenuated and exposure may be completely eliminated or reduced to an acceptable level. The type and amount of shielding needed to achieve a safe working level vary with the type and quantity of radioactive material used. The half-value layer (HVL) may be used as a guide to the thickness of the shielding necessary to block the radiation. The HVL is the thickness of the shielding necessary to reduce the radiation dose rate to half of the original or unshielded dose rate. The HVL is expressed in units of distance (millimeters or centimeters). The HVL is inversely proportional to the attenuation coefficient. The *attenuation coefficient* is a quantity that characterizes how easily a material or medium can be penetrated by a beam of light, sound, particles, or other energy or matter.[4] The HVL can be calculated using the following equation:

$$HVL = \frac{0.693}{\mu},$$

where

μ = linear absorption coefficient for material.

Table 26.4 shows an example of HVL for various sources.
To calculate the approximate effectiveness of a shield, use one of the following equations:
For gamma or x-rays,

$$I = I_o e^{-\mu x},$$

where

I = intensity after shielding
I_o = original intensity
μ = linear absorption coefficient of material
x = shield thickness

TABLE 26.4

Example of Half-Value Thickness (mm)

Material	Source				
	Co 60	Ir 192	Yb 169	100 kV X-rays	250 kV X-rays
Depleted uranium	6.5	2			
Lead	12.4	6.4	2.6	0.24	0.82
Steel	20	13	9.5	6.3	10
Concrete	66	48		19	28

Source: Online Showcase, 2010.http://onlineshowcase.tafensw.edu.au/ndt/content/rad _safety/task4/accessible.htm#topic5.

For example, the intensity at 2 ft from ^{137}Cs source is 35 mrem/h. What is the intensity at this point if a 5-cm lead shield is placed between the source and the detector? (μ for lead [662 keV gamma ray] = 1.23 cm^{-1})

$$I = \left(35\frac{\text{mrem}}{\text{h}}\right)e^{-1.23\times5}$$

$$I = \left(35\frac{\text{mrem}}{\text{h}}\right)(0.002)$$

$$I = 0.075 \text{ mrem/h.}$$

Beta (β) particle radiation is somewhat more complex, in that they are more strongly affected by shielding, since they have charge and mass. Therefore, for beta (β) radiation, use the following equation:

$$I = \beta I_o e^{-\mu x},$$

where
  I = intensity after shielding
  I_o = original intensity
  μ = linear absorption coefficient of material
  x = shield thickness
  β = radiation scatter "buildup" factor (for radiation workers, the Occupational Safety and Health Administration [OSHA] assumes β = 1)

NOTE: Buildup factors can be located in American Nuclear Standard 6.4.3. Using a β of 1 would not make a difference in this calculation.

Personal Protective Equipment

The specific type of protective equipment used depends on the nuclide, level of activity, chemical form, and type of work being performed. Examples of personal protective equipment may include laboratory coats, safety glasses, respirators, or specially designed radiation protection suits. Other equipment may include biological safety cabinets and specially designed ventilation systems.

OSHA Exposure Standards

OSHA standards for ionizing radiation can be found in 29 CFR 1910.1096. Table 26.5 summarizes the exposure limits for ionizing radiation.

TABLE 26.5

Occupational Ionizing Radiation Exposure Limits

Body Part	rem
Whole body; head and trunk; active blood-forming organs; lens of eye; or gonads	1 1/4 rem/quarter
Hands and forearms; feet and ankles	18 3/4 rem/quarter
Skin of whole body	7 1/2 rem/quarter
Whole body	5 rem/year

Note: $D = 5(N - 18)$, where: D = dose; N = age of employee on last birthday. The dose to the whole body, when added to the accumulated occupational dose to the whole body, shall not exceed $5(N - 18)$ rem, where N is equal to the individual's age in years at his last birthday.

Nonionizing Radiation

Nonionizing radiation is described as a series of energy waves composed of oscillating electric and magnetic fields traveling at the speed of light. Nonionizing radiation includes the spectrum of ultraviolet (UV), visible light, infrared (IR), microwave (MW), radio frequency (RF), and extremely low frequency (ELF). Lasers commonly operate in the UV, visible, and IR frequencies. Nonionizing radiation is found in a wide range of occupational settings and can pose a considerable health risk to potentially exposed workers if not properly controlled.[6]

Nonionizing radiation is illustrated in Figure 26.1

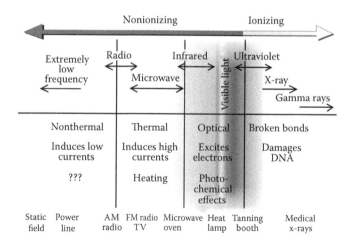

FIGURE 26.1

Types of radiation in the electromagnetic spectrum. (From U.S. Environmental Protection Agency (EPA), http://www.epa.gov/rpdweb00/understand/ionize_nonionize.html.)

UV Radiation

UV radiation energy waves are the range of electromagnetic waves 100 to 400 nm long (between the x-ray and visible light spectrums). The division of UV radiation may be classified as Vacuum UV (100–200 nm), UV-C (200–280 nm), UV-B (280–315 nm), and UV-A (315–400 nm).[8] UV radiation has a high photon energy range and is particularly hazardous because there are usually no immediate symptoms of excess exposure. Sources of UV radiation include the sun, black light, welding arcs, and UV lasers.[6] The effects of UV radiation are primarily limited to the skin and eyes. UV radiation can cause redness or erythema, photosensitization, skin cancer, corneal lesions, photokeratitis (welder's flash), or snowblindness from welder's flash. The controls for UV radiation include solid materials to block the sun and sunscreen applied to the exposed areas.

Visible Light Radiation

The different visible frequencies of the electromagnetic spectrum are "seen" by our eyes as different colors. Good lighting is conducive to increased production and may help prevent incidents related to poor lighting conditions. Excessive visible radiation can damage the eyes and skin.[6] Control measures for visible light include enclosures or filters.

IR Radiation

The skin and eyes absorb IR radiation as heat. Workers normally notice excessive exposure through heat sensation and pain. Sources of IR radiation include furnaces, heat lamps, and IR lasers.[6] Sources of IR radiation include thermal sources such as furnaces, welding, lasers, and incandescent bulbs. Infrared A (0.74–2.5 nm) can penetrate the skin and eyes to the retina. Infrared B (2.5–5 nm) is almost completely absorbed by the upper layers of the skin and eyes. Infrared C (5–300 nm) can cause thermal burns on the skin and cornea and cataracts on the lens. Control measures for IR radiation are enclosures and glass doped without neodymium is opaque to near-IR shielding.

MW Radiation

MW sources include ovens, televisions, and some radars. The effects of MWs include a rise in surface temperature, and longer wavelengths can penetrate and heat deep body tissue and can cause cataracts on the lens of the eyes. Shielding with metals with high dielectric constant and that are transparent is the method to control MW radiation.

Certain behavior characteristics of electromagnetic fields dominate at one distance from the radiating antenna, while a completely different behavior can dominate at another location. Electrical engineers define boundary regions to categorize behavior characteristics of electromagnetic fields as a

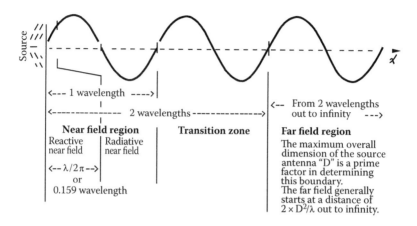

FIGURE 26.2
Diagram of electromagnetic field zones. (From OSHA.[8])

function of distance from the radiating source. These regions are the "near field," "transition zone," and "far field," as illustrated in Figure 26.2.[8]

Calculating MW Radiation (Near Field)

The region located less than one wavelength from the source is called the "near field." Here, the relationship between E and H becomes very complex, and it requires measurement of both E and H to determine the power density. Also, unlike the far field where electromagnetic waves are usually characterized by a single polarization type (horizontal, vertical, circular, or elliptical), all four polarization types can be present in the near field.[8]

To calculate the estimated power density levels in the near field of large aperture circular antennas, use the following equation:

$$W = \frac{16P}{\pi D^2} = \frac{4P}{A},$$

where
W = power density (mW/cm²)
P = antenna power (W)
A = effective antenna area (cm²)
D = diameter of antenna (cm)

Example

Calculate the estimated power density at a diameter of 36 in. (1 in. = 2.54 cm) for 50 W transmitted through an antenna (1 W = 1000 mW).
The first step would be to convert inches into centimeters; 36 in. = 91.44 cm. Now, we can proceed to the following equation:

$$W = \frac{16(50,000 \text{ mW})}{(3.14)(91.44 \text{ cm})^2}$$

$$W = \frac{800,000 \text{ mW}}{26,254.4 \text{ cm}^2}$$

$$W = 30.5 \text{ mW/cm}^2.$$

Calculating MW Radiation (Far Field)

To calculate the estimated power density levels in the far field of large aperture circular antennas, use the following equation:

$$W = \frac{GP}{4\pi r^2} = \frac{AP}{\lambda^2 r^2},$$

where
W = power density (mW/cm²)
P = antenna power (mW)
A = effective antenna area (cm²)
λ = wavelength (cm)
r = distance from antenna (cm)
G = gain $\left(\text{calculated by } G = \frac{4\pi A}{\lambda^2} \right)$. When gain is used in calculations, values of the absolute gain must be used: $G = 10^{\frac{g}{10}} = \text{antilog } \frac{g}{10}$.

Example

Calculate the estimated power density at 150 ft for 10,000 W transmitted through an antenna with a gain of 10. Remember to convert watts into milliwatts and feet into centimeters.

$$W = \frac{GP}{4\pi r^2}$$

$$W = \frac{(10)(10 \text{ mW})}{4(3.14)(4572 \text{ cm})^2}$$

$$W = \frac{100 \text{ mW}}{(4)(3.14)(20,903,184 \text{ cm}^2)}$$

$$W = \frac{100 \text{ mW}}{262,677,157.2 \text{ cm}^2}$$

$$W = 0.000000381 \text{ mW/cm}^2.$$

RF Radiation

RF radiation sources include communication systems, heat sealers, medical equipment, radar, and MW ovens. RFs can cause deep body heating, cataracts, reproductive effects, immune system effects, and endocrine effects. RFs are controlled by enclosures and shielding.

Lasers

LASER is an acronym that stands for *light amplification by stimulated emission of radiation*. It is a mechanism for emitting electromagnetic radiation, usually light or visible light, through stimulated emission. Laser output may be continuous or pulsed. Lasers are used in everyday life including electronics, information technology, military, medicine, and law enforcement. Lasers are classified into four categories, on the basis of their ability to cause injury. Table 26.6 illustrates the four classes of lasers.

Laser Safety-Control Measures

The general control measures for lasers include the following:

- Shielding to reflect radiation
- Shielding to absorb radiation
- Restrict access to the radiation source
- Increase distance from the source
- Limit time of exposure
- Utilize less hazardous radiation

Effective Irradiance

Effective irradiance is the measurement of some radiometers used to measure actinic UV. Effective irradiance is measured in $W/cm^2 \cdot nm$. To calculate the effective irradiance of an instrument, use the following equation:

TABLE 26.6

Laser Classification

Classification	Description
Class I	Low power and low risk.
Class II	Visual system, low power, and low risk. The blinking reflex is typically sufficient to protect employees. However, if staring into the laser is done, retinal damage can occur.
Class III	Medium power and medium risk.
Class IIIA	Limit the exposure to the eyes, visible laser.
Class IIIB	Can cause accidental injury if viewed directly.
Class IV	High power and high risk. There is a concern with specularly reflected beams.

$$E_{\text{eff}} = \sum E_\lambda S_\lambda \Delta_\lambda,$$

where
  E_{eff} = effective irradiance (W/cm²·nm)
  E_λ = spectral irradiance (W/cm²·nm)
  S_λ = relative spectral effectiveness (no unit)
  Δ_λ = bandwidth (nm)

Optical density can be calculated using the following equation:

$$OD = \log\left[\frac{I_{\text{incident}}}{I_{\text{transmitted}}}\right],$$

where
  OD = the measured optical density of the material being evaluated (dimensionless)
  I_{incident} = incident laser beam intensity (W/cm²)
  $I_{\text{transmitted}}$ = transmitted laser beam intensity (W/cm²)

Example

Given a laser having a spectral irradiance of 4 W/cm²·nm, a power density of 15 W/cm², and an optical density of 5.9. What is the effective irradiance?

$$E_{\text{eff}} = \sum\left(4\frac{W}{cm^2}nm\right)\left(15\frac{W}{cm^2}\right)(5.9)$$

$$E_{\text{eff}} = 24.9 \; W/cm^2 \cdot nm.$$

Speed of Light Equation

To calculate the speed of light, use the following equation:

$$C = \lambda f = \frac{\lambda}{T},$$

where
  C = speed of light in a vacuum, which is 2.99792458×10^8 m/s
  λ = the wavelength of the photon (m)
  f = the frequency associated with the photon (cycles/s)

Key Information to Remember on Radiation Safety

1. Ionizing radiation occurs as the result of particles or electromagnetic waves having enough energy to detach electrons from atoms or molecules, thereby causing ionization of the atom.

2. The three types of particle radiation are alpha (α), beta (β), and neutron (n).

3. Alpha radiation is a helium nucleus that has two neutrons and two protons.

4. Beta particles are excess electrons. They are formed when an atom with one excess neutron transforms the neutron to a proton and ejects the extra electron.

5. Neutron radiation is an indirect ionizing radiation, which consists of free neutrons.

6. The two types of electromagnetic radiation are gamma (γ) radiation and x-rays.

7. Gamma rays are released when an atomic nucleus releases excess energy after a decay reaction.

8. X-rays are produced when an atomic nucleus stabilizes itself by taking an electron from an electron cloud.

9. Biological effects of ionizing radiation are attributed to the ionization process that destroys the capacity for cell reproduction or causes cell mutation.

10. Radioactive decay occurs as alpha decay, beta decay, or series chain.

11. The radiological half-life is the time that it takes for one-half of the atoms of that substance to disintegrate into another nuclear form.

12. The three primary controls for radiation are time, distance, and shielding.

13. Reducing the time means less exposure.

14. The dose received by employees is inversely proportional to distance; therefore, the greater the distance from the source means the less the dose.

15. By placing an appropriate shield between the radioactive source and the employee, radiation is attenuated and exposure may be completely eliminated or reduced to an acceptable level.

16. OSHA's ionizing radiation exposure limit is 5 rem/year, 1.25 rem/quarter.

17. Nonionizing radiation is described as a series of energy waves composed of oscillating electric and magnetic fields traveling at the speed of light. Nonionizing radiation includes UV, visible light, IR, MW, RF, and ELF.

References

1. Fleeger, A. and Lillquist, D., 2006. *Industrial Hygiene Reference & Study Guide*, American Industrial Hygiene Association, Fairfax, VA, p. 57.
2. Wikipedia, 2010. Available at http://en.wikipedia.org/wiki/Neutron.
3. U.S. Department of Labor (OSHA), 2010. *Introduction to Ionizing Radiation—Lecture Outline*, p. 4. Available at http://www.osha.gov/SLTC/radiationionizing/introtoionizing/ionizinghandout.html.
4. Wikipedia, 2010. Available at http://en.wikipedia.org/wiki/Attenuation_coefficient.
5. Online Showcase, 2010. Available at http://onlineshowcase.tafensw.edu.au/ndt/content/rad_safety/task4/accessible.htm#topic5.
6. U.S. Department of Labor (OSHA), 2010. *Non-Ionizing Radiation*. Available at http://www.osha.gov/SLTC/radiation_nonionizing/index.html.
7. U.S. Environmental Protection Agency (EPA), 2010. Available at http://www.epa.gov/rpdweb00/understand/ionize_nonionize.html.
8. U.S. Environmental Protection Agency (EPA), 2010. Available at http://www.epa.gov/ogwdw000/mdbp/pdf/alter/chapt_8.pdf.

27

Behavior-Based Safety

Since the 1980s, there have been a wide variety of viewpoints regarding the topic of behavior-based safety programs. They can range from the viewpoint that behavior-based safety programs are the "catch all, end all" solution to accidents to "behavior-based programs are designed strictly to blame the worker without eliminating the hazards." As you read on in this chapter, you will see that behavior-based safety is one tool in the toolbox and has many facets to it.

So, what is behavior-based safety? To answer this question, it may be easier to first tell you what it is not. It is not the "magic bullet" that managers at all levels would like to have that would miraculously eliminate all injuries. It is not, or rather should not be, a method of placing blame on workers for their own injuries. To further attempt to define behavior-based safety, it is necessary to also give a little history of how the safety profession got to the point we are at today in regard to behavior-based safety.

In 1931, Herbert William Heinrich published a book entitled *Industrial Accident Prevention: A Scientific Approach*. Heinrich was an Assistant Superintendent of the Engineering and Inspection Division of Travelers Insurance Company when he published his book. In this book, Heinrich theorized that for every major accidents resulting in injury, there are 29 minor accidents that cause minor injuries and 300 accidents that cause no injuries. This became known as "Heinrich's law" and is represented graphically in Figure 27.1.

Also included in Heinrich's work was the statement that roughly 85%–90% of all accidents occurred as a result of "worker errors." According to Aubrey Daniels in his PM eZine Magazine article, "What Is Behavior-Based Safety?," which discussed Heinrich's conclusion that worker error is the major cause of accidents, it is easy to see how companies began to blame employees for having an accident or causing one. Because of this focus, many of the early safety programs concentrated on stopping unsafe behavior through negative consequences.

With the publication of Heinrich's book, *Industrial Accident Prevention: A Scientific Approach*, companies began to make a more systematic approach to analyzing accident data. However, as far as we can tell from the literature, there was nothing behavioral in his "scientific approach." His interest was in analyzing data and not in changing it. This is not meant to minimize Heinrich's contribution to the systematic study of safety, but he is not in the lineage of modern behavior-based safety.[1]

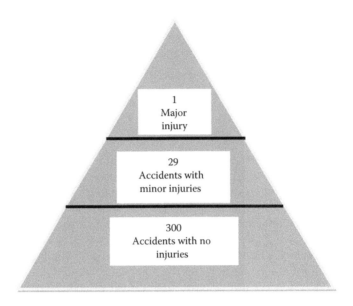

FIGURE 27.1
Heinrich's law.

Quite simply, behavior-based safety has its basis in the study of behaviors. When behaviors are studied or observed specifically for safety, then efforts and resources can be provided in order to convert unsafe behaviors into safe behaviors. Regardless of which specific method you choose to purchase and train the employees on, they all have at their core mission to reduce overall injuries and incidents by modifying behaviors and creating "habit strength" safe behaviors. Thus, if a single definition had to be made, I would suggest that behavior-based safety is a multistage process leading to observation, feedback, and continuous safety improvement.

There are a wide variety of systems available on the market, including Aubrey Daniels, SafeStart, Dupont STOP, Check 6, BST, and others. Each of these programs has the following common elements:

- Identify (or target) behaviors that affect safety
- Define these behaviors precisely enough to measure them reliably
- Develop and implement mechanisms for measuring those behaviors in order to determine their current status and set reasonable goals
- Provide feedback
- Reinforce progress[2]

While there have been substantial changes over the years to "behavior-based safety," the overwhelming data show that if implemented correctly, behavior-based programs greatly assist in the reduction of accidents and injuries. However, if improperly implemented, these same programs can cause great harm to the safety program and to the overall organization as a whole. Some of the issues that have caused serious problems to behavior-based programs include

- Using the program to place blame on employees
- Improper training
- Using behavior-based safety as your whole focused safety program (it is just one tool in the toolbox)
- Using the program to punish employees (this is a very common mistake)
- Not getting initial buy-in from employees
- Not including all employees (management and hourly staff)

Additional Resources

Based on the wide variety of approaches to behavior-based safety, I have provided the following links for your review and consideration:

- http://aubreydaniels.com/
- http://www.training.dupont.com/dupont-stop?pid=ggl.stop& gclid=Cj0KEQjwr-KeBRCMh92Ax9rNgJ8BEiQA1OVm-H1NIX nikcAAfkHR0zhaxZx3MlmLmwTGbABUhwrxGLwaAo4n8P8HAQ
- http://www.safestart.com/
- http://www.bstsolutions.com/en/
- http://www.checksix.com/

Key Points to Remember on Behavior-Based Safety

1. Heinrich's law states that for every 1 major injury, there are 29 minor injuries and 300 accidents with no injuries.
2. Each behavior-based safety program has the following elements in common: identify (or target) behaviors that affect safety, define these behaviors precisely enough to measure them reliably, develop and implement mechanisms for measuring those behaviors in order

to determine their current status and set reasonable goals, provide feedback, and reinforce progress.

3. Behavior-based safety is not a tool that should be used to blame the workers for their own injuries.

References

1. Daniels, A.C., 2013. What is Behavior-Based Safety?: A Look at the History and its Connection to Science. Available at http://aubreydaniels.com/pmezine /what-behavior-based-safety-look-history-and-its-connection-science.
2. Sulzer-Azaroff, B. and Austin, J., 2000. *Professional Safety*, p. 18.

28

Measuring Health and Safety Performance

You can't manage what you can't measure

—Drucker

On the website of the Occupational Safety and Health Administration (OSHA), under the Safety & Health Management Systems eTool link, the author discusses the difference between *responsibility, authority,* and *accountability*. Supervisors and coordinators have responsibility, authority, and, on occasion, accountability. However, the business owner, operator, or manager has overall accountability. Therefore, there must be some form of measurement to determine success or areas in need of improvement. The elements of an effective accountability system include the following:

- *Established standards* in the form of company policies, procedures, or rules that clearly convey standards of performance in safety and health to employees
- *Resources* needed to meet the standards, such as a safe and healthful workplace, effective training, and adequate oversight of work operations
- A *measurement system* that specifies acceptable performance
- *Consequences*, both positive and negative
- *Application* at all levels

When managers and employees are held accountable for their safety and health responsibilities, they are more likely to press for solutions to safety and health problems than to present barriers. By implementing an accountability system, positive involvement in the safety and health program is created.

Examples of measured safety behavior at various levels include the following:

- *Top/mid-level managers*: Measurement at this level includes personal behavior, safety activities, and statistical results, such as following company safety and health rules, enforcing safety and health rules, arranging safety and health training, and workers' compensation costs.
- *Supervisors*: Measurement should include personal safety behavior and safety activities that they are able to control, such as making

sure employees have safe materials and equipment, following and enforcing safety rules, and conducting safety meetings.

- *Employees*: Measurement usually includes personal behavior, such as complying with safety and health rules and reporting injuries and hazards.[1]

With the understanding that not only good management practices but OSHA as well expect a viable measurement system be put into place that holds various parties responsible and accountable, the question then becomes, "How is this to be accomplished?" When measuring safety performance, there are several aspects that must be addressed. As you will learn later in this chapter, there are various types of indicators of performance and relying on old methods and metrics is no longer sufficient to advance the safety program of an organization.

Almost to a person, any CEO, owner, or business director/manager, when asked to describe their company's performance would describe them in terms of profit percentage, return on investment, or their share of the market. The characteristic of each of these measurements is that they are generally positive in nature, reflecting achievement, rather than negative reflecting failure. However, if you were to ask this same group to describe their safety performance, they would inevitably mention the number of injuries or the lack thereof, incident rates, and so on. Some of them may mention their reduction in overall injury/illness rates. The characteristic of each of these measurements is reflective of an injury (or, in some cases, the lack of something bad occurring). Safety metrics in general have been a negative or past performance measurement.[2]

Major Problems with Injury/Illness Health Statistics in General

In general, the problem with traditional measurements of injury/illness health statistics are as follows:

- Underreporting: An emphasis on injury and illness rates as a measure, particularly when related to reward systems, can lead to such events not being reported so as to "maintain" performance.
- Whether a particular event results in an injury is often a matter of chance, so it will not necessarily reflect whether or not a hazard is under control. An organization can have a low injury rate because of luck or fewer people exposed, rather than a good health and safety management program.

- Injury rates often do not reflect the potential severity of an event, merely the consequence. For example, the same failing to adequately guard a machine could result in a cut finger or an amputation.
- People can stay off work for reasons that do not reflect the severity of the event.
- There is evidence to show that there is not necessarily a relationship between "occupational" injury statistics (e.g., slips, trips, and falls) and control of major accident hazards (e.g., loss of containment of flammable or toxic material).
- A low injury rate can lead to complacency.
- A low injury rate results in few data points being available.
- There must have been a failure (i.e., injury or illness), in order to get a data point.
- Injury statistics reflect outcomes not causes.[2]

Some of the aspects or questions that must be answered include the following:

1. Why measure?
2. What do I measure?
3. When to measure?
4. Who should measure?
5. How to measure?

Each of these questions or aspects will be discussed individually.

Why Measure Performance?

If you will recall, one of the management theories presented in Chapter 19, known as the "Deming Cycle," states that there are four basic steps, which includes (1) Plan, (2) Do, (3) Check, and (4) Act. Measurement is one method of checking—to measure, if you will, the overall performance of your safety program. It is equally important to the safety and health profession as it is to the financial, production, or service delivery management.[2] At the beginning of this chapter, there is a quote, by Peter Drucker, which is worth reiterating and states: "You can't manage what you can't measure." Figure 28.1 illustrates where performance measurement fits within the overall safety and health management system.

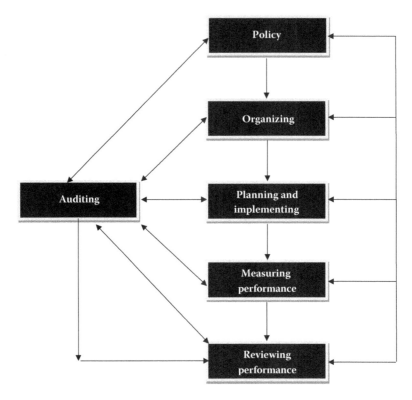

FIGURE 28.1
Health and safety performance management system. (From http://www.hse.gov.uk/opsunit /perfmeas.pdf.)

Therefore, one of the main purposes in measuring performance is to provide information to the decision makers and those accountable for the overall success of the program. The measurement information sustains the operation and development of the health and safety management system, and hence the control of risk, by

- Providing information on how the system operates in practice
- Identifying areas where remedial action is required
- Providing a basis for continued improvement
- Providing feedback and motivation

As stated above, the measurement system helps in deciding

- Where you are relative to where you want to be

- What progress is necessary and reasonable in the circumstances
- How that progress might be achieved against particular restraints (e.g., resources or time)
- The way progress might be achieved
- Priorities and effective use of resources[2]

As has been mentioned earlier in this chapter, one of the overriding purposes of measuring health and safety performance is to hold those in charge accountable for the performance of the program(s). By measuring health and safety performance, those accountable can initially determine where they are, determine where they want to take the program, and then develop a strategic plan on how to achieve those goals and objectives. By measuring, an accountable person can continuously determine where they are relative to the goals and objectives.

What to Measure?

One thing that can be agreed upon by all safety professionals is that health and safety risks must be controlled or eliminated. In order to do this, they must be identified and, therefore, put into the system to be managed. Figure 28.2 graphically illustrates an effective risk control model.

The health and safety management system comprises three levels of control:

- *Level 1*: The key elements of the health and safety management system: the management arrangements (including plans and objectives) necessary to organize, plan, control, and monitor the design and implementation of the risk control system or risk control management system.

Input	Process	Output	Outcome
Uncontrolled	Health and safety management system	Controlled	No injuries
Hazards	Management arrangements (Level 1)	Hazards/risks	No occupational illnesses
The "Hazard burden"	Risk control systems (Level 2)		No incidents
	Workplace precautions (Level 3)		Stakeholder Satisfaction
	Positive H & S culture		

FIGURE 28.2
Effective risk control. (From http://www.hse.gov.uk/opsunit/perfmeas.pdf.)

- *Level 2*: Risk control systems: the basis for ensuring that adequate workplace precautions are provided and maintained.
- *Level 3*: Effective workplace precautions provided and maintained to prevent harm to people at the point of risk.

To effectively answer the question, "What is our health and safety performance?," performance measurement should cover all elements of Figure 28.2. It should be based on a balanced approach that combines the following:

1. *Input*: Monitoring the scale, nature, and distribution of hazards created by the organization's activities—*measures the hazard burden*
2. *Process*: Active monitoring of the adequacy, development, implementation, and deployment of the health and safety management system and the activities to promote a positive health and safety culture—*measures of success*
3. *Outcomes*: Reactive monitoring of adverse outcomes resulting in injuries, illnesses, loss, and accidents with the potential to cause injuries, illness, or loss—*measures of failures*

Measuring the Hazard Burden

An organization has many activities and each of these activities will create hazards, which will vary in nature and significance. Therefore, the range, nature, distribution, and significance of the hazards (*the hazard burden*) will determine the risks that need to be controlled.

Most professionals can agree that the best solution is to eliminate the hazard, either by engineering it out, by replacing it with another method, or simply by no longer carrying out a particular activity. However, this is not always practical or feasible, depending on the situation.

Measuring the hazard burden requires that

- The hazard is first identified
- The hazard be classified according to the significance of the hazard (high, moderate or low risk)
- The hazard be categorized as to whether the significance of the hazard is consistent across the various parts of the organization
- The hazard is determined to be consistent or variable over time
- The hazard reduction or elimination method is successful or not
- A determination is made on what impact business changes or processes are having on the nature and significance of the hazards[2]

Measuring the Health and Safety Management System

The primary and overriding purpose of any safety and health management system is to turn uncontrolled hazards into controlled risks and to ultimately eliminate all injuries, illnesses, and accidents. The key elements of a safety and health management system, graphically illustrated in Figure 28.1, are as follows:

- Policy
- Organizing
- Planning and implementation
- Measuring performance
- Audit and review

Policy

Every organization should have a written health and safety management policy. The measuring process should ensure that this written safety policy[2]

- Exits
- Meets the legal requirements and best practices
- Is up to date
- Assigns specific responsibilities
- Is being implemented effectively

This last bullet point requires more discussion to ensure that it is fully understood. When conducting a self-audit, it is important to verify through whatever means that the policy is fully implemented. What does "fully implemented" mean? Used as a verb, *Merriam Webster Online Dictionary* defines *implement* as follows: "carryout, accomplish; *especially*: to give practical effect to and ensure of actual fulfillment by concrete measures."[3] On the basis of this definition, simply writing a health and safety policy would not be sufficient to "check the box" on the audit form. Some action must be taken to ensure that the plan is (1) communicated to all stakeholders, (2) understood by all stakeholders, and (3) put into action. The individual components of the health and safety plan can be measured or audited separately.

Organizing

Whenever you establish a measurement system, your program should be organized in such a manner to ensure that the following are present, adequate, and implemented:[2]

- That it establishes and maintains management control of health and safety in the organization
- That it promotes effective cooperation and participation of individuals, safety representatives, and relevant groups so that health and safety is a collaborative effort
- That it ensures the effective communication of necessary information throughout the organization
- That it secures the competence of the organization's employees

Planning and Implementation

When measuring the health and safety performance of an organization, it is important to assess the adequacy and implementation of the planning system, which should be able to[2]

- Deliver plans with objectives for developing, maintaining, and improving the health and safety management system
- Design, develop, install, and implement suitable management arrangements, risk control systems, and workplace precautions proportionate to the needs, hazards, and risks of the organization
- Provide effect prioritization of activities based on risk assessment
- Ensure the correct balance of resources and effort is being targeted proportionately according to the hazards/risk profile across the organization (e.g., is disproportionate effort being expended on slips/trips relative to control of major accident hazards or fire safety?)
- Operate, maintain, and improve the system to suit changing needs and process hazards/risks
- Promote a positive health and safety culture

Under the *General Duty Clause*, OSHA states that an employer shall furnish "a place of employment which is free from recognized hazards that are causing or are likely to cause death or serious physical harm to his employees." Where there is no specific standard, OSHA will use the general duty clause for the issuance of citations and fines. The general duty clause can be found in Section 5 (a)(1) of the Occupational Safety and Health Act of 1970.[4] In complying with this requirement, it is necessary to plan and implement an effective health and safety management system.

As mentioned previously, health and safety programs should always include goals and objectives and a viable, sustainable plan to achieve them. While there are many different parts and pieces to an organization, each should be aligned in the overall organizational goals and objectives, to which the Business Manager is to be held accountable. These goals and objectives are linked to the individual components in the various departments of the organization.

It is said that a prerequisite of effective health and safety plans and objectives is that they should be SMART, which is an acronym that stands for[2]

- Specific
- Measureable
- Attainable
- Realistic/Relevant
- Time-bound

Another element is measuring management arrangements and risk control systems. The measurements should address whether the program has the correct capabilities, whether the implementation of the program has been fully and correctly implemented (compliance) and whether it has been properly deployed in order to accomplish the overall goals and objectives.

Capability

In determining whether a system or plan has the capability to deliver the desired outcomes, one must ensure that, first, the system has the necessary physical requirements and resources. Then, just as important as, if not more important than, the physical requirements and resources are the management processes, support, and resources. Without adequate resources in terms of financial, professional resources, and management commitment, the program is doomed to failure from the very beginning.

We discussed the Plan, Do, Check, and Act method earlier in this chapter, but it alone is not sufficient to determine whether a program is in compliance. An example of this is if the system calls for an incident investigation and one is done. Merely doing the investigation would satisfy the requirement on a checklist but would be of no value if the incident investigation did not determine the root cause and then provided some viable recommendations for corrective or preventive actions. In other words, caution should be given that the measurement is not just "technically adequate," but is also practically adequate.

Most importantly, the plan must assign specific responsibilities. For example, some of the key positions within the systems include the following:

- President/CEO
- Department heads
- Safety professionals
- First-line supervisors
- Employees

Each of these key positions has its own particular responsibilities and accountabilities. For example, the President/CEO may have overall responsibility for ensuring that a plan is in place and that the plan is fully funded financially and adequate resources are made available to accomplish the goals and objectives. The safety professional may be responsible for implementing the program, writing the written programs, ensuring that employees receive the appropriate training, and so on.

Compliance

It is often debated as to whether compliance is the same as safe. I have changed my thinking process on this a few times over the years and currently fall into the category that thinks that being compliant with regulatory and corporate requirements lends itself to a more safe and healthy workplace. Not to mention the fact that it is a requirement and, therefore, a vital aspect of any safety professional's job performance. As such, an enormous amount of resources and effort is put into maintaining compliance. The trick is to make sure that compliance merges with safety.

In further discussing the health and safety plans, it is important to point out that the objectives of the program/plan can never be achieved if the individual components of the plan are not being complied with. Therefore, it is vitally important that routine checks of the progress be made to measure the status of the program, and to adjust course, if need be. In a world where there are increasing demands on managers and supervisors to accomplish more with fewer resources, it is incumbent upon them to monitor the most important requirements of their jobs. As the old adage goes, "Inspect what you expect" and the likelihood that it will be accomplished is greatly increased. In doing so, it is also important to ensure that every employee fully understands how his or her organization is expected to operate. Many of us have been in environments where we have had two or more equally important managers to whom we report. For example, we may operationally report to a site manager or manufacturing director, while at the same time reporting to a corporate health and safety director or division vice president or president. Most times, this is not a problem; however, sometimes this gets quite tricky. Furthermore, employees must understand the actual process, such as the individual requirements/programs (energy control program, hearing conservation, etc.) as well as other issues, such as how to report a hazard, who initiates the work order, and so on.

Deployment

As we have discussed, it is not enough that a plan be written and put on a shelf. It must also be implemented, which means that at some point, the plan must have been deployed to every employee. There are a wide variety of methods to deploy a new program or process and each must be tailored to the individual organization. However, when doing so, the deployment process must also be measured as to the status of the deployment.

Figure 28.3 graphically illustrates the three dimensions of measurement that have been discussed above. The main goal of which would be to be in the shaded area of the model. Measuring all three dimensions allows management to gauge a performance that they can apply at different levels, for example, looking at a specific management arrangement (e.g., competence) or risk control system (e.g., entry into confined spaces) or looking at a range of management arrangements and risk control systems at a particular site or across the organization.[2]

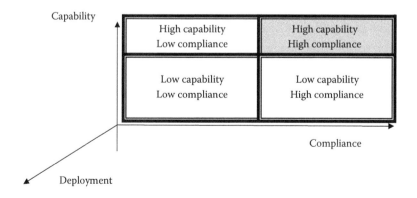

FIGURE 28.3
Three dimensions of measurement. (From http://www.hse.gov.uk/opsunit/perfmeas.pdf.)

When to Measure

As you may have already seen in the preceding sections, measurement of the health and safety program is continuous in nature. If your goals and objectives follow the SMART model, then they should also allow for specific measurement intervals of the milestones. The measurements should be consistent and effective in order to allow design makers to determine if the overall goals and objectives will be met at the end of the final measuring period (i.e., monthly, quarterly, semiannually, or annually). If, for example, the ending period of the current measurement is annually, then maybe monthly measuring periods are the appropriate interval by which to determine if a program is on track to achieve the overall annual goals. If the monthly measurements reveal, at any point, that the overall annual goals and objectives are not going to be met, then adjustments to the design of the system or alteration of the program may become necessary. Some key considerations when determining time intervals are as follows:[2]

- The initial design phase
- Whenever changes are made that could affect the operation of the systems
- When information is obtained that indicates that the system as designed has failed in some way (e.g., when there has been an injury or significant incident)
- When data from the monitoring of the operation of the system indicate that the design is flawed
- The relative importance of the activity or particular precaution relative to the overall control of the risk
- Where intervals for monitoring are prescribed by legislation
- Where there is evidence that there is noncompliance
- Where there is evidence of compliance
- The relative frequency and time at which a particular activity takes place

Who Should Measure

One of the dangers in any organization (and this happens much too often) is where senior management does not understand that there is a problem within their organization unless they are specifically informed. This approach is extremely dangerous and poses significant risks to the organizations.[2]

Therefore, it is necessary that specific systems be put into place (such as the measurement system discussed within this chapter) that indicate whether a program or plan is functional, is properly operational, and ensures that corrective actions are triggered whenever deficiencies are found in the overall plan.

The question as to Who Should Measure is not as straightforward as you might think. For example, one might think that simply because we are discussing the health and safety program, the safety professional assigned would be the logical choice as the person to achieve the measurements. However, it is human nature to report only the positive aspects of any audit. As safety professionals, we understand that the value of a good audit program is to find deficiencies within our programs. However, if the program is tied to a bonus system, then the outcome of the audit may be different if that person were to audit the programs that they were responsible for. That is not to say that there would be no value in a safety professional doing his or her own audits. This occurs with frequent regularity and should. However, the formal audit of the program should be accomplished with both interested and disinterested parties. For example, a good audit or measurement of the system should be objective in nature, meaning that it is truly measureable without bias. Therefore, a measurement system should be verified with disinterested parties, such as an outside auditor, employee representatives, or supervisors.

How to Measure

As we have already discussed in this chapter, the measurement criteria should be an integral part of the overall plan. For example, in your health and safety plan, you may determine that the energy control program will be audited and documented and that a minimum of 10% of all lockout/tagout operations will be audited. Assuming that you have an energy control program that requires a permit system, you could determine how many operations had taken place on a daily or weekly basis and base the 10% on that number. In addition to auditing 10% of the lockout/tagout operations, you will get a detailed report on how the energy control program is operating. By this, I mean that when conducting audits of the energy control program in facilities that I've had experience in, a checklist approach is used that can be quantified to determine overall compliance of the energy control program. For example, if I selected 70 lockout/tagout operations to audit and 9 of them were lacking in some issue, then my overall program would indicate that I am 87% compliant. Delving further into the system, I could determine whether the noncompliance was related to an administrative issue, such as not signing off on the permit, or whether the deficiencies lay in the actual operational aspect of the program, such as failure to isolate a particular valve or electrical source within the system. The operational aspect of the program would be an indication of greater magnitude than that of an administrative error and therefore adjustments to the system can be made.

Oftentimes, many of the programs/plans are audited or measured on an individual basis rather than an overall one-time measurement. As mentioned in the above paragraph, for example, my preference on the energy control program is to conduct weekly audits/measurements of the operations and then roll them up into a monthly and annual audit/measurement report. The frequency and the exact method of how to measure are tailored to the individual organization. As to how to actually perform the measurements, which have already been established in the plan, I prefer the use of forms and checklists that cover the individual aspects of the program.

Leading versus Lagging Indicators

The one thing that I believe every safety professional can agree on is that the overall goal is to ultimately eliminate and, in the interim, to reduce injuries to the lowest possible levels. To do this, we have all searched for the magic tool that will accomplish this ultimate goal. In systematically working our way toward this goal, safety professionals like to quantify information in order to, hopefully, predict our next event. Therefore, we are continually looking for the best safety metrics to monitor.

Ask any safety professional what their safety metrics are and, undoubtedly, almost all, if not everyone, will tell you that they track total case incident rates (TCIRs), severity rates, and days away, restricted or transferred (DARTs). These indicators are, of course, required by both the Mine Safety and Health Administration (MSHA) and OSHA. These metrics are known as lagging indicators, meaning that the events measured have already occurred and are not necessarily predictors of future events. The financial markets have proven this time and time again. If a stock price rose 10% over a 5-year period, then that measurement is no indication of future stock prices.

As Terry L. Mathis, Founder and CEO of ProAct Safety, succinctly put it in an article entitled "5 New Metrics to Transform Safety":[5]

> Safety has historically been measured by its failures. We first measured simply the cost of safety failures and then began to calculate the number and severity of failures. When we realized that numbers did not account for exposure, and thus did not enable comparative metrics, we began to calculate the ratio or rate of our failures per exposure. OSHA selected the magic number of 200,000 hours worked as the standard for calculating accident rates. The standards were then set to define what an accident is and determine what severity of injury constituted the necessity to report the accident.

The question then becomes how valuable to an organizational safety program are lagging indicators? The answer to this question is not an easy one. Besides being required by MSHA and OSHA, most corporate measurements require these to measure how well a safety program has performed over the past year. Have we reduced or eliminated the number of injuries? Have we reduced the severity of injuries? Hence, in this sense, it does provide some measurement of past safety performance. However, these measurements are of little value as predictors of future incidents.

In every interview or discussion that I've had with corporate directors regarding metrics over the past 5–10 years, I've been asked what leading indicators do I use and which ones are the best predictors of future incidents. I will add at this point, if you're looking for an absolute answer in this chapter, you can stop reading at this point. There are no magic bullets and these indicators can vary from company to company and from program to program. In this chapter, I will provide some examples of leading indicators and a discussion of each that I have found over the years.

In an article written by Fred A. Manuele in the December 2009 *Professional Safety* magazine, he states that the selection of leading indicators is largely judgmental and only time will tell whether the indicators selected are the right ones.[6] In other words, even leading indicator information has some lagging indicator qualities. The difference being, that leading indicators do provide some correlation between measured data and reduced incident rates.

What are the characteristics or traits of a good leading indicator? In a report published by the International Council on Mining and Metals,[7] good leading indicators must be reliable, repeatable, consistent, and independent. To be of use in improving safety and health, both leading and lagging indicators should

- Allow accurate and detailed comparisons
- Lead to correct or help avoid erroneous conclusions
- Be well understood by everyone, especially those responsible for implementing change
- Have a quantitative basis (even when measuring a qualitative dimension)
- Measure what they are supposed to, consistently, accurately, and reliably
- Collect information that is relevant to the required management decisions and actions
- Adequately map and identify causal linkages (root causes, precursors, events, and outcomes)
- Prompt an appropriate response, leading to consistent focus on implementing change

TABLE 28.1

Key Differences between the Characteristics of Leading and Lagging Indicators

Leading Indicators	Lagging Indicators
Are actionable, predictive, and relevant to objectives	Are retrospective
Identify hazards before the fact	Identify hazards after the fact
Allow preventative actions before the hazard manifests as an incident	Require corrective actions to prevent another similar incident
Allow response to changing circumstances through implementing control measures before the incident	Indicate that circumstances have changed; control measures can be implemented after the incident
Measure effectiveness of control systems	Measure failures of control system
Measure inputs and conditions	Measure outcomes
Direct toward an outcome that we want or away from an outcome that we don't want	Measure the current outcome without influencing it
Give indications of systems conditions	Measure system failures
Measure what might go wrong and why	Measure what has gone wrong
Provide proactive monitoring of desired state	Provide reactive monitoring of undesired effects
Are useful for internal tracking of performance	Are useful for external benchmarking
Identify weaknesses through risk control system	Identify weaknesses through incidents
Are challenging to identify and measure	Are easy to identify and measure
Evolve as organizational needs change	Are static

Source: ICMM, 2012. http://www.icmm.com/document/4800.

Safety indicators are designed, or rather should be designed, to measure performance in order to identify problem areas, stimulate actions, document management efforts, and reinforce improvements in behavior. The distinguishing feature between leading and lagging indicators is that these objectives are achieved before the potential harm (leading) or after the potential harm (lagging).[7] Table 28.1 illustrates the key difference between the characteristics of leading and lagging indicators.

Continuous Improvement

As mentioned earlier in this chapter, there are no specific indicators that work across the board for every situation and they must be individually selected for each organization and circumstance. With that in mind, whatever indicator that you choose, it must be considered a part of your continual improvement process. You may begin by measuring one aspect of your safety program in order to get to the next step.

Over the span of my 30-plus-year career, I have used numerous kinds of metrics, some of which I currently use and some I have discarded altogether. They are listed here and will be discussed individually:

- Near misses
- Number of safety observations
- Number of participants in the safety observation program
- Number of inspections/audits performed
- Number of safety-related work orders completed
- Number of corrective actions taken
- Employee perceptions of the safety culture
- Quantitative measurements of audit programs (tracking of scores)

Near Misses

The first and most common "leading indicator" that people think of is the number of near misses reported. Whether it is a leading or a lagging indicator can be debated. As Phil La Duke posted on his blog:[8]

> Near misses in themselves aren't leading indicators; they are things that almost killed or injured someone, and most importantly, they are events that happened in the past. Not that anything that happened in the past has to be automatically counted out as a lagging indicator, but unless you still cling to the idea proffered by Heinrich that there is a strict statistical correlation between the number of near misses and fatalities, near misses are no more a leading indicator than your injury rate, lost work days or first aid cases. They simply tell you that something almost happened, and nothing more.

His statement would make you think that near-miss reporting should be discounted as a leading indicator and placed in the lagging indicator category. He goes on to make the statement, which I wholeheartedly believe in, that near-miss reporting is indeed a leading indicator because when people report near misses, (a) they are more actively engaged in safety day to day and (b) the more the individual reports near misses, the better he or she is at identifying hazards. In addition, he states that to gauge the robustness of your safety process, measuring the level of participation in near-miss reporting is a good leading indicator. Just to clarify, it is the participation and not the actual number of near misses that serves as a good leading indicator.

Number of Safety Observations

In my opinion, this metric has become a favorite among many organizations for several reasons. It makes companies feel good when they focus on the number of observations and then they increase their numbers. It is easily quantified and measured. The problem that I've experienced with this metric is that many companies either set a quota for each employee or establish a bonus system for those that submit a certain number. For example, a company may have a policy that says if you complete two safety observations per month, then you will be eligible for a $50 gift card drawing. In addition, for each month you qualify, you will receive one chance at the end of the year for the large drawing, which may be a $1000 award or an item such as a four-wheel ATV. Other problems with this approach are that the employees may be "pencil-whipping" the observations just to become eligible for the awards.

With this said, this metric is not without its value. By increasing the number of observations, there is more focus on the safety aspects of the job. In addition, by increasing the number of observations, employees are at least minimally engaged in the process.

Number of Participants in the Safety Observation Program

Measuring the number of participants over time has become one of my favorite indicators to monitor. Similar to the metrics listed above, there are some real or perceived problems, such as the fact that an employee may be participating in the program just to qualify for the bonus or award. This is a real concern. However, at least in my opinion, the real value in this measurement or metric is that for whatever reason, employees are participating in the program and their focus is on the observation process, at least in the few moments that it takes to conduct the observation. In addition, in my experience, whenever I'm doing observations, I will see discrepancies in an employee's behavior or actions and realize that I have made the same errors while performing that particular task.

Number of Inspections/Audits Performed

A valuable metric that I've found useful, especially in regard to measuring the performance of safety inspectors, professionals, and first-line supervisors is that of the number of inspections or audits conducted in their assigned areas of responsibility. By increasing this number, you achieve several different objectives, as follows:

- Increase visibility of the safety personnel and first-line supervisors, demonstrating and communicating the idea that "Management Cares."
- Supervisors and safety professionals actually get out and know their respective areas.
- Communication is greatly increased when key leadership is involved. (Note: In the organizations that I've been involved in, weekly conversations with intent [focused message, such as wearing of enhanced eyewear] must be held with employees during the audits and therefore conversations with intent are a submeasurement of this metric.)
- Timely corrections of discrepancies can be made.

Quantitative Measurements of Audit Programs (Tracking of Scores)

As previously stated, I prefer the use of checklists for almost every aspect of the safety programs that I've designed, implemented, or been responsible for maintaining. As such, I've developed standardized audit/inspection forms that have quantifiable and weighted measurements for various aspects of the audit (see Chapter 29 for examples). These forms and checklists allow for consistent measurements of individual audits of facilities. One of the metrics that I am partial to is the measurement and tracking of the overall and individual scores, which will, hopefully, show a continual increase in compliance. However, if the opposite holds true and there is a decline, even temporary, immediate corrective measures can be taken to improve the program or correct specific deficiencies.

Percentage of Safety-Related Work Orders Completed

Absolutely, one of my favorite metrics to monitor. The percentage of safety-related work orders completed is a very strong indicator as to how well the maintenance portion of your safety program is functioning. With this being said, an important aspect of this measurement is that there must be a clear definition of what constitutes a safety-related work order. Someone capable of determining whether it is or is not a safety-related work order must be included in the process. Otherwise, this metric can become skewed, as many in the organization will code work orders as safety related, simply to get it moved to a higher priority.

Employee Perceptions of the Safety Culture

This metric can be a valuable tool in evaluating the overall employee perceptions of the safety program. However, it is probably one of the most difficult to measure. The survey itself must be carefully designed with the questions being carefully worded, as to remove all perceived or predetermined bias, which will give the interviewee no real options. Furthermore, it is always interesting to see the results of hourly employees as compared to those of management and first-line supervisors. If this is one of the metrics that you choose to use, I would strongly suggest having the survey questions designed by a professional and then use the same questionnaire over time.

Key Points to Remember on Measuring
Health and Safety Performance

1. According to OSHA, the elements of an effective accountability system include established standards, resources, a measurement system, consequences, and application.

2. Traditional measurements of safety performance, such as TCIRs, DARTs, and so on, have limited or no use in predicting future incidents.

3. Major problems in using injury/illness rates include underreporting, events are a matter-of-chance, actual injury rates are no indication of the severity or potential severity, employees may stay off work for reasons that do not reflect severity of the event, and they are lagging indicators and reflect negative attributes, such as failures or lack of something bad occurring.

4. Elements of an effective health and safety performance management system include policy, organizing, planning and implementing, measuring performance, reviewing performance, and auditing of the performance.

5. An effective measurement system is designed or should be designed to indicate where you currently are and measure periodically as to where you want to be.

6. A prerequisite of effective health and safety goals and objectives should follow the acronym SMART, which represents, Specific, Measurable, Attainable, Realistic/Relevant, and Time-bound.

7. Three dimensions of measurement must include compliance, deployment, and capability of the system to achieve specific and measureable goals and objectives.

8. Continuous improvement of any program must always be built into any sustainable system.

9. The value of any leading indicator must be that it engages employees and supervisors in the safety program.

References

1. http://www.osha.gov/SLTC/etools/safetyhealth/comp1_responsbility.html.
2. http://www.hse.gov.uk/opsunit/perfmeas.pdf.
3. http://www.merriam-webster.com/dictionary/implemented?show=0& t=1406386603.
4. https://www.osha.gov/pls/oshaweb/owadisp.show_document?p_id=3359& p_table=oshact.
5. http://proactsafety.com/uploads/file/articles/5-new-metrics-to-transform -safety.pdf.
6. Professional Safety, 2009. Available at http://www.asse.org/professionalsafety /pastissues/054/12/F2Manuele_1209.pdf.
7. ICMM, 2012. Available at http://www.icmm.com/document/4800.
8. http://philladuke.wordpress.com/2013/06/10/misleading-indicators/.

29

Safety Program Auditing Techniques and Checklists

As previously discussed in Chapter 28, in order to determine whether the goals and objectives of your health and safety program are on track to being met, you must measure them. Chapter 28 provides an overview as to what is expected to measure the program, whereas this chapter will help explain in more detail as to how measurements are accomplished. Examples of some useful tools will also be provided to further help the practicing safety professional.

Purpose of Auditing

The paramount objective of any health and safety audit program is to ensure the health and safety of all employees, contractors, and visitors. This is done by

- Identifying hazards
- Assessing the hazards
- Developing controls to either eliminate or mitigate the hazards
- Implementing the appropriate controls
- Supervising and evaluating the controls to determine whether they are effective or not

In addition to identifying hazards and preventing injuries or illnesses, the health and safety audit program should assist in determining whether or not significant progress is being made in the overall goals and objectives. Remember, one of the main purposes of measuring health and safety performance is to help an organization determine where it currently exists and then to focus the efforts on where it wants to be.

Methodology for Conducting Audits

There are a wide variety of audit techniques in assessing the effectiveness of any health and safety program, but they all boil down to three basic methods according to the Occupational Safety and Health Administration (OSHA). Each of these will be discussed in greater detail throughout the chapter. They are as follows (Note: a major portion of the following sections are taken from Ref. [1]):

- Document review and verification.
- Employee interviews—interviewing employees at all levels will help determine the actual knowledge, awareness, and perceptions.
- Site condition reviews—conducting the actual site visit to conduct a visual audit to determine the existing or potential hazards and finding the weaknesses in the management systems that allowed the hazards to occur or to be "uncontrolled."

Audits can vary from unlimited to limited in scope. For example, an organization-wide health and safety audit can encompass the entire health and safety program, including record keeping, hearing conservation program, confined space entry program, and so on. A limited scope audit may focus on just one of these programs at a time. For example, an audit may be limited to just conducting an audit of the energy control program or just limiting it to one particular area of the facility.

Depending on the scope or level of the audit, one of these audit methods may be used, whereas a more thorough or unlimited audit may require the use of one or more of these methods. For example, whenever I conduct an audit of the energy control program, I will use both employee interviews and document review/verification. While this is a limited-scope audit, the nature of the audit requires that I accomplish two things:

1. To measure the effectiveness of whether the employees have knowledge of the requirements of the energy control program
2. To verify that the procedures listed within the energy control program or isolation procedures are actually being followed

Documentation

Nearly every audit will require that documentation is reviewed. It is a standard technique in auditing. It is particularly useful for understanding whether the tracking of hazards to correction is effective. It can also be used to determine the quality of certain activities, such as self-inspections or routine hazard analysis.

Inspection records can tell the evaluator whether serious hazards are being found, or whether the same hazards are being found repeatedly. If serious hazards are not being found and accidents keep occurring, there may be a need to train inspectors to look for different hazards. If the same hazards are being found repeatedly, the problem may be more complicated. Perhaps the hazards are not being corrected. If so, this would suggest a tracking problem or a problem in accountability for hazard correction.

If certain hazards recur repeatedly after being corrected, someone is not taking responsibility for keeping those hazards under control. Either the responsibility is not clear, or those who are responsible are not being held accountable.[1]

Employee Interviews

Interviewing employees at all levels is a very good technique for determining the quality of health and safety training. If safety and health training is effective, employees will be able to tell you about the hazards they work with and how they protect themselves and others by keeping those hazards controlled. Every employee should also be able to say precisely what he or she is expected to do as part of the program. And all employees should know where to go and the route to follow in an emergency.[1]

When conducting employee interviews, the auditor must be careful in their approach, particularly if you are an outside auditor. If you're an auditor outside of the organization, then there may be some inherent distrust among the employees (i.e., not wanting to say the wrong thing to get their company in trouble). However, if approached correctly, the auditor can quickly relieve the tension between themselves and the employees.

Employee perceptions can provide other useful information. An employee's opinion of how easy it is to report a hazard and get a response will tell you a lot about how well the hazard reporting system is working. If employees indicate that the system for enforcing safety and health rules and safe work practices is inconsistent or confusing, you will know that the system needs improvement.

Interviews should not be limited to hourly employees. Much can be learned from talking with first-line supervisors. It is also helpful to query line managers about their understanding of their safety and health responsibilities.[1]

Site Conditions and Root Causes

It is my opinion that when conducting organization-wide audits, a safety professional must conduct a visual site investigation. Conducting a visual site investigation to look at site conditions is one of the few techniques that will help reveal existing and equally important, potential hazards. Not only can the visual site investigation reveal physical hazards, but it can also help reveal breakdowns in the same management systems meant to prevent or control the hazards.[1] For example, while conducting the visual site visit, you see signage requiring safety glasses and hearing protection. The presence of the signage is obviously a good indication that the communication piece of the health and safety program is being adhered to. However, as you progress in your audit, you see a large percentage of employees in the area not wearing the required personal protective equipment (PPE). The fact that a large percentage of employees are not wearing the required PPE is a very strong indicator that there is a breakdown in what is really being communicated to the employees. By not enforcing the requirement, managers and supervisors are negligent in their responsibilities. One of the benefits of the audit process is to identify hazards, risks, and breakdowns within the system and to take swift action to correct them. Now that this breakdown has been identified, it can be corrected.

Another way to obtain information about safety and health program management is through root analysis of observed hazards. This approach to hazards is much like the most sophisticated accident investigation techniques, in which many contributing factors are located and corrected or controlled.

The audit of the worksite's health and safety program will require the use of one or more of the above methods and will vary with the organization, scope of audit, and the individual auditor. It is now necessary to discuss the individual components of a viable health and safety program, along with discussing useful ways to assess these components.

Elements of an Effective Health and Safety Program

Before beginning to discuss the methods and techniques used to assess the effectiveness of the health and safety program, it is necessary to determine some of the most common elements of a health and safety program.[2] OSHA, in their Voluntary Protection Program, recognizes that effective management of health and safety programs

- Reduces the extent and severity of work-related injuries and illnesses
- Improves employee morale and productivity
- Reduces workers' compensation costs

A common characteristic of an effective health and safety program is that it assigns responsibility to managers, supervisors, and employees. In addition, the program or plan provides for the inspection of facilities and an assessment of the overall program on a regular basis for the presence of hazards in order to more readily control or mitigate them. Another characteristic is that the program allows for the orientation and training of employees to eliminate or avoid hazards.

General Guidelines

An effective health and safety program includes provisions for systematic identification, evaluation, and prevention or control of hazards. It goes beyond specific requirements of the law to address all hazards. OSHA requires that there be a written program. However, it should be understood that it is more important to have an actual, functional health and safety program in reality and that the written program, while required, is less important that what happens in reality. Depending on the size and complexity of the worksite, written programs take on greater importance.[2]

Major Elements

The major elements of an effective health and safety program include the following:

- Management Commitment
- Employee Involvement
- Worksite Analysis
- Hazard Prevention and Control
- Health and Safety Training

Management Commitment

Effective health and safety programs do not just happen. The old adage, "I'd rather be lucky than good," most definitely does not apply to the modern safety professional. For an effective health and safety program to evolve requires leadership and commitment from the very top of an organization. One of the most important aspects of every well-managed company is that management regards worker safety and health as a fundamental value of the organization. Ideally, this means that concern for every aspect of the health and safety of all workers through the facility is demonstrated. Other recommended actions for management leadership include visible management involvement; assigning and communicating responsibility, authority, and resources to responsible parties; and holding those parties accountable. In addition, management should ensure that workers are encouraged to report hazards, symptoms, injuries, and illnesses, and that there are no programs or policies that discourage this reporting.

Visible Leadership

Successful top managers use a variety of techniques that visibly involve them in the health and safety protection of their workers. Managers should look for methods that fit their style and workplace. Some methods include

- Getting out where you can be seen, informally or through formal inspections
- Being accessible
- Being an example, by knowing and following the rules employees are expected to follow
- Being involved by participating on the workplace Safety and Health Committee[3]

This is sometimes referred to as Management by Walking Around (MBWA) or Management by Wandering Around. This term may sound strange or unduly simple. However, the concept behind it is that management should get out of their offices and get involved in what is actually going on within their departments. All too often, we, as managers, are asked to write a policy or procedure, so we set aside the time and "knock it out." With the ever-increasing demands on management to produce more with fewer resources, this can become difficult. However, I would propose the thought that it all becomes a matter of priorities. By way of example, I will use my own experiences with various organizations. As a safety professional, you would like to be involved in every aspect of the organization, including engineering, production, operations, maintenance and distribution, and so on. Combine this with the requirement to be involved in root cause analyses and attending every meeting that you are requested to be in, the task to get out of the office to visit production, operation, and maintenance facilities becomes impossible. Hence, to curb this, I actually fill load my MBWA visits into my weekly calendar and, unless it is an emergency, these times are not interrupted. Sounds simple, doesn't it? The truth of the matter is that this did not occur overnight. It took a very direct and open conversation with my supervisor to allow me to do this. When it was explained that the visibility and focus of the safety department would be greatly improved by doing this, they readily agreed. Now, that is not to say that it is still not a struggle, mainly created by my own desire to be more involved in other aspects, but the plan must be developed and the methods must be complied with to achieve them.

Employee Involvement

Employee involvement provides the means through which workers develop and express their own commitment to safety and health, for both themselves and their fellow workers.[4] OSHA is aware that the growing recognition of the value of employee involvement and the increasing number and variety of employee participation arrangements can raise legal concerns. It makes good sense to consult your labor relations advisor to ensure that your employee involvement program conforms to current legal requirements.

Why should employees be involved? Because it is the right and smart thing to do and here is why:

- Rank-and-file workers are the persons most in contact with potential safety and health hazards. They have a vested interest in effective protection programs.
- Group decisions have the advantage of the group's wider range of experience.
- Employees are more likely to support and use programs in which they have input.

Employees who are encouraged to offer their ideas and whose contributions are taken seriously are more satisfied and productive on the job. What can employees do to be involved? Examples of employee participation include the following:

- Participating on joint labor–management committees and other advisory or specific purpose committees
- Conducting site inspections
- Analyzing routine hazards in each step of a job or process and preparing safe work practices or controls to eliminate or reduce exposure
- Developing and revising the site safety and health rules
- Training both current and newly hired employees
- Providing programs and presentations at safety and health meetings
- Conducting accident/incident investigations
- Reporting hazards
- Fixing hazards within your control
- Supporting your fellow workers by providing feedback on risks and assisting them in eliminating hazards
- Participating in accident/incident investigations
- Performing a preuse or change analysis for new equipment or processes in order to identify hazards up front before use

Worksite Analysis

What is a worksite analysis and how often should it be done? A worksite analysis means that managers and employees analyze all worksite conditions to identify and eliminate existing or potential hazards.[5] There should be a comprehensive, baseline survey, with a system in place for periodic updates.

To help in conducting a worksite analysis, you can

- Request a free OSHA consultation visit
- Become aware of hazards in your industry
- Create safety teams
- Encourage employee reporting of hazards
- Have an adequate system for reporting hazards
- Have trained personnel conduct inspections of the worksite and correct hazards
- Ensure that any changes in process or new hazards are reviewed
- Seek assistance from safety and health experts

Worksite analysis involves a variety of worksite examinations to identify not only existing hazards but also conditions and operations in which changes might create hazards. Effective management actively analyzes the work and the worksite, to anticipate and prevent harmful occurrences.

Here is a suggested plan to identify all worksite hazards:[5]

- Conduct a comprehensive, baseline survey for safety and health and periodic, comprehensive update surveys.
- Change analysis of planned and new facilities, processes, materials, and equipment.
- Perform routine job hazard analyses.
- Conduct periodic and daily safety and health inspections of the workplace.

These four major actions form the basis from which good hazard prevention and control can develop:

Comprehensive Surveys

For small business, OSHA-funded, state-run consultation services can conduct a comprehensive survey at no cost. Many workers' compensation carriers and other insurance companies offer expert services to help their clients evaluate health and safety hazards. Numerous private consultants provide a variety of health expert services. Larger companies may find the needed expertise at the company or corporate level.

For the industrial hygiene survey, at a minimum, all chemicals and hazardous materials in the plant should be inventoried, the hazard communication program should be reviewed, and air samples should be analyzed. For many industries, a survey of noise levels, a review of the respirator program, and a review of ergonomic risk factors are needed.

Change Analysis

Anytime something new is brought into the workplace, whether it is a piece of equipment, different materials, a new process, or an entirely new building, new hazards may unintentionally be introduced. Before considering a change for a worksite, it should be analyzed thoroughly beforehand. Change analysis helps in heading off a problem before it develops. You may find change analysis useful when

- Building or leasing a new facility
- Installing new equipment
- Using new materials
- Starting up new processes
- Staffing changes occur

Hazard Analysis

Hazard analysis techniques can be quite complex. While this is necessary in some cases, frequently a basic, step-by-step review of the operation is sufficient. One of the most commonly used techniques is the Job Hazard Analysis (JHA). Jobs that were initially designed with safety in mind may now include hazards or improper operations. When done for every job, this analysis periodically puts processes back on the safety track.

Other, more sophisticated techniques are called for when there are complex risks involved. These techniques include the following: What-If Checklist, Hazard and Operability Study, Failure Mode and Effect Analysis, and Fault Tree Analysis.

Conducting JHAs does not have to be a complex operation. For example, some of the organizations that I've been involved in use a simple form to be completed for all nonroutine tasks, where there is no written safe work procedure. These forms are simply reminders to the person completing the form to review on the task at hand. Figures 29.1 and 29.2[6,7] are examples of simple JHA forms that can be used.

Health and Safety Inspections

Routine site health and safety inspections are designed to catch hazards missed at other stages. This type of inspection should be done at regular intervals, generally on a weekly basis. Over the course of my career, I find it useful to conduct an overall, comprehensive audit or assessment of each work area on a monthly basis. When addressing larger facilities, I typically break the entire plant into workable zones. For example, if there is a large facility, I may break the facility into 5–10 workable zones and then conduct a weekly audit of one of these zones. The purpose being that by the end of the quarter, I have covered the entire facility and still have time to follow up on the corrective actions identified in the weekly audits. In addition, procedures

Small Business Job Hazard Analysis
(General Example)

Date of analysis: _____   People who participated:

_____

Tasks/jobs where injuries occur, or can occur		
How people get hurt	What causes them to get hurt?	What safe practices or PPE are needed?
Ladders tipping over	• Ladder was not on a level surface • Ladder was on soft ground and the leg sunk in • The person reached out too far • The ladder wasn't high enough to reach up safely – the person stood up near the top of it • Ladder broken or damaged	• Set ladder feet on solid level surfaces • When reaching out, keep belt buckle between the side rails of the ladder • Do not stand on the top of a stepladder or on the first step down from the top • Replace or repair ladder
Lifting heavy objects	• Trying to lift too heavy objects • Bending over at the waist when lifting • Turning (twisting back while lifting	• Use proper lifting practices (bend knees, don't twist) • For very heavy objects, use mechanical devices or get another person to help
Slipping on the floor	• Spilled liquids not cleaned up • Small objects are dropped on the floor and left there • People wear the wrong type of shoes for conditions	• Wipe up all spills, and pick up dropped items, immediately. • Wear sturdy shoes with slip-resistant soles
Using the bench grinder	• Flying particles get in eyes • If grinder wheel breaks, large chunks fly off at high speed • High noise level can injure hearing	• Wear safety glasses and earplugs when using grinder • Keep tongue guards adjusted properly (see sticker on grinder for spacing)

FIGURE 29.1

Sample small business job hazard analysis. (From Google, 2009. http://www.google.com/url ?sa=t&rct=j&q=&esrc=s&frm=1&source=web&cd=15&cad=rja&uact=8&ved=0CFAQFjAEOAo& url=http%3A%2F%2Fwisha-training.lni.wa.gov%2Ftraining%2Fjsa%2Fjsa2009.ppt&ei=5nfVU 4KHAYKRyAS614GYAw&usg=AFQjCNEc4zlFd_XAzfNSfC8x6U99wEeJlw&sig2=dSwRsup3C CeGatFXNhx5Tg.)

that provide a daily inspection of the work area by supervisors and employees should be established. This can be done through the following:

• Using a checklist (I prefer this method)
• Past problem identification/discussion
• Reviewing standards that apply to your particular industry
• Input from everyone involved
• Reviewing the company safety practices or rules

Other important things to remember regarding health and safety inspections include the following:

• Inspections should cover every part of the worksite.
• They should be done at regular intervals.

Task or Step	Hazards	Controls	Personal Protective Equipment (PPE)

FIGURE 29.2
Sample JHA form. (From www.lni.wa.gov/Safety/Topics/AToZ/JHA/PDFs/SampleJHA.)

- In-house inspectors should be trained to recognize and control hazards.
- Identified hazards should be tracked to correction.

Information from inspections/audits should be used to improve the hazard prevention and control program.[5]

Catching Hazards that Escape Controls

After hazards are recognized and controls are put in place, additional analysis tools can help ensure that the controls stay in place and other hazards don't appear. These other tools include the following:

- Employee reports of hazards
- Accident and incident investigations
- Injury and illness trend analysis

Employee Reports of Hazards

Employees play a key role in discovering and controlling hazards that may develop—or that already exist—in the workplace. A reliable system for employee reporting is an important element of an effective safety and health system. The workplace must not only encourage reporting, but must value it.

It is often helpful to establish multiple ways to report hazards so that, depending on comfort level and the nature of the issue, there are several avenues to get concerns addressed. Examples include supervisor chain of command, safety and health committee member, voice mail box, and a suggestion box.

An effective reporting system needs

- A policy that encourages employees to report safety and health concerns
- Timely and appropriate responses to the reporting employee
- Timely and appropriate action where valid concerns exists
- Tracking of required hazard correction
- Protection of reporting employees from any type of reprisal or harassment

Accident/Incident Investigations

Accident/incident investigation is another tool for uncovering hazards that were missed earlier or that slipped by the planned controls. But it's only useful when the process is positive and focuses on finding the root cause, not someone to blame!

All accidents and incidents should be investigated. "Near misses" are considered an incident, because, given a slight change in time or position, injury or damage could have occurred.

Six key questions should be answered in the accident investigation and report: who, what, when, where, why, and how. Thorough interviews with everyone involved are necessary.

The primary purpose of the accident/incident investigation is to prevent future occurrences. Therefore, the results of the investigation should be used to initiate corrective action.

Trend Analysis

The final action recommended under Worksite Analysis is analysis of injury and illness trends over time, so that patterns with common causes can be identified and prevented. Review of the OSHA injury and illness forms is the most common form of pattern analysis, but other records of hazards can be analyzed for patterns. Examples are inspection records and employee hazard reporting records.

- Injury and Illness Records Analysis:
  - Since there must be enough information for patterns to emerge, small sites may require a review of 3–5 years of records. Larger sites may find useful trends yearly, quarterly, or monthly.
  - When analyzing injury and illness records, look for similar injuries and illnesses. These generally indicate a lack of hazard controls. Look for where the injury or illness occurred, what type of work was being done, time of day, or type of equipment.

- Analysis of Other Records:
    - Repeat hazards, just like repeat injuries or illnesses, mean that controls are not working, and patterns in hazard identification records can show up over shorter periods than accidents or incidents. Upgrading a control may involve something as basic as improving communication or accountability.

Hazards found during worksite analysis should be reviewed to determine what failure in the safety and health system permitted the hazard to occur. The system failure should then be corrected to ensure that similar hazards do not reoccur.[5]

Health and Safety Training

How do we know what the hazards are? In order to know what to look for, it's important that everyone in the workplace is properly trained.[8] This includes the floor worker to the supervisors, managers, contractors, and part-time and temporary workers.

Other suggestions related to training include the following:

- Only properly authorized and instructed employees should be allowed to do any job.
- Make sure no one does a job that appears unsafe.
- Hold emergency preparedness drills.
- Pay particular attention when new operations are being learned so that everyone has the proper job skills and awareness of hazards.

Supervisors and managers should be trained to recognize hazards and understand their responsibilities.

Does everyone in the workplace know

- The workplace plan in case of a fire or other emergency?
- When and where PPE is required?
- The types of chemicals used in the workplace?
- The precautions when handling them?

Training can help develop the knowledge and skills needed to understand workplace hazards and safe procedures. It is most effective when integrated into a company's overall training in performance requirements and job practices.

The content of a company's training program and the methods of presentation should reflect the needs and characteristics of the particular workforce. Therefore, identification of needs is an important early step in training design. Involving everyone in this process and in the subsequent teaching can be highly effective.

The five principles of teaching and learning should be followed to maximize program effectiveness. They are as follows:

- Trainees should understand the purpose of the training.
- Information should be organized to maximize effectiveness.
- People learn best when they can immediately practice and apply newly acquired knowledge and skills.
- As trainees practice, they should get feedback.

People learn in different ways, so an effective program will incorporate a variety of training methods.

Some types of health and safety training needed are as follows:

- Orientation training for site workers and contractors
- JSA, SOPs, and other hazard recognition training
- Training required by OSHA standards, including the Process Safety Management standard
- Training for emergency response people
- Accident investigation training
- Training on new processes or new equipment

Who needs training? Training should target new hires, contract workers, employees who wear PPE, and workers in high-risk areas. Managers and supervisors should also be included in the training plan. Training for managers should emphasize the importance of their role in visibly supporting the safety and health program and setting a good example. Supervisors should receive training in company policies and procedures, as well as hazard detection and control, accident investigation, handling of emergencies, and how to train and reinforce training.

The long-term worker whose job changes as a result of new processes or materials should not be overlooked. And the entire workforce needs periodic refresher training in responding to emergencies.

Plan to evaluate the training program when initially designing the training. If the evaluation is done right, it can identify your program's strengths and weaknesses and provide a basis for future program changes.

Keeping training records will help ensure that everyone who should get training does. A simple form can document the training record for each employee. OSHA has developed voluntary training guidelines to assist in the design and implementation of effective training programs.

Assessing the Effectiveness of the Overall Health and Safety Program

Now that we have a basic understanding of what is expected or required under the OSHA standards in regard to an effective health and safety program, we can now discuss some methods and techniques for assessing the individual components of the program. (Note: A major portion of the following sections are taken from Ref. [1])

Assessing the Key Components of Leadership, Participation, and Line Accountability

Worksite Policy on Safe and Healthful Working Conditions

Documentation

One of the first areas to look is the written health and safety program/plan/policy. If there is a written policy, does it clearly declare the priority of worker safety and health over other organizational values, such as production? Does it clearly define areas of responsibility for key players, including management, supervisors, and employees?

Interviews

When asked, can employees at all levels express the worksite policy on worker safety and health? If the policy is written, can hourly employees tell you where they have seen it? Can employees at all levels explain the priority of worker safety and health over other organizational values, as the policy intends?

Site Conditions and Root Causes of Hazards

Have injuries occurred because employees at any level did not understand the importance of safety precautions in relation to other organizational values, such as production?

Goal and Objectives for Worker Safety and Health

Documentation

If there is a written goal for safety and health program, is it updated annually? If there are written objectives, such as an annual plan to reach that goal, are they clearly stated? If managers and supervisors have written objectives, do these documents include objectives for the safety and health program?

Interviews

Do managers and supervisors have a clear idea of their objectives for worker safety and health? Do hourly employees understand the current objectives of the safety and health program?

Site Conditions and Root Causes of Hazards (Only helpful in a general sense.)

Visible Top Management Leadership

Documentation

Are there one or more written programs that involve top-level management in safety and health activities? For example, top management can receive and sign off on inspection reports either after each inspection or in a quarterly summary. These reports can then be posted for employees to see. Top management can provide "open door" times each week or each month for employees to come in to discuss safety and health concerns. Top management can reward the best safety suggestions each month or at other specified intervals.

Interviews

Can hourly employees describe how management officials are involved in safety and health activities? Do hourly employees perceive that managers and supervisors follow safety and health rules and work practices, such as wearing appropriate PPE?

Site Conditions and Root Causes of Hazards

When employees are found not wearing required PPE or not following safe work practices, have any of them said that managers or supervisors also did not follow these rules?[1]

Employee Participation

Documentation

Are there one or more written programs that provide for employee participation in decisions affecting their safety and health? Is there documentation of these activities, for example, employee inspection reports, minutes of joint employee–management, or employee committee meetings? Is there written documentation of any management response to employee safety and health program activities? Does the documentation indicate that employee safety and health activities are meaningful and substantive? Are there written guarantees of employee protection from harassment resulting from safety and health program involvement?

Interviews

Are employees aware of ways they can participate in decisions affecting their safety and health? Do employees appear to take pride in the achievements of the worksite safety and health program? Are employees comfortable answering questions about safety and health programs and conditions at the site? Do employees feel they have the support of management for their safety and health activities?

Site Conditions and Root Causes of Hazards (Not applicable.)

Assignment of Responsibility

Documentation

Are responsibilities written out so that they can be clearly understood?

Interviews

Do employees understand their own responsibilities and those of others?

Site Conditions and Root Causes of Hazards

Are hazards caused in part because no one was assigned the responsibility to control or prevent them? Are hazards allowed to exist in part because someone in management did not have the clear responsibility to hold a lower-level manager or supervisor accountable for carrying out assigned responsibilities?

Adequate Authority and Resources

Documentation (Only generally applicable.)

Interviews

Do safety staff members or any other personnel with responsibilities for ensuring safe operation of production equipment have the authority to shut down that equipment or to order maintenance or parts? Do employees talk about not being able to get safety or health improvements because of cost? Do employees mention the need for more safety or health personnel or expert consultants?

Site Conditions and Root Causes of Hazards

Do recognized hazards go uncorrected because of lack of authority or resources? Do hazards go unrecognized because greater expertise is needed to diagnose them?

Accountability of Managers, Supervisors, and Hourly Employees

Documentation

Do performance evaluations for all line managers and supervisors include specific criteria relating to safety and health protection? Is there documented evidence of employees at all levels being held accountable for safety and health responsibilities, including safe work practices? Is accountability accomplished through either performance evaluations affecting pay and promotions or disciplinary actions?

Interviews

When you ask employees what happens to people who violate safety and health rules or safe work practices, do they indicate that rule breakers are clearly and consistently held accountable? Do hourly employees indicate that supervisors and managers genuinely care about meeting safety and health responsibilities? When asked what happens when rules are broken, do hourly employees complain that supervisors and managers do not follow rules and are never disciplined for infractions?

Site Conditions and Root Causes of Hazards

Are hazards occurring because employees, supervisors, or managers are not being held accountable for their safety and health responsibilities? Are identified hazards not being corrected because those persons assigned the responsibility are not being held accountable?

Evaluation of Contractor Programs

Documentation

Are there written policies for onsite contractors? Are contractor rates and safety and health programs reviewed before selection? Do contracts require the contractor to follow site safety and health rules? Are there means for removing a contractor who violates the rules?

Interviews

Do employees describe hazardous conditions created by contract employees? Are employees comfortable reporting hazards created by contractors? Do contract employees feel they are covered by the same, or the same quality, safety and health program as regular site employees?

Site Conditions and Root Causes of Hazards

Do areas where contractors are working appear to be in the same condition as areas where regular site employees are working? Better? Worse? Does the working relationship between site and contract employees appear cordial?

Assessing the Key Components of Worksite Analysis

Comprehensive Surveys, Change Analysis, and Routine Hazard Analysis

Documentation

Are there documents that provide comprehensive analysis of all potential safety and health hazards of the worksite? Are there documents that provide both the analysis of potential safety and health hazards for each new facility, equipment, material, or process and the means for eliminating or controlling such hazards? Does documentation exist of the step-by-step analysis of the hazards in each part of each job, so that you can clearly discern the evolution of decisions on safe work procedures?

If complicated processes exist, with a potential for catastrophic impact from an accident but low probability of such accident (as in nuclear power or chemical production), are there documents analyzing the potential hazards in each part of the processes and the means to prevent or control them? If there are processes with a potential for catastrophic impact from an accident but low probability of an accident, have analyses such as "fault tree" or "what if?" been documented to ensure enough backup systems for worker protection in the event of multiple control failure?

Interviews

Do employees complain that new facilities, equipment, materials, or processes are hazardous? Do any employees say they have been involved in job safety analysis or process review and are satisfied with the results? Does the safety and health staff indicate ignorance of existing or potential hazards at the worksite? Does the occupational nurse/doctor or other health care provider understand the potential occupational diseases and health effects in this worksite?

Site Conditions and Root Causes of Hazards

Have hazards appeared where no one in management realized there was potential for their development? Where workers have faithfully followed job procedures, have accidents or near misses occurred because of hidden hazards? Have hazards been discovered in the design of new facilities, equipment, materials, and processes after use has begun? Have accidents or near misses occurred when two or more failures in the hazard control system occurred at the same time, surprising everyone?

Regular Site Safety and Health Inspections

Documentation

If inspection reports are written, do they show that inspections are done on a regular basis? Do the hazards found indicate good ability to recognize those hazards typical of this industry? Are hazards found during inspections tracked to complete correction?

What is the relationship between hazards uncovered during inspections and those implicated in injuries or illness?

Interviews

Do employees indicate that they see inspections being conducted, and that these inspections appear thorough?

Site Conditions and Root Causes of Hazards

Are the hazards discovered during accident investigations ones that should have been recognized and corrected by the regular inspection process?

Employee Reports of Hazards

Documentation

Is the system for written reports being used frequently? Are valid hazards that have been reported by employees tracked to complete correction? Are the responses timely and adequate?

Interviews

Do employees know whom to contact and what to do if they see something they believe to be hazardous to themselves or coworkers? Do employees think that responses to their reports of hazards are timely and adequate? Do employees say that sometimes when they report a hazard, they hear nothing further about it? Do any employees say that they or other workers are being harassed, officially or otherwise, for reporting hazards?

Site Conditions and Root Causes of Hazards

Are hazards ever found where employees could reasonably be expected to have previously recognized and reported them? When hazards are found, is there evidence that employees had complained repeatedly but to no avail?

Accident and Near-Miss Investigations

Documentation

Do accident investigation reports show a thorough analysis of causes, rather than a tendency automatically to blame the injured employee? Are near misses (property damage or close calls) investigated using the same techniques as accident investigations? Are hazards that are identified as contributing to accidents or near misses tracked to correction?

Interviews

Do employees understand and accept the results of accident and near-miss investigations? Do employees mention a tendency on management's part to blame the injured employee? Do employees believe that all hazards contributing to accidents are corrected or controlled?

Site Conditions and Root Causes of Hazards

Are accidents sometimes caused at least partly by factors that might also have contributed to previous near misses that were not investigated or accidents that were too superficially investigated?

Injury and Illness Pattern Analysis

Documentation

In addition to the required OSHA log, are careful records kept of first-aid injuries and illnesses that might not immediately appear to be work related? Is there any periodic, written analysis of the patterns of near misses, injuries, or illnesses over time, seeking previously unrecognized connections between them that indicate unrecognized hazards needing correction or control? Looking at the OSHA 200 log and, where applicable, first-aid logs, are there patterns of illness or injury that should have been analyzed for previously undetected hazards? If there is an occupational nurse/doctor on the worksite, or if employees suffering from ordinary illness are encouraged to see a nearby health care provider, are the lists of those visits analyzed for clusters of illness that might be work related?

Interviews

Do employees mention illnesses or injuries that seem work related to them but that have not been analyzed for previously undetected hazards?

Site Conditions and Root Causes of Hazards (Not generally applicable.)

Figure 29.3 is an example of a worksite analysis form that I developed for use in industrial and manufacturing facilities. It can easily be adopted to serve your individual needs. I prefer checklists that are weighted and provide an overall measurement.

Assessing the Key Components of Hazard Prevention and Control

(**NOTE**: A major portion of the following sections are taken from Ref. [1])

Appropriate Use of Engineering Controls, Work Practices, Personal Protective Equipment, and Administrative Controls

Documentation

If there are documented comprehensive surveys, are they accompanied by a plan for systematic prevention or control of hazards found? If there is a written plan, does it show that the best method of hazard protection was chosen? Are there written safe work procedures? If respirators are used, is there a written respirator program?

Interviews

Do employees say they have been trained in and have ready access to reliable, safe work procedures? Do employees say they have difficulty accomplishing their work because of unwieldy controls meant to protect them? Do

Industrial Safety Assessment Team (ISAT) Checklist

Business Unit: _____   Date: _____

Area Coordinator: _____   Safety
                                     Rep.: _____

Section I: BBS Participation (5%)				
Item	**Weight**	**Raw Score**	**Weighted Score**	**Score**
Are area hourly personnel participating in the SWI program? (0-25% = 25 points; 26-50% = 50 points; 51-75% = 75 points and 75-100% = 100 points)	50%	75	37.5	
Are area salaried personnel participating in the SWI program? (0-25% = 25 points; 26-50% = 50 points; 51-75% = 75 points and 75-100% = 100 points)	50%	50	25	
	100%		62.5	
Total Section I Score				**3.125**

Section II: Personal Protective Equipment (5%)				
Item	**Weight**	**Raw Score**	**Weighted Score**	**Score**
Is PPE appropriate for this work area and/or assigned job tasks? (If yes = 100 points, If no = 0 points)	15%	0	0.00	
Is PPE functional and in good repair? (If yes = 100 points, If no = 0 points)	15%	100	15.00	
Have employees been properly trained and can they demonstrate acceptable proficiency in the care, use and limitations of PPE? If yes = 100 points, If no = 0 points)	40%	100	40.00	

FIGURE 29.3
Example of an industrial safety assessment team checklist.

(Continued)

employees ever mention PPE, work procedures, or engineering controls as interfering with their ability to work safely?

Do employees who use PPE understand why they use it and how to maintain it? Do employees who use PPE indicate that the rules for PPE use are consistently and fairly enforced?

Safety glasses, ear protection , and other protection available and their use enforced? (If yes = 100 points, If no = 0 points)	15%	100	15.00	
Are approved respirators provided for regular or emergency use where needed? Have employees been trained and fit-tested in their use? (If yes = 100 points, If no = 0 points)	15%	100	15.00	
	100%		85.00	
Total Section II Score				**4.25**

Section III: Housekeeping, Egress & Walking Surfaces (10%)				
Item	**Weight**	**Raw Score**	**Weighted Score**	**Score**
Work areas, storage rooms, penthouse, corridors & stairways are clear of clutter, debris, equipment & obstacles? (If yes = 100 points, If no = 0 points)	7%	100	6.66	
Evacuation plans exist to assist employees with disabilities? (If yes = 100 points, If no = 0 points)	7%	100	6.66	
Material is not allowed to be stored on top of cabinets where it could fall? (If yes = 100 points, If no = 0 points)	7%	0	0	
Are changes of direction or elevation readily identifiable? (If yes = 100 points, If no = 0 points)	7%	100	6.66	
Major exits and routes are unobstructed and marked with illuminated "EXIT" signs? (If yes = 100 points, If no = 0 points)	7%	0	0	

FIGURE 29.3 (CONTINUED)
Example of an industrial safety assessment team checklist.

(Continued)

Do employees indicate that safe work procedures are fairly and consistently enforced?

Site Conditions and Root Causes of Hazards

Are controls meant to protect workers actually putting them at risk or not providing enough protection? Are employees engaging in unsafe practices or creating unsafe conditions because rules and work practices are not fairly and consistently enforced? Are employees in areas designated for PPE

Are floors clean and dry? Is drainage maintained and are gratins, mats or raised platforms used where wet processes are performed? (If yes = 100 points, If no = 0 points)	7%	0	0
Are permanent isles and passageways appropriately marked? (If yes = 100 points, If no = 0 points)	7%	100	6.66
Are floor load rating limits marked on plates and conspicuously posted? (If yes = 100 points, If no = 0 points)	7%	0	0
Are combustible scrap and debris removed at regular intervals? (If yes = 100 points, If no = 0 points)	7%	100	6.66
Are holes in the floor, sidewalk or other walking surface repaired properly, covered or otherwise made safe? (If yes = 100 points, If no = 0 points)	7%	0	0
Is there safe clearance for walking in aisles where motorized or mechanical handling equipment is operating? (If yes = 100 points, If no = 0 points)	7%	100	6.66
Is material or equipment stored in such a way that sharp projections will not interfere with the walkway? (If yes = 100 points, If no = 0 points)	7%	100	6.66
Are steps on stairs and stairways designed or provided with a surface that renders them slip resistant? (If yes = 100 points, If no = 0 points)	7%	100	6.66
Are toe boards installed around the edges of permanent floor openings (where persons may pass below the openings?) (If yes = 100 points, If no = 0 points)	7%	100	6.66
Heavy storage cabinets, bookcases and file cabinets secured from tipping? (If yes = 100 points, If no = 0 points)	7%	100	6.66

FIGURE 29.3 (CONTINUED)
Example of an industrial safety assessment team checklist.

(Continued)

		Raw	Weighted	
	100%		66.601	
Total Section III Score				**6.6601**
Section IV: Emergency & Fire Protection (10%)				
Item	**Weight**	**Raw Score**	**Weighted Score**	**Score**
Emergency lighting is provided, maintained, and periodically tested? (If yes = 100 points, If no = 0 points)	7.69%	100	7.69	
Employees have received emergency evacuation instructions and know where to find a copy of their Area/Building Emergency Action Plans? (If yes = 100 points, If no = 0 points)	7.69%	100	7.69	
Are all extinguishers fully charged and in their designated places? (If yes = 100 points, If no = 0 points)	7.69%	100	7.69	
Is access to extinguishers free from obstructions or blockage? (If yes = 100 points, If no = 0 points)	7.69%	100	7.69	
Do temporary office heaters have "tip over" cutoff switches and are UL approved? (If yes = 100 points, If no = 0 points)	7.69%	100	7.69	
Are items not allowed to be stored within 18 inches of sprinkler heads? (If yes = 100 points, If no = 0 points)	7.69%	100	7.69	
Are fire doors prevented from being propped open? (If yes = 100 points, If no = 0 points)	7.69%	100	7.69	
Is fire extinguishing equipment available when welding cutting, burning or grinding? (If yes = 100 points, If no = 0 points)	7.69%	100	7.69	

FIGURE 29.3 (CONTINUED)
Example of an industrial safety assessment team checklist.

(Continued)

	Weight	Raw Score	Weighted Score	Score
Are materials, equipment, and trash cans kept out of the way of fire extinguishers or other emergency equipment? (If yes = 100 points, If no = 0 points)	7.69%	100	7.69	
Is emergency equipment easy to identify, well marked and easy to access? (If yes = 100 points, If no = 0 points)	7.69%	100	7.69	
Are calibrations and inspections up to date? Is it tested periodically? (If yes = 100 points, If no = 0 points)	7.69%	100	7.69	
Are personnel trained in the use and inspection of emergency equipment in their work area? (If yes = 100 points, If no = 0 points)	7.69%	100	7.69	
Are emergency lights installed in stairwells and means of egress areas? (If yes = 100 points, If no = 0 points)	7.69%	100	7.69	
	100%		99.9703	
Total Section IV Score				**9.99703**

Section V: Machine and Tool Safety (10%)				
Item	**Weight**	**Raw Score**	**Weighted Score**	**Score**
Bench-top grinders have properly adjusted tool rests and tongue guards? (If yes = 100 points, If no = 0 points)	8.33%	100	8.33	

FIGURE 29.3 (CONTINUED)
Example of an industrial safety assessment team checklist.

(*Continued*)

Are all cord-connected, electrically operated tools and equipment effectively grounded, or of the approved double insulated type? (If yes = 100 points, If no = 0 points)	8.33%	100	8.33	
Power tools are used with the correct shield, guard, or attachment, recommended by the manufacturer? (If yes = 100 points, If no = 0 points)	8.33%	100	8.33	
Are portable fans provided with full guards or screens having openings ½ inch or less? (If yes = 100 points, If no = 0 points)	8.33%	100	8.33	
Is documentation kept for hoses and equipment periodic inspections? (If yes = 100 points, If no = 0 points)	8.33%	100	8.33	
Are all machines with moving parts properly guarded? (If yes = 100 points, If no = 0 points)	8.33%	0	0	
Are load capacities and operating speeds, conspicuously posted for machinery? (If yes = 100 points, If no = 0 points)	8.33%	0	0	
Is equipment and machinery securely placed and anchored to prevent moving? (If yes = 100 points, If no = 0 points)	8.33%	100	8.33	
Is sufficient Clearance provided around and between machines to allow for safe operations, setup, serving, material handling and waste removal? (If yes = 100 points, If no = 0 points)	8.33%	100	8.33	
Are all portable and fixed ladders inspected periodically for defects? (If yes = 100 points, If no = 0 points)	8.33%	100	8.33	
Are pneumatic power tools secured to the hose or whip in a positive manner to prevent accidental disconnection? (If yes = 100 points, If no = 0 points)	8.33%	100	8.33	

FIGURE 29.3 (CONTINUED)
Example of an industrial safety assessment team checklist.

(*Continued*)

Are pneumatic and hydraulic hoses on power operated tools checked regularly for deterioration or damage? (If yes = 100 points, If no = 0 points)	8.33%	100	8.33	
	100%		83.3004	
Total Section V Score				**8.33004**

Section VI: Flammable and Hazardous Substances and Storage (10%)				
Item	**Weight**	**Raw Score**	**Weighted Score**	**Score**
Flammable wastes (e.g., oil or solvent rags) are discarded into an approved flammable waste container? (If yes = 100 points, If no = 0 points)	6.66%	100	6.66	
Safety Data Sheets (SDS) are available for each hazardous substance in the work area in either hard copy or electronically? (If yes = 100 points, If no = 0 points)	6.66%	100	6.66	
Flammable liquid quantities >10 gallons are stored in UL/FM-approved flammable liquid cabinets? (If yes = 100 points, If no = 0 points)	6.66%	100	6.66	
Emergency eye washes and showers are readily accessible wherever corrosive or toxic materials are present? (If yes = 100 points, If no = 0 points)	6.66%	100	6.66	

FIGURE 29.3 (CONTINUED)
Example of an industrial safety assessment team checklist.

(Continued)

Are hazardous materials and flammable liquids stored in proper containers/cabinets? Are containers stored away from means or egress or exits? (If yes = 100 points, If no = 0 points)	6.66%	100	6.66
Are storage spaces or rooms for hazardous materials marked? (If yes = 100 points, If no = 0 points)	6.66%	100	6.66
Have personnel who handle chemicals been trained on the hazards involved? (If yes = 100 points, If no = 0 points)	6.66%	100	6.66
Are bulk drums of flammable liquids grounded and bonded to containers during dispensing? (If yes = 100 points, If no = 0 points)	6.66%	100	6.66
Are "NO SMOKING" signs posted where appropriate in areas where flammable or combustible materials are used or stored? (If yes = 100 points, If no = 0 points)	6.66%	100	6.66
Is there a list of hazardous substances used in your workplace? (If yes = 100 points, If no = 0 points)	6.66%	100	6.66
Is there a written Hazard Communication Program dealing with Safety Data Sheets (SDS), labeling, and employee training? (If yes = 100 points, If no = 0 points)	6.66%	100	6.66
Is there an employee worksite orientation/training program before allowing employees to use hazardous substances? (If yes = 100 points, If no = 0 points)	6.66%	100	6.66
Is each container for a hazardous substance labeled with product identity and a hazard warning (communication of the specific health hazards and physical hazards)? (If yes = 100 points, If no = 0 points)	6.66%	100	6.66

FIGURE 29.3 (CONTINUED)
Example of an industrial safety assessment team checklist.

(*Continued*)

Item	Weight	Raw Score	Weighted Score	Score
Are incompatible materials banned from being stored together? (If yes = 100 points, If no = 0 points)	6.66%	100	6.66	
Is there a review procedure in place to ensure that banned materials are not entering the site, in accordance with F-02? (If yes = 100 points, If no = 0 points)	6.66%	100	6.66	
	100%		99.901	
Total Section VI Score				**9.9901**

Section VII: Electrical Safety (15%)				
Item	Weight	Raw Score	Weighted Score	Score
Are electrical enclosures such as switches, receptacles, and junction boxes provided with tight fitting covers or plates? (If yes = 100 points, If no = 0 points)	5.26%	100	5.26	
Is each motor disconnecting switch or circuit breaker located within sight of the motor control device? (If yes = 100 points, If no = 0 points)	5.26%	100	5.26	
Are employees prohibited from working alone on energized lines or equipment over 600 volts? (If yes = 100 points, If no = 0 points)	5.26%	100	5.26	
Three feet of clear space is maintained in front of electrical panels? (If yes = 100 points, If no = 0 points)	5.26%	100	5.26	
Are GFI provided near wet, damp, or conductive locations? (If yes = 100 points, If no = 0 points)	5.26%	100	5.26	
Are breaker or motor control center panel doors closed and latched? (If yes = 100 points, If no = 0 points)	5.26%	100	5.26	

FIGURE 29.3 (CONTINUED)
Example of an industrial safety assessment team checklist.

(*Continued*)

Are employees who may reasonably face the risk of shock trained in electrical safety? (If yes = 100 points, If no = 0 points)	5.26%		100	5.26
Are the necessary voltage, wattage or current ratings labeled on equipment? (If yes = 100 points, If no = 0 points)	5.26%		100	5.26
Is the insulation on electrical wires free from fraying or cuts? (If yes = 100 points, If no = 0 points)	5.26%		100	5.26
Is out of order equipment shut down, segregated and tagged until repaired? (If yes = 100 points, If no = 0 points)	5.26%		100	5.26
Extension cords are not used in place of permanent wiring (not to exceed 6-feet in length)? (If yes = 100 points, If no = 0 points)	5.26%		100	5.26
Extension cords are not frayed, spliced, cut, or otherwise in poor condition? (If yes = 100 points, If no = 0 points)	5.26%		100	5.26
Do extension cords being used have a grounding conductor? (If yes = 100 points, If no = 0 points)	5.26%		100	5.26
Circuit breakers are labeled to indicate which breaker controls which circuit? (If yes = 100 points, If no = 0 points)	5.26%		100	5.26
Are multiple plug adaptors prohibited?	5.26%		100	5.26
Are GFCI's installed on each temporary 15 or 20 ampere, 120 volt AC circuit at locations where construction/modifications are being performed?	5.26%		100	5.26
Are flexible cords and cables free of splices or taps? (If yes = 100 points, If no = 0 points)	5.26%		100	5.26

FIGURE 29.3 (CONTINUED)
Example of an industrial safety assessment team checklist.

(*Continued*)

Item	Weight		Raw Score		Weighted Score	Score
Are cords routed to avoid overloading outlets? (If yes = 100 points, If no = 0 points)	5.26%		100		5.26	
Is the use of metal ladders prohibited in areas where they could come in contact with energized parts of equipment, fixtures or circuit conductors? (If yes = 100 points, If no = 0 points)	5.26%		100		5.26	
	100%				99.9406	
						14.99109

Section VIII: Control of Hazardous Energy (15%)						
Item	Weight		Raw Score		Weighted Score	Score
Are provisions made to prevent machines from automatically starting when power is restored after a power failure or shutdown? (If yes = 100 points, If no = 0 points)	9.09%		100		9.09	
Are suspended loads or potential energy (such as compressed springs, hydraulics or jacks) controlled to prevent hazards? (If yes = 100 points, If no = 0 points)	9.09%		100		9.09	
Does the organization have a written Control of Hazardous Energy Program that meets the requirements of 29CFR 1910.146? (If yes = 100 points, If no = 0 points)	9.09%		100		9.09	
Are tagout procedures used if an energy isolating device cannot be locked out? (If yes = 100 points, If no = 0 points)	9.09%		100		9.09	
Are employees trained and do they understand the organization's lockout and tagout program? (If yes = 100 points, If no = 0 points)	9.09%		100		9.09	

FIGURE 29.3 (CONTINUED)
Example of an industrial safety assessment team checklist.

(*Continued*)

		Raw Score	Weighted Score	Score
To the extent possible, are live parts deenergized, locked out and tagged out before personnel work on or near parts? (If yes = 100 points, If no = 0 points)	9.09%	100	9.09	
Are all equipment control valve handles provided with a means for locking-out? (If yes = 100 points, If no = 0 points)	9.09%	100	9.09	
Are appropriate employees provided with individually keyed personal safety locks? (If yes = 100 points, If no = 0 points)	9.09%	100	9.09	
Is it required that only the employee exposed to the hazard place or remove the safety lock? (If yes = 100 points, If no = 0 points)	9.09%	100	9.09	
In the event that equipment or lines cannot be shut down, locked-out and tagged, is a safe job procedure established and rigidly followed? (If yes = 100 points, If no = 0 points)	9.09%	100	9.09	
Is there an audit of lockout/tagout procedures completed at least annually? (If yes = 100 points, If no = 0 points) Last audit date: _____	9.09%	100	9.09	
	100%		99.9901	
				14.998515

Section IX: Compressed Gases and Compressed Air (5%)				
Item	Weight	Raw Score	Weighted Score	Score
Compressed gas cylinders are secured? (If yes = 100 points, If no = 0 points)	20.00%	100	20	

FIGURE 29.3 (CONTINUED)
Example of an industrial safety assessment team checklist.

(*Continued*)

Compressed gas cylinders not in use have valve protection covers in place? (If yes = 100 points, If no = 0 points)	20.00%	100	20
Incompatible gas cylinders are adequately separated (e.g., flammables separated from oxidizers)? (If yes = 100 points, If no = 0 points)	20.00%	100	20
Compressed air nozzles are provided with pressure reducing devices that restrict nozzle pressure to less than 30 psi? (If yes = 100 points, If no = 0 points)	20.00%	100	20
Do cutting torches have flash back preventors installed on the downstream side? (If yes = 100 points, If no = 0 points)	20.00%	100	20
	100%		100
			5

Section VIII: Control of Hazardous Energy (15%)				
Item	Weight	Raw Score	Weighted Score	Score
Does the organization have a written Confined Spaces Safety Program that meets the requirements of 29CFR 1910.147? (If yes = 100 points, If no = 0 points)	20.00%	100	20	
Are Permit Required confined spaces identified by signs to warn personnel prior to entry? (If yes = 100 points, If no = 0 points)	20.00%	100	20	

FIGURE 29.3 (CONTINUED)
Example of an industrial safety assessment team checklist.

(Continued)

When entered, is a standby person assigned outside the space and in constant communication? (If yes = 100 points, If no = 0 points)	20.00%		100		20		
Are appropriate atmospheric tests performed to check for oxygen deficiency, toxic substances and explosive concentrations in the confined space before entry? (If yes = 100 points, If no = 0 points)	20.00%		100		20		
Is all portable electrical equipment used inside confined spaces either grounded and insulated, or equipped with ground fault protection? (If yes = 100 points, If no = 0 points)	20.00%		100		20		
	100%				100		
							15

Overall Assessment Score   92.34

COMMENTS:

FIGURE 29.3 (CONTINUED)
Example of an industrial safety assessment team checklist.

wearing it properly, with no exceptions? Are hazards that could feasibly be controlled through improved design being inadequately controlled by other means?

Facility and Equipment Preventive Maintenance

Documentation

Is there a preventive maintenance schedule that provides for timely maintenance of the facilities and equipment? Is there a written or computerized record of performed maintenance that shows the schedule has been followed? Do maintenance request records show a pattern of certain facilities or equipment needing repair or breaking down before maintenance was scheduled or actually performed? Do any accident/incident investigations list facility or equipment breakdown as a major cause?

Interviews

Do employees mention difficulty with improperly functioning equipment or facilities in poor repair? Do maintenance employees believe that the preventive maintenance system is working well? Do employees believe that hazard controls needing maintenance are properly cared for?

Site Conditions and Root Causes of Hazards

Is poor maintenance a frequent source of hazards? Are hazard controls in good working order? Does equipment appear to be in good working order?

Establishing a Medical Program

Documentation

Are good, clear records kept of medical testing and assistance?

Interviews

Do employees say that test results were explained to them? Do employees feel that more first-aid or CPR-trained personnel should be available? Are employees satisfied with the medical arrangements provided at the site or elsewhere? Does the occupational health care provider understand the potential hazards of the worksite, so that occupational illness symptoms can be recognized?

Site Conditions and Root Causes of Hazards

Have further injuries or worsening of injuries occurred because proper medical assistance (including trained first-aid and CPR providers) was not readily available? Have occupational illnesses possibly gone undetected because no one with occupational health specialty training reviewed employee symptoms as part of the medical program?

Emergency Planning and Preparation

Documentation

Are there clearly written procedures for every likely emergency, with clear evacuation routes, assembly points, and emergency telephone numbers?

Interviews

When asked about any kind of likely emergency, can employees tell you exactly what they are supposed to do and where they are supposed to go?

Site Conditions and Root Causes of Hazards

Have hazards occurred during actual or simulated emergencies because of confusion about what to do? In larger worksites, are emergency evacuation routes clearly marked?

Are emergency telephone numbers and fire alarms in prominent, easy-to-find locations?

Assessing the Key Components of Safety and Health Training

Ensuring that all Employees Understand Hazards

(NOTE: A major portion of the following sections are taken from Ref. [1])

Documentation

Does the written training program include complete training for every employee in emergency procedures and in all potential hazards to which employees may be exposed? Do training records show that every employee received the planned training? Do the written evaluations of training indicate that the training was successful and that the employees learned what was intended?

Interviews

Can employees tell you what hazards they are exposed to, why those hazards are a threat, and how they can help protect themselves and others? If PPE is used, can employees explain why they use it and how to use and maintain it properly? Do employees feel that health and safety training is adequate?

Site Conditions and Root Causes of Hazards

Have employees been hurt or made ill by hazards of which they were completely unaware, or whose dangers they did not understand, or from which they did not know how to protect themselves? Have employees or rescue workers ever been endangered by employees not knowing what to do or where to go in a given emergency situation? Are there hazards in the workplace that exist, at least in part, because one or more employees have not received adequate hazard control training? Are there any instances of employees not wearing required PPE properly because they have not received proper training? Or because they simply don't want to and the requirement is not enforced?

Ensuring that Supervisors Understand Their Responsibilities

Documentation

Do training records indicate that all supervisors have been trained in their responsibilities to analyze work under their supervision for unrecognized hazards, to maintain physical protections, and to reinforce employee training through performance feedback and, where necessary, enforcement of safe work procedures and safety and health rules?

Interviews

Are supervisors aware of their responsibilities? Do employees confirm that supervisors are carrying out these duties?

Site Conditions and Root Causes

Has a supervisor's lack of understanding of safety and health responsibilities played a part in creating hazardous activities or conditions?

Ensuring that Managers Understand Their Safety and Health Responsibilities

Documentation

Do training plans for managers include training in safety and health responsibilities? Do records indicate that all line managers have received this training?

Interviews

Do employees indicate that managers know and carry out their safety and health responsibilities?

Site Conditions and Root Causes of Hazards

Has an incomplete or inaccurate understanding by management of its safety and health responsibilities played a part in the creation of hazardous activities or conditions?

Key Points to Remember on Health and Safety Program Auditing

1. In addition to identifying hazards and preventing illnesses and injuries, the health and safety program audit should assist in determining whether or not significant progress is being made in the overall goals and objectives of your health and safety program.
2. The three basic methods used to conduct health and safety program audits are document review/verification, employee interviews, and site conditions.
3. Inspection records can tell the evaluator whether serious hazards are being found or whether the same hazards are being found repeatedly. If serious hazards are not being found and accidents keep occurring, there may be a need to train inspectors to look for different hazards.
4. Employee interviews are extremely useful in determining the quality of health and safety training.
5. Site conditions and root causes are useful in identifying present and potential hazards in the workplace.

6. The major elements of an effective health and safety program include management commitments, employee involvement, worksite analysis, hazard prevention and control, and health and safety training.

7. One of the best methods used by managers and supervisors to continuously evaluate the health and safety effectiveness of their program is known as Management by Walking Around.

8. Employee involvement provides the means through which workers develop and express their own commitment to health and safety, for both themselves and for their fellow employees.

9. Worksite analysis means that managers and employees alike analyze all worksite conditions to identify and eliminate existing or potential hazards.

10. One of the most commonly used techniques in conducting hazard analysis is the use of a Job Hazard Analysis form.

References

1. https://www.osha.gov/doc/outreachtraining/htmlfiles/evaltool.html.
2. https://www.osha.gov/dte/library/safety_health_program/index.html.
3. https://www.osha.gov/SLTC/etools/safetyhealth/comp1_mgt_lead.html.
4. https://www.osha.gov/SLTC/etools/safetyhealth/comp1_empl_envolv.html.
5. https://www.osha.gov/SLTC/etools/safetyhealth/comp2.html.
6. Google, 2009. Available at http://www.google.com/url?sa=t&rct=j&q=&esrc=s&frm=1&source=web&cd=15&cad=rja&uact=8&ved=0CFAQFjAEOAo&url=http%3A%2F%2Fwisha-training.lni.wa.gov%2Ftraining%2Fjsa%2Fjsa2009.ppt&ei=5nfVU4KHAYKRyAS614GYAw&usg=AFQjCNEc4zlFd_XAzfNSfC8x6U99wEeJlw&sig2=dSwRsup3CCeGatFXNhx5Tg.
7. www.lni.wa.gov/Safety/Topics/AToZ/JHA/PDFs/SampleJHA.
8. https://www.osha.gov/SLTC/etools/safetyhealth/comp4.html.

30

Environmental Management

As a safety professional or future safety professional reading this book, you may be wondering what "Environmental Management" has to do with safety. It has been my experience that these two disciplines have as much in common as they are different. Therefore, I've included this chapter as an introduction on environmental management by having a working knowledge of many of the main environmental regulations.

History and Evolution of US Environmental Policies

I'll begin with a history and evolution of US environmental policies. Formally, I will begin the history in 1970, when formal laws were passed that are directly attributable to the protection of the environment. However, just to be clear, this is not the beginning of the environmental or conservation movement, which began almost 100 years prior to this.

The National Environmental Policy Act (NEPA) (42 U.S.C. 4321)[1]

The National Environmental Policy Act (NEPA) was the beginning of modern-day environmental policies within the United States. It was signed into law on January 1, 1970, by President Richard M. Nixon. The Act establishes national environmental policy and goals for the protection, maintenance, and enhancements of the environment and provides a process for implementing these goals within the federal agencies. The Act also establishes the Council on Environmental Quality (CEQ).

NEPA Requirements

Title I of NEPA contains a Declaration of National Environmental Policy that requires the federal government to use all practicable means to create and maintain conditions under which man and nature can exist in productive harmony. Section 102 requires federal agencies to incorporate environmental considerations in their planning and decision making through a systematic interdisciplinary approach. Specifically, all federal agencies are to prepare detailed statements assessing the environmental impact of and alternatives to major federal actions significantly affecting the environment. These statements are commonly referred to as environmental impact statements (EISs).

Title II of NEPA establishes the CEQ.

Oversight of NEPA

The CEQ, which is headed by a full-time Chair, oversees NEPA. A staff assists the Council. The duties and functions of the Council are listed in Title II, Section 204 of NEPA and include the following:

- Gathering information on the conditions and trends in environmental quality
- Evaluating federal programs in light of the goals established in Title I of the Act
- Developing and promoting national policies to improve environmental quality
- Conducting studies, surveys, research, and analyses relating to ecosystems and environmental quality

Implementation

In 1978, CEQ promulgated regulations (40 CFR Parts 1500–1508) implementing NEPA, which are binding on all federal agencies. The regulations address the procedural provisions of NEPA and the administration of the NEPA process, including preparation of EISs. To date, the only change in the NEPA regulations occurred on May 27, 1986, when CEQ amended Section 1502.22 of its regulations to clarify how agencies are to carry out their environmental evaluations in situations where information is incomplete or unavailable.

CEQ has also issued guidance on various aspects of the regulations including an information document on "Forty Most Asked Questions Concerning CEQ's National Environmental Policy Act," Scoping Guidance, and Guidance Regarding NEPA Regulations. Additionally, most federal agencies have promulgated their own NEPA regulations and guidance, which generally follow the CEQ procedures but are tailored for the specific mission and activities of the agency.

The NEPA Process

The NEPA process consists of an evaluation of the environmental effects of a federal undertaking including its alternatives. There are three levels of analysis:

categorical exclusion determination, preparation of an environmental assessment/finding of no significant impact (EA/FONSI), and preparation of an EIS.

- *Categorical Exclusion*: At the first level, an undertaking may be categorically excluded from a detailed environmental analysis if it meets certain criteria that a federal agency has previously determined as having no significant environmental impact. A number of agencies have developed lists of actions that are normally categorically excluded from environmental evaluation under their NEPA regulations.
- *EA/FONSI*: At the second level of analysis, a federal agency prepares a written EA to determine whether or not a federal undertaking would significantly affect the environment. If the answer is no, the agency issues a FONSI. The FONSI may address measures that an agency will take to mitigate potentially significant impacts.
- *EIS*: If the EA determines that the environmental consequences of a proposed federal undertaking may be significant, an EIS is prepared. An EIS is a more detailed evaluation of the proposed action and alternatives. The public, other federal agencies, and outside parties may provide input into the preparation of an EIS and then comment on the draft EIS when it is completed.

If a federal agency anticipates that an undertaking may significantly affect the environment, or if a project is environmentally controversial, a federal agency may choose to prepare an EIS without having to first prepare an EA.

After a final EIS is prepared and at the time of its decision, a federal agency will prepare a public record of its decision addressing how the findings of the EIS, including consideration of alternatives, were incorporated into the agency's decision-making process.

EAs and EIS Components

An EA is described in Section 1508.9 of the CEQ NEPA regulations. Generally, an EA includes brief discussions of the following:

- The need for the proposal
- Alternatives (when there is an unresolved conflict concerning alternative uses of available resources)
- The environmental impacts of the proposed action and alternatives
- A listing of agencies and persons consulted

An EIS, which is described in Part 1502 of the regulations, should include the following:

- Discussions of the purpose of and need for the action
- Alternatives

- The affected environment
- The environmental consequences of the proposed action
- Lists of preparers, agencies, organizations, and persons to whom the statement is sent
- An index
- An appendix (if any)

Federal Agency Role

The role of a federal agency in the NEPA process depends on the agency's expertise and relationship to the proposed undertaking. The agency carrying out the federal action is responsible for complying with the requirements of NEPA.

- *Lead Agency*: In some cases, there may be more than one federal agency involved in an undertaking. In this situation, a lead agency is designated to supervise preparation of the environmental analysis. Federal agencies, together with state, tribal, or local agencies, may act as joint lead agencies.
- *Cooperating Agency*: A federal, state, tribal, or local agency having special expertise with respect to an environmental issue or jurisdiction by law may be a cooperating agency in the NEPA process. A cooperating agency has the responsibility to assist the lead agency by participating in the NEPA process at the earliest possible time; by participating in the scoping process; in developing information and preparing environmental analyses including portions of the EIS concerning which the cooperating agency has special expertise; and in making available staff support at the lead agency's request to enhance the lead agency's interdisciplinary capabilities.
- *CEQ*: Under Section 1504 of CEQ's NEPA regulations, federal agencies may refer to CEQ on interagency disagreements concerning proposed federal actions that might cause unsatisfactory environmental effects. CEQ's role, when it accepts a referral, is generally to develop findings and recommendations, consistent with the policy goals of Section 101 of NEPA.

Environmental Protection Agency's Role

The Environmental Protection Agency (EPA), like other federal agencies, prepares and reviews NEPA documents. However, EPA has a unique responsibility in the NEPA review process. Under Section 309 of the Clean Air Act (CAA), EPA is required to review and publicly comment on the environmental impacts of major federal actions, including actions that are the subject of

EISs. If EPA determines that the action is environmentally unsatisfactory, it is required by Section 309 to refer the matter to CEQ.

In accordance with a Memorandum of Agreement between EPA and CEQ, EPA carries out the operational duties associated with the administrative aspects of the EIS filing process. The Office of Federal Activities in EPA has been designated the official recipient in EPA of all EISs prepared by federal agencies.

The Public's Role

The public has an important role in the NEPA process, particularly during scoping, in providing input on what issues should be addressed in an EIS and in commenting on the findings in an agency's NEPA documents. The public can participate in the NEPA process by attending NEPA-related hearings or public meetings and by submitting comments directly to the lead agency. The lead agency must take into consideration all comments received from the public and other parties on NEPA documents during the comment period.

Resource Conservation and Recovery Act

The Resource Conservation and Recovery Act—commonly referred to as RCRA—is our nation's primary law governing the disposal of solid and hazardous waste. Congress passed RCRA on October 21, 1976, to address the increasing problems the nation faced from our growing volume of municipal and industrial waste. RCRA, which amended the Solid Waste Disposal Act of 1965, set national goals for

- Protecting human health and the environment from the potential hazards of waste disposal
- Conserving energy and natural resources
- Reducing the amount of waste generated
- Ensuring that wastes are managed in an environmentally sound manner

To achieve these goals, RCRA established three distinct, yet interrelated, programs:

- The solid waste program, under RCRA Subtitle D, encourages states to develop comprehensive plans to manage nonhazardous industrial solid waste and municipal solid waste (MSW), sets criteria for MSW landfills and other solid waste disposal facilities, and prohibits the open dumping of solid waste.

- The hazardous waste program, under RCRA Subtitle C, establishes a system for controlling hazardous waste from the time it is generated until its ultimate disposal—in effect, from "cradle to grave."
- The underground storage tank (UST) program, under RCRA Subtitle I, regulates USTs containing hazardous substances and petroleum products.

RCRA banned all open dumping of waste, encouraged source reduction and recycling, and promoted the safe disposal of municipal waste. RCRA also mandated strict controls over the treatment, storage, and disposal of hazardous waste. The first RCRA regulations, "Hazardous Waste and Consolidated Permit Regulations," published in the Federal Register on May 19, 1980 (45 FR 33066; May 19, 1980), established the basic "cradle to grave" approach to hazardous waste management that exists today.

RCRA was amended and strengthened by Congress in November 1984 with the passing of the Federal Hazardous and Solid Waste Amendments (HSWA). These amendments to RCRA required phasing out land disposal of hazardous waste. Some of the other mandates of this strict law include increased enforcement authority for EPA, more stringent hazardous waste management standards, and a comprehensive UST program.

RCRA has been amended on two occasions since HSWA:

1. Federal Facility Compliance Act of 1992—strengthened enforcement of RCRA at Federal facilities
2. Land Disposal Program Flexibility Act of 1996 (5 pp, 24 kB)—provided regulatory flexibility for land disposal of certain wastes

RCRA focuses only on active and future facilities and does not address abandoned or historical sites that are managed under the Comprehensive Environmental Response, Compensation, and Liability Act—commonly known as Superfund.[2]

Regulations promulgated pursuant to Subtitle C of RCRA (40 CFR Parts 260–299) establish a "cradle-to-grave" system governing hazardous waste from the point of generation to disposal. RCRA hazardous wastes include the specific materials listed in the regulations (commercial chemical products, designated with the code "P" or "U"; hazardous wastes from specific industries/sources, designated with the code "K"; hazardous wastes from nonspecific sources, designated with the code "F") and materials that exhibit a hazardous waste characteristic (ignitability, corrosivity, reactivity, or toxicity) designated with the code "D."

Regulated entities that generate hazardous waste are subject to waste accumulation, manifesting, and record-keeping standards. Facilities that treat, store, or dispose of hazardous waste must obtain a permit, either from EPA or from a state agency that EPA has authorized to implement the permitting program.

Subtitle C permits contain general facility standards such as contingency plans, emergency procedures, record-keeping and reporting requirements, financial assurance mechanisms, and unit-specific standards. RCRA also contains provisions (40 CFR Part 264 Subpart S and Part 264.10) for conducting corrective actions that govern the cleanup of releases of hazardous waste or constituents from solid waste management units at RCRA-regulated facilities.

Although RCRA is a federal statute, many states implement the RCRA program. Currently, EPA has delegated its authority to implement various provisions of RCRA to 46 of the 50 states.

Most RCRA requirements are not industry specific but apply to any company that generates, transports, treats, stores, or disposes of hazardous waste. Here are some important RCRA regulatory requirements:

- Solid Waste and Hazardous Waste
- Universal Waste
- Used Oil Management Standards
- USTs

Solid Waste and Hazardous Waste[3]

Solid waste means any garbage or refuse; sludge from a wastewater treatment plant, water supply treatment plant, or air pollution control facility; and other discarded material, including solid, liquid, semisolid, or contained gaseous material resulting from industrial, commercial, mining, and agricultural operations, and from community activities. Solid wastes include both hazardous and nonhazardous waste.

A waste may be considered hazardous if it is ignitable (i.e., burns readily), corrosive, or reactive (e.g., explosive). Waste may also be considered hazardous if it contains certain amounts of toxic chemicals. In addition to these characteristic wastes, EPA has also developed a list of more than 500 specific hazardous wastes. Hazardous waste takes many physical forms and may be solid, semisolid, or even liquid.

Acute hazardous wastes contain such dangerous chemicals that they could pose a threat to human health and the environment even when properly managed. These wastes are fatal to humans and animals even in low doses.

Subtitle C of the RCRA creates a cradle-to-grave management system for hazardous waste to ensure proper treatment, storage, and disposal in a manner protective of human health and the environment.

Identification of Solid and Hazardous Wastes

This regulation (40 CFR Part 261) lays out the procedure every generator should follow to determine whether the material created is considered a hazardous waste or a solid waste or is exempted from regulation.

Standards for Generators of Hazardous Waste

This regulation (40 CFR Part 262) establishes the responsibilities of hazardous waste generators, including obtaining an identification number, preparing a manifest, ensuring proper packaging and labeling, meeting standards for waste accumulation units, and record-keeping and reporting requirements. Generators can accumulate hazardous waste for up to 90 days (or 180 days depending on the amount of waste generated) without obtaining a permit for being a treatment, storage, and disposal (TSD) facility.

Land Disposal Restrictions

Land disposal restrictions (LDRs) are regulations prohibiting the disposal of hazardous waste on land without prior treatment. Under 40 CFR 268, materials must meet treatment standards before placement in an RCRA land disposal unit (landfill, land treatment unit, waste pile, or surface impoundment). Wastes subject to the LDR include solvents, electroplating wastes, heavy metals, and acids. Generators of waste subject to these restrictions must notify the designated TSD facility to ensure proper treatment before disposal.

Tanks and Containers

Tanks and containers used to store hazardous waste with a high volatile organic concentration must meet emission standards under RCRA. Regulations (40 CFR Parts 264 and 265, Subpart CC) require generators to test the waste to determine the concentration of the waste, to satisfy tank and container emissions standards, and to inspect and monitor regulated units. These regulations apply to all facilities that store such waste, including generators operating under the 90-day accumulation rule.

Hazardous Waste and Agriculture

Irrigation return flows are not considered hazardous waste. Agricultural producers disposing of waste pesticides from their own use are exempt from hazardous waste requirements as long as (1) they triple rinse the emptied containers in accordance with the labeling to facilitate removal of the chemical from the container and (2) they dispose of the pesticide residue on their own agricultural establishment in a manner consistent with the disposal instructions on the pesticide label.

Disposal of hazardous waste on an agricultural establishment could subject the agricultural producer to significant responsibility, including closure and postclosure care. Offsite disposal of hazardous waste could subject agricultural producers to hazardous waste generator requirements.

Universal Waste[3]

The universal waste rule is designed to reduce the amount of hazardous waste items in the MSW stream, encourage recycling and proper disposal of certain common hazardous wastes, and reduce the regulatory burden on businesses that generate these wastes. Universal wastes include the following:

- *Batteries* such as nickel–cadmium (Ni–Cd) and small sealed lead-acid batteries, which are found in many common items in the business and home setting, including electronic equipment, mobile telephones, portable computers, and emergency backup lighting.
- *Agricultural pesticides* that have been recalled or banned from use are obsolete, have become damaged, or are no longer needed because of changes in cropping patterns or other factors. They often are stored for long periods in sheds or barns.
- *Thermostats*, which can contain as much as 3 g of liquid mercury and are located in almost any building, including commercial, industrial, agricultural, community, and household buildings.

Universal wastes are generated by small and large businesses that are regulated under RCRA and have been required to handle these materials as hazardous wastes. The universal waste rule eases the regulatory burden on businesses that generate these wastes. Specifically, it streamlines the requirements related to notification, labeling, marking, prohibitions, accumulation time limits, employee training, response to releases, offsite shipments, tracking, exports, and transportation. For example, the rule extends the amount of time that businesses can accumulate these materials on site. It also allows companies to transport them with a common carrier, instead of a hazardous waste transporter, and no longer requires companies to obtain a manifest.

Universal Waste and Agriculture

The universal waste rule does not apply to businesses (such as many agricultural establishments and other agribusinesses) that generate less than 100 kg of universal wastes per month (Conditionally Exempt Small Quantity Generators). EPA encourages these businesses to participate voluntarily in collection and recycling programs by bringing these wastes to collection centers for proper treatment and disposal.

Used Oil Management Standards[3]

Used oil management standards (40 CFR Part 279) impose management requirements affecting the storage, transportation, burning, processing, and re-refining of the used oil. For facilities that merely generate used oil, the regulations establish storage standards. A facility that is considered a used oil marketer (one who generates and sells off-specification used oil directly to a used oil burner) must satisfy additional tracking and paperwork requirements.

EPA's regulatory definition of used oil is as follows: Used oil is any oil (either synthetic or refined from crude oil) that has been used and, as a result of such use, is contaminated by physical or chemical impurities. Simply put, used oil is exactly what its name implies—any petroleum-based or synthetic oil that has been used. During normal use, impurities such as dirt, metal scrapings, water, or chemicals can get mixed in with the oil, so that in time the oil no longer performs well. Eventually, this used oil must be replaced with virgin or re-refined oil to do the job at hand.

EPA's used oil management standards include a three-pronged approach to determine if a substance meets the definition of used oil. To meet EPA's definition of used oil, a substance must meet each of the following three criteria:

- *Origin*—the first criterion for identifying used oil is based on the origin of the oil. Used oil must have been refined from crude oil or made from synthetic materials. Animal and vegetable oils are excluded from EPA's definition of used oil.

- *Use*—the second criterion is based on whether and how the oil is used. Oils used as lubricants, hydraulic fluids, heat transfer fluids, buoyants, and for other similar purposes are considered used oil. Unused oils, such as bottom clean-out waste from virgin fuel oil storage tanks or virgin fuel oil recovered from a spill, do not meet EPA's definition of used oil because these oils have never been "used." EPA's definition also excludes products used as cleaning agents or solely for their solvent properties, as well as certain petroleum-derived products like antifreeze and kerosene.

- *Contaminants*—the third criterion is based on whether or not the oil is contaminated with either physical or chemical impurities. In other words, to meet EPA's definition, used oil must become contaminated as a result of being used. This aspect of EPA's definition includes residues and contaminants generated from handling, storing, and processing used oil. Physical contaminants could include metal shavings, sawdust, or dirt. Chemical contaminants could include solvents, halogens, or salt water.

Used Oil and Agriculture

Agricultural producers who generate an average of 25 gal or less per month from vehicles or machinery per calendar year are exempt from these regulations. Those exceeding 25 gal are required to store it in tanks meeting underground or above-ground technical requirements and use transporters with EPA authorization numbers for removal from the agricultural establishment. Storage in unlined surface impoundments (defined as wider than they are deep) is banned.

Underground Storage Tanks[3]

A UST system is a tank and any underground piping connected to the tank that has at least 10% of its combined volume underground. USTs containing petroleum and hazardous substances are regulated under Subtitle I of RCRA. Subtitle I regulations (40 CFR Part 280) contain tank design and release detection requirements, as well as financial responsibility and corrective action standards for USTs. The UST program also establishes increasingly stringent standards, including upgrade requirements for existing tanks, that were to be met by 1998.

In 1984, Congress responded to the increasing threat to groundwater posed by leaking USTs by adding Subtitle I to the RCRA. Subtitle I required EPA to develop a comprehensive regulatory program for USTs storing petroleum or certain hazardous substances.

In 1986, Congress amended Subtitle I of RCRA and created the Leaking Underground Storage Tank Trust Fund and established financial responsibility requirements. Congress directed EPA to publish regulations that would require UST owners and operators to demonstrate that they are financially capable of cleaning up releases and compensating third parties for resulting damages.

The following USTs are excluded from regulation and, therefore, do not need to meet federal requirements for USTs:

- Farm and residential tanks of 1100 gal or less capacity holding motor fuel used for noncommercial purposes
- Tanks storing heating oil used on the premises where it is stored
- Tanks on or above the floor of underground areas, such as basements or tunnels
- Septic tanks and systems for collecting storm water and wastewater
- Flow-through process tanks
- Tanks of 110 gal or less capacity
- Emergency spill and overfill tanks

Subtitle I of RCRA allows state UST programs approved by EPA to operate in lieu of the federal program, and EPA's state program approval regulations set standards for state programs to meet. States may have more stringent regulations than the federal requirements. People who are interested in requirements for USTs should contact their state UST program for information on state requirements.

USTs and Agriculture

For agricultural establishments, USTs and their associated piping holding less than 1100 gal of motor fuel for noncommercial purposes, tanks holding less than 110 gal, tanks holding heating oil used on the premises, and septic tanks are excluded from regulations. All newly installed regulated USTs are required to meet regulations related to construction, monitoring, operating, reporting to state or federal regulatory agencies, owner record keeping, and financial responsibility. Requirements for regulated USTs installed before December 22, 1988, were phased in through December 2, 1989.

Toxic Substances Control Act (TSCA) (15 U.S.C. §2601 et seq. (1976))[4]

The Toxic Substances Control Act of 1976 provides EPA with authority to require reporting, record-keeping and testing requirements, and restrictions relating to chemical substances or mixtures. Certain substances are generally excluded from TSCA, including, among others, food, drugs, cosmetics, and pesticides.

TSCA addresses the production, importation, use, and disposal of specific chemicals including polychlorinated biphenyls, asbestos, radon, and lead-based paint.

Various sections of TSCA provide authority to

- Require, under Section 5, premanufacture notification for "new chemical substances" before manufacture.
- Require, under Section 4, testing of chemicals by manufacturers, importers, and processors where risks or exposures of concern are found.
- Issue Significant New Use Rules, under Section 5, when it identifies a "significant new use" that could result in exposures to, or releases of, a substance of concern.
- Maintain the TSCA Inventory, under Section 8, which contains more than 83,000 chemicals. As new chemicals are commercially manufactured or imported, they are placed on the list.
- Require those importing or exporting chemicals, under Sections 12(b) and 13, to comply with certification reporting and other requirements.
- Require, under Section 8, reporting and record keeping by persons who manufacture, import, process, or distribute chemical substances in commerce.

- Require, under Section 8(e), that any person who manufactures (including imports), processes, or distributes in commerce a chemical substance or mixture and who obtains information that reasonably supports the conclusion that such substance or mixture presents a substantial risk of injury to health or the environment to immediately inform EPA, except where EPA has been adequately informed of such information. EPA screens all TSCA b§8(e) submissions as well as voluntary "For Your Information" submissions. The latter are not required by law but are submitted by industry and public interest groups for a variety of reasons.

What Does It Mean for a Chemical to Be on the TSCA Inventory?

Substances on the TSCA Inventory are considered "existing" chemicals in US commerce, and substances not on the TSCA Inventory are considered "new" chemicals. If a substance is determined to be a "new" chemical substance for TSCA purposes, it is subject to TSCA Section 5 Premanufacture Notice (PMN) requirements, unless the substance meets a TSCA reporting exclusion (e.g., is a naturally occurring material) or is exempt from PMN reporting (e.g., is an exempted polymer). (The TSCA Inventory must be consulted to determine if a specific substance is "new" or "existing.") For substances that are "existing" chemical substances in US commerce, the TSCA Inventory can be used to determine if there are restrictions on manufacture or use.[5]

How Are Chemicals Added to the TSCA Inventory?

After PMN review has been completed, the company that submitted the PMN must provide a Notice of Commencement of Manufacture or Import (NOC) (EPA Form 7710-56) to EPA within 30 calendar days of the date the substance is first manufactured or imported for nonexempt commercial purposes. A chemical substance is considered to be on the TSCA Inventory and becomes an existing chemical as soon as a complete NOC is received by EPA. The Agency receives between 500 and 1000 NOCs each year; thus, the TSCA Inventory changes daily.

Non-PMN submissions (Low Volume Exemptions, Low Release/Low Exposure Exemptions, and Test Market Exemptions) and exempt uses not subject to submission (R&D) do not require an NOC and are not listed on the TSCA Inventory.[5]

How to Get a Determination from EPA on whether a Chemical Is on the Inventory

If an intended manufacturer or importer of a chemical substance is unsure of the TSCA Inventory status of the chemical (e.g., cannot find that substance on one of the public sources of nonconfidential TSCA Inventory data), the company or representative can obtain a written determination from EPA if it can demonstrate a "genuine intent." It can do this by submitting a Bona Fide Intent to Manufacture or Import Notice pursuant to the procedures at 40 CFR Section 720.25.[5]

Bona Fide Intent to Manufacture or Import Notice

In a bona fide notice, a submitter must

- Provide specific chemical identification data including the Chemical Abstracts Index name as well as information about the substances' manufacture or importation. Proper CA Index names can be obtained from the Chemical Abstracts Service's Inventory Expert Service.
- Possibly submit a letter of support (see below). This may be necessary if some information is withheld from the submitter by, for example, a supplier.
- Certify intent to manufacture or import the intended substance for a commercial purpose.
- Provide all of the other information required in support of a bona fide notice.

In some cases, a potential manufacturer may be intending to use reactants whose specific chemical identities are held confidential by their suppliers. In certain other cases, a potential importer may be intending to bring into the United States a substance whose identity is known only to its foreign manufacturer. In these cases, a letter of support from the domestic or foreign manufacturer of the confidential substances can be provided directly to EPA and should include specific chemical identity information. When using a Branded Material of Confidential Composition, users will need information from their suppliers to ensure that they are and remain in compliance.

Instructions on Notices of Bona Fide Intent to Manufacture are available from the TSCA hotline.[5]

Letter of Support

Specific chemical identity information is required for Bona Fide Intent to Manufacture or Import Notices, PMNs, and other purposes under TSCA. EPA must be notified of any confidential chemical identity information (e.g., a reactant only known by a trade name is used in the manufacture of a chemical substance that is the subject of a bona fide notice or PMN). Information that has been withheld from the submitter by a third party should be submitted directly to EPA by that third party (e.g., usually a domestic or foreign supplier or manufacturer). In its letter of support, the third party must provide chemical identity information for the confidential substance as specified in the amended regulation at 40 CFR Section 720.45(a).

If confidential substances are involved and require a third-party letter of support, a bona fide notice or PMN submitter must keep in mind that all supporting material must be received by EPA for a bona fide notice or PMN to be considered complete. A submitter should also have an agreement with its supplier to ensure being informed of any changes in composition that can change the chemical identity of the confidential substance.[5]

Branded Materials of Confidential Composition

Manufacturers and importers whose reportable substances are manufactured with branded materials that have confidential components should take steps to be informed in a timely manner if the branded materials change in composition. EPA does not use brand names in listing substances on the TSCA Inventory, in part because branded material formulations can change and in part because the TSCA Inventory identifies and lists specific chemical substances and not formulations.[5]

Federal Insecticide, Fungicide, and Rodenticide Act

The objective of the Federal Insecticide, Fungicide, and Rodenticide Act (FIFRA) is to provide federal control of pesticide distribution, sale, and use. All pesticides used in the United States must be registered (licensed) by EPA. Registration assures that pesticides will be properly labeled and that, if used in accordance with specifications, they will not cause unreasonable harm to the environment. Use of each registered pesticide must be consistent with use directions contained on the label or labeling.

The first pesticide control law was enacted in 1910. This law was primarily aimed at protecting consumers from ineffective products and deceptive labeling. When the FIFRA was first passed in 1947, it established procedures for registering pesticides with the US Department of Agriculture and established labeling provisions. The law was still, however, primarily concerned with the efficacy of pesticides and did not regulate pesticide use.

FIFRA was essentially rewritten in 1972 when it was amended by the Federal Environmental Pesticide Control Act (FEPCA). The law has been amended numerous times since 1972, including some significant amendments in the form of the Food Quality Protection Act of 1996. In its current form, FIFRA mandates that EPA regulate the use and sale of pesticides to protect human health and preserve the environment.

Since the FEPCA amendments, EPA is specifically authorized to (1) strengthen the registration process by shifting the burden of proof to the chemical manufacturer, (2) enforce compliance against banned and unregistered products, and (3) promulgate the regulatory framework missing from the original law.

FIFRA provides EPA with the authority to oversee the sale and use of pesticides. However, because FIFRA does not fully preempt state/tribal or local law, each state/tribe and local government may also regulate pesticide use.[6]

Tolerances and Exemptions

Before EPA can register a pesticide that is used on raw agricultural products, it must grant a tolerance or exemption. A tolerance is the maximum amount of a pesticide that can be on a raw product when it is used and still be considered safe. Under the Food, Drug, and Cosmetic Act (FDCA), a raw agricultural product is deemed unsafe if it contains a pesticide residue, unless the residue is within the limits of a tolerance established by EPA or is exempt from the requirement. The FDCA requires EPA to establish these residue tolerances.[6]

Tolerances and Agriculture

Food or feed residues that lack tolerances or have residues exceeding the established tolerances are subject to seizure, and the applicators or agricultural producers are subject to prosecution under FIFRA if misuse is found.[6]

Registration of New Pesticides

Under FIFRA Section 3, all new pesticides (with minor exceptions) used in the United States must be registered by the Administrator of EPA. Pesticide registration is very specific; it is not valid for all uses of a particular chemical. Each registration specifies the crops/sites on which it may be applied, and each use must be supported by research data. Ordinarily, the manufacturer (domestic or foreign) of the pesticide files an application for registration. The application process often requires the submission of extensive environmental, health, and safety data.

Conditional registration may be given to some new pesticides while EPA obtains the data needed to make a full analysis of the product. After a pesticide is registered, the registrant must also notify EPA of any newly uncovered facts concerning adverse environmental effects (FIFRA Section 6(a)(2)). In addition, EPA is required to periodically review pesticide registrations, with a goal of review every 15 years. Variations of the registration requirements exist for "minor use pesticides," "antimicrobial pesticides," and "reduced risk pesticides."

Data Requirements for Registration

The categories of data required include the product's chemistry, environmental fate, residue chemistry, dietary and nondietary hazards to humans, and hazards to domestic animals and nontarget organisms (40 CFR Part 158).

Under the "product chemistry" category, applicants must supply technical information describing the product's active and inert ingredients, manufacturing or formulating processes, and physical and chemical characteristics.

Data from "environmental fate" studies are used to assess the effects of pesticide residues on the environment, including the effect on nontarget organisms and their habitat.

Residue chemistry information includes the expected frequency, amounts, and time of application, and test results of residue remaining on treated food or feed.

Information under "hazards to humans, domestic animals, and nontarget organisms" includes specific test data assessing acute, subchronic, and chronic toxicity; skin and eye irritation potential; and potential exposure by various routes (i.e., oral, dermal, or inhalation).

All studies must be conducted under conditions that meet Good Laboratory Practice standards (40 CFR Part 160). Guidelines for studies of product chemistry, residue chemistry, environmental chemistry, hazard evaluation, and occupational and residential exposure can be found in 40 CFR Part 158.

Registration Criteria

To register a pesticide, the Administrator must find the following to be true:

- Its composition is such as to warrant the proposed claims for it.
- Its labeling and other material required to be submitted comply with the requirements of the Act.
- It will perform its intended function without unreasonable adverse effects on the environment.
- When used in accordance with widespread and commonly recognized practice, it will not generally cause unreasonable adverse effects on the environment.

Unreasonable Adverse Effects on the Environment

FIFRA defines an "unreasonable adverse effect on the environment" as "(1) any unreasonable risk to man or the environment, taking into account the economic, social, and environmental costs and benefits of the use of the pesticide, or (2) a human dietary risk from residues that result from a use of a pesticide in or on any food inconsistent with the standard under Section 408 of the Federal Food, Drug, and Cosmetic Act (21U.S.C.346a)."

Clean Air Act of 1970 (42 U.S.C. §7401 et seq. (1970))

The CAA is the comprehensive federal law that regulates air emissions from stationary and mobile sources. Among other things, this law authorizes EPA to establish National Ambient Air Quality Standards (NAAQS) to protect public health and public welfare and to regulate emissions of hazardous air pollutants.

One of the goals of the Act was to set and achieve NAAQS in every state by 1975 in order to address the public health and welfare risks posed by certain widespread air pollutants. The setting of these pollutant standards was coupled with directing the states to develop state implementation plans (SIPs), applicable to appropriate industrial sources in the state, in order to achieve these standards. The Act was amended in 1977 and 1990 primarily to set new goals (dates) for achieving attainment of NAAQS since many areas of the country had failed to meet the deadlines.

Section 112 of the CAA addresses emissions of hazardous air pollutants. Prior to 1990, CAA established a risk-based program under which only a few standards were developed. The 1990 CAA Amendments revised Section 112 to first require issuance of technology-based standards for major sources and certain area sources. "Major sources" are defined as a stationary source or group of stationary sources that emit or have the potential to emit 10 tons per year or more of a hazardous air pollutant or 25 tons per year or more of a combination of hazardous air pollutants. An "area source" is any stationary source that is not a major source.

For major sources, Section 112 requires that EPA establish emission standards that require the maximum degree of reduction in emissions of hazardous air pollutants. These emission standards are commonly referred to as "maximum achievable control technology" or "MACT" standards. Eight years after the technology-based MACT standards are issued for a source category, EPA is required to review those standards to determine whether any residual risk exists for that source category and, if necessary, revise the standards to address such risk.[7]

Clean Air Act of 1990[8]

In June 1989, President Bush proposed sweeping revisions to the CAA. Building on Congressional proposals advanced during the 1980s, the President proposed legislation designed to curb three major threats to the nation's environment and to the health of millions of Americans: acid rain, urban air pollution, and toxic air emissions. The proposal also called for establishing a national permits program to make the law more workable and an improved enforcement program to help ensure better compliance with the Act.

By large votes, both the House of Representatives (401–21) and the Senate (89–11) passed Clean Air bills that contained the major components of the President's proposals. Both bills also added provisions requiring the phase-out of ozone-depleting chemicals, roughly according to the schedule outlined in international negotiations (Revised Montreal Protocol). The Senate and House bills also added specific research and development provisions, as well as detailed programs to address accidental releases of toxic air pollutants.

A joint conference committee met from July to October 1990 to iron out differences in the bills and both Houses overwhelmingly voted out the package recommended by the Conferees. The President received the Bill from Congress on November 14, 1990, and signed it on November 15, 1990.

Several progressive and creative new themes are embodied in the Amendments; themes necessary for effectively achieving the air quality goals and regulatory reform expected from these far-reaching amendments. Specifically, the new law

- Encourages the use of market-based principles and other innovative approaches, like performance-based standards and emission banking and trading
- Provides a framework from which alternative clean fuels will be used by setting standards in the fleet and California pilot program that can be met by the most cost-effective combination of fuels and technology
- Promotes the use of clean, low-sulfur coal and natural gas, as well as innovative technologies to clean high-sulfur coal through the acid rain program
- Reduces enough energy waste and creates enough of a market for clean fuels derived from grain and natural gas to cut dependency on oil imports by one million barrels/day
- Promotes energy conservation through an acid rain program that gives utilities flexibility to obtain needed emission reductions through programs that encourage customers to conserve energy

With these themes providing the framework for the CAA amendments and with our commitment to implement the new law quickly, fairly, and efficiently, Americans will get what they asked for: a healthy, productive environment, linked to sustainable economic growth and sound energy policy.

Title I: Provisions for Attainment and Maintenance of NAAQS

Although the CAA of 1977 brought about significant improvements in our nation's air quality, the urban air pollution problems of ozone (smog), carbon monoxide (CO), and particulate matter (PM-10) persist. Currently, over 100 million Americans live in cities that are out of attainment with the public health standards for ozone.

The most widespread and persistent urban pollution problem is ozone. The causes of this and the lesser problem of carbon monoxide (CO) and particulate matter (PM-10) pollution in our urban areas are largely attributed to the diversity and number of urban air pollution sources. One component of urban smog—hydrocarbons—comes from automobile emissions, petroleum refineries, chemical plants, dry cleaners, gasoline stations, house painting, and printing shops. Another key component—nitrogen oxides—comes from the combustion of fuel for transportation, utilities, and industries.

While there are other reasons for continued high levels of ozone pollution, such as growth in the number of stationary sources of hydrocarbons and continued growth in automobile travel, perhaps the most telling reason is that the remaining sources of hydrocarbons are also the most difficult to control. These are the small sources—generally those that emit less than 100 tons of hydrocarbons per year. These sources, such as auto body shops and dry cleaners, may individually emit less than 10 tons per year, but collectively emit many hundreds of tons of pollution.

The CAA Amendments of 1990 create a new, balanced strategy for the nation to attack the problem of urban smog. Overall, the new law reveals the Congress' high expectations of the states and the federal government. While it gives states more time to meet the air quality standard—up to 20 years for ozone in Los Angeles—it also requires states to make constant formidable progress in reducing emissions. It requires the federal government to reduce emissions from cars, trucks, and buses; from consumer products such as hair spray and window washing compounds; and from ships and barges during loading and unloading of petroleum products. The federal government must also develop the technical guidance that states need to control stationary sources.

The new law addresses the urban air pollution problems of ozone (smog), carbon monoxide (CO), and particulate matter (PM-10). Specifically, it clarifies how areas are designated and redesignated "attainment." It also allows EPA to define the boundaries of "nonattainment" areas: geographical areas whose air quality does not meet federal air quality standards designed to protect public health.

The new law also establishes provisions defining when and how the federal government can impose sanctions on areas of the country that have not met certain conditions.

For the pollutant ozone, the new law establishes nonattainment area classifications ranked according to the severity of the areas' air pollution problem. These classifications are marginal, moderate, serious, severe, and extreme. EPA assigns each nonattainment area one of these categories, thus triggering varying requirements the area must comply with in order to meet the ozone standard.

As mentioned, nonattainment areas will have to implement different control measures, depending upon their classification. Marginal areas, for example, are the closest to meeting the standard. They will be required to conduct an inventory of their ozone-causing emissions and institute a permit program. Nonattainment areas with more serious air quality problems must implement various control measures. The worse the air quality, the more controls areas will have to implement.

The new law also establishes similar programs for areas that do not meet the federal health standards for the pollutants carbon monoxide and particulate matter. Areas exceeding the standards for these pollutants will be divided into "moderate" and "serious" classifications. Depending upon the degree to which they exceed the carbon monoxide standard, areas will be required to implement programs introducing oxygenated fuels and enhanced emission inspection programs, among other measures. Depending upon their classification, areas exceeding the particulate matter standard will have to implement either reasonably available control measures or best available control measures, among other requirements.

Title II: Provisions Relating to Mobile Sources

While motor vehicles built today emit fewer pollutants (60% to 80% less, depending on the pollutant) than those built in the 1960s, cars and trucks still account for almost half the emissions of the ozone precursors VOCs and NOx, and up to 90% of the CO emissions in urban areas. The principal reason for this problem is the rapid growth in the number of vehicles on the roadways and the total miles driven. This growth has offset a large portion of the emission reductions gained from motor vehicle controls.

In view of the unforeseen growth in automobile emissions in urban areas combined with the serious air pollution problems in many urban areas, the Congress has made significant changes to the motor vehicle provisions on the 1977 CAA.

The CAA of 1990 establishes tighter pollution standards for emissions from automobiles and trucks. These standards will reduce tailpipe emissions of hydrocarbons, carbon monoxide, and nitrogen oxides on a phased-in basis beginning in model year 1994. Automobile manufacturers will also be required to reduce vehicle emissions resulting from the evaporation of gasoline during refueling.

Title III: Air Toxics

Toxic air pollutants are those pollutants that are hazardous to human health or the environment but are not specifically covered under another portion of the CAA. These pollutants are typically carcinogens, mutagens, and reproductive toxins. The CAA Amendments of 1977 failed to result in substantial reductions of the emissions of these very threatening substances. In fact, over the history of the air toxics program, only seven pollutants have been regulated.

We know that the toxic air pollution problem is widespread. Information generated from The Superfund "Right to Know" rule (SARA Section 313) indicates that more than 2.7 billion pounds of toxic air pollutants are emitted annually in the United States. EPA studies indicate that exposure to such quantities of air toxics may result in 1000 to 3000 cancer deaths each year.

The CAA of 1990 offers a comprehensive plan for achieving significant reductions in emissions of hazardous air pollutants from major sources. Industry reports in 1987 suggest that an estimated 2.7 billion pounds of toxic air pollutants were emitted into the atmosphere, contributing to approximately 300–1500 cancer fatalities annually. The new law will improve EPA's ability to address this problem effectively and it will dramatically accelerate progress in controlling major toxic air pollutants.

The new law includes a list of 189 toxic air pollutants of which emissions must be reduced. EPA must publish a list of source categories that emit certain levels of these pollutants within 1 year after the new law is passed. The list of source categories must include (1) major sources emitting 10 tons/year of any one, or 25 tons/year of any combination of those pollutants; and (2) area sources (smaller sources, such as dry cleaners).

EPA then must issue MACT standards for each listed source category according to a prescribed schedule. These standards will be based on the best demonstrated control technology or practices within the regulated industry, and EPA must issue the standards for 40 source categories within 2 years of passage of the new law. The remaining source categories will be controlled according to a schedule that ensures all controls will be achieved within 10 years of enactment. Companies that voluntarily reduce emissions according to certain conditions can get a 6-year extension from meeting the MACT requirements.

Eight years after MACT is installed on a source, EPA must examine the risk levels remaining at the regulated facilities and determine whether additional controls are necessary to reduce unacceptable residual risk.

The new law also establishes a Chemical Safety Board to investigate accidental releases of chemicals. Further, the new law requires EPA to issue regulations controlling air emissions from municipal, hospital, and other commercial and industrial incinerators.

Title IV: Acid Deposition Control

As many know, acid rain occurs when sulfur dioxide and nitrogen oxide emissions are transformed in the atmosphere and return to the earth in rain, fog, or snow. Approximately 20 million tons of SO_2 are emitted annually in the United States, mostly from the burning of fossil fuels by electric utilities. Acid rain damages lakes, harms forests and buildings, contributes to reduced visibility, and is suspected of damaging health.

The new law also includes specific requirements for reducing emissions of nitrogen oxides, based on EPA regulations to be issued not later than mid-1992 for certain boilers and 1997 for all remaining boilers.

Title V: Permits

The new law introduces an operating permits program modeled after a similar program under the Federal National Pollution Elimination Discharge System law. The purpose of the operating permits program is to ensure compliance with all applicable requirements of the CAA and to enhance EPA's ability to enforce the Act. Air pollution sources subject to the program must obtain an operating permit, states must develop and implement the program, and EPA must issue permit program regulations, review each state's proposed program, and oversee the state's efforts to implement any approved program. EPA must also develop and implement a federal permit program when a state fails to adopt and implement its own program.

This program—in many ways the most important procedural reform contained in the new law—will greatly strengthen enforcement of the CAA. It will enhance air quality control in a variety of ways. First, adding such a program updates the CAA, making it more consistent with other environmental statutes. The Clean Water Act, the RCRA, and the FIFRA all require permits. The 1977 Clean Air laws also require a construction permit for certain pollution sources, and about 35 states have their own laws requiring operating permits.

The new program clarifies and makes more enforceable a source's pollution control requirements. Currently, a source's pollution control obligations may be scattered throughout numerous hard-to-find provisions of state and federal regulations, and in many cases, the source is not required under the applicable SIP to submit periodic compliance reports to EPA or the states. The permit program will ensure that all of a source's obligations with respect to its pollutants will be contained in one permit document and that the source will file periodic reports identifying the extent to which it has complied with those obligations. Both of these requirements will greatly enhance the ability of federal and state agencies to evaluate its air quality situation.

In addition, the new program will provide a ready vehicle for states to assume administration, subject to federal oversight, of significant parts of the air toxics program and the acid rain program, and, through the permit fee provisions, required under Title V, the program will greatly augment a state's resources to administer pollution control programs by requiring sources of pollution to pay their fair share of the costs of a state's air pollution program.

Under the new law, EPA must issue program regulations within 1 year of enactment. Within 3 years of enactment, each state must submit to EPA a permit program meeting these regulatory requirements. After receiving the state submittal, EPA has 1 year to accept or reject the program. EPA must levy sanctions against a state that does not submit or enforce a permit program.

Each permit issued to a facility will be for a fixed term of up to 5 years. The new law establishes a permit fee whereby the state collects a fee from the permitted facility to cover reasonable direct and indirect costs of the permitting program.

All sources subject to the permit program must submit a complete permit application within 12 months of the effective date of the program. The state permitting authority must determine whether or not to approve an application within 18 months of the date it receives the application.

EPA has 45 days to review each permit and to object to permits that violate the CAA. If EPA fails to object to a permit that violates the Act or the implementation plan, any person may petition EPA to object within 60 days following EPA's 45-day review period, and EPA must grant or deny the permit within 60 days. Judicial review of EPA's decision on a citizen's petition can occur in the federal court of appeals.

Title VI: Stratospheric Ozone and Global Climate Protection

The new law builds on the market-based structure and requirements currently contained in EPA's regulations to phase out the production of substances that deplete the ozone layer. The law requires a complete phaseout of CFCs and halons with interim reductions and some related changes to the existing Montreal Protocol, revised in June 1990.

Under these provisions, EPA must list all regulated substances along with their ozone depletion potential, atmospheric lifetimes, and global warming potentials within 60 days of enactment.

In addition, EPA must ensure that Class I chemicals be phased out on a schedule similar to that specified in the Montreal Protocol—CFCs, halons, and carbon tetrachloride by 2000; methyl chloroform by 2002—but with more stringent interim reductions. Class II chemicals (hydrochlorofluorocarbons) will be phased out by 2030. Regulations for Class I chemicals will be required within 10 months, and Class II chemical regulations were required by December 31, 1999.

The law also requires EPA to publish a list of safe and unsafe substitutes for Class I and II chemicals and to ban the use of unsafe substitutes.

The law requires nonessential products releasing Class I chemicals to be banned within 2 years of enactment. In 1994, a ban went into effect for aerosols and noninsulating foams using Class II chemicals, with exemptions for flammability and safety. Regulations for this purpose will be required within 1 year of enactment, to become effective 2 years afterward.

Title VII: Provisions Relating to Enforcement

The CAA of 1990 contains a broad array of authorities to make the law more readily enforceable, thus bringing it up to date with the other major environmental statutes.

EPA has new authorities to issue administrative penalty orders up to $200,000 and field citations up to $5000 for lesser infractions. Civil judicial penalties are enhanced. Criminal penalties for knowing violations are upgraded from misdemeanors to felonies, and new criminal authorities for knowing and negligent endangerment will be established.

In addition, sources must certify their compliance, and EPA has authority to issue administrative subpoenas for compliance data. EPA will also be authorized to issue compliance orders with compliance schedules of up to 1 year.

The citizen suit provisions have also been revised to allow citizens to seek penalties against violators, with the penalties going to a US Treasury fund for use by EPA for compliance and enforcement activities. The government's right to intervene is clarified and citizen plaintiffs will be required to provide the United States with copies of pleadings and draft settlements.

Other Titles

The CAA Amendments of 1990 continue the federal acid rain research program and contain several new provisions relating to research, development, and air monitoring. They also contain provisions to provide additional unemployment benefits through the Job Training Partnership Act to workers laid off as a consequence of compliance with the CAA. The Act also contains provisions to improve visibility near National Parks and other parts of the country.

Key Points to Remember on Environmental Management

1. The National Environmental Protection Act (NEPA) was signed into law on January 1, 1970. Its major focus was to require the federal government to use all practicable means to create and maintain conditions under which man and nature can exist in productive harmony.

2. NEPA specifically requires all federal agencies to prepare detailed environmental impact statements (EISs) that assess the environmental impact of and alternatives to major federal actions significantly affecting the environment.

3. NEPA establishes the Council on Environmental Quality (CEQ), which is specifically appointed to oversee the provisions of NEPA.

4. The Resource Conservation and Recovery Act (RCRA) is the primary policy of the United States governing the disposal of solid and hazardous wastes.

5. RCRA specifically covers solid and hazardous waste, universal waste, used oil management, and underground storage tanks.

6. Solid waste means any garbage or refuse; sludge from a wastewater treatment plant, water supply treatment plant, or air pollution control facility; and other discarded material, including solid, liquid, semisolid, or contained gaseous material resulting from industrial, commercial, mining, and agricultural operations, and from community activities. Solid wastes include both hazardous and nonhazardous waste.

7. Universal wastes include items such as batteries, agricultural pesticides, and thermostats.

8. The Toxic Substances Control Act (TSCA) provides EPA with the authority to require reporting, record-keeping and testing requirements, and restrictions relating to chemical substances and mixtures.

9. The purpose of the Federal Insecticide, Fungicide and Rodenticide Act (FIFRA) is to provide federal control of pesticide distribution, sale, and use.

10. The Clean Air Act of 1990 was designed to curb three major threats to the nation's environment, including acid rain, urban air pollution, and toxic air emissions.

References

1. http://www.epa.gov/compliance/basics/nepa.html.
2. http://www.epa.gov/osw/laws-regs/rcrahistory.htm.
3. http://www.epa.gov/oecaagct/lrca.html#About.
4. http://www2.epa.gov/laws-regulations/summary-toxic-substances-control-act.
5. http://www.epa.gov/oppt/existingchemicals/pubs/tscainventory/basic.html#what.
6. http://www.epa.gov/agriculture/lfra.html.
7. http://www2.epa.gov/laws-regulations/summary-clean-air-act.
8. http://epa.gov/oar/caa/caaa_overview.html.

31

OSHA's Laboratory Safety Standard (29 CFR 1910.1450)

More than a half million workers are employed in laboratories across the United States. These same laboratory workers are exposed to numerous potential hazards, including chemical, biological, physical, and radioactive hazards, as well as musculoskeletal stresses. There have been individual aspects over the years that the Occupational Safety and Health Administration (OSHA) has governed or provided guidance on to make laboratories increasingly safe for employees. The Occupational Exposure to Hazardous Chemicals in Laboratories standard (29 CFR 1910.1450) was created specifically for nonproduction laboratories. Additional OSHA standards provide rules that protect workers, including those who work in laboratories, from chemical hazards as well as biological, physical, and safety hazards. For those hazards that are not covered by a specific OSHA standard, OSHA often provides guidance on protecting workers from these hazards. The majority of this chapter was derived from https://www.osha.gov/Publications/laboratory/OSHA3404laboratory-safety-guidance.pdf.[1]

Scope and Applicability

The Laboratory Standard applies to all individuals engaged in laboratory use of hazardous chemicals. Work with hazardous chemicals outside of laboratories is covered under the Hazardous Communication Standard (29 CFR 1910.1200). Laboratory uses of chemicals that provide no potential for exposure (e.g., chemically impregnated test media or prepared kits for pregnancy testing) are not covered by the Laboratory Standard.

OSHA's Occupational Exposure to Hazardous Chemicals in Laboratories standard (29 CFR 1910.1450), referred to as the Laboratory Standard, covers laboratories where chemical manipulation generally involves small amounts of a limited variety of chemicals. This standard applies to all hazardous chemicals meeting the definition of "laboratory use" and having the potential for worker exposure.

Elements of the Laboratory Standard

This standard applies to employers engaged in laboratory use of hazardous chemicals.

- "Laboratory" means a facility where the "laboratory use of hazardous chemicals" occurs. It is a workplace where relatively small quantities of hazardous chemicals are used on a nonproduction basis.
- "Laboratory use of hazardous chemicals" means handling or use of such chemicals in which all of the following conditions are met:
  - Chemical manipulations are carried out on a "laboratory scale" (i.e., work with substances in which the containers used for reactions, transfers, and other handling of substances is designed to be easily handled by one person).
  - Multiple chemical procedures or chemicals are used.
  - The procedures involved are not part of a production process, nor do they in any way simulate a production process.
  - "Protective laboratory practices and equipment" are available and in common use to minimize the potential for worker exposure to hazardous chemicals.
- Any hazardous chemical use that does not meet this definition is regulated under other standards. This includes other hazardous chemical use within a laboratory. For instance:
  - Chemicals used in building maintenance of a laboratory are not covered under the Laboratory Standard.
  - The production of a chemical for commercial sale, even in small quantities, is not covered by the Laboratory Standard.
  - Quality control testing of a product is not covered under the Laboratory Standard.
- If the Laboratory Standard applies, employers must develop a chemical hygiene plan (CHP). A CHP is the laboratory's program that addresses all aspects of the Laboratory Standard.
  - The employer is required to develop and carry out the provisions of a written CHP.
  - A CHP must address virtually every aspect of the procurement, storage, handling, and disposal of chemicals in use in a facility.
- The primary elements of a CHP include the following:
  - Minimizing exposure to chemicals by establishing standard operating procedures, requirements for personal protective

equipment (PPE), engineering controls (e.g., chemical fume hoods, air handlers, etc.) and waste disposal procedures.

- For some chemicals, the work environment must be monitored for levels that require action or medical attention.
- Procedures to obtain free medical care for work-related exposures must be stated.
- The means to administer the plan must be specified.
- Responsible persons must be designated for procurement and handling of safety data sheets (SDSs), organizing training sessions, monitoring employee work practices, and annual revision of the CHP.

NOTE: The scope of the Formaldehyde standard (29 CFR 1910.1048) is not affected in most cases by the Laboratory Standard. The Laboratory Standard specifically does not apply to formaldehyde use in histology, pathology, and human or animal anatomy laboratories; however, if formaldehyde is used in other types of laboratories that are covered by the Laboratory Standard, the employer must comply with 29 CFR 1910.1450.[2]

Required Elements of the CHP

OSHA's Occupational Exposure to Hazardous Chemicals in Laboratories standard (29 CFR 1910.1450), referred to as the Laboratory Standard, specifies the mandatory requirements of a CHP to protect laboratory workers from harm attributed to hazardous chemicals. The CHP is a written program stating the policies, procedures, and responsibilities that protect workers from the health hazards associated with the hazardous chemicals used in that particular workplace. The requirements of this plan are as follows:

1. Standard operating procedures relevant to safety and health considerations for each activity involving the use of hazardous chemicals.
2. Criteria that the employer will use to determine and implement control measures to reduce exposure to hazardous materials (i.e., engineering controls, the use of PPE, and hygiene practices) with particular attention given to selecting control measures for extremely hazardous materials.
3. A requirement to ensure that fume hoods and other protective equipment are functioning properly and identify the specific measures the employer will take to ensure proper and adequate performance of such equipment.

4. Information to be provided to laboratory personnel working with hazardous substances include the following:

   a. The contents of the Laboratory Standard and its appendices

   b. The location and availability of the employer's CHP

   c. The permissible exposure limits (PELs) for OSHA-regulated substances or recommended exposure limits for other hazardous chemicals where there is no applicable OSHA standard

   d. The signs and symptoms associated with exposures to hazardous chemicals used in the laboratory

   e. The location and availability of known reference materials on the hazards, safe handling, storage, and disposal of hazardous chemicals found in the laboratory including, but not limited to, the SDSs received from the chemical supplier

5. The circumstances under which a particular laboratory operation, procedure, or activity requires prior approval from the employer or the employer's designee before being implemented.

6. Designation of personnel responsible for implementing the CHP, including the assignment of a Chemical Hygiene Officer (CHO) and, if appropriate, establishment of a Chemical Hygiene Committee.

7. Provisions for additional worker protection for work with particularly hazardous substances. These include "select carcinogens," reproductive toxins, and substances that have a high degree of acute toxicity. Specific consideration must be given to the following provisions and shall be included where appropriate:

   a. Establishment of a designated area

   b. Use of containment devices such as fume hoods or glove boxes

   c. Procedures for safe removal of contaminated waste

   d. Decontamination procedures

8. The employer must review and evaluate the effectiveness of the CHP at least annually and update it as necessary.

   Worker training must include the following:

   a. Methods and observations that may be used to detect the presence or release of a hazardous chemical (such as monitoring conducted by the employer, continuous monitoring devices, visual appearance or odor of hazardous chemicals when being released, etc.)

   b. The physical and health hazards of chemicals in the work area

   c. The measures workers can take to protect themselves from these hazards, including specific procedures the employer has implemented to protect workers from exposure to hazardous

chemicals, such as appropriate work practices, emergency procedures, and PPE to be used

d. The applicable details of the employer's written CHP

Medical Exams and Consultation

The employer must provide all personnel who work with hazardous chemicals an opportunity to receive medical attention, including any follow-up
examinations that the examining physician determines to be necessary,
under the following circumstances:

- Whenever a worker develops signs or symptoms associated with a
  hazardous chemical to which the worker may have been exposed
  in the laboratory, the worker must be provided an opportunity to
  receive an appropriate medical examination.

- Where exposure monitoring reveals an exposure level routinely
  above the action level (or in the absence of an action level, the PEL)
  for an OSHA-regulated substance for which there are exposure
  monitoring and medical surveillance requirements, medical surveillance must be established for the affected worker(s) as prescribed by
  the particular standard.

- Whenever an event takes place in the work area such as a spill, leak,
  explosion, or other occurrence resulting in the likelihood of a hazardous
  exposure, the affected worker(s) must be provided an opportunity for a
  medical consultation to determine the need for a medical examination.

- All medical examinations and consultations must be performed by
  or under the direct supervision of a licensed physician and be provided without cost to the worker, without loss of pay, and at a reasonable time and place.

OSHA Standards Applying to Laboratories

As mentioned in the introduction, there are other standards that have been
implemented over the years to protect the health and safety of laboratory
workers prior to the implementation of the Laboratory Standards. These standards still apply to the laboratories in addition to the Laboratory Standard.
Therefore, it is necessary for us to mention the primary ones here.

The Hazard Communication Standard (29 CFR 1910.1200)

This standard, which is sometimes called the HazCom standard, is a set of requirements first issued in 1983 by OSHA. The standard requires evaluating the potential hazards of chemicals and communicating information concerning those hazards and appropriate protective measures to employees. The standard includes provisions for developing and maintaining a written hazard communication program for the workplace, including lists of hazardous chemicals present; labeling of containers of chemicals in the workplace, as well as of containers of chemicals being shipped to other workplaces; preparation and distribution of SDSs to workers and downstream employers; and development and implementation of worker training programs regarding hazards of chemicals and protective measures. This OSHA standard requires manufacturers and importers of hazardous chemicals to provide material SDSs to users of the chemicals describing potential hazards and other information. They must also attach hazard warning labels to containers of the chemicals. Employers must make SDSs available to workers. They must also train their workers in the hazards caused by the chemicals workers are exposed to and the appropriate protective measures that must be used when handling the chemicals.

The Bloodborne Pathogens Standard (29 CFR 1910.1030)

The Bloodborne Pathogens standard, including changes mandated by the *Needlestick Safety and Prevention Act of 2001*, requires employers to protect workers from infection with human bloodborne pathogens in the workplace. The standard covers all workers with "reasonably anticipated" exposure to blood or other potentially infectious materials (OPIM). It requires that information and training be provided before the worker begins work that may involve occupational exposure to bloodborne pathogens, annually thereafter, and before a worker is offered hepatitis B vaccination. The Bloodborne Pathogens standard also requires advance information and training for all workers in research laboratories who handle human immunodeficiency virus (HIV) or hepatitis B virus (HBV). The standard was issued as a performance standard, which means that the employer must develop a written exposure control plan (ECP) to provide a safe and healthy work environment, but is allowed some flexibility in accomplishing this goal. Among other things, the ECP requires employers to make an exposure determination, establish procedures for evaluating incidents, and determine a schedule for implementing the standard's requirements, including engineering and work practice controls. The standard also requires employers to provide and pay for appropriate PPE for workers with occupational exposures. Although this standard only applies to bloodborne pathogens, the protective measures in this standard (e.g., ECP, engineering and work practice controls, administrative controls, PPE, housekeeping, training, and postexposure medical follow-up) are the same measures for effectively controlling exposure to other biological agents.

The PPE Standard (29 CFR 1910.132)

The PPE standard requires that employers provide and pay for PPE and ensure that it is used wherever "hazards of processes or environment, chemical hazards, radiological hazards, or mechanical irritants are encountered in a manner capable of causing injury or impairment in the function of any part of the body through absorption, inhalation or physical contact" [29 CFR 1910.132(a) and 1910.132(h)]. In order to determine whether and what PPE is needed, the employer must "assess the workplace to determine if hazards are present, or are likely to be present, which necessitate the use of [PPE]" [29 CFR 1910.132(d)(1)]. On the basis of that assessment, the employer must select appropriate PPE (e.g., protection for eyes, face, head, extremities; protective clothing; respiratory protection; shields and barriers) that will protect the affected worker from the hazard [29 CFR 1910.132 (d)(1)(i)], communicate selection decisions to each affected worker [29 CFR 1910.132 (d)(1)(ii)], and select PPE that properly fits each affected employee [29 CFR 1910.132(d)(1) (iii)]. Employers must provide training for workers who are required to use PPE that addresses when and what PPE is necessary, how to wear and care for PPE properly, and the limitations of PPE [29 CFR 1910.132(f)].

The Eye and Face Protection Standard (29 CFR 1910.133)

The Eye and Face Protection standard requires employers to ensure that each affected worker uses appropriate eye or face protection when exposed to eye or face hazards from flying particles, molten metal, liquid chemicals, acids or caustic liquids, chemical gases or vapors, or potentially injurious light radiation [29 CFR 1910.133(a)].

The Respiratory Protection Standard (29 CFR 1910.134)

The Respiratory Protection standard requires that a respirator be provided to each worker when such equipment is necessary to protect the health of such individuals. The employer must provide respirators that are appropriate and suitable for the purpose intended, as described in 29 CFR 1910.134(d) (1). The employer is responsible for establishing and maintaining a respiratory protection program, as required by 29 CFR 1910.134(c), which includes, but is not limited to, the following: selection of respirators for use in the workplace; medical evaluations of workers required to use respirators; fit testing for tight-fitting respirators; proper use of respirators during routine and emergency situations; procedures and schedules for cleaning, disinfecting, storing, inspecting, repairing, and discarding of respirators;

procedures to ensure adequate air quality, quantity, and flow of breathing air for atmosphere-supplying respirators; training of workers in respiratory hazards that they may be exposed to during routine and emergency situations; training of workers in the proper donning and doffing of respirators, and any limitations on their use and maintenance; and regular evaluation of the effectiveness of the program.

The Hand Protection Standard (29 CFR 1910.138)

The Hand Protection standard requires employers to select and ensure that workers use appropriate hand protection when their hands are exposed to hazards such as those from skin absorption of harmful substances, severe cuts or lacerations, severe abrasions, punctures, chemical burns, thermal burns, and harmful temperature extremes [29 CFR 1910.138(a)]. Further, employers must base the selection of the appropriate hand protection on an evaluation of the performance characteristics of the hand protection relative to the task(s) to be performed, conditions present, duration of use, and the hazards and potential hazards identified [29 CFR 1910.138(b)].

The Control of Hazardous Energy Standard (29 CFR 1910.147)

The Control of Hazardous Energy standard, often called the "Lockout/ Tagout" standard, establishes basic requirements for locking or tagging out equipment while installation, maintenance, testing, repair, or construction operations are in progress. The primary purpose of the standard is to protect workers from the unexpected energization or start-up of machines or equipment, or release of stored energy. The procedures apply to the shutdown of all potential energy sources associated with machines or equipment, including pressures, flows of fluids and gases, electrical power, and radiation. In addition to the standards listed above, other OSHA standards that pertain to electrical safety (29 CFR 1910 Subpart S—Electrical); fire safety (Portable Fire Extinguishers standard, 29 CFR 1910.157); and slips, trips, and falls (29 CFR 1910 Subpart D—Walking–Working Surfaces, Subpart E—Means of Egress, and Subpart J—General Environmental Controls) are discussed at pages 25–28. These standards pertain to general industry, as well as laboratories. When laboratory workers are using large analyzers and other equipment, their potential exposure to electrical hazards associated with this equipment must be assessed by employers and appropriate precautions must be taken. Similarly, worker exposure to wet floors or spills and clutter can lead to slips/trips/falls and other possible injuries and employers must assure that these hazards are minimized. While large laboratory fires are rare, there is the potential for small bench-top fires, especially in laboratories using flammable solvents. It is the responsibility of employers to implement appropriate protective measures to assure the safety of workers.[1]

Laboratory Standard (29 CFR 1910.1450)

Now that the description and overall general requirements of the Laboratory Standard have been discussed, we can now move on to discuss the more intricate details of the standard. The Laboratory Standard consists of five major elements:

- Hazard identification
- Chemical hygiene plan[3]
- Information and training
- Exposure monitoring
- Medical consultation and examinations

One very important aspect of any program is that the laboratories covered under the standard must appoint a CHO to develop and implement a CHP, as previously discussed. The CHO is responsible for duties such as monitoring processes, procuring chemicals, helping project directors upgrade facilities, and advising administrators on improved chemical hygiene policies and practices. A worker designated as the CHO must be qualified, by training and experience, to provide technical guidance in developing and implementing the provisions of the CHP.

Hazard Identification

Each laboratory must identify which hazardous chemicals will be encountered by its workers. All containers for chemicals must be clearly labeled. An employer must ensure that workers do not use, store, or allow any other person to use or store any hazardous substance in his or her laboratory if the container does not meet the labeling requirements outlined in the Hazard Communication standard [29 CFR 1910.1200(f)(4)]. Labels on chemical containers must not be removed or defaced.[1]

Chemical Hygiene Plan

The purpose of the CHP is to provide guidelines for prudent practices and procedures for the use of chemicals in the laboratory. The Laboratory Standard requires that the CHP set forth procedures, equipment, PPE, and work practices capable of protecting workers from the health hazards presented by chemicals used in the laboratory. The following information must be included in each CHP.

Standard Operating Procedures
Prudent laboratory practices that must be followed when working with chemicals in a laboratory. These include general and laboratory-specific procedures for work with hazardous chemicals.

Criteria for Exposure Control Measures

Criteria used by the employer to determine and implement control measures to reduce worker exposure to hazardous chemicals including engineering controls, the use of PPE, and hygiene practices.

Adequacy and Proper Functioning of Fume Hoods and other Protective Equipment

Specific measures that must be taken to ensure proper and adequate performance of protective equipment, such as fume hoods.

Information and Training

The employer must provide information and training required to ensure that workers are apprised of the hazards of chemicals in their work areas and related information.

Requirement of Prior Approval of Laboratory Procedures

The circumstances under which certain laboratory procedures or activities require approval from the employer or employer's designee before work is initiated.

Medical Consultations and Examinations

Provisions for medical consultation and examination when exposure to a hazardous chemical has or may have taken place.

CHO Designation

Identification of the laboratory CHO and outline of his or her role and responsibilities, and, where appropriate, establishment of a Chemical Hygiene Committee.

Particularly Hazardous Substance

Outlines additional worker protections for work with particularly hazardous substances. These include select carcinogens, reproductive toxins, and substances that have a high degree of acute toxicity.[1]

Information and Training

Laboratory workers must be provided with information and training relevant to the hazards of the chemicals present in their laboratory. The training must be provided at the time of initial assignment to a laboratory and prior to assignments involving new exposure situations.

The employer must inform workers about the following:

- The content of the OSHA Laboratory Standard and its appendices (the full text must be made available)
- The location and availability of the CHP

- PELs for OSHA-regulated substances, or recommended exposure levels for other hazardous chemicals where there is no applicable standard
- Signs and symptoms associated with exposure to hazardous chemicals in the laboratory
- The location and availability of reference materials on the hazards, safe handling, storage, and disposal of hazardous chemicals in the laboratory, including, but not limited to, SDSs

Training must include the following:

- Methods and observations used to detect the presence or release of a hazardous chemical. These may include employer monitoring, continuous monitoring devices, and familiarity with the appearance and odor of the chemicals
- The physical and health hazards of chemicals in the laboratory work area
- The measures that workers can take to protect themselves from these hazards, including protective equipment, appropriate work practices, and emergency procedures
- Applicable details of the employer's written CHP
- Retraining, if necessary[1]

Exposure Determination

OSHA has established PELs, as specified in 29 CFR 1910, subpart Z, for hundreds of chemical substances. A PEL is the chemical-specific concentration in inhaled air that is intended to represent what the average, healthy worker may be exposed to daily for a lifetime of work without significant adverse health effects. The employer must ensure that workers' exposures to OSHA-regulated substances do not exceed the PEL. However, most of the OSHA PELs were adopted soon after the Agency was first created in 1970 and were based upon scientific studies available at that time. Since science has continued to move forward, in some cases, there may be health data that suggest a hazard to workers below the levels permitted by the OSHA PELs. Other agencies and organizations have developed and updated recommended occupational exposure limits (OELs) for chemicals regulated by OSHA, as well as other chemicals not currently regulated by OSHA. Employers should consult other OELs, in addition to the OSHA PEL, to make a fully informed decision about the potential health risks to workers associated with chemical exposures. The American Conference of Governmental Industrial Hygienists, the American Industrial Hygiene Association, the National Institute for Occupational Safety and Health, and some chemical manufacturers have established OELs to assess safe exposure limits for various chemicals.

Employers must conduct exposure monitoring, through air sampling, if there is reason to believe that workers may be exposed to chemicals above the action level or, in the absence of an action level, the PEL. Periodic exposure monitoring should be conducted in accord with the provisions of the relevant standard. The employer should notify workers of the results of any monitoring within 15 working days of receiving the results. Some OSHA chemical standards have specific provisions regarding exposure monitoring and worker notification. Employers should consult relevant standards to see if these provisions apply to their workplace.

Medical Consultations and Examinations

Employers must do the following:

- Provide all exposed workers with an opportunity to receive medical attention by a licensed physician, including any follow-up examinations that the examining physician determines to be necessary.
- Provide an opportunity for a medical consultation by a licensed physician whenever a spill, leak, explosion, or other occurrence results in the likelihood that a laboratory worker experienced a hazardous exposure in order to determine whether a medical examination is needed.
- Provide an opportunity for a medical examination by a licensed physician whenever a worker develops signs or symptoms associated with a hazardous chemical to which he or she may have been exposed in the laboratory.
- Establish medical surveillance for a worker as required by the particular standard when exposure monitoring reveals exposure levels routinely exceeding the OSHA action level or, in the absence of an action level, the PEL for an OSHA-regulated substance.
- Provide the examining physician with the identity of the hazardous chemical(s) to which the individual may have been exposed, and the conditions under which the exposure may have occurred, including quantitative data, where available, and a description of the signs and symptoms of exposure the worker may be experiencing.
- Provide all medical examinations and consultations without cost to the worker, without loss of pay, and at a reasonable time and place.

The examining physician must complete a written opinion that includes the following information:

- Recommendations for further medical follow-up.
- The results of the medical examination and any associated tests.

- Any medical condition revealed in the course of the examination that may place the individual at increased risk as a result of exposure to a hazardous chemical in the workplace.
- A statement that the worker has been informed of the results of the consultation or medical examination and any medical condition that may require further examination or treatment. However, the written opinion must not reveal specific findings of diagnoses unrelated to occupational exposure.

A copy of the examining physician's written opinion must be provided to the exposed worker.[1]

Record Keeping

Employers must also maintain an accurate record of exposure monitoring activities and exposure measurements as well as medical consultations and examinations, including medical tests and written opinions. Employers generally must maintain worker exposure records for 30 years and medical records for the duration of the workers' employment plus 30 years, unless one of the exemptions listed in 29 CFR 1910.1020(d)(1)(i)(A)–(C) applies. Such records must be maintained, transferred, and made available, in accord with 29 CFR 1910.1020, to an individual's physician or made available to the worker or his or her designated representative upon request.

Roles and Responsibilities in Implementing the Laboratory Standard

The following are the National Research Council's recommendations concerning the responsibilities of various individuals for chemical hygiene in laboratories.

Chief Executive Officer
- Bears ultimate responsibility for chemical hygiene within the facility
- Provides continuing support for institutional chemical hygiene

Chemical Hygiene Officer
- Develops and implements appropriate chemical hygiene policies and practices
- Monitors procurement, use, and disposal of chemicals used in the laboratory
- Ensures that appropriate audits are maintained
- Helps project directors develop precautions and adequate facilities

- Knows the current legal requirements concerning regulated substances
- Seeks ways to improve the chemical hygiene program

Laboratory Supervisors

- Have overall responsibility for chemical hygiene in the laboratory
- Ensure that laboratory workers know and follow the chemical hygiene rules
- Ensure that protective equipment is available and in working order
- Ensure that appropriate training has been provided
- Provide regular, formal chemical hygiene and housekeeping inspections, including routine inspections of emergency equipment
- Know the current legal requirements concerning regulated substances
- Determine the required levels of PPE and equipment
- Ensure that facilities and training for use of any material being ordered are adequate

Laboratory Workers

- Plan and conduct each operation in accord with the facility's chemical hygiene procedures, including use of PPE and engineering controls, as appropriate
- Develop good personal chemical hygiene habits
- Report all accidents and potential chemical exposures immediately[1]

Specific Chemical Hazards

Air Contaminants Standard (29 CFR 1910.1000)

The Air Contaminants standard provides rules for protecting workers from airborne exposure to more than 400 chemicals. Several of these chemicals are commonly used in laboratories and include toluene, xylene, and acrylamide. Toluene and xylene are solvents used to fix tissue specimens and rinse stains. They are primarily found in histology, hermatology, microbiology, and cytology laboratories. These three chemicals, along with their exposure routes, symptoms, and target organs are listed in Tables 31.1, 31.2, and 31.3.

Formaldehyde Standard (29 CFR 1910.1048)

As mentioned previously, formaldehyde has its own standard. Formaldehyde is used as a fixative and is commonly found in most laboratories. The

TABLE 31.1

Toluene Exposure Routes, Symptoms, and Target Organs

Toluene		
Exposure Routes	**Symptoms**	**Target Organs**
Inhalation	Irritation of eyes, nose	Eyes
Ingestion	Weakness, exhaustion, confusion,	Skin
Skin and/or eye contact	euphoria, headache	Respiratory system
Skin absorption	Dilated pupils, tearing	Central nervous
	Anxiety	system
	Muscle fatigue	Liver
	Insomnia	Kidneys
	Tingling, prickling, or numbness of skin	
	Dermatitis	
	Liver, kidney damage	

Source: https://www.osha.gov/Publications/laboratory/OSHA3404laboratory-safety-guidance.pdf.

TABLE 31.2

Acrylamide Exposure Routes, Symptoms, and Target Organs

Acrylamide		
Exposure Routes	**Symptoms**	**Target Organs**
Inhalation	Irritation of eyes, nose	Eyes
Ingestion	Ataxia (staggering gait), numb limbs,	Skin
Skin and/or eye contact	tingling, pricking, or numbness of skin	Respiratory system
Skin absorption	Muscle weakness	
	Absence of deep tendon reflex	
	Hand sweating	
	Tearing, drowsiness	
	Reproductive effects	
	Potential occupational carcinogen	

Source: https://www.osha.gov/Publications/laboratory/OSHA3404laboratory-safety-guidance.pdf.

TABLE 31.3

Xylene Exposure Routes, Symptoms, and Target Organs

Xylene		
Exposure Routes	**Symptoms**	**Target Organs**
Inhalation	Irritation of eyes, nose, throat	Eyes
Ingestion	Dizziness, excitement, drowsiness,	Skin
Skin and/or eye contact	incoherence, staggering gait	Respiratory system
Skin absorption	Corneal vacuolization (cell debris)	Central nervous system
	Anorexia, nausea, vomiting, abdominal pain	Gastrointestinal tract
	Dermatitis	Blood
		Liver
		Kidneys

Source: https://www.osha.gov/Publications/laboratory/OSHA3404laboratory-safety-guidance.pdf.

TABLE 31.4

Formaldehyde Exposure Routes, Symptoms, and Target Organs

Formaldehyde		
Exposure Routes	**Symptoms**	**Target Organs**
Inhalation	Irritation of eyes, skin, nose, throat,	Eyes
Ingestion	respiratory system	Skin
Skin and/or eye contact	Tearing	Respiratory system
	Coughing	
	Wheezing	
	Dermatitis	
	Potential occupational nasal carcinogen	

Source: https://www.osha.gov/Publications/laboratory/OSHA3404laboratory-safety-guid ance.pdf.

employer must ensure that no worker is exposed to an airborne concentration of formaldehyde that exceeds 0.75 ppm, as an 8-h time-weighted average [29 CFR 1910.1048(c)(1)].

Table 31.4 provides exposure routes, symptoms, and target organs of formaldehyde.

Chemical Fume Hoods

The fume hood is often the primary control device for protecting laboratory workers when working with flammable or toxic chemicals. OSHA's Laboratory Standard requires that fume hoods be maintained and function properly when used [29 CFR 1910.1450(e)(3)(iii)].

Before using a fume hood

- Make sure that you understand how the hood works.
- You should be trained to use it properly.
- Know the hazards of the chemical you are working with; refer to the chemical's SDS if you are unsure.
- Ensure that the hood is on.
- Make sure that the sash is open to the proper operating level, which is usually indicated by arrows on the frame.
- Make sure that the air gauge indicates that the air flow is within the required range.

When using a fume hood

- Never allow your head to enter the plane of the hood opening. For example, for vertical rising sashes, keep the sash below your face; for horizontal sliding sashes, keep the sash positioned in front of you and work around the side of the sash.

- Use appropriate eye protection.
- Be sure that nothing blocks the airflow through the baffles or through the baffle exhaust slots.
- Elevate large equipment (e.g., a centrifuge) at least 2 in. off the base of the hood interior.
- Keep all materials inside the hood at least 6 in. from the sash opening. When not working in the hood, close the sash.
- Do not permanently store any chemicals inside the hood.
- Promptly report any hood that is not functioning properly to your supervisor. The sash should be closed and the hood "tagged" and taken out of service until repairs can be completed.
- When using extremely hazardous chemicals, understand your laboratory's action plan in case an emergency, such as a power failure, occurs.[4]

Biological Hazards

Biological Agents (Other than Bloodborne Pathogens) and Biological Toxins

Many laboratory workers encounter daily exposure to biological hazards. These hazards are present in various sources throughout the laboratory such as blood and body fluids, culture specimens, body tissue and cadavers, and laboratory animals, as well as other workers.

A number of OSHA's Safety and Health Topics Pages mentioned below have information on select agents and toxins. These are federally regulated biological agents (e.g., viruses, bacteria, fungi, and prions) and toxins that have the potential to pose a severe threat to public health and safety, to animal or plant health, or to animal or plant products. The agents and toxins that affect animal and plant health are also referred to as high-consequence livestock pathogens and toxins, nonoverlap agents and toxins, and listed plant pathogens. Select agents and toxins are defined by lists that appear in Section 73.3 of Title 42 of the Code of Federal Regulations (HHS/CDC *Select Agent Regulations*), Sections 121.3 and 121.4 of Title 9 of the Code of Federal Regulations (USDA/APHIS/VS Select Agent Regulations), and Section 331.3 of Title 7 of the Code of Federal Regulations (plants—USDA/APHIS/PPQ *Select Agent Regulations*) and Part 121, Title 9, Code of Federal Regulations (animals—USDA/APHIS). Select agents and toxins that are regulated by both HHS/CDC and USDA/APHIS are referred to as "overlap" select agents and toxins (see 42 CFR Section 73.4 and 9 CFR 121.4).

Employers may use the list below as a starting point for technical and regulatory information about some of the most virulent and prevalent biological agents and toxins. The OSHA Safety and Health Topics Page entitled Biological Agents can be accessed at www.osha.gov/SLTC/biological agents/index.html.

Anthrax. Anthrax is an acute infectious disease caused by a spore-forming bacterium called *Bacillus anthracis*. It is generally acquired following contact with anthrax-infected animals or anthrax-contaminated animal products. *B. anthracis* is an HHS and USDA select agent.

Avian Flu. Avian influenza is caused by Influenza A viruses. These viruses normally reside in the intestinal tracts of water fowl and shore birds, where they cause little, if any, disease. However, when they are passed on to domestic birds, such as chickens, they can cause a deadly contagious disease, highly pathogenic avian influenza (HPAI). HPAI viruses are considered USDA/APHIS select agents.

Botulism. Cases of botulism are usually associated with consumption of preserved foods. However, botulinum toxins are currently among the most common compounds explored by terrorists for use as biological weapons. Botulinum neurotoxins, the causative agents of botulism, are HHS/CDC select agents.

Foodborne Disease. Foodborne illnesses are caused by viruses, bacteria, parasites, toxins, metals, and prions (microscopic protein particles). Symptoms range from mild gastroenteritis to life-threatening neurologic, hepatic, and renal syndromes.

Hantavirus. Hantaviruses are transmitted to humans from the dried droppings, urine, or saliva of mice and rats. Animal laboratory workers and persons working in infested buildings are at increased risk to this disease.

Legionnaires' Disease. Legionnaires' disease is a bacterial disease commonly associated with water-based aerosols. It is often the result of poorly maintained air conditioning cooling towers and potable water systems.

Molds and Fungi. Molds and fungi produce and release millions of spores small enough to be air-, water-, or insect-borne, which may have negative effects on human health including allergic reactions, asthma, and other respiratory problems.

Plague. The World Health Organization reports 1000 to 3000 cases of plague every year. A bioterrorist release of plague could result in a rapid spread of the pneumonic form of the disease, which could have devastating consequences. *Yersinia pestis*, the causative agent of plague, is an HHS/CDC select agent.

Ricin. Ricin is one of the most toxic and easily produced plant toxins. It has been used in the past as a bioterrorist weapon and remains a serious threat. Ricin is an HHS/CDC select toxin.

Severe Acute Respiratory Syndrome (SARS). SARS is an emerging, some-times fatal, respiratory illness. According to the Centers for Disease Control and Prevention (CDC), the most recent human cases of SARS were reported in China in April 2004 and there is currently no known transmission anywhere in the world.

Smallpox. Smallpox is a highly contagious disease unique to humans. It is estimated that no more than 20% of the population has any immu-nity from previous vaccination. Variola major virus, the causative agent for smallpox, is an HHS/CDC select agent.

Tularemia. Tularemia is also known as "rabbit fever" or "deer fly fever" and is extremely infectious. Relatively few bacteria are required to cause the disease, which is why it is an attractive weapon for use in bioterrorism. *Francisella tularensis*, the causative agent for tularemia, is an HHS/CDC select agent.

Viral Hemorrhagic Fevers (VHFs). Hemorrhagic fever viruses are among the agents identified by the CDC as the most likely to be used as biological weapons. Many VHFs can cause severe, life-threatening disease with high fatality rates. Many VHFs are HHS/CDC select agents; for example, Marburg virus, Ebola viruses, and the Crimean–Congo hemorrhagic fever virus.

Pandemic Influenza. A pandemic is a global disease outbreak. An influenza pandemic occurs when a new influenza virus emerges for which there is little or no immunity in the human population, begins to cause serious illness, and then spreads easily person to person worldwide.[1]

Specific Laboratory Practices

Bloodborne Pathogens

- Bloodborne pathogens and the bloodborne pathogen standard have been discussed in general descriptions, earlier in this chapter and in Chapter 2. For those laboratories who present a potential exposure to blood or OPIM, additional actions must be taken to mitigate the hazard. OSHA defines blood to mean human blood, human blood components, and products made from human blood. OPIM means the following human body fluids: semen, vaginal secretions, cerebro-spinal fluid, synovial fluid, pleural fluid, pericardial fluid, peritoneal fluid, amniotic fluid, saliva in dental procedures, any body fluid that is visibly contaminated with blood, and all body fluids in situations where it is difficult or impossible to differentiate between body fluids.

- Any unfixed tissue or organ (other than intact skin) from a human (living or dead).
- HIV- or HBV-containing cell or tissue cultures, organ cultures, and HIV- or HBV-containing culture medium or other solutions; and blood, organs, or other tissues from experimental animals infected with HIV or HBV.[1]

Engineering Controls and Work Practices for All HIV/HBV Laboratories

Employers must ensure that

- All activities involving OPIM are conducted in Biological Safety Cabinets (BSCs) or other physical-containment devices; work with OPIM must not be conducted on the open bench [29 CFR 1910.1030(e)(2)(ii)(E)].
- Certified BSCs or other appropriate combinations of personal protection or physical containment devices, such as special protective clothing, respirators, centrifuge safety cups, sealed centrifuge rotors, and containment caging for animals, be used for all activities with OPIM that pose a threat of exposure to droplets, splashes, spills, or aerosols [29 CFR 1910.1030(e)(2)(iii)(A)].
- Each laboratory contains a facility for hand washing and an eye-wash facility that is readily available within the work area [29 CFR 1910.1030(e)(3)(i)].
- Each work area contains a sink for washing hands and a readily available eyewash facility. The sink must be foot, elbow, or automatically operated and must be located near the exit door of the work area [29 CFR 1910.1030(e)(4)(iii)].

Additional BBP Standard Requirements Applied to HIV and HBV Research Laboratories

Requirements include the following:

- Waste materials:
  - All regulated waste must be either incinerated or decontaminated by a method such as autoclaving known to effectively destroy bloodborne pathogens [29 CFR 1910.1030(e)(2)(i)].
  - Contaminated materials that are to be decontaminated at a site away from the work area must be placed in a durable, leakproof, labeled, or color-coded container that is closed before being removed from the work area [29 CFR 1910.1030(e)(2)(ii)(B)].

- Access:
  - Laboratory doors must be kept closed when work involving HIV or HBV is in progress [29 CFR 1910.1030(e)(2)(ii)(A)].
  - Access to the production facilities' work area must be limited to authorized persons. Written policies and procedures must be established whereby only persons who have been advised of the potential biohazard, who meet any specific entry requirements, and who comply with all entry and exit procedures must be allowed to enter the work areas and animal rooms [29 CFR 1910.1030(e)(2)(ii)(C)].
  - Access doors to the production facilities' work area or containment module must be self-closing [29 CFR 1910.1030(e)(4)(iv)].
  - Work areas must be separated from areas that are open to unrestricted traffic flow within the building. Passage through two sets of doors must be the basic requirement for entry into the work area from access corridors or other contiguous areas. Physical separation of the high-containment work area from access corridors or other areas or activities may also be provided by a double-doored clothes-change room (showers may be included), airlock, or other access facility that requires passing through two sets of doors before entering the work area [29 CFR 1910.1030(e)(4)(i)].
  - The surfaces of doors, walls, floors, and ceilings in the work area must be water resistant so that they can be easily cleaned. Penetrations in these surfaces must be sealed or capable of being sealed to facilitate decontamination [29 CFR 1910.1030(e)(4)(ii)]. (These requirements do not apply to clinical or diagnostic laboratories engaged solely in the analysis of blood, tissue, or organs [29 CFR 1910.1030(e)(1)].[4])

Research Animals

Laboratory research conducted on animals should only be done by trained personnel wearing the appropriate PPE. By wearing the appropriate PPE, employees can minimize the likelihood of being bitten, scratched, or exposed to animal body fluids and tissues. In addition to these hazards, working with animals can result in sprains, strains, bites, and allergies.

Zoonotic Diseases

Infectious agents that can be transmitted to humans by animals causing illness are known as *zoonotic diseases*. The most common routes of exposure to these infectious agents are inhalation, inoculation, ingestion, and contamination of skin and mucous membranes. Various procedures can result in contamination of humans such as splashes, lacerations, and so on.

Specific Engineering Control—BSCs

Properly maintained BSCs, when used in conjunction with good microbiological techniques, provide an effective containment system for safe manipulation of moderate- and high-risk infectious agents (Biosafety Level 2 [BSL 2] and 3 [BSL 3] agents). BSCs protect laboratory workers and the immediate environment from infectious aerosols generated within the cabinet.

Biosafety Cabinet Certifications

BSCs must be certified when installed, whenever they are moved and at least annually [29 CFR 1030(e)(2)(iii)(B)].[1]

Radiation

Radiation used in the laboratory can be either ionizing or nonionizing. Each is discussed as follows.

Ionizing Radiation

OSHA's Ionizing Radiation standard (29 CFR 1910 1096) sets forth the limitations on exposure to radiation from atomic particles. Ionizing radiation sources are found in a wide range of occupational settings, including laboratories. These radiation sources can pose a considerable health risk to affected workers if not properly controlled. Any laboratory possessing or using radioactive isotopes must be licensed by the Nuclear Regulatory Commission (NRC) or by a state agency that has been approved by the NRC (10 CFR 31.11 and 10 CFR 35.12). The fundamental objectives of radiation protection measures are (1) to limit entry of radionuclides into the human body (via ingestion, inhalation, absorption, or through open wounds) to quantities as low as reasonably achievable and always within the established limits, and (2) to limit exposure to external radiation to levels that are within established dose limits and as far below these limits as is reasonably achievable.

The OSHA Ionizing Radiation standard requires precautionary measures and personnel monitoring for workers who are likely to be exposed to radiation hazards. Personnel monitoring devices (film badges, thermoluminescent dosimeters, pocket dosimeters, etc.) must be supplied and used if required to measure an individual's radiation exposure from gamma, neutron, energetic beta, and x-ray sources. The standard monitoring device is a clip-on badge or ring badge bearing the individual assignee's name, date of the monitoring period, and a unique identification number. The badges are provided, processed, and reported through a commercial service company that meets current requirements of the National Institute of Standards and Technology's National Voluntary Laboratory Accreditation Program.

Nonionizing Radiation

Nonionizing radiation is described as a series of energy waves composed of oscillating electric and magnetic fields traveling at the speed of light. Nonionizing radiation includes the spectrum of ultraviolet (UV), visible light, infrared (IR), microwave (MW), radio frequency (RF), and extremely low frequency (ELF). Lasers commonly operate in the UV, visible, and IR frequencies. Nonionizing radiation is found in a wide range of occupational settings and can pose a considerable health risk to potentially exposed workers if not properly controlled.

ELF Radiation

ELF radiation at 60 Hz is produced by power lines, electrical wiring, and electrical equipment. Common sources of intense exposure include ELF induction furnaces and high-voltage power lines.

RF and MW Radiation

MW radiation is absorbed near the skin, while RF radiation may be absorbed throughout the body. At high enough intensities, both will damage tissue through heating. Sources of RF and MW radiation include radio emitters and cell phones.

IR Radiation

The skin and eyes absorb IR radiation as heat. Workers normally notice excessive exposure through heat sensation and pain. Sources of IR radiation include heat lamps and IR lasers.

Visible Light Radiation

The different visible frequencies of the electromagnetic spectrum are "seen" by our eyes as different colors. Good lighting is conducive to increased production, and may help prevent incidents related to poor lighting conditions. Excessive visible radiation can damage the eyes and skin.

UV Radiation

UV radiation has a high photon energy range and is particularly hazardous because there are usually no immediate symptoms of excessive exposure. Sources of UV radiation in the laboratory include black lights and UV lasers.

Laser Hazards

Lasers typically emit optical (UV, visible light, and IR) radiations and are primarily an eye and skin hazard. Common lasers include CO_2 IR laser, helium–neon, neodymium YAG, and ruby visible lasers, and the Nitrogen UV laser. LASER is an acronym that stands for light amplification by stimulated emission of radiation. The laser produces an intense, highly directional beam of light. The most common cause of laser-induced tissue damage is thermal in nature, where the tissue proteins are denatured owing to the temperature rise following absorption of laser energy.

The human body is vulnerable to the output of certain lasers, and under certain circumstances, exposure can result in damage to the eye and skin. Research relating to injury thresholds of the eye and skin has been carried out in order to understand the biological hazards of laser radiation. It is now widely accepted that the human eye is almost always more vulnerable to injury than human skin.[1]

Noise

In an area that exceeds 85 dBA over an 8-h work shift, the employer should follow the requirements of OSHA's Hearing Conservation Program outlined in Chapter 2.

Autoclaves and Sterilizers

Workers should be trained to recognize the potential for exposure to burns or cuts that can occur from handling or sorting hot sterilized items or sharp instruments when removing them from autoclaves/sterilizers or from steam lines that service the autoclaves. In order to prevent injuries from occurring, employers must train workers to follow good work practices.

Centrifuges

Centrifuges, because of the high speed at which they operate, have great potential for injuring users if not operated properly. Unbalanced centrifuge rotors can result in injury, even death. Sample container breakage can generate aerosols that may be harmful if inhaled. The majority of all centrifuge accidents are the result of user error. In order to prevent injuries or exposure to dangerous substances, employers should train workers to follow good work practices.

Employers should instruct workers when centrifuging infectious materials that they should wait 10 min after the centrifuge rotor has stopped before opening the lid. Workers should also be trained to use appropriate decontamination and cleanup procedures for the materials being centrifuged if a spill occurs and to report all accidents to their supervisor immediately.

Compressed Gases

According to OSHA's Laboratory Standard, a "compressed gas" (1) is a gas or mixture of gases in a container having an absolute pressure exceeding 40 pounds per square inch (psi) at 70°F (21.1°C), or (2) is a gas or mixture of gases having an absolute pressure exceeding 104 psi at 130°F (54.4°C) regardless of the pressure at 70°F (21.1°C), or (3) is a liquid having a vapor pressure exceeding 40 psi at 100°F (37.8°C) as determined by ASTM (American Society for Testing and Materials) D-323-72 [29 CFR 1910.1450(c)(1)–(3)].

Within laboratories, compressed gases are usually supplied either through fixed piped gas systems or individual cylinders of gases. Compressed gases can be toxic, flammable, oxidizing, corrosive, or inert. Leakage of any of these gases can be hazardous.

Leaking inert gases (e.g., nitrogen) can quickly displace air in a large area creating an oxygen-deficient atmosphere; toxic gases can create poisonous atmospheres; and flammable (oxygen) or reactive gases can result in fire and exploding cylinders.

In addition, there are hazards from the pressure of the gas and the physical weight of the cylinder. A gas cylinder falling over can break containers and crush feet. The gas cylinder can itself become a missile if the cylinder valve is broken off. Laboratories must include compressed gases in their inventory of chemicals in their CHP. Compressed gases contained in cylinders vary in chemical properties, ranging from inert and harmless to toxic and explosive. The high pressure of the gases constitutes a serious hazard in the event that gas cylinders sustain physical damage or are exposed to high temperatures.

- Store, handle, and use compressed gases in accord with OSHA's Compressed Gases standard (29 CFR 1910.101) and Pamphlet P-1-1965 from the Compressed Gas Association.
- All cylinders whether empty or full must be stored upright.
- Secure cylinders of compressed gases. Cylinders should never be dropped or allowed to strike each other with force.
- Transport compressed gas cylinders with protective caps in place and do not roll or drag the cylinders.[4]

Cryogens and Dry Ice

Cryogens, substances used to produce very low temperatures [below –153°C (–243°F)], such as liquid nitrogen (LN2), which has a boiling point of –196°C (–321°F), are commonly used in laboratories. Although not a cryogen, solid carbon dioxide or dry ice, which converts directly to carbon dioxide gas at –78°C (–109°F), is also often used in laboratories.

Shipments packed with dry ice, samples preserved with liquid nitrogen, and, in some cases, techniques that use cryogenic liquids, such as cryogenic grinding of samples, present potential hazards in the laboratory.

Overview of Cryogenic Safety Hazards

The safety hazards associated with the use of cryogenic liquids are categorized as follows:

- *Cold contact burns*: Liquid or low-temperature gas from any cryogenic substance will produce effects on the skin similar to a burn.
- *Asphyxiation*: Degrees of asphyxia will occur when the oxygen content of the working environment is less than 20.9% by volume. This decrease in oxygen content can be caused by a failure/leak of a cryogenic vessel or transfer line and subsequent vaporization of the cryogen. Effects from oxygen deficiency become noticeable at levels below approximately 18% and sudden death may occur at approximately 6% oxygen content by volume.
- *Explosion—Pressure*: Heat flux into the cryogen from the environment will vaporize the liquid and potentially cause pressure buildup in cryogenic containment vessels and transfer lines. Adequate pressure relief should be provided to all parts of a system to permit this routine outgassing and prevent explosion.
- *Explosion—Chemical*: Cryogenic fluids with a boiling point below that of liquid oxygen are able to condense oxygen from the atmosphere. Repeated replenishment of the system can thereby cause oxygen to accumulate as an unwanted contaminant. Similar oxygen enrichment may occur where condensed air accumulates on the exterior of cryogenic piping. Violent reactions, for example, rapid combustion or explosion, may occur if the materials that make contact with the oxygen are combustible.

Employer Responsibility

It is the responsibility of the employer, specifically the supervisor in charge of an apparatus, to ensure that the cryogenic safety hazards are minimized. This will entail (1) a safety analysis and review for all cryogenic facilities; (2) cryogenic safety and operational training for relevant workers; (3) appropriate maintenance of cryogenic systems in their original working order, that is, the condition in which the system was approved for use; and (4) upkeep of inspection schedules and records.

Employers must train workers to use the appropriate PPE. Whenever handling or transfer of cryogenic fluids might result in exposure to the cold liquid, boil-off gas, or surface, protective clothing must be worn. This includes

- Face shield or safety goggles
- Safety gloves
- Long-sleeved shirts, laboratory coats, and aprons

Eye protection is required at all times when working with cryogenic fluids. When pouring a cryogen, working with a wide-mouth Dewar flask or around the exhaust of cold boil-off gas, use of a full face shield is recommended. Hand protection is required to guard against the hazard of touching cold surfaces. It is recommended that Cryogen Safety Gloves be used by the worker.[1]

Fire

Fire is the most common serious hazard that one faces in a typical laboratory. While proper procedures and training can minimize the chances of an accidental fire, laboratory workers should still be prepared to deal with a fire emergency should it occur. In dealing with a laboratory fire, all containers of infectious materials should be placed into autoclaves, incubators, refrigerators, or freezers for containment. Small bench-top fires in laboratory spaces are not uncommon. Large laboratory fires are rare. However, the risk of severe injury or death is significant because fuel load and hazard levels in laboratories are typically very high. Laboratories, especially those using solvents in any quantity, have the potential for flash fires, explosion, rapid spread of fire, and high toxicity of products of combustion (heat, smoke, and flame).

Employers Should Ensure That Workers Are Trained to Do the Following in Order to Prevent Fires

- Plan work. Have a written emergency plan for your space and operation.
- Minimize materials. Have present in the immediate work area and use only the minimum quantities necessary for work in progress. Not only does this minimize fire risk, it reduces costs and waste.
- Observe proper housekeeping. Keep work areas uncluttered, and clean frequently. Put unneeded materials back in storage promptly. Keep aisles, doors, and access to emergency equipment unobstructed at all times.
- Observe restrictions on equipment (i.e., keeping solvents only in an explosion-proof refrigerator).
- Keep barriers in place (shields, hood doors, laboratory doors). Wear proper clothing and PPE.
- Avoid working alone.

- Store solvents properly in approved flammable liquid storage cabinets.
- Shut door behind you when evacuating.
- Limit open flames use to under fume hoods and only when constantly attended.
- Keep combustibles away from open flames.
- Do not heat solvents using hot plates.
- Remember the "RACE" rule in case of a fire.
  - R = Rescue/remove all occupants
  - A = Activate the alarm system
  - C = Confine the fire by closing doors
  - E = Evacuate/extinguish

Employers Should Ensure That Workers Are Trained in the Following Emergency Procedures

- Know what to do. You tend to do under stress what you have practiced or preplanned. Therefore, planning, practice, and drills are essential.
- Know where things are: The nearest fire extinguisher, fire alarm box, exit(s), telephone, emergency shower/eyewash, first-aid kit, and so on.
- Be aware that emergencies are rarely "clean" and will often involve more than one type of problem. For example, an explosion may generate medical, fire, and contamination emergencies simultaneously.
- Train workers and exercise the emergency plan.
- Learn to use the emergency equipment provided.

Employers must be knowledgeable about OSHA's Portable Fire Extinguishers standard (29 CFR 1910.157) and train workers to be aware of the different fire extinguisher types and how to use them. OSHA's Portable Fire Extinguishers standard (29 CFR 1910.157) applies to the placement, use, maintenance, and testing of portable fire extinguishers provided for the use of workers. This standard requires that a fire extinguisher be placed within 75 ft for Class A fire risk (ordinary combustibles; usually fuels that burn and leave "ash") and within 50 ft for high-risk Class B fire risk (flammable liquids and gases; in the laboratory, many organic solvents and compressed gases are fire hazards). The two most common types of extinguishers in the chemistry laboratory are pressurized dry chemical (Type BC or ABC) and carbon dioxide. In addition, you may also have a specialized Class D dry powder extinguisher for use on flammable metal fires. Water-filled extinguishers are not acceptable for laboratory use.

Employers Should Train Workers to Remember the "PASS" Rule for Fire Extinguishers

PASS summarizes the operation of a fire extinguisher.

- P = Pull the pin
- A = Aim extinguisher nozzle at the base of the fire
- S = Squeeze the trigger while holding the extinguisher upright
- S = Sweep the extinguisher from side to side; cover the fire with the spray

Employers Should Train Workers on Appropriate Procedures in the Event of a Clothing Fire

- If the floor is not on fire, STOP, DROP, and ROLL to extinguish the flames or use a fire blanket or a safety shower if not contraindicated (i.e., there are no chemicals or electricity involved).
- If a coworkers' clothing catches fire and he or she runs down the hallway in panic, tackle him or her and smother the flames as quickly as possible, using appropriate means that are available (e.g., fire blanket or fire extinguisher).

Key Points to Remember on the Laboratory Safety Standard

1. The Laboratory Standard consists of five major elements: hazard identification, chemical hygiene plan, information and training, exposure monitoring, and medical consultation and examinations.
2. The primary elements of a chemical hygiene plan include minimizing exposure to chemicals by establishing standard operating procedures, monitoring the work environment for certain chemicals, and establishing procedures to obtain free medical care for work-related exposures.
3. Air contaminant PELs are covered in the Air Contaminant Standard (29 CFR 1910.1000).
4. Toluene primarily affects the eyes, skin, respiratory system, central nervous system, liver, and kidneys.
5. Acrylamide affects the eyes, skin, and respiratory system.
6. Xylene affects the eyes, skin, respiratory system, central nervous system, gastrointestinal tract, blood, liver, and kidneys.

7. Formaldehyde is covered by the Formaldehyde standard (29 CFR 1910.1048).

8. The target organs for formaldehyde are the eyes, skin, and respiratory system.

9. OSHA's Laboratory Standard requires that fume hoods be maintained and function properly when used [29 CFR 1910.1450.(e)(3)(iii)].

10. Compressed gases must be stored, handled, and used in accordance with the OSHA Compressed Gases Standard (29 CFR 1910.101) and Pamphlet P-1-1965 from the Compressed Gas Association.

References

1. https://www.osha.gov/Publications/laboratory/OSHA3404laboratory-safety-guidance.pdf.

2. https://www.osha.gov/Publications/laboratory/OSHAfactsheet-laboratory-safety-osha-lab-standard.pdf.

3. https://www.osha.gov/Publications/laboratory/OSHAfactsheet-laboratory-safety-chemical-hygiene-plan.pdf.

4. https://www.osha.gov/Publications/laboratory/OSHAquickfacts-lab-safety-chemical-fume-hoods.pdf.

32

OSHA's Process Safety Management Standard

Process safety management is a system of tools, processes, and procedures that when properly implemented and followed will greatly reduce the potential for accidental release of highly hazardous chemicals. The Occupational Safety and Health Administration (OSHA) has regulated the implementation of such programs in their Process Safety Management Standard 29 CFR 1910.119. The primary purpose of the standard is to prevent unwanted releases of hazardous chemicals especially into locations that could expose employees and others to serious hazards. An effective process safety management program requires a systematic approach to evaluating the whole chemical process. Using this approach, the process design, process technology, process changes, operational and maintenance activities and procedures, nonroutine activities and procedures, emergency preparedness plans and procedures, training programs, and other elements that affect the process are all considered in the evaluation.[1]

The process safety management standard targets highly hazardous chemicals that have the potential to cause a catastrophic incident. This standard as a whole is to aid employers in their efforts to prevent or mitigate episodic chemical releases that could lead to a catastrophe in the workplace and possibly to the surrounding community. To control these types of hazards, employers need to develop the necessary expertise, experiences, judgment, and proactive initiative within their workforce properly as environed in the OSHA standard. The OSHA standard is required by the Clean Air Act Amendments, as is the Environmental Protection Agency's Risk Management Plan. Employers who merge the two sets of requirements into their process safety management program will better assure full compliance with each as well as enhance their relationship with the local community.[2]

Applicability

This standard applies to all of the following:

- A process that involves chemical at or above the specified threshold quantities listed in Appendix A of 29 CFR 1910.119

- A process that involves a Category 1 flammable gas [as defined in 29 CFR 1910.1200 (c)] or a flammable liquid with a flashpoint below 100°F (37.8°C) on site in one location, in a quantity of 10,000 lb (4535.9 kg) or more except for
  - Hydrocarbon fuels used solely for workplace consumption as a fuel (e.g., propane used for comfort heating, gasoline for vehicle refueling), if such fuels are not a part of a process containing another highly hazardous chemical covered by 29 CFR 1910.119
  - Flammable liquids stored in atmospheric tanks or transferred, which are kept below their normal boiling point without benefit of chilling or refrigeration

This standard does not apply to

- Retail facilities
- Oil or gas well drilling or servicing operations
- Normally unoccupied remote facilities[2]

Definitions

Before moving forward in our discussion on process safety management, it is necessary to provide a clear definition of many of the terms:

Atmospheric tank means a storage tank that has been designed to operate at pressures from atmospheric through 0.5 psig (pounds per square inch gauge, 3.45 kPa).

Boiling point means the boiling point of a liquid at a pressure of 14.7 pounds per square inch absolute (psia) (760 mm). For the purposes of this section, where an accurate boiling point is unavailable for the material in question, or for mixtures that do not have a constant boiling point, the 10% point of a distillation performed in accordance with the Standard Method of Test for Distillation of Petroleum Products, ASTM D-86-62, which is incorporated by reference as specified in Section 1910.6, may be used as the boiling point of the liquid.

Catastrophic release means a major uncontrolled emission, fire, or explosion, involving one or more highly hazardous chemicals, that presents serious danger to employees in the workplace.

Facility means the buildings, containers, or equipment that contain a process.

Highly hazardous chemical means a substance possessing toxic, reactive, flammable, or explosive properties and specified by paragraph (a)(1) of this section.

Hot work means work involving electric or gas welding, cutting, brazing, or similar flame or spark-producing operations.

Normally unoccupied remote facility means a facility that is operated, maintained, or serviced by employees who visit the facility only periodically to check its operation and to perform necessary operating or maintenance tasks. No employees are permanently stationed at the facility. Facilities meeting this definition are not contiguous with and must be geographically remote from all other buildings, processes, or persons.

Process means any activity involving a highly hazardous chemical including any use, storage, manufacturing, handling, or the on-site movement of such chemicals, or combination of these activities. For purposes of this definition, any group of vessels that are interconnected and separate vessels that are located such that a highly hazardous chemical could be involved in a potential release shall be considered a single process.

Replacement in kind means a replacement that satisfies the design specification.

Trade secret means any confidential formula, pattern, process, device, information, or compilation of information that is used in an employer's business and that gives the employer an opportunity to obtain an advantage over competitors who do not know or use it. See Appendix E to § 1910.1200—Definition of a Trade Secret (which sets out the criteria to be used in evaluating trade secrets).

Employee Participation

Employers shall develop a written plan of action regarding the implementation of the employee participation. Employers shall consult employees and their representatives on the conduct and development of process hazard analyses and on the development of the other elements of process safety management. Employers shall also provide to employees and their representatives access to process hazard analyses and to all other information required to be developed under this standard.

Process Safety Information

In accordance with the schedule set forth in paragraph (e)(1) of this standard, the employer shall complete a compilation of written process safety information before conducting any process hazard analysis required by the standard. The compilation of written process safety information is to enable the employer and the employees involved in operating the process to identify and understand the hazards posed by those processes involving highly hazardous chemicals. This process safety information shall include information pertaining to the hazards of the highly hazardous chemicals used or produced by the process, information pertaining to the technology of the process, and information pertaining to the equipment in the process.

Information pertaining to the hazards of the highly hazardous chemicals in the process shall consist of at least the following:

- Toxicity information
- Permissible exposure limits
- Physical data
- Reactivity data
- Corrosivity data
- Thermal and chemical stability data
- Hazardous effects of inadvertent mixing of different materials that could foreseeably occur

NOTE: Safety data sheets meeting the requirements of 29 CFR 1910.1200(g) may be used to comply with this requirement to the extent they contain the information required by this standard.

Information Pertaining to the Technology of the Process

A block flow diagram is used to show the major process equipment and interconnecting process flow lines and show flow rates, stream composition, temperatures, and pressures when necessary for clarity. The block flow diagram is a simplified diagram.

Process flow diagrams are more complex and will show all main flow streams including valves to enhance the understanding of the process, as well as pressures and temperatures on all feed and product lines within all major vessels, in and out of headers and heat exchangers, and points of pressure and temperature control. Also, materials of construction information, pump capacities and pressure heads, compressor horsepower, and vessel design pressures and temperatures are shown when necessary for clarity. In addition, major components of control loops are usually shown along with key utilities on process flow diagrams.

Piping and instrument diagrams (P&IDs) may be the more appropriate type of diagrams to show some of the above details and to display the information for the piping designer and engineering staff. The P&IDs are to be used to describe the relationships between equipment and instrumentation as well as other relevant information that will enhance clarity. Computer

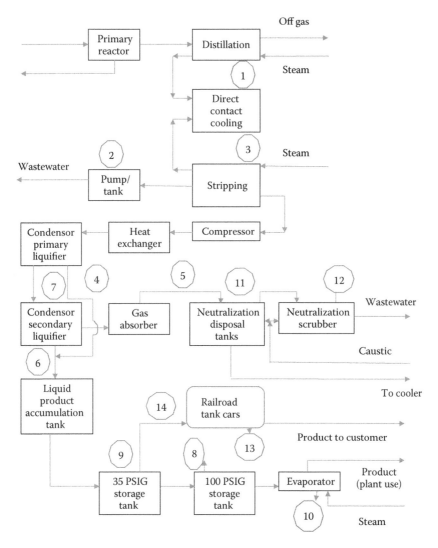

FIGURE 32.1
Example of a block flow diagram.

software programs that do P&IDs or other diagrams useful to the information package may be used to help meet this requirement.[3]

Information concerning the technology of the process shall include at least the following:

- A block flow diagram or simplified process flow diagram (see Figures 32.1 and 32.2)
- Process chemistry
- Maximum intended inventory
- Safe upper and lower limits for such items as temperatures, pressures, flows, or compositions
- An evaluation of the consequences of deviations, including those affecting the safety and health of employees

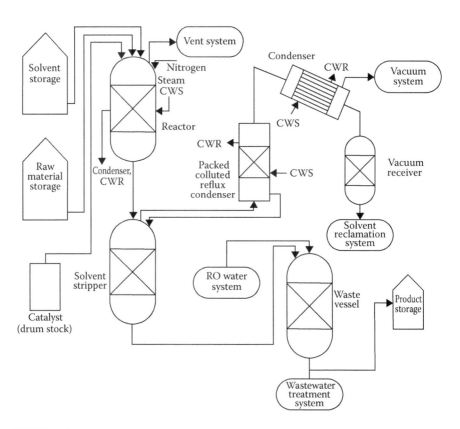

FIGURE 32.2
Example of a process flow diagram.

Where the original technical information no longer exists, such information may be developed in conjunction with the process hazard analysis in sufficient detail to support the analysis.

Information Pertaining to the Equipment in the Process

Information pertaining to the equipment in the process shall include

- Materials of construction
- P&IDs
- Electrical classification
- Relief system design and design basis
- Ventilation system design
- Design codes and standards employed
- Material and energy balances for processes built after May 26, 1992
- Safety systems (e.g., interlocks, detection or suppression systems)

The employer shall document that equipment complies with recognized and generally accepted good engineering practices. For existing equipment designed and constructed in accordance with codes, standards, or practices that are no longer in general use, the employer shall determine and document that the equipment is designed, maintained, inspected, tested, and operating in a safe manner.

Process Hazard Analysis

The employer shall perform an initial process hazard analysis (hazard evaluation) on processes covered by this standard. The process hazard analysis shall be appropriate to the complexity of the process and shall identify, evaluate, and control the hazards involved in the process. Employers shall determine and document the priority order for conducting process hazard analyses on the basis of a rationale that includes such considerations as extent of the process hazards, number of potentially affected employees, age of the process, and operating history of the process. The process hazard analysis shall be conducted as soon as possible, but not later than the following schedule:

- No less than 25% of the initial process hazard analyses shall be completed by May 26, 1994.
- No less than 50% of the initial process hazard analyses shall be completed by May 26, 1995.
- No less than 75% of the initial process hazard analyses shall be completed by May 26, 1996.
- All initial process hazard analyses shall be completed by May 26, 1997.

Process hazard analyses completed after May 26, 1987, that meet the requirements of this paragraph are acceptable as initial process hazard analyses. These process hazard analyses shall be updated and revalidated, on the basis of their completion date, in accordance with paragraph (e)(6) of this standard.

The employer shall use one or more of the following methodologies that are appropriate to determine and evaluate the hazards of the process being analyzed:

- What-If
- Checklist
- What-If/Checklist
- Hazard and Operability Study
- Failure Mode and Effects Analysis
- Fault Tree Analysis
- An appropriate equivalent methodology

The process hazard analysis shall address

- The hazards of the process
- The identification of any previous incident that had a likely potential for catastrophic consequences in the workplace
- Engineering and administrative controls applicable to the hazards and their interrelationships such as appropriate application of detection methodologies to provide early warning of releases (acceptable detection methods might include process monitoring and control instrumentation with alarms and detection hardware such as hydrocarbon sensors)
- Consequences of failure of engineering and administrative controls
- Facility siting
- Human factors
- A qualitative evaluation of a range of the possible safety and health effects of failure of controls on employees in the workplace

The process hazard analysis shall be performed by a team with expertise in engineering and process operations, and the team shall include at least one employee who has experience and knowledge specific to the process being evaluated. Also, one member of the team must be knowledgeable in the specific process hazard analysis methodology being used.

The employer shall establish a system to promptly address the team's findings and recommendations; assure that the recommendations are resolved in a timely manner and that the resolution is documented; document what actions are to be taken; complete actions as soon as possible; develop a written schedule of when these actions are to be completed; communicate the actions to operating, maintenance, and other employees whose work assignments are in the process and who may be affected by the recommendations or actions.

At least every 5 years after the completion of the initial process hazard analysis, the process hazard analysis shall be updated and revalidated by a team meeting the requirements in paragraph (e)(4) of this section, to assure that the process hazard analysis is consistent with the current process.

Employers shall retain process hazard analyses and updates or revalidations for each process covered by this section, as well as the documented resolution of recommendations described in paragraph (e)(5) of this section for the life of the process.

Operating Procedures

The employer shall develop and implement written operating procedures that provide clear instructions for safely conducting activities involved in each covered process consistent with the process safety information and shall address at least the following elements.

Steps for each operating phase:
- Initial start-up
- Normal operations
- Temporary operations
- Emergency shutdown including the conditions under which emergency shutdown is required, and the assignment of shutdown responsibility to qualified operators to ensure that emergency shutdown is executed in a safe and timely manner
- Emergency operations
- Normal shutdown
- Start-up following a turnaround, or after an emergency shutdown

Operating limits:

- Consequences of deviation
- Steps required to correct or avoid deviation
- Safety and health considerations
- Properties of, and hazards presented by, the chemicals used in the process
- Precautions necessary to prevent exposure, including engineering controls, administrative controls, and personal protective equipment
- Control measures to be taken if physical contact or airborne exposure occurs
- Quality control for raw materials and control of hazardous chemical inventory levels
- Any special or unique hazards
- Safety systems and their functions

Operating procedures shall be readily accessible to employees who work in or maintain a process. The operating procedures shall be reviewed as often as necessary to assure that they reflect current operating practice, including changes that result from changes in process chemicals, technology, and equipment, and changes to facilities. The employer shall certify annually that these operating procedures are current and accurate.

Operating procedures describe tasks to be performed, data to be recorded, operating conditions to be maintained, samples to be collected, and safety and health precautions to be taken. The procedures need to be technically accurate, understandable to employees, and revised periodically to ensure that they reflect current operations. The process safety information package is to be used as a resource to better assure that the operating procedures and practices are consistent with the known hazards of the chemicals in the process and that the operating parameters are accurate. Operating procedures should be reviewed by engineering staff and operating personnel to ensure that they are accurate and provide practical instructions on how to actually carry out job duties safely.

Operating procedures will include specific instructions or details on what steps are to be taken or followed in carrying out the stated procedures. These operating instructions for each procedure should include the applicable safety precautions and should contain appropriate information on safety implications. For example, the operating procedures addressing operating parameters will contain operating instructions about pressure limits, temperature ranges, flow rates, what to do when an upset condition occurs, what alarms and instruments are pertinent if an upset condition occurs, and other subjects. Another example of using operating instructions to properly implement operating procedures is in starting up or shutting down the process. In these cases, different parameters will be required from those of normal operation. These operating instructions need to clearly indicate the distinctions between start-up and normal operations such as the appropriate allowances for heating up a unit to reach the normal operating parameters. Also, the operating instructions need to describe the proper method for increasing the temperature of the unit until the normal operating temperature parameters are achieved.

Computerized process control systems add complexity to operating instructions. These operating instructions need to describe the logic of the software as well as the relationship between the equipment and the control system; otherwise, it may not be apparent to the operator.

Operating procedures and instructions are important for training operating personnel. The operating procedures are often viewed as the standard operating practices for operations. Control room personnel and operating staff, in general, need to have a full understanding of operating procedures. If workers are not fluent in English, then procedures and instructions need to be prepared in a second language understood by the workers. In addition, operating procedures need to be changed when there is a change in the process as a result of the management of change procedures. The consequences of operating procedure changes need to be fully evaluated and the information needs to be conveyed to the personnel. For example, mechanical changes to the process made by the maintenance department (like changing a valve from steel to brass or other subtle changes) need to be evaluated to determine if operating procedures and practices also need to be changed. All management of change actions must be coordinated and integrated with current operating procedures and operating personnel must be oriented to the changes in procedures before the change is made. When the process is shut down in order to make a change, then the operating procedures must be updated before start-up of the process.[3]

The employer shall develop and implement safe work practices to provide for the control of hazards during operations such as lockout/tagout; confined space entry; opening process equipment or piping; and control over entrance into a facility by maintenance, contractor, laboratory, or other support personnel. These safe work practices shall apply to employees and contractor employees.[2]

Training

Initial Training

Each employee presently involved in operating a process, and each employee before being involved in operating a newly assigned process, shall be trained in an overview of the process and in the operating procedures as specified in paragraph (f) of this section. The training shall include emphasis on the specific safety and health hazards, emergency operations including shutdown, and safe work practices applicable to the employee's job tasks.

In lieu of initial training for those employees already involved in operating a process on May 26, 1992, an employer may certify in writing that the employee has the required knowledge, skills, and abilities to safely carry out the duties and responsibilities as specified in the operating procedures.

Refresher Training

Refresher training shall be provided at least every 3 years, and more often if necessary, to each employee involved in operating a process to assure that the employee understands and adheres to the current operating procedures of the process. The employer, in consultation with the employees involved in operating the process, shall determine the appropriate frequency of refresher training.

Training Documentation

The employer shall ascertain that each employee involved in operating a process has received and understood the training required by this paragraph. The employer shall prepare a record that contains the identity of the employee, the date of training, and the means used to verify that the employee understood the training.[2]

All employees, including maintenance and contractor employees, involved with highly hazardous chemicals need to fully understand the safety and health hazards of the chemicals and processes they work with for the protection of themselves, their fellow employees, and the citizens of nearby communities. Training conducted in compliance with 1910.1200, the Hazard Communication standard, will help employees to be more knowledgeable about the chemicals they work with as well as familiarize them with reading and understanding SDSs. However, additional training in subjects such as operating procedures and safety work practices, emergency evacuation and response, safety procedures, routine and nonroutine work authorization activities, and other areas pertinent to process safety and health will need to be covered by an employer's training program.

In establishing their training programs, employers must clearly define the employees to be trained and what subjects are to be covered in their training. Employers in setting up their training program will need to clearly establish the goals and objectives they wish to achieve with the training that they provide to their employees. The learning goals or objectives should be written in clear measurable terms before the training begins. These goals and objectives need to be tailored to each of the specific training modules or segments. Employers should describe the important actions and conditions under which the employee will demonstrate competence or knowledge as well as what is acceptable performance.

Hands-on training where employees are able to use their senses beyond listening will enhance learning. For example, operating personnel who will work in a control room or at control panels would benefit by being trained at a simulated control panel or panels. Upset conditions of various types could be displayed on the simulator, and then the employee could go through the proper operating procedures to bring the simulator panel back to the normal operating parameters. A training environment could be created to help the trainee feel the full reality of the situation but, of course, under controlled conditions. This realistic type of training can be very effective in teaching employees correct procedures while allowing them to also see the consequences of what might happen if they do not follow established operating procedures. Other training techniques using videos or on-the-job training can also be very effective for teaching other job tasks, duties, or other important information. An effective training program will allow the employee to fully participate in the training process and to practice their skill or knowledge.

Employers need to periodically evaluate their training programs to see if the necessary skills, knowledge, and routines are being properly understood and implemented by their trained employees. The means or methods for evaluating the training should be developed along with the training program goals and objectives. Training program evaluation will help employers determine the amount of training their employees understood and whether the desired results were obtained. If, after the evaluation, it appears that the trained employees are not at the level of knowledge and skill that was expected, the employer will need to revise the training program, provide retraining, or provide more frequent refresher training sessions until the deficiency is resolved. Those who conducted the training and those who received the training should also be consulted as to how best to improve the training process. If there is a language barrier, the language known to the trainees should be used to reinforce the training messages and information.

Careful consideration must be given to assure that employees including maintenance and contract employees receive current and updated training. For example, if changes are made to a process, affected employees must be trained in the changes and understand the effects of the changes on their job tasks (e.g., any new operating procedures pertinent to their tasks). Additionally, as already discussed, the evaluation of the employee's absorption of training will certainly influence the need for training.

Contractors

Application

This paragraph applies to contractors performing maintenance or repair, turnaround, major renovation, or specialty work on or adjacent to a covered process. It does not apply to contractors providing incidental services that do not influence process safety, such as janitorial work, food and drink services, laundry, delivery, or other supply services.

Employer Responsibilities

The employer, when selecting a contractor, shall obtain and evaluate information regarding the contract employer's safety performance and programs. The employer shall inform contract employers of the known potential fire, explosion, or toxic release hazards related to the contractor's work and the process. The employer shall explain to contract employers the applicable provisions of the emergency action plan required by the standard.

The employer shall develop and implement safe work practices consistent with paragraph (f)(4) of the standard to control the entrance, presence, and exit of contract employers and contract employees in covered process areas.

The employer shall periodically evaluate the performance of contract employers in fulfilling their obligations as specified in the standard. The employer shall maintain a contract employee injury and illness log related to the contractor's work in process areas.

Contract Employer Responsibilities

The contract employer shall assure that each contract employee is trained in the work practices necessary to safely perform his or her job. The contract employer shall assure that each contract employee is instructed in the known potential fire, explosion, or toxic release hazards related to his or her job and the process, and the applicable provisions of the emergency action plan.

The contract employer shall document that each contract employee has received and understood the training required by this paragraph. The contract employer shall prepare a record that contains the identity of the contract employee, the date of training, and the means used to verify that the employee understood the training.

The contract employer shall assure that each contract employee follows the safety rules of the facility including the safe work practices required by the standard.

The contract employer shall advise the employer of any unique hazards presented by the contract employer's work or of any hazards found by the contract employer's work.

Pre-Start-Up Safety Review

The employer shall perform a pre-start-up safety review for new facilities and for modified facilities when the modification is significant enough to require a change in the process safety information. The pre-start-up safety review shall confirm that prior to the introduction of highly hazardous chemicals to a process

- Construction and equipment is in accordance with design specifications.
- Safety, operating, maintenance, and emergency procedures are in place and are adequate.
- For new facilities, a process hazard analysis has been performed and recommendations have been resolved or implemented before start-up, and modified facilities meet the requirements contained in management of change.
- Training of each employee involved in operating a process has been completed.

Mechanical Integrity

Application

This area shall apply to the following process equipment:

- Pressure vessels and storage tanks
- Piping systems (including piping components such as valves)
- Relief and vent systems and devices
- Emergency shutdown systems
- Controls (including monitoring devices and sensors, alarms, and interlocks)
- Pumps

Written Procedures

The employer shall establish and implement written procedures to maintain the ongoing integrity of process equipment.

Training for Process Maintenance Activities

The employer shall train each employee involved in maintaining the ongoing integrity of process equipment in an overview of that process and its hazards and in the procedures applicable to the employee's job tasks to assure that the employee can perform the job tasks in a safe manner.

Inspection and Testing

Inspections and tests shall be performed on process equipment. Inspection and testing procedures shall follow recognized and generally accepted good engineering practices. The frequency of inspections and tests of process equipment shall be consistent with applicable manufacturers' recommendations and good engineering practices, and more frequently if determined to be necessary by prior operating experience.

    The employer shall document each inspection and test that has been performed on process equipment. The documentation shall identify the date of the inspection or test, the name of the person who performed the inspection or test, the serial number or other identifier of the equipment on which the inspection or test was performed, a description of the inspection or test performed, and the results of the inspection or test.

Equipment Deficiencies

The employer shall correct deficiencies in equipment that are outside acceptable limits (defined by the process safety information in paragraph (d) of this section) before further use or in a safe and timely manner when necessary means are taken to assure safe operation.

Quality Assurance

In the construction of new plants and equipment, the employer shall assure that equipment as it is fabricated is suitable for the process application for which it will be used. Appropriate checks and inspections shall be performed to assure that equipment is installed properly and consistent with design specifications and the manufacturer's instructions. The employer shall assure that maintenance materials, spare parts, and equipment are suitable for the process application for which they will be used.

Hot Work Permit

The employer shall issue a hot work permit for hot work operations conducted on or near a covered process. The permit shall document that the fire prevention and protection requirements in 29 CFR 1910.252(a) have been implemented prior to beginning the hot work operations; it shall indicate the date(s) authorized for hot work and identify the object on which hot work is to be performed. The permit shall be kept on file until completion of the hot work operations.

Management of Change

The employer shall establish and implement written procedures to manage changes (except for "replacements in kind") to process chemicals, technology, equipment, and procedures, and changes to facilities that affect a covered process. The procedures shall assure that the following considerations are addressed prior to any change:

- The technical basis for the proposed change
- Impact of change on safety and health
- Modifications to operating procedures
- Necessary period for the change
- Authorization requirements for the proposed change

Employees involved in operating a process and maintenance and contract employees whose job tasks will be affected by a change in the process shall be informed of, and trained in, the change prior to start-up of the process or affected part of the process. If a change covered by this paragraph results in a change in the process safety information required by paragraph (d) of this section, such information shall be updated accordingly. If a change covered by this paragraph results in a change in the operating procedures or practices required by paragraph (f) of this section, such procedures or practices shall be updated accordingly.

Incident Investigation

The employer shall investigate each incident that resulted in, or could reasonably have resulted in, a catastrophic release of highly hazardous chemical in the workplace. An incident investigation shall be initiated as promptly as possible, but not later than 48 h following the incident. An incident investigation team shall be established and consist of at least one person knowledgeable in the process involved, including a contract employee if the incident involved work of the contractor, and other persons with appropriate knowledge and experience to thoroughly investigate and analyze the incident.

A report shall be prepared at the conclusion of the investigation, which includes at a minimum

- Date of incident
- Date investigation began
- A description of the incident
- The factors that contributed to the incident
- Any recommendations resulting from the investigation

The employer shall establish a system to promptly address and resolve the incident report findings and recommendations. Resolutions and corrective actions shall be documented. The report shall be reviewed with all affected personnel whose job tasks are relevant to the incident findings including contract employees where applicable. *Incident investigation reports shall be retained for 5 years.*

Emergency Planning and Response

The employer shall establish and implement an emergency action plan for the entire plant in accordance with the provisions of 29 CFR 1910.38. In addition, the emergency action plan shall include procedures for handling small releases. Employers covered under this standard may also be subject to the hazardous waste and emergency response provisions contained in 29 CFR 1910.120 (a), (p), and (q).

Compliance Audits

Employers shall certify that they have evaluated compliance with the provisions of this section at least every 3 years to verify that the procedures and practices developed under the standard are adequate and are being followed. The compliance audit shall be conducted by at least one person knowledgeable in the process. A report of the findings of the audit shall be developed. The employer shall promptly determine and document an appropriate response to each of the findings of the compliance audit and document that deficiencies have been corrected. *Employers shall retain the two most recent compliance audit reports.*

An effective audit includes a review of the relevant documentation and process safety information, inspection of the physical facilities, and interviews with all levels of plant personnel. Utilizing the audit procedure and checklist developed in the preplanning stage, the audit team can systematically analyze compliance with the provisions of the standard and any other corporate policies that are relevant. For example, the audit team will review all aspects of the training program as part of the overall audit. The team will review the written training program for adequacy of content, frequency of training, effectiveness of training in terms of its goals and objectives, and how it fits into meeting the standard's requirements, documentation, and so on. Through interviews, the team can determine the employee's knowledge and awareness of the safety procedures, duties, rules, emergency response assignments, and so forth. During the inspection, the team can observe actual practices such as safety and health policies, procedures, and work authorization practices. This approach enables the team to identify deficiencies and determine where corrective actions or improvements are necessary.

An audit is a technique used to gather sufficient facts and information, including statistical information, to verify compliance with standards. Auditors should select as part of their preplanning a sample size sufficient to give a degree of confidence that the audit reflects the level of compliance with the standard. The audit team, through this systematic analysis, should document areas that require corrective action as well as those areas where the process safety management system is effective and working in an effective manner. This provides a record of the audit procedures and findings and serves as a baseline of operation data for future audits. It will assist future auditors in determining changes or trends from previous audits.

Corrective action is one of the most important parts of the audit. It includes not only addressing the identified deficiencies but also planning, follow-up, and documentation. The corrective action process normally begins with a management review of the audit findings. The purpose of this review is to determine what actions are appropriate, as well as to establish priorities, timetables, resource allocations, and requirements and responsibilities. In some cases, corrective action may involve a simple change in procedure or minor maintenance effort to remedy the concern. Management of change procedures need to be used, as appropriate, even for what may seem to be a minor change. Many of the deficiencies can be acted on promptly, while some may require engineering studies or in-depth review of actual procedures and practices. There may be instances where no action is necessary and this is a valid response to an audit finding. All actions taken, including an explanation where no action is taken on a finding, need to be documented as to what was done and why.[3]

It is important to assure that each deficiency identified is addressed, the corrective action to be taken noted, and the audit person or team responsible be properly documented by the employer. To control the corrective action process, the employer should consider the use of a tracking system. This tracking system might include periodic status reports shared with affected levels of management, specific reports such as completion of an engineering study, and a final implementation report to provide closure for audit findings that have been through management of change, if appropriate, and then shared with affected employees and management. This type of tracking system provides the employer with the status of the corrective action. It also provides the documentation required to verify that appropriate corrective actions were taken on deficiencies identified in the audit.

Trade Secrets

Employers shall make all information necessary to comply with the section available to those persons responsible for compiling the process safety information, those assisting in the development of the process hazard analysis, those responsible for developing the operating procedures, and those involved in incident investigations, emergency planning, and response and compliance audits without regard to possible trade secret status of such information.

Nothing in this standard shall preclude the employer from requiring the persons to whom the information is made available to enter into confidentiality agreements not to disclose the information as set forth in 29 CFR

1910.1200. Subject to the rules and procedures set forth in 29 CFR 1910.1200(i) (1) through 1910.1200(i)(12), employees and their designated representatives shall have access to trade secret information contained within the process hazard analysis and other documents required to be developed by this standard.

Key Points to Remember on OSHA's Process Safety Management Standard

1. Process safety management is a system of tools, processes, and procedures that when properly implemented and followed will greatly reduce the potential for accidental release of highly hazardous chemicals.

2. The process safety management standard targets highly hazardous chemicals that have the potential to cause a catastrophic incident.

3. Employers are required to develop a written plan of action regarding the implementation of the employee participation.

4. Information pertaining to the hazards of the highly hazardous chemicals in the process shall consist of the following: (a) toxicity information, (b) permissible exposure limits, (c) physical data, (d) reactivity data, (e) corrosivity data, (f) thermal and chemical stability data, and (g) hazardous effects of inadvertent mixing of different materials that could foreseeably occur.

5. The employer shall perform an initial process hazard analysis on processes covered by this standard.

6. The employer shall develop and implement written operating procedures that provide clear instructions for safely conducting activities involved in each covered process consistent with the process safety information.

7. The information required in #6 above shall address at least the following elements: initial start-up, normal operations, temporary operations, emergency shutdown, emergency operations, normal shutdown, and start-up following a turnaround, or after an emergency shutdown.

8. Each employee presently involved in operating a process and each employee before being involved in operating a new process shall be trained in an overview of the process and in the operating procedures.

9. Refresher training shall be provided at least every 3 years, and more often if necessary.

10. The employer shall prepare a record that contains the identity of the employee, the date of training, and the means used to verify that the employee understood the training.

References

1. https://www.osha.gov/Publications/osha3133.html.
2. https://www.osha.gov/pls/oshaweb/owadisp.show_document%3Fp_table%3DSTANDARDS%26p_id%3D9760.
3. https://www.osha.gov/pls/oshaweb/owadisp.show_document?p_table=STANDARDS&p_id=9763.

33

BCSP Code of Ethics

Professional ethics are essential for any professional organization and profession. They are a set of codes that establish operating guidelines for any profession, which the members agree to abide by. These codes serve as overriding principles that provide professional organizations and their members with moral principles when serving the public. The Board of Certified Safety Professionals (BCSP) has an established Code of Ethics that every certifcate holder must abide by. This code is republished in the following and is testable information on both the Associate Safety Professional and Certified Safety Professional examinations.[1]

BCSP Code of Ethics

This code sets forth the code of ethics and professional standards to be observed by holders of documents of certification conferred by the BCSP. Certificants shall, in their professional activities, sustain and advance the integrity, honor, and prestige of the profession by adherence to these standards.

Standards

1. Hold paramount the safety and health of people, the protection of the environment, and protection of property in the performance of professional duties and exercise their obligation to advise employers, clients, employees, the public, and appropriate authorities of danger and unacceptable risks to people, the environment, or property.

2. Be honest, fair, and impartial; act with responsibility and integrity. Adhere to high standards of ethical conduct with balanced care for the interests of the public, employers, clients, employees, colleagues, and the profession. Avoid all conduct or practice that is likely to discredit the profession or deceive the public.

3. Issue public statements only in an objective and truthful manner and only when founded upon knowledge of the facts and competence in the subject matter.

4. Undertake assignments only when qualified by education or experience in the specific technical fields involved. Accept responsibility for their continued professional development by acquiring and maintaining competence through continuing education, experience, and professional training.

5. Avoid deceptive acts that falsify or misrepresent their academic or professional qualifications. Not misrepresent or exaggerate their degree of responsibility in or for the subject matter of prior assignments. Presentations incident to the solicitation of employment shall not misrepresent pertinent facts concerning employers, employees, associates, or past accomplishments with the intent and purpose of enhancing their qualifications and their work.

6. Conduct their professional relations by the highest standards of integrity and avoid compromise of their professional judgment by conflicts of interest.

7. Act in a manner free of bias with regard to religion, ethnicity, gender, age, national origin, sexual orientation, or disability.

8. Seek opportunities to be of constructive service in civic affairs and work for the advancement of the safety, health, and well-being of their community and their profession by sharing their knowledge and skills.

BCSP Code of Ethics and Professional Conduct

Interpretation #1—Approved October 28, 2006

Subject: Use of US degrees that are not awarded by accredited schools and use of degrees from schools recognized by US federal or state governments as diploma mills.

Interpretation: Use of degrees from colleges and universities not holding accreditation from a body recognized by the US Department of Education or the Council for Higher Education Accreditation or degrees from colleges and universities identified by the US government or any US state government as a diploma mill or similar unacceptable institution when used to establish or demonstrate professional qualifications shall be deemed in violation of Standard #5 of the BCSP Code of Ethics and Professional Conduct, which states: Avoid deceptive acts that falsify or misrepresent their academic or professional qualifications. BCSP will rely on information found in these and other reference lists:

http://ope.ed.gov/accreditation/

http://ope.ed.gov/accreditation/search.asp

http://www.ed.gov/admins/finaid/accred/index.html

http://www.ed.gov/students/prep/college/diplomamills/index.html

http://www.chea.org

http://www.chea.org/search/default.asp

http://www.chea.org/degreemills/frmStates.htm

http://www.michigan.gov/documents/Non-accreditedSchools_78090_7
.pdf

http://www.osac.state.or.us/oda/unaccredited.html

Effective Date: This interpretation is effective January 1, 2007.

Implementation: Should BCSP receive a complaint after the effective date from anyone regarding an individual holding a certification or status with BCSP that includes suitable evidence of use of a degree defined above, BCSP will act to remove the certification or status from the individual in accordance with Article XIV of the BCSP Bylaws.

Filing a Complaint: The complaint must name the individual holding a certification or other status with BCSP and include supporting evidence. The evidence must show that (a) the individual named in the complaint uses the unacceptable degree defined above on a business card, in a resume, curriculum vitae, promotional brochure, or other document that presents to the public credentials or any such academic degree or in some other way uses the degree, and (b) the degree named was awarded by an institution included in the definition above or the individual admitted the degree source on their own document. A complaint that does not provide both elements of evidence is not considered a valid complaint under this interpretation.

BCSP Procedures:

1. Upon receipt of a valid complaint that includes the required evidence, BCSP will notify the named individual and request that the individual

    a. Provide evidence that the complaint is not true

    b. State in writing whether the individual wishes to retain the BCSP certification or status and agrees to discontinue use of the degree

    BCSP may require the individual to sign a written agreement to comply with Interpretation #1 and to provide such evidence as may be required from time to time to satisfy BCSP that the individual remains in compliance.

2. If there is no suitable response within 30 days of BCSP forwarding a copy of the complaint to the individual named in the valid complaint, BCSP may institute disciplinary action in accordance with Article XIV of the BCSP Bylaws.

3. If BCSP receives a second valid complaint for the same person relating to the same or a different degree falling under Interpretation #1, BCSP may, upon giving notice to the individual, immediately take disciplinary action against the individual as provided in Article XIV of the BCSP Bylaws.

Reference

1. http://www.bcsp.org/pdf/ethics.pdf.

Appendix A: BCSP Supplied Equations

Mechanics

$$F = \mu N$$

F = frictional force (newtons)
μ = coefficient of friction
N = newtons

$$F_1 D_1 = F_2 D_2$$

F = force (newtons)
D = distance

$$v = v_o + at$$

v = velocity
v_o = original velocity at the start of acceleration
a = acceleration
t = time (seconds)

$$s = v_o t + \frac{at^2}{2}$$

s = displacement of the object (change in position—normally described in distance from its original position)
v_o = initial velocity
t = time
a = acceleration

$$v^2 = v_o^2 + 2as$$

v = final velocity
v_o = initial velocity
a = acceleration of the object (meters per second)

s = displacement of the object (normally described in distance from original position)

$$K.E. = \frac{mv^2}{2}$$

K.E. = kinetic energy (newtons)
m = mass of the object
v = speed of the object (velocity)

$$P.E. = mgh$$

P.E. = potential energy (joules)
m = mass of the object (kilograms)
g = gravitation acceleration of the earth (9.8 m/s²)
h = height above earth's surface (meters)

$$P.E. = \frac{kx^2}{2}$$

P.E. = potential energy (elastic)(joules)
k = spring constant (N/m²)
x = amount of compression (distance in meters)

$$\rho = mv$$

ρ = momentum
m = mass (kilograms)
v = velocity (meters per second)

$$F = ma$$

F = amount of force
m = mass (kilograms)
a = acceleration (meters per second squared)

$$W = mg$$

W = amount of work done on or to an object due to gravity
m = mass (kilograms)
g = gravity (9.8 m/s²) (constant)

$$W = Fs$$

W = work done on or to a system (usually in joules or N/m²) (1 J = 1 N × 1 m)
F = amount of force (newtons)
s = distance (usually meters or feet)

Ergonomic (Revised NIOSH Lifting Equations)

$$\text{Lifting Index (LI)} = \frac{L}{\text{RWL}}$$

LI = Lifting Index
L = object weight
RWL = recommended weight limit

$$\text{RWL} = \text{LC} \times \text{HM} \times \text{VM} \times \text{DM} \times \text{AM} \times \text{FM} \times \text{CM}$$

RWL = recommended weight limit
LC = load constant
HM = horizontal multiplier
VM = vertical multiplier
DM = distance multiplier
AM = asymmetric multiplier
FM = frequency multiplier
CM = coupling multiplier

Heat Stress and Relative Humidity

Indoor (no solar load)

$$0.7 \text{ WB} + 0.3 \text{ GT}$$

Outdoors (with solar heat load)

$$0.7 \text{ WB} + 0.2 \text{ GT} = 0.1 \text{ DB}$$

WB = wet-bulb temperature
GT = globe temperature
DB = dry-bulb temperature

Concentrations of Vapors and Gases

$$\text{ppm} = \frac{\frac{\text{mg}}{\text{m}^3} \times 24.45}{\text{MW}}$$

ppm = parts per million
mg/m^3 = measured mg/m^3 of the contaminant
MW = molecular weight of the contaminant
24.45 = constant = 1 g-mol

$$\text{TLV}_\text{m} = \frac{1}{\left(\dfrac{f_1}{\text{TLV}_1} + \dfrac{f_2}{\text{TLV}_2} + \cdots \dfrac{f_n}{\text{TLV}_n} \right)}$$

f = fraction of chemical (weight percent of liquid mixture)
TLV = threshold limit value of the chemical

$$\text{LFL}_\text{m} = \frac{1}{\left(\dfrac{f_1}{\text{LFL}_1} + \dfrac{f_2}{\text{LFL}_2} + \cdots \dfrac{f_n}{\text{LFL}_n} \right)}$$

f = fraction of chemical in the mixture
LFL = lower flammability limit

$$PV = nRT$$

P = absolute pressure (atm)
V = volume (liters)
n = number of molecules (moles)
R = universal gas constant (derived from table)
T = temperature (Rankine or Kelvin)

$$\frac{P_1 V_1}{T_1} = \frac{P_2 V_2}{T_2}$$

P = absolute pressure (atm)
V = volume of gas (liters)
T = temperature of gas (Kelvin temperature scale)

Ventilation

$$Q = VA$$

Q = volumetric flow rate (cfm)
V = velocity of the air (fpm)
A = cross-sectional area of the duct (sf)

$$V = 4005 \, C_e \sqrt{SP_h}$$

V = velocity (fpm)
C_e = coefficient of entry loss
SP_h = static pressure of the hood ("wg)

$$SP_{fan} = SP_{out} - SP_{in} - VP_{in}$$

SP = static pressure ("wg)

$$SP_h = VP + h_e$$

SP_h = static pressure of the hood ("wg)
VP = duct velocity pressure ("wg)
h_e = overall hood entry loss ("wg)

NOTE: When calculating for SP_h, it is understood that the static pressure of the hood is always positive; therefore, in this equation, the SP_h should be interpreted as the absolute value.

$$h_e = \frac{\left(1 - C_e^2\right) VP}{C_e^2}$$

h_e = hood entry loss ("wg)
C_e = coefficient of entry loss
VP = velocity pressure of duct ("wg)

$$Q = \frac{403 \times 10^6 \times SG \times ER \times K}{MW \times C}$$

Q = actual ventilation rate (cfm)
SG = specific gravity of volatile liquid
ER = evaporation rate of liquid (pints per minute)
K = design distribution constant to allow for incomplete mixing of contaminant air (1–10)
MW = molecular weight of liquid
C = desired concentration of gas or vapor at time t (ppm)—normally the TLV or PEL

$$V = 4005\sqrt{VP}$$

V = velocity of air (fpm)
VP = velocity pressure ("wg)

$$TP = SP + VP$$

TP = total pressure ("wg)
SP = static pressure ("wg)
VP = velocity pressure ("wg)

$$V = \frac{Q}{10x^2 + A}$$

V = velocity (fpm)
Q = flow rate (cfm)
x = source distance from hood opening (feet) (the equation is only accurate for a limited distance of 1.5 times the diameter of a round duct or the side of a rectangle or square duct)
A = area (sf)

$$C_e = \sqrt{\frac{VP}{SP_h}}$$

C_e = coefficient of entry loss
VP = velocity pressure of the duct ("wg)
SP_h = static pressure of the hood ("wg)

$$Q' = \frac{G}{C}$$

Q' = the effective rate of ventilation corrected for incomplete mixing (cfm)
($Q' = Q/K$)
K = design distribution constant to allow for incomplete mixing of contaminant air (1–10)
G = generation rate (cfm)

$$G = \frac{\text{constant} \times \text{specific gravity} \times \text{evaporation rate}}{\text{molecular weight}}$$

$$C = \frac{G}{Q'}\left(1 - e^{-\frac{Nt}{60}}\right)$$

C = concentration at a give time (ppm)
G = rate of generation of contaminant (cfm)
$Q' = (Q/K)$
K = design distribution constant (1–10)
Q = flow rate (cfm)
Nt = number of air changes

$$\ln\left(\frac{C_2}{C_1}\right) = -\frac{Q'}{V}(t_2 - t_1)$$

C_1 = the measured concentration
C_2 = the desired concentration
$Q' = (Q/K)$
K = design distribution constant (1–10)
Q = flow rate (cfm)
$t_2 - t_1$ = time interval or Δt
V = volume of space (ft³)

$$\ln\left(\frac{G - Q'C_2}{G - Q'C_1}\right) = -\frac{Q'(t_2 - t_1)}{V}$$

G = rate of generation of contaminant (cfm)
$Q' = (Q/K)$
$t_2 - t_1$ = time interval or Δ
C_1 = the measured concentration
C_2 = the desired concentration
V = volume of space (ft³)

$$C = \frac{P_v \times 10^6}{P_b}$$

C = concentration (ppm)
P_v = pressure of chemical (mm Hg)
P_b = barometric pressure (mm Hg)

Engineering Economy

$$F = P(1 + i)^n$$

F = future value of money
P = present value of money (principal)
i = interest rate (decimal)
n = number of years

$$P = F(1 + i)^{-n}$$

P = present worth of money (principal)
F = future worth (or savings)
i = interest rate (decimal)
n = number of years

$$F = A\left(\frac{(1+i)^n - 1}{i}\right)$$

F = future value of money
A = each payment ($)
i = interest rate (decimal)
n = number of periods

$$A = F\left(\frac{i}{(1+i)^n - 1}\right)$$

A = yearly payment of loan
F = future value of loan
i = interest

$$P = A\left(\frac{(1+i)^n - 1}{i(1+i)^n}\right)$$

P = present worth of money (\$)
A = period payment amount (\$)
i = annual percentage rate (decimal)
n = number of periods

$$A = P\left(\frac{i(1+i)^n}{(1+i)^n - 1}\right)$$

A = period payment amount (\$)
P = present worth of money
i = annual percentage rate (decimal)
n = number of periods

Reliability

$$P_f = 1 - R(t)$$

P_f = probability of failure
$R(t)$ = reliability

$$R(t) = e^{-\lambda t}$$

$R(t)$ = reliability
t = time in which reliability is measured
λ = number of failures divided by number of time units during which all items were exposed

$$P_f = (1 - P_s)$$

P_f = probability of failure
P_s = probability of success or reliability of the system

Noise

$$I = \frac{p^2}{\rho c}$$

I = sound intensity (W/m²)
p = sound pressure level (N/m²)
ρ = the density of the medium (in air, 1.2 kg/m²)
c = the speed of sound (in air, 344 m/s)

NOTE: The equation shown on the exam reference sheet does not show the RMS sound pressure squared. Therefore, it is necessary for you to remember the correction in this equation if performing sound intensity level calculations.

$$L_{pt} = 10\log\left[\sum_{i=1}^{N} 10^{\left(\frac{L_{pi}}{10}\right)}\right]$$

L_{pt} = combined sound pressure level
L_{pi} = individual measured sound pressure level

$$L_w = 10\log_{10}\frac{W}{W_o}$$

L_w = sound power level (dB)
W = acoustic power (W)
W_o = reference acoustic power (10^{-12} W)

$$L_p = 20\log_{10}\frac{p}{p_o}\,dB$$

L_p = sound pressure level (dB)
p = measured sound pressure level (N/m²)
p_o = reference sound pressure level (0.00002 N/m²)

$$T = \frac{8}{2^{(L-90)/5}}$$

T = time allowed for exposure
L = sound pressure level (dB)

$$D = 100 \left[\sum_{i=1}^{N} \frac{C_i}{T_i} \right]$$

D = dosage (or effective dose)
C_i = actual exposure time
T_i = allowed exposure time

$$dB_1 = dB_o - 20 \log_{10} \left(\frac{d_o}{d_1} \right)$$

dB_o = the original sound level measurement
dB_1 = the calculated sound level measurement at another distance
d_o = the original distance where noise measurement was taken
d_1 = the second distance that you would like to calculate the sound level reading

$$TWA = 16.61 \log_{10} \left[\frac{D}{100} \right] + 90$$

TWA = time weighted average
D = dose

$$dB = 10 \log_{10} \left(\frac{A_2}{A_1} \right)$$

dB = noise reduction in decibels
A_1 = total number of absorption units (sabins) in the room before treatment
A_2 = total number of absorption units (sabins) in the room after treatment

$$NR = \frac{12.6 P \alpha^{1.4}}{A} \, dB/ft$$

NR = noise reduction (dB per foot of length)
P = perimeter of the duct (inches)
α = absorption coefficient of the lining material at the frequency of interest
A = cross-sectional area of the duct (square inches)

Radiation

Ionizing

$$I_2 = I_1 \frac{(d_1)^2}{(d_2)^2}$$

I = intensity of source
d = distance from source

$$S \cong 6CE$$

$$I = I_0 e^{-\mu x}$$

I = intensity
I_0 = initial intensity
μ = linear attenuation coefficient (cm^{-1})
x = shielding thickness (cm)

$$I = \beta I_0 e^{-\mu x}$$

I = intensity on opposite side of shield
I_0 = initial intensity
μ = linear attenuation coefficient (cm^{-1})
x = shielding thickness (cm)
β = radiation scatter "buildup" factor (assume β = 1)

Nonionizing

$$W = \frac{16P}{\pi D^2} = \frac{4P}{A}$$

W = power density (mW/cm^2)
P = antenna power (mW)

A = effective antenna area (cm²)
D = diameter of antenna (cm)
A = cross-sectional area of antenna (cm²)

$$W = \frac{GP}{4\pi r^2} = \frac{AP}{\lambda^2 r^2}$$

W = power density (mW/cm²)
P = antenna power (mW)
A = effective antenna area (cm²)
λ = wavelength (cm)
r = distance from antenna (cm)

G = gain $\left(\text{calculated by } G = \dfrac{4\pi A}{\lambda^2}\right)$. When gain is used in calculations,

values of the absolute gain must be used: $G = 10^{\frac{g}{10}} = \text{antilog } \dfrac{g}{10}$.

$$E_{eff} = E_\lambda S_\lambda \Delta_\lambda$$

E_{eff} = effective irradiance
E_λ = spectral irradiance (W/cm²·nm)
S_λ = power density (W/m²)
Δ_λ = optical density (dimensionless)

$$c = \lambda f = \frac{\lambda}{T}$$

c = speed of light (3 × 10¹⁰ cm/s)
λ = wavelength (cm)
f = frequency (cycles per second)

Appendix B: Conversions and Standards

Length

2.54 cm = 1 inch
25.4 mm = 1 inch
12 inches = 1 foot
3 feet = 1 yard
5280 feet = 1 mile
1760 yards = 1 mile
1609 meters = 1 mile
0.868 nautical miles = 1 US mile

Area

1 sq. in. = 6.452 sq. cm
1 sq. in. = 6452 sq. mm
1 sq. ft = 144 sq. in.
1 sq. ft = 0.09290 sq. m
1 sq. yd. = 9 sq. ft

Volume

1 cu. cm = 1 mL
1 cu. in. = 16.39 cu. cm
1 cu. ft = 1728 cu. in.
1 cu. ft = 0.02832 cu. m
1 L = 1.057 quarts (US)
0.9463 L = 1 quart
1 US gallon = 0.1337 cu. ft
1 US gallon = 0.8327 Imperial gallons

Time

60 s = 1 min
60 min = 1 h
24 h = 1 day
3600 s = 1 h

Mass and Weight

16 ounces = 1 lb
1 g = 1000 mg
1 lb = 2.205 kg
1 ton (US) = 2000 lb
1 tonne (Metric ton) = 1000 kg
1 dram = 1771.85 mg

Energy

1 cal = 0.003968 Btu
1 cal = 4.187 J
1 Btu = 252 cal
1 J = 0.2388 cal

Velocity

1 ft/s = 0.3048 m/s
1 ft/s = 0.8618 mi/h
1 ft/s = 0.5921 knots
1 m/s = 3.281 ft/s
1 m/s = 2.237 mi/h
1 mi/h = 0.4470 m/s
1 mi/h = 1.467 ft/s
1 knot = 0.5148 m/s
1 knot = 1.151 mi/h
1 knot = 1.689 ft/s

Density

1 lb/cu. ft = 0.01602 cu. cm
1 g/cu. cm = 62.42 lb/cu. ft
1 g/cu. cm = 1000 kg/cu. m
1 kg/cu. m = 0.001 g/cu. cm

Pressure

1 psia = 6.895 kN/m^2
1 psia = 0.0680 atm
1 psia = 27.67 in. H_2O
1 psia = 51.72 mm Hg
1 psig = ADD 14.7 psia
1 mm Hg (Torr) = 0.01934 psia
1 mm Hg (Torr) = 0.1333 kN/m^2
1 in. H_2O = 0.2491 kN/m^2
1 kg/cm^2 = 735.6 mm Hg (Torr)
1 in. H_2O = 0.3614 psia
1 kg/cm^2 = 0.9678 atm
1 atm = 101.3 kN/m^2
1 kg/cm^2 = 14.22 psia
1 atm = 14.70 psia

1 bar = 100 kN/m²
1 kN/m² = 0.1450 psia
1 bar = 0.9869 atm
1 kN/m² = 0.009869 atm
1 bar = 1.020 kg/cm²

Heat Capacity
1 Btu/lb/°F = 1 cal/g/°C
1 Btu/lb/°F = 4187 J/kg·K
1 J/kg·K = 0.0002388 Btu/lb/°F

1 cal/g/°C = 1 Btu/lb/°F

Concentration
1 ppm = 1 mg/L
1 mg/L = 1 ppm
1 mg/m³ = 1 × 10⁻⁹ g/cm³
1 g/cm³ = 1 × 10⁹ mg/m³
1 g/m³ = 62.42 lb/ft³
1 lb/ft³ = 0.01602 g/cm³

SI Conversion Factors

From	Multiply By	To Find
Length		
Inches	25.4	Millimeters
Feet	0.305	Meters
Yards	0.914	Meters
Miles	1.61	Kilometers
Area		
Square inches	645.2	Square millimeters
Square feet	0.093	Square meters
Square yards	0.836	Square meters
Acres	0.405	Hectares
Square miles	2.59	Square kilometers
Volume		
Fluid ounce	29.57	Milliliters
Gallons	3.785	Liters
Cubic feet	0.028	Cubic meters
Cubic yards	0.765	Cubic meters
Mass		
Ounce	28.35	Grams
Pounds	0.454	Kilograms
Temperature		
Fahrenheit	$5(°F - 32)/9$	Celsius
Celsius	$9(°C + 32)/5$	Fahrenheit
Celsius	$°C + 273$	Kelvin
Fahrenheit	$°F + 460$	Rankine

Force, Pressure, or Stress		
Pound-force	4.45	Newtons
Pound-force/square inch	6.89	Kilopascals
Illumination		
Foot-candle	10.76	Lux
Foot-lambert	3.426	Candela/square meter

Standards and Constants

Physical Constants

Acceleration of gravity = 32.2 ft/s^2 = 9.8 m/s^2

Velocity of light = 3.0×10^8 m/s

Planck's constant = 6.626×10^{-34} J·s

Avogadro's number = 6.024×10^{23}/g-mol

Radiation

1 rad = 10^{-2} gray

1 rem = 10^{-2} sievert

1 curie = 3.7×10^{10} becquerel

1 becquerel = 1 disintegration/s

Density of Water

1 g/cm^3 = 1.94 slugs/ft^3

weight density = 62.4 lb/ft^3

1 US gallon of H$_2$O = 8.345 lb

Angles

1 radian = $180°/\pi$

Light

1 candela = 1 lumen/steradian

1 foot-candle = 10.76 candela/m^2 = 10.76 lux

Magnetic Fields

1 tesla = 10,000 gauss

Energy

1 British Thermal Unit (Btu) = 1055 J

1 faraday = 9.65×10^4 coulombs

1 g-cal = 4.19 J

1 g-mol at 0°C and 1 atm = 22.4 L

1 g-mol at 25°C and 1 atm = 24.45 L

1 ampere-hour = 3600 coulombs

1 W = 1 J/s = 1 A × 1 V

1 kwh = 3.6×10^6 J

Standards

STP (physical science) = 0°C and 1 atm

STP (ventilation) = 70°F and 1 atm

Air density = 0.075 lb/ft^3 at 70°F and 1 atm

STP (industrial hygiene) = 25°C and 1 atm

Appendix C: OSHA Regional and Area Offices

OSHA Regional Offices

Region 1 (ME, NH, MA, RI, CT, VT)
JFK Federal Building, Room E340
Boston, MA 02203
(617) 565-9860
(617) 565-9827 FAX

Region 2 (NY, NJ, PR, VI)
201 Varick Street, Room 670
New York, NY 10014
(212) 337-2378
(212) 337-2371 FAX

Region 3 (DC, DE, MD, PA, VA, WV)
US Department of Labor/OSHA
The Curtis Center-Suite 740 West
170 S. Independence Mall West
Philadelphia, PA 19106-3309
TEL: (215) 861-4900
FAX: (215) 861-4904

Region 4 (KY, TN, NC, SC, GA, AL, MS, FL)
61 Forsyth Street, SW
Room 6T50
Atlanta, GA 30303
(404) 562-2300
(404) 562-2295 FAX

Region 5 (IL, IN, MI, MN, OH, WI)
230 South Dearborn Street,
Room 3244
Chicago, IL 60604
(312) 353-2220
(312) 353-7774 FAX

Region 6 (AR, LA, NM, OK, TX)
525 Griffin Street, Suite 602
Dallas, TX 75202
(972) 850-4145
(972) 850-4149 FAX
(972) 850-4150 FSO FAX

Region 7 (IA, KS, MO, NE)
Two Pershing Square Building
2300 Main Street, Suite 1010
Kansas City, MO 64108-2416
Phone: (816) 283-8745
Voice: (816) 283-0545
FAX: (816) 283-0547

Region 8 (CO, MT, ND, SD, UT, WY)
1999 Broadway, Suite 1690
Denver, CO 80202
720-264-6550
720-264-6585 FAX

Region 9 (CA, NV, AZ)
90 7th Street, Suite 18100
San Francisco, CA 94103
(415) 625-2547 (Main
    Public—8:00 a.m. to
    4:30 p.m. Pacific)
(800) 475-4019 (for Technical
    Assistance)
(800) 475-4020 (for
    Complaints—Accidents/
    Fatalities)
Note: The 800 number for
    Complaints—Accidents/
    Fatalities is Regional only.
(800) 475-4022 (for Publication
    Requests)
(415) 625-2534 FAX

Region 10 (AK, ID, OR, WA)
1111 Third Avenue, Suite 715
Seattle, WA 98101-3212
(206) 553-5930
(206) 553-6499 FAX

OSHA Area Offices

Alabama

Birmingham Area Office
Medical Forum Building
950 22nd Street North, Room 1500
Birmingham, AL 35203
(205) 731-1534
(205) 731-0504 FAX

Mobile Area Office
1141 Montlimar Drive, Suite 1006
Mobile, AL 36609
(251) 441-6131
(251) 441-6396 FAX

Alaska

Anchorage Area Office
Scott Ketcham, Area Director
U.S. Department of Labor—OSHA
222 W. 8th Avenue, Room A14
Anchorage, AK 99513
Mailing Address:
Anchorage Area Office
Scott Ketcham, Area Director
U.S. Department of Labor—OSHA
222 W. 7th Avenue, Box 22
Anchorage, AK 99513
Comm. Phone: (907) 271-5152
Facsimile Number: (907) 271-4238

American Samoa

Region IX Federal Contact Numbers
90 7th Street, Suite 18100
San Francisco, CA 94103
(415) 625-2547
(Main Public—8:00 a.m. to 4:30 p.m. Pacific)
(800) 475-4019 (for Technical Assistance)
(800) 475-4020 (for Complaints—Accidents/ Fatalities)
Note: The 800 number for Complaints—Accidents/ Fatalities is Regional only.
(800) 475-4022 (for Publication Requests)
(415) 625-2534 FAX

Arizona

Industrial Commission of Arizona (ICA)
800 W. Washington Street
Phoenix, AZ 85007
Laura L. McGrory, Director & State Designee
Arizona Division of Occupational Safety and Health (ADOSH)

Phoenix Office
U.S. Department of Labor—OSHA
Phoenix Federal Building
230 N. 1st Avenue, Suite 202
Phoenix, AZ 85003
Phone: (602) 514-7250
Fax: (602) 514-7251
Darin Perkins, Director
Bill Wright, Assistant Director
Babak Emami, Consultation & Training Manager
PH: (602) 542-1769

Tucson Office
2675 E. Broadway Blvd. #239
Tucson, AZ 85716
PH: (520) 628-5478
Fax: (520) 322-8008
Mark Norton, Assistant
Director

Arkansas

Little Rock Area Office
10810 Executive Center Dr
Danville Bldg #2; Ste 206
Little Rock, AR 72211
Phone: 501-224-1841
Fax: 501-224-4431

California

Department of Industrial Relations
Office of the Director
455 Golden Gate Avenue
San Francisco, CA 94102
PH: (415) 703-5050
John Duncan, Director and
State Designee

Division of Occupational Safety and Health
1515 Clay Street Suite 1901
Oakland, CA 94612
PH: (510) 286-7000
FAX: (510) 286-7037
Len Welsh, Chief
Chris Lee, Deputy Chief of
Enforcement
Cal/OSHA Consultation
Services
2424 Arden Way, Suite 485
Sacramento, CA 95825
PH: (916) 263-5765
1 (800) 963-9424

Occupational Safety and Health Standards Board
2520 Venture Oaks Way, Suite 350
Sacramento, CA 95833
PH: (916) 274-5721
FAX: (916) 274-5743
Send mail to the OSHSB

Occupational Safety and Health Appeals Board
2520 Venture Oaks Way, Suite 300
Sacramento, CA 95833
PH: (916) 274-5751
FAX: (916) 274-5785
Send mail to Appeals Board
Michael Wimberly, Executive
Officer
Division of Labor Standards
Enforcement
Discrimination Complaint
Investigation Unit
2031 Howe Avenue, Suite 100
Sacramento, CA 95825
PH: (916) 263-1811
FAX: (916) 916-5378
Angela Bradstreet, Labor
Commissioner

Colorado

Denver Area Office
1391 Speer Boulevard, Suite 210
Denver, CO 80204-2552
(303) 844-5285
(303) 844-6676 FAX
The Denver Area Office also
oversees the federal pro-
gram for Utah.
Contact: Herb Gibson, Area
Director, Denver Area
Office, (303) 844-5285,
Ext. 106

Englewood Area Office
7935 East Prentice Avenue,
Suite 209
Englewood, CO 80111-2714
(303) 843-4500
(303) 843-4515 FAX

Connecticut

Bridgeport Area Office
Clark Building
1057 Broad Street, 4th Floor
Bridgeport, CT 06604
(203) 579-5581
Fax: (203) 579-5516

Hartford Area Office
Federal Building
450 Main Street, Room 613
Hartford, CT 06103
(860) 240-3152
Fax: (860) 240-3155

Delaware

Wilmington Area Office
Mellon Bank Building, Suite
900
919 Market Street
Wilmington, DE 19801-3319
(302) 573-6518
(302) 573-6532 FAX

District of Columbia

**Baltimore/Washington, D.C. Area
Office**
OSHA Area Office
U.S. Department of
Labor—OSHA
1099 Winterson Road, Suite 140
Linthicum, MD 21090
Phone: (410) 865-2055/2056
Fax: (410) 865-2068

Florida

Fort Lauderdale Area Office
8040 Peters Road, Building
H-100
Fort Lauderdale, FL 33324
(954) 424-0242
(954) 424-3073 FAX

Jacksonville Area Office
Ribault Building, Suite 227
1851 Executive Center Drive
Jacksonville, FL 32207
(904) 232-2895
(904) 232-1294 FAX

Tampa Area Office
5807 Breckenridge Parkway,
Suite A
Tampa, FL 33610-4249
(813) 626-1177
(813) 626-7015 FAX

Georgia

Atlanta East Area Office
LaVista Perimeter Office Park
2183 N. Lake Parkway,
Building 7
Suite 110
Tucker, GA 30084-4154
(770) 493-6644
(770) 493-7725 FAX

Atlanta West Area Office
2400 Herodian Way, Suite 250
Smyrna, GA 30080-2968
(770) 984-8700
(770) 984-8855 FAX

Savannah Area Office
450 Mall Boulevard, Suite J
Savannah, GA 31406
(912) 652-4393
(912) 652-4329 FAX

Guam

Region IX Federal Contact Numbers
90 7th Street, Suite 18100
San Francisco, CA 94103
(415) 625-2547 (Main
Public—8:00 a.m. to 4:30
p.m. Pacific)
(800) 475-4019 (for Technical
Assistance)
(800) 475-4020 (for
Complaints—Accidents/
Fatalities)
Note: The 800 number for
Complaints—Accidents/
Fatalities is Regional only.
(800) 475-4022 (for Publication
Requests)
(415) 625-2534 FAX

Hawaii

Department of Labor & Industrial Relations
830 Punchbowl Street, Suite 321
Honolulu, HI 96813
PH: (808) 586-8844
Darwin Ching, Director of
Department of Labor &
Industrial Relations
James Hardway, Special
Assistant to the Director

HIOSH (Enforcement & Consultation)
830 Punchbowl Street, Suite 425
Honolulu, HI 96813
Jamesner A. Dumlao,
Operations Manager
PH: (808) 586-9078

Consultation
PH: (808) 586-9100

Accident Reporting Line
PH: (808) 586-9102

Complaints
PH: (808) 586-9092
FAX: (808) 586-9104

Idaho

Boise Area Office
1150 North Curtis Road, Suite
201
Boise, ID 83706
(208) 321-2960
(208) 321-2966 Fax

Illinois

Calumet City Area Office
1600 167th Street, Suite 9
Calumet City, IL 60409
(708) 891-3800
(708) 862-9659 FAX

Chicago North Area Office
701 Lee Street—Suite 950
Des Plaines, IL 60016
(847) 803-4800
(847) 390-8220 FAX

Fairview Heights District Office
11 Executive Drive, Suite 11
Fairview Heights, IL 62208
(618) 632-8612
(618) 632-5712 FAX

North Aurora Area Office
365 Smoke Tree Plaza
North Aurora, IL 60542
(630) 896-8700
(630) 892-2160 FAX

Peoria Area Office
2918 W. Willows Knolls Road
Peoria, IL 61614
(309) 589-7033
(309) 589-7326 FAX

Indiana

Indianapolis Area Office
46 East Ohio Street, Room 453
Indianapolis, IN 46204
(317) 226-7290
(317) 226-7292 FAX

Iowa

**U.S. Department of Labor
Occupational Safety and Health
Administration**
210 Walnut St., Rm 815
Des Moines, IA 50309-2015
(515) 284-4794
(515) 284-4058 FAX

Kansas

Wichita Area Office
271 W. 3rd Street North, Room
400
Wichita, KS 67202
(316) 269-6644
(316) 269-6646 Voice Mail
(316) 269-6185 FAX
Toll Free (Kansas Residents
Only): 1-800-362-2896

Kentucky

Frankfort Area Office
John C. Watts Federal Office
Building
330 West Broadway, Room 108
Frankfort, KY 40601-1922
(502) 227-7024
(502) 227-2348 FAX

Louisiana

Baton Rouge Area Office
9100 Bluebonnet Centre Blvd,
Suite 201
Baton Rouge, LA 70809
(225) 298-5458
(225) 298-5457 FAX

Maine

Bangor District Office
382 Harlow Street
Bangor, ME 04401
(207) 941-8177
Fax: (207) 941-8179
Augusta Area Office
E.S. Muskie Federal Bldg
40 Western Ave., Room G-26
Augusta, ME 04330
(207) 626-9160
Fax: (207) 622-8213

Maryland

**Baltimore/Washington, D.C. Area
Office**
OSHA Area Office
U.S. Department of
Labor—OSHA
1099 Winterson Road, Suite 140
Linthicum, MD 21090
Phone: (410) 865-2055/2056
Fax: (410) 865-2068

Massachusetts

North Boston Area Office
Shattuck Office Center
138 River Road, Suite 102
Andover, MA 01810
(978) 837-4460
Fax: 978-837-4455

South Boston Area Office
639 Granite Street, 4th Floor
Braintree, MA 02184
(617) 565-6924
Fax: (617) 565-6923

Springfield Area Office
1441 Main Street, Room 550
Springfield, MA 01103-1493
(413) 785-0123
Fax: (413) 785-0136

Michigan

Lansing Area Office

U.S. Department of Labor
Occupational Safety and
Health Administration
315 West Allegan Street, Suite
207
Lansing, MI 48933
(517) 487-4996
(517) 487-4997 FAX

Minnesota

Eau Claire Area Office

1310 W. Clairemont Avenue
Eau Claire, WI 54701
(715) 832-9019
(715) 832-1147 FAX

Mississippi

Jackson Area Office

3780 I-55 North, Suite 210
Jackson, MS 39211-6323
(601) 965-4606
(601) 965-4610 Fax

Missouri

Kansas City Area Office

2300 Main Street, Suite 168
Kansas City, MO 64108
(816) 483-9531
(816) 483-9724 Fax
Toll Free (Missouri Residents
Only): 1-800-892-2674

St. Louis Area Office

1222 Spruce Street, Room 9.104
St. Louis, MO 63103
(314) 425-4249
(314) 425-4255 Voice Mail
(314) 425-4289 Fax
Toll Free (Missouri Residents
Only): 1-800-392-7743

Montana

Billings Area Office

2900 4th Avenue North, Suite
303
Billings, MT 59101
(406) 247-7494
(406) 247-7499 FAX

Nebraska

Omaha Area Office

Overland-Wolf Building
6910 Pacific Street, Room 100
Omaha, NE 68106
(402) 553-0171
(402) 551-1288 FAX
Toll Free (Nebraska Residents
Only): 1-800-642-8963

Nevada

Division of Industrial Relations

Department of Business &
Industry
400 W. King Street, Suite 400
Carson City, NV 89703
Donald Jayne, Director & State
Designee
Nevada OSHA
1301 N. Green Valley Parkway,
Suite 200
Henderson, NV 89074
PH: (702) 486-9044
Fax: (702) 990-0365
Steve Coffield, Chief
Administrative Officer

Reno Office

4600 Kietzke Lane, Suite F-153
Reno, NV 89502
PH: (775) 824-4600
Fax: (775) 688-1378

Nevada Safety Consultation and Training Section (SCATS)
1301 N. Green Valley Parkway,
    Suite 200
Henderson, NV 89074
PH: (702) 486-9140
Fax: (702) 990-0362
Jan Rosenberg, Chief
    Administrative Officer

New Hampshire
Concord Area Office
J.C. Cleveland Federal Bldg
53 Pleasant Street, Room 3901
Concord, NH 03301
(603) 225-1629
Fax: (603) 225-1580

New Jersey

Avenel Area Office
1030 St. Georges Avenue
Plaza 35, Suite 205
Avenel, NJ 07001
(732) 750-3270
(732) 750-4737 FAX

Hasbrouck Heights Area Office
500 Route 17 South
2nd Floor
Hasbrouck Heights, NJ 07604
(201) 288-1700
(201) 288-7315 FAX

Marlton Area Office
Marlton Executive Park,
    Building 2
701 Route 73 South, Suite 120
Marlton, NJ 08053
(856) 396-2594
(856) 396-2593 FAX

Parsippany Area Office
299 Cherry Hill Road, Suite 103
Parsippany, NJ 07054
(973) 263-1003
(973) 299-7161 FAX

New Mexico

Lubbock Area Office
1205 Texas Avenue, Room 806
Lubbock, TX 79401
(806) 472-7681
(806) 472-7686 FAX

New York
Albany Area Office
401 New Karner Road, Suite 300
Albany, NY 12205-3809
(518) 464-4338
(518) 464-4337 FAX

Queens District Office of the Manhattan Area Office
45-17 Marathon Parkway
Little Neck, NY 11362
(718) 279-9060
(718) 279-9057 FAX

Buffalo Area Office
U.S. Dept. of Labor/OSHA
130 S. Elmwood Avenue, Suite
    500
Buffalo, NY 14202-2465
(716) 551-3053
(716) 551-3126 FAX

Long Island Area Office
1400 Old Country Road
Suite 208
Westbury, NY 11590
(516) 334-3344
(516) 334-3326 FAX

Manhattan Area Office
201 Varick Street RM. 908
New York, NY 10014
(212) 620-3200
(212) 620-4121 (FAX)

Syracuse Area Office
3300 Vickery Road
North Syracuse, NY 13212
(315) 451-0808
(315) 451-1351 FAX

Tarrytown Area Office
660 White Plains Road, 4th
Floor
Tarrytown, NY 10591-5107
(914) 524-7510
(914) 524-7515 FAX

North Carolina

Raleigh Area Office
4407 Bland Road
Somerset Park Suite 210
Raleigh, NC 27609
(919) 790-8096
(919) 790-8224 FAX

North Dakota

Bismarck Area Office
Federal Office Building
1640 East Capitol Avenue
Bismarck, ND 58501
(701) 250-4521
(701) 250-4520 FAX

Ohio

Cincinnati Area Office
36 Triangle Park Drive
Cincinnati, OH 45246
(513) 841-4132
(513) 841-4114 FAX

Cleveland Area Office
1240 East 9th Street, Room 899
Cleveland, OH 44199
(216) 615-4266
(216) 615-4234 FAX

Columbus Area Office
200 North High Street, Room
620
Columbus, OH 43215
(614) 469-5582
(614) 469-6791 FAX

Toledo Area Office
420 Madison Avenue, Suite 600
Toledo, OH 43604
(419) 259-7542
(419) 259-6355 FAX

Oklahoma

Oklahoma City Area Office
55 North Robinson—Suite 315
Oklahoma City, OK 73102-9237
(405) 278-9560
(405) 278-9572 FAX

Oregon

Portland Area Office
Federal Office Building
1220 Southwest 3rd Avenue,
Room 640
Portland, OR 97204
(503) 326-2251
(503) 326-3574 FAX

Pennsylvania

Allentown Area Office
850 North 5th Street
Allentown, PA 18102
(610) 776-0592
(610) 776-1913 FAX

Erie Area Office
  1128 State Street, Suite 200
  Erie, PA 16501
  (814) 461-1492
  (814) 461-1498 FAX

Harrisburg Area Office
  Progress Plaza
  49 North Progress Avenue
  Harrisburg, PA 17109-3596
  (717) 782-3902
  (717) 782-3746 FAX

Philadelphia Area Office
  US Custom House, Room 242
  Second & Chestnut Street
  Philadelphia, PA 19106-2902
  (215) 597-4955
  (215) 597-1956 FAX

Pittsburgh Area Office
  U.S. Department of
    Labor—OSHA
  William Moorhead Federal
    Building, Room 905
  1000 Liberty Avenue
  Pittsburgh, PA 15222
  (412) 395-4903
  (412) 395-6380 FAX

Wilkes-Barre Area Office
  The Stegmaier Building, Suite
    410
  7 North Wilkes-Barre
    Boulevard
  Wilkes-Barre, PA 18702-5241
  (570) 826-6538
  (570) 821-4170 FAX

Puerto Rico

Puerto Rico Area Office
  Triple S Building
  1510 FD Roosevelt Avenue,
    Suite 5B
  Guaynabo, PR 00968
  (787) 277-1560
  (787) 277-1567 FAX

Rhode Island

Providence Area Office
  Federal Office Building
  380 Westminster Mall, Room
    543
  Providence, RI 02903
  (401) 528-4669
  Fax: (401) 528-4663

South Carolina

Columbia Area Office
  Strom Thurmond Federal
    Building
  1835 Assembly Street, Room
    1472
  Columbia, SC 29201-2453
  (803) 765-5904
  (803) 765-5591 FAX

South Dakota

Bruce Beelman, Area Director
  U.S. Department of Labor
  Occupational Safety and
    Health Administration
  Bismarck Area Office
  1640 East Capitol Avenue
  Bismarck, ND 58501
  (701) 250-4521

Tennessee

Nashville Area Office
51 Century Boulevard Suite 340
Nashville, TN 37214
(615) 232-3803
(615) 232-3827 FAX

Texas

Austin Area Office
La Costa Green Bldg.
1033 La Posada Dr. Suite 375
Austin, TX 78752-3832
(512) 374-0271
(512) 374-0086 FAX

Corpus Christi Area Office
Wilson Plaza
606 N Carancahua, Ste. 700
Corpus Christi, TX 78476
(361) 888-3420
(361) 888-3424 FAX

Dallas Area Office
8344 East RL Thornton Freeway
Suite 420
Dallas, TX 75228
(214) 320-2400
(214) 320-2598 FAX

El Paso District Office
U.S. Dept. of Labor—OSHA
4849 N. Mesa, Suite 200
El Paso, TX 79912-5936
(915) 534-6251
(915) 534-6259 FAX

Fort Worth Area Office
North Starr II, Suite 302
8713 Airport Freeway
Fort Worth, TX 76180-7610
(817) 428-2470
(817) 581-7723 FAX

Houston North Area Office
507 North Sam Houston
   Parkway East
Suite 400
Houston, TX 77060
(281) 591-2438
(281) 999-7457 FAX

Houston South Area Office
17625 El Camino Real, Suite
   400
Houston, TX 77058
(281) 286-0583
(281) 286-6352 FAX
Toll Free: (800) 692-4202

Lubbock Area Office
1205 Texas Avenue, Room
   806
Lubbock, TX 79401
(806) 472-7681 (7685)
(806) 472-7686 FAX

San Antonio District Office
Washington Square Blvd, Suite
   203
800 Dolorosa Street
San Antonio, TX 78207-4559
(210) 472-5040
(210) 472-5045 FAX

Utah

Herb Gibson, Area Director
   U.S. Department of Labor
   Occupational Safety and
      Health Administration
   1391 Speer Blvd, Suite 210
   Denver, CO 80204-2552
   (303) 844-5285, Ext. 106
   Fax (303) 844-6676

Vermont

Vermont Department of Labor
   5 Green Mountain Drive
   P.O. Box 488
   Montpelier, VT 05601-0488
   PH: (802) 828-4000
   FAX: (802) 888-4022
   Patricia Moulton Powden,
      Commissioner
   PH: (802) 828-4301
   Workers' Compensation and
      Safety Division

J. Stephen Monahan, Director
   PH: (802) 828-2138
   VOSHA
   Robert McLeod, Manager
   PH: (802) 828-5084

Virginia

Norfolk Area Office
   Federal Office Building, Room
      614
   200 Granby Street
   Norfolk, VA 23510-1811
   (757) 441-3820
   (No direct lines to staff)
   (757) 441-3594 FAX

Virgin Islands

Virgin Islands Department of Labor
   Albert Bryan, Jr., Commissioner
   PH: (340) 773-1994
   Division of Occupational Safety
      and Health (VIDOSH)
   3012 Golden Rock
   Christiansted, St. Croix VI 00890
   PH: (340) 772-1315
   FAX: (340) 772-4323
   Jannette Barbosa, Acting
      Director
   PH: (340) 772-1315

Washington

Bellevue Area Office
   505 106th Avenue NE, Suite 302
   Bellevue, WA 98004
   Comm. Phone: (425) 450-5480
   Facsimile Number: (425) 450-5483

West Virginia

Charleston Area Office
   405 Capitol Street, Suite 407
   Charleston, WV 25301-1727
   (304) 347-5937
   (No direct lines to staff)
   (304) 347-5275 FAX

Wisconsin

Appleton Area Office
   1648 Tri Park Way
   Appleton, WI 54914
   (920) 734-4521
   (920) 734-2661 FAX
   Eau Claire Area Office
   1310 W. Clairemont Avenue
   Eau Claire, WI 54701
   (715) 832-9019
   (715) 832-1147 FAX

Madison Area Office
4802 E. Broadway
Madison, WI 53716
(608) 441-5388
(608) 441-5400 FAX

Milwaukee Area Office
310 West Wisconsin Avenue,
Room 1180
Milwaukee, WI 53203
(414) 297-3315
(414) 297-4299 FAX

Wyoming

Herb Gibson, Area Director
U.S. Department of Labor
Occupational Safety and
Health Administration
1391 Speer Blvd, Suite 210
Denver, CO 80204-2552
(303) 844-5285, Ext. 106

Index

Page numbers followed by f and t indicate figures and tables, respectively.